The Quantum Vacuum

An Introduction to Quantum Electrodynamics

The Quantum Vacuum

An Introduction to Quantum Electrodynamics

Peter W. Milonni
Los Alamos, New Mexico

Academic Press
San Diego New York Boston
London Sydney Tokyo Toronto

ACADEMIC PRESS, INC.
A Division of Harcourt Brace & Company
525 B Street, Suite 1900, San Diego, California 92101-4495

United Kingdom Edition published by
ACADEMIC PRESS LIMITED
24–28 Oval Road, London NW1 7DX

Library of Congress Cataloging-in-Publication Data

Milonni, Peter W.
    The quantum vacuum : an introduction to quantum electrodynamics /
Peter W. Milonni.
        p.   cm.
    Includes bibliographical references and index.
    ISBN 0-12-498080-5
    1. Quantum electrodynamics.   I. Title.
QC688.M553   1993
537.6'7—dc20   93–29780
                                                              CIP

PRINTED IN THE UNITED STATES OF AMERICA
  95 96 97 EB 9 8 7 6 5 4 3 2

For Shwu-Fang, Moe, Nin, Carol Ann, and Betty

For no man can write anything who does not think that what he writes is for the time the history of the world; or do anything well who does not esteem his work to be of importance. My work may be of none, but I must not think of it as none, or I shall not do it with impunity.

In like manner, there is throughout nature something mocking, something that leads us on and on, but arrives nowhere; keeps no faith with us. All promise outruns the performance. We live in a system of approximations.

Ralph Waldo Emerson, Essay on *Nature* (1844).

Contents

Preface

According to present ideas there is no vacuum in the ordinary sense of tranquil nothingness. There is instead a fluctuating quantum vacuum. One purpose of this book is to survey some of our most important ideas about the quantum vacuum. A second is to describe, based on fundamental vacuum processes, the physical concepts of quantum electrodynamics (QED).

Why bother? Few people doubt the reality or significance of vacuum field fluctuations, and the formalism for perturbative QED calculations can already be found in many books. My answer is that, if QED is indeed the nonpareil physical theory, and if "the vacuum holds the key to a full understanding of the forces of nature,"[1] then it is worthwhile to look carefully at the *physical* ideas underlying QED vacuum effects, including not only such things as mass and charge renormalization, Lamb shifts and Casimir effects, but even more "elementary" things such as spontaneous emission, van der Waals forces, and the fundamental linewidth of a laser. Phenomena of the latter type, primarily nonrelativistic, are basic to quantum optics and other aspects of modern, applied QED. All of them involve the vacuum electromagnetic field in one way or another. All of them, furthermore, can be described physically in ways that involve *source* fields. A third purpose of this book is to exhibit and explain the relation between vacuum and source fields.

The modern view of the vacuum is closely related to zero-point energy, the energy associated with motion persisting even at the absolute zero of temperature, where classically all motion ceases. The idea of zero-point energy arose with the work of Planck and Einstein on the blackbody problem, and is connected to early, vague premonitions of wave–particle duality. Many physicists, including Pauli and Kramers, have been uncomfortable with zero-point energy, and even today the concept retains a peculiar flavor. We review the origins of the concept of zero-point energy in the first chapter.

---

[1] P. C. W. Davies, *Superforce* (Simon and Schuster, New York, 1985), p. 104.

Chapters 2-8 are devoted to various aspects of the electromagnetic vacuum in nonrelativistic theory. In Chapter 2 we review the quantization of the electromagnetic field in the simplest way, by exploiting the mathematical equivalence of the field to a collection of harmonic oscillators. (A more sophisticated path is taken in Chapter 10.) The zero-point field energy associated with the source-free, vacuum field is a consequence of the zero-point energy of the simple harmonic oscillator. This zero-point energy is often ignored on the grounds that it is a constant addition to the field Hamiltonian and can therefore be eliminated by simply redefining the zero of energy. However, deletion of the zero-point field energy from the Hamiltonian does not eliminate it once and for all from consideration, for it "re-emerges" as the homogeneous solution of the operator Maxwell equations in the Heisenberg picture. In fact the vacuum field, with its zero-point energy, is *required* for the formal consistency of QED, and in particular to preserve canonical commutation relations.

In Chapter 2 we also introduce the Casimir force between two perfectly conducting plates in the standard way, by considering the change in the electromagnetic zero-point energy due to the presence of the plates. The Casimir effect was proposed at the same time (1948) that the Lamb shift and the anomalous magnetic moment of the electron were being interpreted in terms of zero-point field energy and fluctuations by Welton and others. I often wondered whether Casimir was at all influenced by these other developments, and in early 1992 I wrote to him with this question. Dr. Casimir has kindly given me permission to quote from his answer:[2]

> No, I was not at all familiar with the work of Welton and others. I went my own, somewhat clumsy way ...
>
> A point I should have mentioned in later publications: Summer or autumn 1947 (but I am not absolutely certain that it was not somewhat earlier or later) I mentioned my results to Niels Bohr, during a walk. That is nice, he said, that is something new. I told him that I was puzzled by the extremely simple form of the expressions for the interaction at very large distances and he mumbled something about zero-point energy. That was all, but it put me on a new track ...
>
> I found that calculating changes of zero-point energy really leads to the same results as the calculations of Polder and myself ...
>
> I do not think there were outside influences apart from those mentioned above. I did not myself contribute to further developments, nor to experimental confirmations (apart from proposing a crazy model for an electron, which did not lead to a value of the fine structure constant).

---

[2]H. B. G. Casimir, private communication, 12 March 1992.

In Chapter 3 we discuss various "vacuum fluctuation effects" — spontaneous emission, the (nonrelativistic) Lamb shift and anomalous moment, the Casimir and Casimir–Polder forces, and the van der Waals interaction — in elementary terms. These effects can be described and derived in different ways in terms of *source* fields. Different interpretations and derivations result from an *arbitrary* choice of ordering of photon annihilation and creation operators, as discussed in Chapter 4. In Chapter 5 we briefly review the nonrelativistic theory of radiation reaction, which underlies the source-field interpretation of "vacuum fluctuation effects."

Chapter 6 shows how the vacuum field appears in various problems of quantum optics, cavity QED, and laser physics, including the fundamental quantum limit to the laser linewidth.

The Casimir effect plays such a prominent role in discussions of the quantum vacuum that we devote Chapters 7 and 8 to a rather detailed treatment of the forces between both conducting and dielectric bodies. We derive the Lifshitz expression for the force between two dielectric slabs from the perspectives of both vacuum and source fields, and show how various approaches are related. The macroscopic approach pioneered by Lifshitz is justified using the Ewald–Oseen extinction theorem. We discuss the experimental status of Casimir effects and some theoretical elaborations. We also describe Casimir's "crazy model for an electron."

Chapters 9 and 10 review some elementary features of the first-quantized Dirac equation and quantum field theory, respectively, including the relation between causality and spin statistics and the propagators for various quantum fields. We emphasize that *all quantum fields have zero-point energies and vacuum fluctuations.* In particular, the Dirac vacuum of electrons and positrons has a negative (infinite) zero-point energy, and this gives rise to Casimir effects analogous to those for the pure electromagnetic field.

Chapter 11 is an old-fashioned approach to relativistic self-energies and mass and charge renormalization. Some of the most important physical ideas of QED, including vacuum polarization, emerge rather clearly in such an approach. The modern approach, of course, is based principally on covariant perturbation theory and Feynman diagrams. In Chapter 12 we discuss the construction and utility of the diagrams, and again the calculation of basic things such as the electron self-energy, using both quantum field theory and the intuitive way of Feynman. The identification of some of the main physical ideas of relativistic QED in Chapter 11, *before* the introduction of Feynman diagrams, seems to me to have the advantage of avoiding confusion between the diagrams and the *physics*, although the distinction between the two is admittedly not easy to make when perturbative calculations are taken to high orders. Chapters 9, 10, and 12 provide a painless short course, I think, for those crossing the relativistic Rubicon for the first or second time.

Originally I hoped to cover more ground, including chapters on the vacuum in quantum chromodynamics and cosmology. However, I gradually realized that in a book of the intended size I could not do this without sacrificing the rather considerable detail I wanted for various QED effects, particularly van der Waals and Casimir interactions. Since these things are not treated in any sort of detail in other books I know of, I decided to omit the somewhat more speculative aspects of vacuum theory and to focus on quantum electrodynamics.

Writing this book has given me much pleasure. I wish to thank especially Mei-Li Shih and Gordon W. F. Drake. Mei-Li not only was relentless in her encouragement but also checked many of the calculations and made good suggestions. In my opinion, but not hers, she is an author of this book. Professor Drake served as a reviewer for Academic Press and, by suggesting changes and identifying errors and non sequiturs, helped greatly to improve the quality of the manuscript.

Discussions relating to this book with Richard J. Cook, Joseph H. Eberly, Walter T. Grandy, Jr., Michael Lieber, David Nesbitt, M.D., Edwin A. Power, David H. Sharp, and Larry Spruch were enlightening and often encouraging. Professor Eberly supervised the Ph.D thesis work from which this book gradually evolved.

I think it is also appropriate to thank Gabriel Barton, Timothy H. Boyer, Trevor W. Marshall, and again, Edwin Power; their work has been partly responsible for my continued fascination with the subject.

Needless to say, H. B. G. Casimir has given us a lot to think about, and I thank him for his kind replies to my letters and for permission to quote from them.

Finally I wish to acknowledge the inestimable help I have received from the editors at Academic Press. I hope this book lives up to their high standards and expectations.

Peter W. Milonni
Los Alamos, New Mexico

Chapter 1

Zero-Point Energy in Early Quantum Theory

The existence of a zero-point energy of size $\frac{1}{2}h\nu$ [is] probable.
— Albert Einstein and Otto Stern (1913)

1.1 Introduction

The importance of the blackbody problem in the development of quantum theory is recognized by every serious student of modern physics. What is not so widely known is that blackbody theory led also to the concept of zero-point energy, which was later to appear naturally in the mathematics of quantum theory. The relation of this energy to early premonitions of wave-particle dualism is similarly not widely appreciated. This chapter is a discussion of these roots of the concept of zero-point energy. We do not proffer any sort of rigorous historical analysis, but only a glimpse into some of the early physics of energy at the absolute zero of temperature.

1.2 The Blackbody Problem

In 1860 Kirchhoff derived a general relation between the radiative and absorptive strengths of a body held at a fixed temperature T. According to Kirchhoff's law the ratio of the radiative strength to the absorption coefficient for radiation of wavelength λ is the same for all bodies at temperature T, and defines a universal function $F(\lambda, T)$. This led to the abstraction of an ideal *blackbody* for which the absorption coefficient is unity at every

1

wavelength, corresponding to total absorption. Thus $F(\lambda, T)$ characterizes the radiative strength at wavelength λ of a blackbody at temperature T. The problem was to determine the universal function $F(\lambda, T)$.

An important step was taken in 1884 by Boltzmann, who invoked several aspects of Maxwell's electromagnetic theory. The most important of these for the present discussion is the result that isotropic radiation exerts on a perfectly reflecting surface a pressure $u/3$, where u is the energy density of the radiation.[1] Boltzmann considered blackbody radiation confined in a cylinder of volume V, one end of which is a perfectly reflecting piston. The radiation pressure on the piston increases the volume by dV, and in order to maintain a constant temperature an amount of heat

$$dQ = dU + PdV = d(uV) + \frac{1}{3}udV = Vdu + \frac{4}{3}udV \qquad (1.1)$$

must be added, according to the first law of thermodynamics. Kirchhoff's law implies that the total energy density u over all wavelengths is a function only of T, so that

$$dQ = V\frac{du}{dT}dT + \frac{4}{3}udV. \qquad (1.2)$$

Associated with the expansion of the cylinder is an increase in entropy by

$$dS = \frac{1}{T}dQ = \frac{V}{T}\frac{du}{dT}dT + \frac{4}{3}\frac{u}{T}dV, \qquad (1.3)$$

which, according to the second law of thermodynamics, is an exact differential. Thus

$$\frac{\partial S}{\partial T} = \frac{V}{T}\frac{du}{dT}, \quad \frac{\partial S}{\partial V} = \frac{4}{3}\frac{u}{T} \qquad (1.4)$$

and

$$\frac{\partial^2 S}{\partial T \partial V} = \frac{\partial^2 S}{\partial V \partial T} = \frac{1}{T}\frac{du}{dT} = \frac{4}{3}\frac{d}{dT}\left(\frac{u}{T}\right), \qquad (1.5)$$

from which it follows that $du/dT = 4u/T$ and

$$u = bT^4 \quad \text{(Stefan–Boltzmann law)}, \qquad (1.6)$$

where b is a universal constant. Stefan in 1879 had in fact suggested such a relation from an analysis of experimental data.

---

[1] A plane wave exerts a pressure $2u$ on a reflecting surface on which it is normally incident. (See, for instance, W. K. H. Panofsky and M. Phillips, *Classical Electricity and Magnetism* (Addison–Wesley, Reading, Mass., 1962), p. 193.) For plane waves propagating with equal intensities in both directions normal to the surface, this is reduced to u, or $u/3$ if the radiation is isotropic.

The Stefan–Boltzmann law stands in conflict with elementary classical models of equilibrium between radiation and matter. Consider the classical oscillator model of an atom, where an electron is assumed to be bound by an elastic restoring force. If $\rho(\nu)d\nu$ denotes the energy per unit volume of radiation in the frequency interval $[\nu, \nu + d\nu]$, then the rate at which the atom absorbs energy from the radiation field may be shown to be given by the formula (see Appendix A)

$$\dot{W}_A = \frac{\pi e^2}{3m}\rho(\nu_o), \tag{1.7}$$

where W_A is the electron energy, e and m are its charge and mass, respectively, and ν_o is the natural oscillation frequency of the electron in the atom. The rate at which the electron radiates electromagnetic energy W_{EM} is given by the well-known classical Larmor formula:

$$\dot{W}_{EM} = \frac{2e^2 a^2}{3c^3}, \tag{1.8}$$

where a is the acceleration of the electron. For oscillation at frequency $\nu_o = \omega_o/2\pi$, $a = -\omega_o^2 x$ and

$$\dot{W}_{EM} = \left(\frac{32\pi^4 e^2 \nu_o^4}{3c^3}\right) x^2, \tag{1.9}$$

where x is the electron displacement from its equilibrium position in the classical oscillator model of the atom. Now according to the virial theorem of classical mechanics the average potential energy $\frac{1}{2}m\omega_o^2 x^2$ of the (one-dimensional) electron oscillator is equal to the average kinetic energy, and their sum is the total oscillator energy U. In a state of equilibrium between radiation and matter, furthermore, the energy absorption rate (1.7) should equal the emission rate (1.9). Thus

$$\rho(\nu_o) = \frac{8\pi\nu_o^2}{c^3}(m\omega_o^2 x^2) = \frac{8\pi\nu_o^2}{c^3}U, \tag{1.10}$$

or more generally

$$\rho(\nu) = \frac{8\pi\nu^2}{c^3}U \tag{1.11}$$

for a blackbody, which absorbs at all frequencies ν. Finally the equipartition theorem of classical statistical mechanics demands that the average value of U in thermal equilibrium is kT, where k is Boltzmann's constant, so that the spectral energy density of thermal radiation must be

$$\rho(\nu) = \left(\frac{8\pi\nu^2}{c^3}\right) kT \quad \text{(Rayleigh–Jeans distribution).} \tag{1.12}$$

The total electromagnetic energy density

$$u = \int_0^\infty \rho(\nu)d\nu \qquad (1.13)$$

violates the Stefan–Boltzmann law. Furthermore the Rayleigh–Jeans law suffers from the ultraviolet catastrophe: u diverges when (1.12) is used for $\rho(\nu)$.

Equation (1.11) was derived by Planck and, as we shall see, played a very important role in his work on the blackbody problem.

Equation (1.12) for the spectral energy density of blackbody radiation was first deduced in a less explicit form by Rayleigh in 1900.[2] Although the derivation just outlined might be criticized for its reliance on a particular model of an atomic electron, it is easy to derive the Rayleigh–Jeans distribution on more general classical grounds. An electromagnetic field mode of frequency ν is basically just a linear harmonic oscillator (see Chapter 2) that, according to the classical equipartition theorem, has an average energy kT at thermal equilibrium. Since the number of modes per unit volume in the frequency interval $[\nu, \nu + d\nu]$ is $(8\pi\nu^2/c^3)d\nu$, the electromagnetic energy per unit volume in this frequency interval should be $(8\pi\nu^2/c^3)(kT)d\nu = \rho(\nu)d\nu$, which is the Rayleigh–Jeans law, independent of any particular model for the atoms with which the radiation is in thermal equilibrium. From this perspective the failure of classical theory, according to Kelvin and Rayleigh, must lie in its equipartition theorem.

Another classical result, due to Wien in 1893, must be mentioned. Wien basically followed Boltzmann's model of radiation contained in a cylinder with a piston, but included the Doppler shift of radiation reflected by the moving piston. This allowed radiant energy to be exchanged among different frequencies. Wien showed that the spectral energy density must follow the general form

$$\rho(\nu) = \nu^3 \phi_1(\nu/T) \qquad \text{(Wien displacement law)}, \qquad (1.14)$$

or, in terms of wavelength,

$$\rho(\lambda) = \rho(\nu)\left|\frac{d\nu}{d\lambda}\right| = \lambda^{-5}\phi_2(\lambda T) \qquad \text{(Wien displacement law)}, \qquad (1.15)$$

where ϕ_1 and ϕ_2 are undetermined functions. The Rayleigh–Jeans distribution obviously obeys Wien's "displacement law" (1.14).

---

[2] Motivated by Wien's work, Rayleigh also allowed for the possiblility that a factor $e^{-(\text{const})\nu/T}$ should be included, thus avoiding the ultraviolet catastrophe.

A few years later Wien presented arguments in support of the distribution

$$\rho(\lambda) = \alpha \lambda^{-5} e^{-\beta/\lambda T} \quad \text{(Wien distribution)}, \quad (1.16)$$

where α and β are constants. A similar distribution function, with the factor λ^{-5} replaced by $\lambda^{-\gamma}$, had just been proposed by Paschen as a fit to his experimental data. Paschen's data indicated that γ was between 5 and 6, thus providing some support for the displacement law. Further measurements showed that γ was indeed close to 5.

Wien's arguments for (1.16) seem to have been guided more by the desired result than by physics. To wit, he made the peculiar assumption that the wavelength and intensity of the radiation from a given atom (or molecule) are determined only by that atom's velocity. This allowed him to adduce the exponential term in (1.16) from the factor $\exp(-mv^2/2kT)$ in the Maxwell–Boltzmann velocity distribution function. In any case the Wien distribution was soon to find a more secure provenance in Planck's work.

1.3   Planck's First Theory

Given that Planck was an expert in thermodynamics, it is not surprising that his work on the blackbody problem emphasized the concept of entropy. In a series of papers in the late 1890s, Planck produced a derivation of the Wien distribution from general thermodynamical considerations plus the *assumption* that the entropy of a collection of radiators depends only on their total energy. An important result was the following relation between the entropy S and average energy U of an elementary radiator (or "molecule" for our purposes) in thermal equilibrium with radiation at temperature T :

$$\frac{\partial^2 S}{\partial U^2} = -\frac{A}{U}, \quad (1.17)$$

where for a given radiator A is a constant. From this equation and the general relation $\partial S/\partial U = 1/T$ it follows that

$$U = B e^{-1/AT}, \quad (1.18)$$

where B is another constant that, like A, may depend on the frequency of a given radiator. This result, together with (1.11), yields the radiation spectral energy density

$$\rho(\nu) = f(\nu) e^{-1/AT}, \quad (1.19)$$

where $f(\nu)$ is some function of ν. Wien's displacement law implies that $f(\nu)$ and A are proportional to ν^3 and ν^{-1}, respectively, so that

$$\rho(\nu) = C\nu^3 e^{-D\nu/T} \quad (C, D \text{ constants}) \tag{1.20}$$

or

$$\rho(\lambda) = \alpha\lambda^{-5} e^{-\beta/\lambda T} \quad (\alpha, \beta \text{ constants}), \tag{1.21}$$

which is the Wien distribution.

The Wien distribution, however, was soon found to be incorrect as experimentalists extended their spectral measurements to higher wavelengths. This was accomplished by the "residual rays" method, whereby longer wavelengths were isolated by multiple reflections off an appropriate crystal. In February 1900 Lummer and Pringsheim reported data that deviated from the Wien distribution by 40–50% for wavelengths between 12 and 18 μm, and in October similar conclusions were reported by Rubens and Kurlbaum.

It was the work of his friend Rubens that led Planck to his formula for the spectral energy density of thermal radiation. In particular, the data indicated that $\rho(\nu)$ was proportional to the temperature T for small ν and large T. Planck found a formula with that behavior at small ν and which approximated the Wien distribution for large ν.

In a paper delivered at a meeting on 19 October Planck presented his formula and provided some justification for it.[3] For small ν and large T, the experimental result $\rho(\nu) \propto T$ and equation (1.11) imply $U \propto T$ and therefore, since $\partial S/\partial U = T^{-1}$, $\partial^2 S/\partial U^2 \propto U^{-2}$ and $S \propto \log U$. On the other hand (1.17) leads to the Wien distribution, which has the correct form for large ν and small T. Planck proposed the interpolation

$$\frac{\partial^2 S}{\partial U^2} = \frac{-A}{U(B + U)} \quad (A, B \text{ constants}). \tag{1.22}$$

According to Planck, equation (1.22) "is the simplest by far of all the expressions which yield S as a logarithmic function of U (a condition which probability theory suggests) and which besides coincides with the Wien law for small values of U." Using again the relation $\partial S/\partial U = 1/T$, equation (1.11), and the Wien displacement law, one obtains from (1.22) the spectral energy density

$$\rho(\lambda) = \frac{\alpha\lambda^{-5}}{e^{\beta/\lambda T} - 1} \quad (\alpha, \beta \text{ constants}). \tag{1.23}$$

This formula was found to agree with all the existing data. In order to give it "a real physical meaning," Planck began what he later described as "a

---

[3] See the books by Kuhn and Pais and the articles by Klein cited at the end of the chapter.

few weeks of the most strenuous work of my life." The culmination of that work was the birth of quantum theory.

Planck's reasoning may be glibly summarized as follows. Consider N radiators of frequency ν and total energy $U_N = NU = P\epsilon$, where P is a large integer and ϵ is some finite element of energy. The entropy $S_N = NS = k \log W_N$, where W_N is the number of ways in which the P energy elements can be distributed among the N radiators. If $N = P = 2$, for instance, then the different partitions of the energy between the two radiators are $(2\epsilon, 0)$, (ϵ, ϵ), and $(0, 2\epsilon)$ *if the energy elements are assumed to be indistinguishable.* Under this assumption we have, in general,

$$W_N = \frac{(N - 1 + P)!}{P!(N - 1)!}, \qquad (1.24)$$

which is the number of ways in which P indistinguishable balls can be put into N distinguishable boxes. Stirling's approximation ($\log M! \cong M \log M - M$ for large M) then gives, for $N, P \gg 1$,

$$
\begin{aligned}
S &= \frac{k}{N} \log \frac{(N - 1 + P)!}{P!(N - 1)!} \\
&\cong k\left[\left(1 + \frac{P}{N}\right)\log\left(1 + \frac{P}{N}\right) - \frac{P}{N}\log\frac{P}{N}\right] \\
&= k\left[\left(1 + \frac{U}{\epsilon}\right)\log\left(1 + \frac{U}{\epsilon}\right) - \frac{U}{\epsilon}\log\frac{U}{\epsilon}\right].
\end{aligned}
\qquad (1.25)
$$

Thus

$$\frac{\partial S}{\partial U} = \frac{1}{T} = \frac{k}{\epsilon}\log\left(1 + \frac{\epsilon}{U}\right) \qquad (1.26)$$

or

$$U = \frac{\epsilon}{e^{\epsilon/kT} - 1} \qquad (1.27)$$

for the average energy of each radiator. The excellent agreement between (1.23) and experiment, together with equation (1.11), suggests that ϵ is inversely proportional to the wavelength, or directly proportional to the frequency of the oscillator:

$$\epsilon = h\nu. \qquad (1.28)$$

Then

$$U = \frac{h\nu}{e^{h\nu/kT} - 1} \qquad (1.29)$$

and (1.11) implies

$$\rho(\nu) = \frac{8\pi h\nu^3/c^3}{e^{h\nu/kT} - 1} \quad \text{(Planck spectrum)} \qquad (1.30)$$

for the spectral energy density of thermal radiation.

The expression (1.25) for S satisfies equation (1.22) with $A = k$ and $B = \epsilon$. Once (1.25) is obtained, therefore, one is led to the form (1.23) for the spectral energy density. The great success of (1.23) in fitting the experimental data led Planck to what he later called an "act of desperation" needed to derive (1.25).

One aspect of this desperate act is the way Planck counted the number of ways, or "complexions," in which P energy elements could be distributed among N radiators. His counting procedure was totally at odds with classical statistical methods in its treatment of the energy elements as fundamentally *indistinguishable*. In one sense Planck was following Boltzmann in regarding all complexions as equally likely, but of course his way of counting the number of complexions was radically different. His "energy elements" obeyed what would much later be recognized as Bose–Einstein statistics.

Another revolutionary (nonclassical) aspect of Planck's calculation, of course, is the physical significance it attaches to the "energy elements" of size ϵ, and the relation (1.28) between ϵ and the frequency ν of a *material* oscillator. Boltzmann had also employed "energy elements" in his counting of complexions, but in his calculations ϵ had no particular significance and in fact could ultimately be taken to be zero once a formula for W_N had been obtained. If Planck had taken the limit $\epsilon \to 0$ in equation (1.26), however, then $\partial S/\partial U \to k/U$ and $\partial^2 S/\partial U^2 \to -k/U^2$, which leads to the Rayleigh–Jeans distribution. In Planck's derivation of his spectrum, therefore, the quantization of energy was absolutely essential.

This is the traditional view of Planck's innovation. It should be noted, however, that Kuhn (1978) has concluded that Planck did not in 1900 introduce any physical quantization of either radiation or material radiators. He argues that Planck's radiators were simply "a device for bringing radiation to equilibrium, and it was justified, not by knowledge of the physical processes involved, but by Kirchhoff's law, which made the equilibrium field independent of the equilibrium-producing material."

Until about 1905 Planck's formula was regarded as little more than a superb fit to the experimental data. Its true significance began to be appreciated only when it was realized that the Rayleigh–Jeans law was an inevitable consequence of classical physics and the equipartition theorem, and therefore that the blackbody experiments had uncovered a fundamental failure of known (classical) theory.

A curious circumstance relating to zero-point energy, which was noted by Einstein and Stern (1913), is worth mentioning. Consider the classical limit $kT \gg h\nu$ of the expression (1.29) for the average energy of an oscillator in thermal equilibrium with radiation:

$$U = \frac{h\nu}{e^{h\nu/kT} - 1} \cong \frac{h\nu}{1 + \frac{h\nu}{kT} + \frac{1}{2}(\frac{h\nu}{kT})^2 - 1} = \frac{kT}{1 + \frac{1}{2}\frac{h\nu}{kT}} \cong kT - \frac{1}{2}h\nu. \quad (1.31)$$

Thus U contains a first order temperature-independent correction to kT, the energy predicted by the equipartition theorem in the classical limit. But

$$U + \frac{1}{2}h\nu = \frac{h\nu}{e^{h\nu/kT} - 1} + \frac{1}{2}h\nu, \quad (1.32)$$

which includes the zero-point energy $\frac{1}{2}h\nu$, does not have a first-order correction to kT in the classical limit. In Planck's "second theory" U was in fact replaced by $U + \frac{1}{2}h\nu$.

1.4   Planck's Zero-Point Energy

It was mentioned earlier that it took several years for the profound significance of Planck's distribution to be appreciated. Planck himself was unsatisfied with the largely ad hoc theory he had used to derive his spectrum, and for many years he explored alternative hypotheses that might lead to it.

In 1912 Planck published his "second theory." The absorption of radiation was assumed to proceed according to classical theory, whereas emission of radiation occurred discontinuously in discrete quanta of energy. Assume that an oscillator can radiate only after it has (continuously) absorbed an energy $h\nu$. Let P_n be the probability that it has energy between $(n-1)h\nu$ and $nh\nu$. When, as a result of absorption of radiation, its energy reaches $nh\nu$, there is a probability p that it will lose *all* its energy in the form of radiation, and a probability $1-p$ that it continues to absorb without emission of radiation. Thus $P_2 = P_1(1-p)$, $P_3 = P_2(1-p) = P_1(1-p)^2$, ..., $P_n = P_1(1-p)^{n-1}$, and

$$\sum_{n=1}^{\infty} P_n = 1 = \sum_{n=1}^{\infty} P_1(1-p)^{n-1} = P_1/p \quad (1.33)$$

or $P_1 = p$ is the probability that an oscillator in equilibrium with radiation has energy between 0 and $h\nu$, $P_2 = p(1-p)$ is the probability that it has energy between $h\nu$ and $2h\nu$, and $P_n = p(1-p)^{n-1}$ is the probability that it has energy between $(n-1)h\nu$ and $nh\nu$. Following Boltzmann, Planck defines the oscillator entropy as

$$S = -k\sum_{n=1}^{\infty} P_n \log P_n = -k\sum_{n=1}^{\infty} p(1-p)^{n-1} \log[p(1-p)^{n-1}]$$

$$= -k \left[\frac{1}{p} \log p + (\frac{1}{p} - 1) \log(\frac{1}{p} - 1) \right]. \tag{1.34}$$

Planck now assumes that all energies between $(n-1)h\nu$ and $nh\nu$ are equally likely, so that the average energy of the oscillators with energy between $(n-1)h\nu$ and $nh\nu$ is $\frac{1}{2}(n+n-1)h\nu = (n - \frac{1}{2})h\nu$. The average oscillator energy is then

$$U = \sum_{n=1}^{\infty}(n - \frac{1}{2})h\nu P_n = h\nu \sum_{n=1}^{\infty}(n - \frac{1}{2})p(1-p)^{n-1} = (\frac{1}{p} - \frac{1}{2})h\nu \tag{1.35}$$

or $1/p = U/h\nu + \frac{1}{2}$. From (1.34), therefore,

$$S = k \left[(\frac{U}{h\nu} + \frac{1}{2}) \log(\frac{U}{h\nu} + \frac{1}{2}) - (\frac{U}{h\nu} - \frac{1}{2}) \log(\frac{U}{h\nu} - \frac{1}{2}) \right]. \tag{1.36}$$

Using once again the relation $\partial S/\partial U = 1/T$, Planck obtained

$$U = \frac{1}{2}h\nu \frac{e^{h\nu/kT}+1}{e^{h\nu/kT}-1} = \frac{h\nu}{e^{h\nu/kT}-1} + \frac{1}{2}h\nu. \tag{1.37}$$

This implies that $U \neq 0$ when $T \rightarrow 0$: when $T \rightarrow 0$, $U \rightarrow \frac{1}{2}h\nu$. *Planck's equation (1.37) marked the birth of the concept of zero-point energy.*

To derive $\rho(\nu)$ Planck could not resort to equation (1.11), since the derivation of that equation assumed continuous absorption and emission processes. Instead he made the assumption that the ratio of the probability that an oscillator does not emit radiation, to the probability that it does, is proportional to $\rho(\nu)$: $(1-p)/p = C\rho(\nu)$, or $1/p = C\rho(\nu) + 1$, where C is a constant of proportionality. This assumption is plausible in that, the greater the radiation intensity, the more absorption should dominate emission. (Planck, of course, was not at this time aware of the possibility of *stimulated* emission!) Then, from (1.35), $U = [C\rho(\nu) + \frac{1}{2}]h\nu$ or

$$\rho(\nu) = \frac{1}{C} \frac{1}{e^{h\nu/kT}-1} . \tag{1.38}$$

To determine C Planck appeals to the classical limit, where the Rayleigh–Jeans law should apply: for $kT >> h\nu$, $\rho(\nu)$ should reduce to (1.12), which requires that $1/C = 8\pi h\nu^3/c^3$ and therefore that

$$\rho(\nu) = \frac{8\pi h\nu^3/c^3}{e^{h\nu/kT}-1} . \tag{1.39}$$

It is interesting that, in deducing C in this way, Planck was employing what would soon be called the *correspondence principle*. Furthermore

Planck's probability p might well be regarded as the first example of a quantum transition probability.

It is also noteworthy that in Planck's second theory the material oscillators have zero-point energy but the electromagnetic field does not: $\rho(\nu) \to 0$ for $T \to 0$. Had Planck simply used equation (1.11) to relate $\rho(\nu)$ and U, he would have obtained from (1.37) the spectral energy density

$$\rho'(\nu) = \rho(\nu) + 4\pi h\nu^3/c^3 = \frac{8\pi h\nu^3/c^3}{e^{h\nu/kT} - 1} + 4\pi h\nu^3/c^3, \qquad (1.40)$$

which, as we will see later, turns out to be the correct spectrum from the standpoint of modern quantum electrodynamics. The zero-point energy appearing in Planck's expression (1.37) is also perfectly correct according to modern theory, even though Planck's route to it is not.

By 1914 Planck was convinced that zero-point energy would be of no experimental consequence. However, the concept attracted much attention, and soon came to play a major role in the work of Einstein.

1.5   The Einstein–Hopf Model

"Concerning a Heuristic Point of View Toward the Emission and Transformation of Light," Einstein (1905) deduced that radiation satisfying the Wien distribution "behaves thermodynamically as though it consisted of a number of independent energy quanta of magnitude $[h\nu]$." Based on this viewpoint he *predicted* the linear relation between radiation frequency and stopping potential in the photoelectric effect, a prediction confirmed by Millikan's experiments in 1916. In 1906 he argued that "in emission and absorption the energy of a [Planck oscillator] changes by jumps which are integral multiples of $h\nu$." These were the beginnings of the photon concept.

Einstein struggled with the blackbody problem for more than ten years after he introduced his heuristic viewpoint concerning energy quanta of radiation. In one important paper Einstein and Hopf (1910b) studied a simple model for the thermal equilibrium between oscillating dipoles and electromagnetic radiation. Imagine each dipole to consist of a particle of mass m and charge e, bound by an elastic restoring force to a mass $M(\gg m)$ of opposite charge. The equation of motion for a linear dipole oscillator is then (see Appendix A)

$$\frac{d^2z}{dt^2} + \omega_o^2 z - \tau\,\ddot{z} = \frac{e}{m}E_z(t), \qquad (1.41)$$

where $\omega_o(= 2\pi\nu_o)$ is the natural oscillation frequency, $E_z(t)$ is the z-component of the external electric field acting on the particle, $\tau\,\dddot{z}$ is the radiation reaction term, and $\tau = 2e^2/3mc^3$. The two oppositely charged particles define an electric dipole moment $ez(t)$. Implicit in equation (1.41) is the *electric dipole approximation* of neglecting any spatial variation of $E_z(t)$ over the distance separating the particles. It is also assumed that the interaction of the dipole with the magnetic field is negligible.

Equation (1.41) is essentially the same equation used earlier by Planck to derive equation (1.11) (Appendix A). In the Einstein–Hopf model, however, the dipole oscillators of mass $M + m \cong M$ are allowed to move; for simplicity they are constrained to move only along the x axis. Einstein and Hopf showed that there is a retarding force on a moving dipole as a result of its interaction with the field. This force acts to decrease its kinetic energy. Due to recoil associated with emission and absorption, however, the field also acts to increase the kinetic energy of a dipole. The condition for equilibrium is that the increase in kinetic energy due to recoil balances the decrease in kinetic energy associated with the retarding force.

Assuming $v/c << 1$, Einstein and Hopf showed that the retarding force due to motion through a thermal field of spectral energy density $\rho(\omega_o)$ is

$$F = -Rv, \tag{1.42}$$

where

$$R = \frac{4\pi^2 e^2}{5mc^2}\left[\rho(\omega_o) - \frac{\omega_o}{3}\frac{d\rho}{d\omega_o}\right] \tag{1.43}$$

and v is the velocity of the dipole. Essentially this same result is derived in Appendix B.

Consider now a dipole with linear momentum $Mv(t)$ at time t. After a short time δt its momentum is

$$Mv(t + \delta t) = Mv(t) + \Delta - Rv(t)\delta t, \tag{1.44}$$

where Δ is the impulse imparted to the dipole in the time interval δt as a result of recoil associated with emission and absorption of radiation. Then

$$M^2v^2(t + \delta t) - M^2v^2(t) = \Delta^2 - 2MRv^2(t)\delta t + (2M - R\delta t)v(t)\Delta \tag{1.45}$$

when δt is taken to be small enough (or M large enough) that terms quadratic in δt are negligible. Now take the equilibrium ensemble average of both sides of (1.45):

$$2M[\langle\frac{1}{2}Mv^2(t + \delta t)\rangle - \langle\frac{1}{2}Mv^2(t)\rangle] = 0 = \langle\Delta^2\rangle - 4R\delta t\langle\frac{1}{2}Mv^2(t)\rangle. \tag{1.46}$$

In writing this expression we have used the fact that $\langle v(t)\Delta\rangle = 0$, since Δ is equally likely to be positive or negative in the time interval from t

to $t + \delta t$. In thermal equilibrium, furthermore, the equipartition theorem requires the average kinetic energy to be $\langle \frac{1}{2} M v^2(t) \rangle = \frac{1}{2} kT$. The condition for thermal equilibrium is therefore

$$(\delta t)^{-1} \langle \Delta^2 \rangle = 2RkT. \tag{1.47}$$

It remains to determine $\langle \Delta^2 \rangle$.

The force on an electric dipole moment $e \hat{z} z(t)$ in the Einstein–Hopf model, where the dipole points in the z direction and is free to move only along x, may for our purposes be taken to be $F_x = e \hat{z} z(t) \partial E_z(t) / \partial x$.[4] The impulse imparted to the particle during the time interval from $t = 0$ to $t = \delta t$ is thus

$$\Delta = e \int_0^{\delta t} dt\, z(t) \frac{\partial E_z(t)}{\partial x} . \tag{1.48}$$

Einstein and Hopf write the electric field as a superposition of plane waves with independent random phases $\theta_{\mathbf{k}\lambda}$:

$$\mathbf{E}(\mathbf{r}, t) = i \sum_{\mathbf{k}\lambda} [A_{\mathbf{k}\lambda} e^{-i(\omega_k t + \theta_{\mathbf{k}\lambda})} - A^*_{\mathbf{k}\lambda} e^{i(\omega_k t + \theta_{\mathbf{k}\lambda})}] \mathbf{e}_{\mathbf{k}\lambda}, \tag{1.49}$$

where $\mathbf{e}_{\mathbf{k}\lambda}$ is a unit polarization vector for a plane wave with wave vector $\mathbf{k}$ and linear polarization index λ ($= 1, 2$). The steady-state solution of equation (1.41) is then

$$z(t) = \frac{ie}{m} \sum_{\mathbf{k}\lambda} [F_{\mathbf{k}\lambda} e^{-i(\omega_k t + \theta_{\mathbf{k}\lambda})} - F^*_{\mathbf{k}\lambda} e^{i(\omega_k t + \theta_{\mathbf{k}\lambda})}], \tag{1.50}$$

where the origin of coordinates has been chosen to be at the position of the dipole and $F_{\mathbf{k}\lambda} = -A_{\mathbf{k}\lambda} e_{\mathbf{k}\lambda z} [\omega_k^2 - \omega_o^2 + i\tau\omega_k^3]^{-1}$, where $e_{\mathbf{k}\lambda z}$ is the z-component of $\mathbf{e}_{\mathbf{k}\lambda}$. In a separate paper Einstein and Hopf (1910a) show that $E_z(t)$ and $\partial E_z / \partial x$ must be treated as independent random variables in the time integral (1.48). It then follows from (1.48)–(1.50) by straightforward manipulations that $\langle \Delta \rangle = 0$ and

$$(\delta t)^{-1} \langle \Delta^2 \rangle = \left(\frac{4\pi^4 c^4 \tau}{5\omega_o^2} \right) \rho^2(\omega_o), \tag{1.51}$$

where the ensemble average is taken over the random phases $\theta_{\mathbf{k}\lambda}$.

Equation (1.47), together with (1.43) and (1.51), now gives a differential equation that must be satisfied by the spectral energy density of thermal radiation:

$$\rho(\omega) - \frac{\omega}{3} \frac{d\rho}{d\omega} = \left(\frac{\pi^2 c^3}{3\omega^2 kT} \right) \rho^2(\omega). \tag{1.52}$$

---

[4] See P. W. Milonni and M.-L. Shih, *Am. J. Phys.* **59**, 684 (1991).

The solution of this equation satisfying $\rho(0) = 0$ is

$$\rho(\omega) = \frac{\omega^2 kT}{\pi^2 c^3}, \tag{1.53}$$

which is seen to be just the Rayleigh–Jeans law when we recall that $\omega = 2\pi\nu$ and $\rho(\omega) = \rho(\nu)/2\pi$.

The beautifully cogent arguments of Einstein and Hopf provide further evidence that the Rayleigh–Jeans law is an inexorable consequence of classical physics. However, we shall see that their results are dramatically altered when zero-point energy is postulated.

1.6   Einstein and Stern's Zero-Point Energy

In 1913 Einstein and Stern noted that an ad hoc postulate about zero-point energy in the Einstein–Hopf model would lead to the Planck spectrum. First let us note that equation (1.11) allows us to write (1.52) in a form in which the average dipole energy U appears explicitly:

$$\rho(\omega) - \frac{\omega}{3}\frac{d\rho}{d\omega} = \frac{1}{3kT}\rho(\omega)U. \tag{1.54}$$

Now suppose the average oscillator energy U is replaced by $U + \hbar\omega$. This means that the dipole oscillators are now assumed to have a zero-point energy $\hbar\omega$. Equation (1.54) is then replaced by

$$\begin{aligned}
\rho(\omega) - \frac{\omega}{3}\frac{d\rho}{d\omega} &= \frac{1}{3kT}\rho(\omega)U + \frac{\hbar\omega}{3kT}\rho(\omega) \\
&= \frac{\pi^2 c^3}{3\omega^2 kT}\rho^2(\omega) + \frac{\hbar\omega}{3kT}\rho(\omega) \\
&= \frac{\pi^2 c^3}{3\omega^2 kT}[\rho^2(\omega) + \frac{\hbar\omega^3}{\pi^2 c^3}\rho(\omega)].
\end{aligned} \tag{1.55}$$

The solution of this equation satisfying $\rho(0) = 0$ is

$$\rho(\omega) = \frac{\hbar\omega^3/\pi^2 c^3}{e^{\hbar\omega/kT} - 1} \quad \text{(Planck spectrum).} \tag{1.56}$$

In other words, if it is assumed that the dipole oscillators in the Einstein–Hopf model have a zero-point energy $\hbar\omega$, then the equilibrium spectrum of radiation is found to be the Planck spectrum.

The oscillator zero-point energy postulated by Einstein and Stern is twice that found earlier by Planck. Since we now know that Planck's zero-point energy $\frac{1}{2}\hbar\omega$ is the correct one, it is interesting to see how Einstein and Stern arrived at the correct spectrum using the wrong zero-point energy.

According to quantum theory a field mode of frequency ω, like a material oscillator, has a zero-point energy $\frac{1}{2}\hbar\omega$ (see Chapter 2). The total zero-point energy of a linear dipole oscillator of frequency ω and a field mode of the same frequency is therefore $\frac{1}{2}\hbar\omega + \frac{1}{2}\hbar\omega = \hbar\omega$. Einstein and Stern's zero-point energy $\hbar\omega$ is just this, *but they attributed it solely to the material dipole oscillators.*

Suppose we include in the Einstein–Hopf model a zero-point energy $\frac{1}{2}\hbar\omega$ for a dipole oscillator *and* a zero-point energy $\frac{1}{2}\hbar\omega$ for each field mode. Since there are $(8\pi\nu^2/c^3)d\nu = (\omega^2/\pi^2 c^3)d\omega$ field modes per unit volume in the frequency interval $[\omega, \omega + d\omega]$, the spectral energy density of the zero-point field is

$$\rho_0(\omega) = (\omega^2/\pi^2 c^3)\frac{1}{2}\hbar\omega = \frac{\hbar\omega^3}{2\pi^2 c^3} . \tag{1.57}$$

If we replace $\rho(\omega)$ in (1.54) by $\rho(\omega) + \rho_0(\omega)$, the left side is unchanged:

$$[\rho(\omega) + \rho_0(\omega)] - \frac{\omega}{3}\frac{d}{d\omega}[\rho(\omega) + \rho_0(\omega)] = \rho(\omega) - \frac{\omega}{3}\frac{d\rho}{d\omega} . \tag{1.58}$$

If we also account for the zero-point energy of the dipole oscillators by replacing U by $U + \frac{1}{2}\hbar\omega$, the product $\rho(\omega)U$ on the right side of (1.54) is changed to

$$
\begin{aligned}
[\rho(\omega) \quad + \quad &\rho_0(\omega)][U + \frac{1}{2}\hbar\omega] = \rho(\omega)U + \frac{1}{2}\hbar\omega\rho(\omega) \\
&+ \rho_0(\omega)U + \frac{1}{2}\hbar\omega\rho_0(\omega) \\
= \quad &\frac{\pi^2 c^3}{\omega^2}[\rho^2(\omega) + \rho_0(\omega)\rho(\omega) + \frac{\hbar\omega^3}{2\pi^2 c^3}\rho(\omega)] + \frac{1}{2}\hbar\omega\rho_0(\omega) \\
= \quad &\frac{\pi^2 c^3}{\omega^2}[\rho^2(\omega) + \frac{\hbar\omega^3}{\pi^2 c^3}\rho(\omega)] + \frac{1}{2}\hbar\omega\rho_0(\omega), \tag{1.59}
\end{aligned}
$$

where we have used (1.57) and (1.11) in the form $U = (\pi^2 c^3/\omega^2)\rho(\omega)$.

The term $\frac{1}{2}\hbar\omega\rho_0(\omega)$ in (1.59) results from a coupling of the zero-point motion of a dipole oscillator to the zero-point oscillations of the field. In quantum theory, in effect, no such coupling arises: an oscillator in its ground state in the absence of any applied field remains in its ground state. We shall see later how this comes about, but for now let us just accept it and drop the term $\frac{1}{2}\hbar\omega\rho_0(\omega)$ in (1.59):

$$[\rho(\omega) + \rho_0(\omega)][U + \frac{1}{2}\hbar\omega] \rightarrow \frac{\pi^2 c^3}{\omega^2}[\rho^2(\omega) + \frac{\hbar\omega^3}{\pi^2 c^3}\rho(\omega)]. \tag{1.60}$$

From (1.54), (1.58), and (1.60), then, we have

$$\rho(\omega) - \frac{\omega}{3}\frac{d\rho}{d\omega} = \frac{\pi^2 c^3}{3\omega^2 kT}[\rho^2(\omega) + \frac{\hbar\omega^3}{\pi^2 c^3}\rho(\omega)], \tag{1.61}$$

which is exactly the Einstein–Stern equation (1.55). The complete spectrum $\rho(\omega) + \rho_0(\omega)$ is then given by equation (1.40).

This route to the Planck spectrum may be summarized as follows. We modified the Einstein–Hopf model to include a zero-point energy $\frac{1}{2}\hbar\omega$ for a dipole oscillator and a zero-point energy $\frac{1}{2}\hbar\omega$ for each mode of the electromagnetic field, and anticipated a result of quantum theory that there is no contribution from the coupling of the zero-point oscillations of the dipole and the field. This led to the Einstein–Stern equation (1.55). Einstein and Stern, however, did not invoke any zero-point energy of the field, and to arrive at the Planck spectrum their dipole oscillators had to have a zero-point energy $\frac{1}{2}\hbar\omega$ plus what we now know to be the zero-point energy of a field mode of the same frequency.

Why did Einstein and Stern not assume zero-point energy for the field? After all, one might have thought that the relation (1.11) between $\rho(\omega)$ and U would have made it obvious that, if either the dipole oscillator or the field has a zero-point energy, then so must the other. If Planck's zero-point energy $\frac{1}{2}\hbar\omega$ is added to U in (1.11), for instance, then for consistency we must add the spectral energy density $\rho_0(\omega)$ of the zero-point field to $\rho(\omega)$:

$$\rho(\omega) + \rho_0(\omega) = \frac{\omega^2}{\pi^2 c^3}\left(U + \frac{1}{2}\hbar\omega\right), \qquad (1.62)$$

or again $\rho_0(\omega) = \hbar\omega^3/2\pi^2 c^3$, which in turn implies that each field mode has a zero-point energy $\frac{1}{2}\hbar\omega$.

However, such a "consistency" argument rests on the usual acuity of hindsight. The fact is that at various stages in Einstein's long efforts to understand the Planck spectrum he seriously doubted the general validity of Planck's equation (1.11). This is not surprising, for if Planck had simply invoked equipartition of energy and used $U = kT$ in (1.11), he would have obtained the Rayleigh–Jeans spectrum. It is not clear whether Planck was even aware at the time of the classical equipartition theorem. If he had known and believed the equipartition theorem, as Einstein later remarked, "he would probably not have made his great discovery."[5]

There is another reason why Einstein and Stern might have been unwilling to attribute a zero-point energy to the field: if $\rho(\omega)$ and U are replaced by $\rho(\omega) + \rho_0(\omega)$ and $U + \frac{1}{2}\hbar\omega$, respectively, in the Einstein–Hopf model, then one obtains the Rayleigh–Jeans spectrum for the total spectral density $\rho(\omega) + \rho_0(\omega)$. Crucial to the derivation of the Planck spectrum is the omission of the term $\frac{1}{2}\hbar\omega\rho_0(\omega)$ in (1.59). This omission occurs automatically in the quantum theory of the Einstein–Hopf model, as we shall see in the next

---

[5] *Albert Einstein: Philosopher-Scientist*, ed. P. A. Schilpp (Tudor, New York, 1949), p. 43.

chapter. Without this consequence of quantum theory available to them, Einstein and Stern may have simply discounted the possibility of zero-point electromagnetic energy. Indeed, the first suggestion that there might be a zero-point electromagnetic field is due not to Planck or Einstein and Stern, but to Nernst (1916).

1.7   Einstein's Fluctuation Formula

Prior to his work with Hopf and Stern, Einstein (1909) had derived a formula for the energy fluctuations of thermal radiation. Denoting the variance in energy in the volume V and in the frequency interval $[\omega, \omega + d\omega]$ by $\langle \Delta E_\omega^2 \rangle$, we may write the Einstein fluctuation formula as

$$\langle \Delta E_\omega^2 \rangle = [\hbar\omega\rho(\omega) + \frac{\pi^2 c^3}{\omega^2}\rho^2(\omega)]V\,d\omega. \tag{1.63}$$

The importance of this formula lies in Einstein's interpretation of it. The first term in brackets, according to Einstein, may be obtained "if radiation were to consist of independently moving pointlike quanta of energy $h\nu$":

$$\langle \Delta E_\omega^2 \rangle_{\text{particles}} = \hbar\omega\rho(\omega)V\,d\omega, \tag{1.64}$$

whereas the second term follows when the field is treated as a superposition of independently fluctuating *waves*:

$$\langle \Delta E_\omega^2 \rangle_{\text{waves}} = \frac{\pi^2 c^3}{\omega^2}\rho^2(\omega)V\,d\omega. \tag{1.65}$$

Thus $\langle \Delta E_\omega^2 \rangle$ has both wave and particle contributions. The Einstein fluctuation formula was the earliest indicator of the wave-particle dualism in quantum theory.

The "wave" term (1.65) may be derived from the superposition (1.49) of waves with independent random phases. For instance,

$$\begin{aligned}
\langle \mathbf{E}^2(\mathbf{r}, t) \rangle &= -2\text{Re}\sum_{\mathbf{k}_1\lambda_1}\sum_{\mathbf{k}_2\lambda_2}[A_{\mathbf{k}_1\lambda_1}A_{\mathbf{k}_2\lambda_2}e^{-i(\omega_{k_1}+\omega_{k_2})t}e^{i(\mathbf{k}_1+\mathbf{k}_2)\cdot\mathbf{r}} \\
&\quad \times \langle e^{-i(\theta_{\mathbf{k}_1\lambda_1}+\theta_{\mathbf{k}_2\lambda_2})}\rangle - A_{\mathbf{k}_1\lambda_1}A_{\mathbf{k}_2\lambda_2}^* e^{-i(\omega_{k_1}-\omega_{k_2})t} \\
&\quad \times e^{i(\mathbf{k}_1-\mathbf{k}_2)\cdot\mathbf{r}}\langle e^{-i(\theta_{\mathbf{k}_1\lambda_1}-\theta_{\mathbf{k}_2\lambda_2})}\rangle]\mathbf{e}_{\mathbf{k}_1\lambda_1}\cdot\mathbf{e}_{\mathbf{k}_2\lambda_2}, \quad (1.66)
\end{aligned}$$

where again the average is over the phases $\theta_{\mathbf{k}\lambda}$, which are assumed to be independent, uniformly distributed random variables on the interval $[0, 2\pi]$.

Thus

$$\langle \mathbf{E}^2(\mathbf{r}, t) \rangle = 2 \sum_{\mathbf{k}\lambda} |A_{\mathbf{k}\lambda}|^2 \tag{1.67}$$

and similarly

$$\langle \mathbf{E}^4(\mathbf{r}, t) \rangle = 8(\sum_{\mathbf{k}\lambda} |A_{\mathbf{k}\lambda}|^2)^2, \tag{1.68}$$

so that

$$\langle \mathbf{E}^4(\mathbf{r}, t) \rangle - \langle \mathbf{E}^2(\mathbf{r}, t) \rangle^2 = 4(\sum_{\mathbf{k}\lambda} |A_{\mathbf{k}\lambda}|^2)^2 = \langle \mathbf{E}^2(\mathbf{r}, t) \rangle^2. \tag{1.69}$$

Since the electromagnetic energy density is proportional to $\langle \mathbf{E}^2 \rangle$, it follows from (1.69) that the variance in energy associated with frequency ω is proportional to $\rho^2(\omega)$. We omit the trivial details of the derivation, which leads directly to equation (1.65).

The "particle" term (1.64) in the Einstein fluctuation formula is of far less obvious origin, and to derive it we temporarily assume the field energy can be written as

$$E = \sum_{\mathbf{k}\lambda} n_{\mathbf{k}\lambda} \hbar \omega_k, \tag{1.70}$$

so that its variance is

$$\langle \Delta E^2 \rangle = \sum_{\mathbf{k}\lambda} \langle \Delta n_{\mathbf{k}\lambda} \rangle^2 \hbar^2 \omega_k{}^2, \tag{1.71}$$

where the $n_{\mathbf{k}\lambda}$ are integers. Thus we are assuming that the field energy is comprised of discrete quanta of energy $\hbar \omega_k$, and that the numbers of quanta associated with different modes fluctuate independently. We assume Poisson statistics for these quanta, so that

$$\langle \Delta n_{\mathbf{k}\lambda}^2 \rangle = \langle n_{\mathbf{k}\lambda} \rangle \tag{1.72}$$

and

$$\langle \Delta E^2 \rangle = \sum_{\mathbf{k}\lambda} \langle n_{\mathbf{k}\lambda} \rangle \hbar^2 \omega_k{}^2. \tag{1.73}$$

Since $\rho(\omega)$ is proportional to the average number of photons at frequency ω, equation (1.73) leads easily to the particle term (1.64) in the Einstein fluctuation formula.

The Einstein fluctuation formula is derived more thoroughly in the next chapter. For the present discussion we simply note that *we can obtain both* the "wave" and "particle" terms using the classical wave picture with zero-point energy. That is, if we replace $\rho(\omega)$ in (1.65) by $\rho(\omega) + \rho_o(\omega)$, where the spectral energy density $\rho_o(\omega)$ of the zero-point field is given by (1.57), we have

$$
\begin{aligned}
\langle \Delta E_\omega^2 \rangle_{\text{waves}} \quad &\rightarrow \quad \frac{\pi^2 c^3}{\omega^2} [\rho^2(\omega) + 2\rho_o(\omega)\rho(\omega) + \rho_o^2(\omega)] V \, d\omega \\
&= \quad [\frac{\pi^2 c^3}{\omega^2} \rho^2(\omega) + \hbar\omega\rho(\omega)] V \, d\omega + \frac{\pi^2 c^3}{\omega^2} \rho_o^2(\omega) V \, d\omega \\
&= \quad \langle \Delta E_\omega^2 \rangle_{\text{waves}} + \langle \Delta E_\omega^2 \rangle_{\text{particles}} + \frac{\pi^2 c^3}{\omega^2} \rho_o^2(\omega) V \, d\omega \\
&= \quad \langle \Delta E_\omega^2 \rangle_{\text{waves}} + \langle \Delta E_\omega^2 \rangle_{\text{particles}} + \frac{1}{2} \hbar\omega \rho_o(\omega) V \, d\omega.
\end{aligned}
$$
(1.74)

The "extra" (third) term in this expression does not appear in the Einstein fluctuation formula. Indeed we shall see in the following chapter that it does not appear at all in quantum theory, for the same reason that the term $\frac{1}{2}\hbar\omega\rho_o(\omega)$ in equation (1.59) is absent in quantum theory.

But aside from this spurious "extra" term, we have obtained the Einstein fluctuation formula from a classical wave perspective that includes zero-point field energy. Obviously the argument is essentially the same as in our approach to the Einstein–Stern theory, and suggests that *the particle term in the Einstein fluctuation formula may be regarded as a consequence of zero-point field energy.*

The particle term was in fact the novel element in Einstein's fluctuation formula, and Einstein emphasized that this term was incompatible with classical wave theory (without zero-point energy). If there were only classical wave fluctuations in thermal radiation, we could ignore the term proportional to $\rho(\omega)$ in equation (1.61). The result is

$$
\rho(\omega) - \frac{\omega}{3} \frac{d\rho}{d\omega} = \frac{\pi^2 c^3}{3\omega^2 kT} \rho^2(\omega)
$$
(1.75)

and the solution is the Rayleigh–Jeans spectrum, $\rho(\omega) = (\omega^2/\pi^2 c^3)kT$. Without the wave term, on the other hand, (1.61) becomes

$$
\rho(\omega) - \frac{\omega}{3} \frac{d\rho}{d\omega} = \frac{\hbar\omega}{3kT} \rho(\omega)
$$
(1.76)

and the solution of this equation is $\rho(\omega) = (\hbar\omega^3/\pi^2 c^3)e^{-\hbar\omega/kT}$, the Wien distribution. This is consistent with the fact that in 1905 Einstein had deduced his "heuristic point of view" concerning radiation energy quanta by considering only radiation satisfying the Wien distribution.

1.8 Einstein's A and B Coefficients

Einstein wrote to his friend Besso in November 1916 that "A splendid light has dawned on me about the absorption and emission of radiation." He was referring to his new insight into his "heuristic principle" of 1905, and the basis it provided for an "astonishingly simple" derivation of the Planck spectrum.

For the sake of completeness we summarize the argument here. Einstein assumes that an atom (or molecule) has discrete energy levels. Let N_1 and N_2 be the numbers of atoms in energy levels E_1 and E_2, respectively, with $E_2 > E_1$. (For simplicity we ignore the possibility of level degeneracies, which does not affect the result for the spectral density of thermal radiation.) The rate at which N_1 changes due to the absorption of radiation, with the atom making an upward transition to the level E_2, is assumed to be proportional to N_1 and the spectral energy density $\rho(\omega_o)$ at the Bohr transition frequency $\omega_o = (E_2 - E_1)/\hbar$:

$$(\dot{N}_1)_{\text{absorption}} = -B_{12}N_1\rho(\omega_o). \tag{1.77}$$

Einstein proposes two kinds of emission processes by which an atom can jump from level E_2 to E_1 with the emission of radiation of frequency ω_o. One is spontaneous emission, which can occur in the absence of any radiation and is described by the rate constant A_{21} :

$$(\dot{N}_1)_{\text{spontaneous emission}} = A_{21}N_2. \tag{1.78}$$

The other is *stimulated* emission, which is assumed to proceed at a rate proportional to both N_2 and $\rho(\omega_o)$:

$$(\dot{N}_1)_{\text{stimulated emission}} = B_{21}N_2\rho(\omega_o). \tag{1.79}$$

The condition for equilibrium is

$$(\dot{N}_1)_{\text{absorption}} + (\dot{N}_1)_{\text{spontaneous emission}} + (\dot{N}_1)_{\text{stimulated emission}} = 0 \tag{1.80}$$

or

$$A_{21}N_2 + B_{21}N_2\rho(\omega_o) = B_{12}N_1\rho(\omega_o), \tag{1.81}$$

$$\rho(\omega_o) = \frac{A_{21}/B_{21}}{(B_{12}/B_{21})(N_1/N_2) - 1} = \frac{A_{21}/B_{21}}{(B_{12}/B_{21})e^{\hbar\omega_o/kT} - 1}, \tag{1.82}$$

since $N_2/N_1 = e^{-(E_2 - E_1)/kT} = e^{-\hbar\omega_o/kT}$ in thermal equilibrium. We are using Bohr's postulate (1913) that $E_2 - E_1 = \hbar\omega_o$, but it is worth noting that this relation in fact emerged naturally from Einstein's analysis once the assumption of discrete energy levels was made and the Wien displacement law was invoked.

At very high temperatures $\rho(\omega_o)$ becomes so large that spontaneous emission is much less probable than stimulated emission. Then from (1.81) we must have $B_{21} = B_{12}$ and, from (1.82),

$$\rho(\omega_o) = \frac{A_{21}/B_{21}}{e^{\hbar\omega_o/kT} - 1}. \tag{1.83}$$

For $kT \gg \hbar\omega_o$, furthermore,

$$\rho(\omega_o) \cong \frac{A_{21}}{B_{21}} \frac{kT}{\hbar\omega_o}. \tag{1.84}$$

This is the limit where the radiation energy quanta are so small compared with kT that the classical Rayleigh-Jeans law should be applicable. This requires $(A_{21}/B_{21})(kT/\hbar\omega_o) = (\omega_o^2/\pi^2 c^3)kT$, or

$$\frac{A_{21}}{B_{21}} = \frac{\hbar\omega_o^3}{\pi^2 c^3} \tag{1.85}$$

and equation (1.83) then yields the Planck spectrum for $\rho(\omega)$.

This derivation of the Planck spectrum joined aspects of Einstein's earlier work on radiation quanta with the theories of Planck and Bohr. But in it Einstein had made several profoundly important theoretical advances, and he suggested that "The simplicity of the hypotheses makes it seem probable ... that these will become the basis of the future theoretical description." He was absolutely correct: none of the developments since 1917 has required any modification of Einstein's derivation of the blackbody spectrum.

One major consequence of Einstein's work, of course, was the introduction of the concept of stimulated emission. Without the stimulated emission term, (1.81) and (1.82) are replaced by

$$A_{21}N_2 = B_{12}N_1\rho(\omega_o), \tag{1.86}$$

$$\rho(\omega_o) = \frac{A_{21}}{B_{12}} \frac{N_2}{N_1} = \frac{\hbar\omega_o^3}{\pi^2 c^3} e^{-\hbar\omega_o/kT}. \tag{1.87}$$

Without stimulated emission, therefore, Einstein would have obtained the Wien distribution.

Einstein's work was also the first to reveal atomic radiation in the form of spontaneous emission as a nonclassical process in which "God plays dice": there is nothing to tell us exactly *when* the atom will make a spontaneous jump to a state of lower energy. Einstein later wrote to Born that "That business about causality causes me a lot of trouble ... Can the quantum absorption and emission of light ever be understood in the sense of the complete causality requirement, or would a statistical residue remain? ... I would be very unhappy to renounce complete causality." That displeasure prevented Einstein from ever accepting quantum theory as a complete description of Nature.

Another novel aspect of Einstein's work was that it brought out the fact that photons carry linear momentum $h\nu/c$ as well as energy $h\nu$.[6] This part of Einstein's work of 1917 is not nearly as widely known as the derivation of the Planck spectrum just reviewed. According to Einstein, however, "a theory [of thermal radiation] can only be regarded as justified when it is able to show that the impulses transmitted by the radiation field to matter lead to motions that are in accordance with the theory of heat." Einstein showed that the momentum transfers accompanying emission and absorption are consistent with statistical mechanics *if the thermal radiation follows the Planck distribution.*

Consider the interaction with radiation of an atom initially at rest in the laboratory frame of reference. After a time δt it acquires some linear momentum Δ due to emission and absorption of radiation. Each emission or absorption process imparts to the atom a linear momentum λ_i, which may be positive or negative. If n emission and absorption processes occur during the time interval δt, then

$$\Delta = \sum_{i=1}^{n} \lambda_i \qquad (1.88)$$

and, assuming the λ_i to be independent random variables of zero mean,

$$\langle \Delta^2 \rangle = \sum_{i=1}^{n} \langle \lambda_i^2 \rangle \rightarrow \frac{1}{3} \left(\frac{\hbar \omega_o}{c} \right)^2 n \qquad (1.89)$$

if we associate with each process of emission or absorption a momentum transfer (photon momentum) $\hbar \omega_o / c$. We have also included a factor of 1/3 because, as in the Einstein–Hopf model, the atoms are assumed to move in only one direction. The average number n of emission and absorption events occuring in the time interval δt is given, according to the foregoing analysis, by

$$n = N_2 A_{21} \delta t + (N_1 + N_2) B_{12} \rho(\omega_o) \delta t, \qquad (1.90)$$

---

[6] The term *photon* for radiation quanta was coined in 1926 by Gilbert Lewis, a physical chemist.

so that

$$(\delta t)^{-1}\langle \Delta^2 \rangle = \frac{1}{3}\left(\frac{\hbar\omega_o}{c}\right)^2 [N_2 A_{21} + (N_1 + N_2)B_{12}\rho(\omega_o)]$$

$$= \frac{2}{3}\left(\frac{\hbar\omega_o}{c}\right)^2 N_1 B_{12}\rho(\omega_o), \qquad (1.91)$$

where we have used the equilibrium condition (1.81).

This result shows that an atom interacting with radiation will continually gain kinetic energy unless there is some retarding force to maintain the fixed average kinetic energy $\langle \frac{1}{2}mv^2 \rangle = \frac{1}{2}kT$ demanded by statistical mechanics. The origin of this retarding force is the same as in the Einstein–Hopf model, except that now we must express it in terms of quantities characteristic of an *atom* rather than a classical dipole oscillator. As shown in Appendix B, this force is given by the formula

$$F = -Rv = -\left(\frac{\hbar\omega_o}{c^2}\right)(N_1 - N_2)B_{12}\left[\rho(\omega_o) - \frac{\omega_o}{3}\frac{d\rho}{d\omega_o}\right]v. \qquad (1.92)$$

As in the classical Einstein–Hopf model the condition for thermal equilibrium is $\langle \Delta^2 \rangle/\delta t = 2RkT$ or, from (1.91) and (1.92),

$$\rho(\omega_o) - \frac{\omega_o}{3}\frac{d\rho}{d\omega_o} = \left(\frac{\hbar\omega_o}{3kT}\right)\left(\frac{N_1}{N_1 - N_2}\right)\rho(\omega_o)$$

$$= \left[\frac{\hbar\omega_o/3kT}{1 - e^{-\hbar\omega_o/kT}}\right]\rho(\omega_o). \qquad (1.93)$$

The solution of this equation is the Planck spectrum. Thus Einstein showed that in his theory of thermal radiation, "the impulses transmitted by the radiation field to matter lead to motions that are in accordance with the theory of heat."

1.9 Discussion

In Section 1.6 we alluded to the fact that an oscillator (or atom) in its ground state does not absorb zero-point electromagnetic radiation. The reason for this is discussed in Chapter 4. The question arises whether an *excited* atom undergoes *stimulated emission* due to the zero-point field.

Let us suppose that it does. Then, according to the Einstein theory described in the preceding section, the rate at which an atom in level 2 is stimulated by the zero-point field to drop to level 1 should be given by

$$(\dot{N}_2)_{\text{stimulated emission}}^{(\text{o})} = -B_{21}\rho_{\text{o}}(\omega_o)N_2 = -B_{21}\left(\frac{\hbar\omega_o^3}{2\pi^2 c^3}\right)N_2, \qquad (1.94)$$

where we have used equation (1.57) for the spectral energy density $\rho_o(\omega)$ of the zero-point field. Using (1.85), therefore, we have

$$(\dot{N}_2)^{(o)}_{\text{stimulated emission}} = -B_{21}\left(\frac{A_{21}}{2B_{21}}\right)N_2 = -\frac{1}{2}A_{21}N_2$$

$$= \frac{1}{2}(\dot{N}_2)_{\text{spontaneous emission}}. \qquad (1.95)$$

Thus we can almost interpret spontaneous emission as stimulated emission due to the zero-point field — *almost* because we calculate within this interpretation only half the correct A coefficient for spontaneous emission. In spite of this discrepancy, one repeatedly hears and reads statements to the effect that "spontaneous emission is induced by the zero-point electromagnetic field." We attempt to clarify the situation in the following chapters.

The result (1.95), however, does suggest that spontaneous emission has something to do with zero-point radiation, even if it is not simply emission induced by this radiation. Another way to infer this is to use the equation

$$\frac{N_1}{N_1 - N_2} = 1 + \frac{B_{21}}{A_{21}}\rho(\omega_o), \qquad (1.96)$$

which follows from (1.81), in equation (1.93):

$$\rho(\omega_o) - \frac{\omega_o}{3}\frac{d\rho}{d\omega_o} = \frac{\hbar\omega_o}{3kT}\left[1 + \frac{B_{21}}{A_{21}}\rho(\omega_o)\right]\rho(\omega_o)$$

$$= \frac{\pi^2 c^3}{3\omega_o^2 kT}\left[\rho^2(\omega_o) + \frac{A_{21}}{B_{21}}\rho(\omega_o)\right]. \qquad (1.97)$$

The identity (1.85) shows that this result is equivalent to (1.61). But now it is evident that the second term in brackets is associated with spontaneous emission. In other words, *the particle term in the Einstein fluctuation formula is a consequence of spontaneous emission.* The fact that the particle term may also be related as in Section 1.7 to the zero-point field thus suggests again some connection between spontaneous emission and the zero-point field. This connection will be explored in Chapter 4.

We noted in Section 1.7 that the particle term was the nonclassical feature of the Einstein fluctuation formula. In fact this term, which we have just related to the existence of spontaneous emission and zero-point radiation, led Einstein in 1917 to conclude that "Outgoing radiation in the form of spherical waves does not exist."

To understand this conclusion, let us first note that the recoil associated with spontaneous emission contributes only to the particle term in $(\delta t)^{-1}\langle\Delta^2\rangle$, not to the wave term. Now the wave term has contributions from both absorption and stimulated emission (neither of which contributes to the particle term), and it is obvious from the classical wave picture that absorption and stimulated emission must cause the atom to recoil, simply because the field carries linear momentum. But why does spontaneous emission not contribute likewise to the recoil associated with the wave term?

The reason is simple. In a classical wave description of spontaneous emission, the radiation is a wave with inversion symmetry about the position $\mathbf{r} = 0$ of the atom. Thus any recoil associated with radiation propagating in the direction $\mathbf{r}$ from the atom is cancelled by the contribution from the radiation in the direction $-\mathbf{r}$. The classical wave pattern associated with spontaneous emission is, loosely speaking, "everywhere at once," and its inversion symmetry precludes any possibility of atomic recoil. In the quantum-electrodynamical description of spontaneous emission, however, the radiated field amplitude has the same spatial distribution predicted classically, but it represents a *probability amplitude for directional photon emission*. The expectation value of the net recoil vanishes because there is no preferred direction of emission, just as predicted by the classical wave picture. But contrary to the classical wave picture, there *is* a nonvanishing mean-square momentum transfer to the atom that, for radiation of frequency ω_o, is $(\hbar\omega_o/c)^2$. It is in this sense that the classical picture of outgoing waves fails.

It is perhaps worth noting that the recoil of a spontaneously emitting atom is an experimental fact, as are the recoils associated with the absorption and stimulated emission of radiation. In absorption the recoil is in the same direction as the incoming (absorbed) photon, whereas in stimulated emission the recoil is in the direction opposite to that of the incoming (stimulating) photon; these are simple consequences of the conservation of linear momentum. In spontaneous emission the direction of recoil cannot be predicted, since the direction of the emitted photon is unpredictable. Recoil accompanying spontaneous emission was inferred experimentally by Frisch in 1933, and has in recent years been confirmed more accurately.

We conclude the present discussion with a tribute to the unsung experimentalists who so painstakingly measured blackbody spectra: when Planck fit his formula to their data he obtained $h = 6.55 \times 10^{-27}$ erg-sec for his constant, within 1% of the modern value $h = 6.63 \times 10^{-27}$. For the Boltzmann constant Planck obtained $k = 1.35 \times 10^{-16}$erg/K, the modern value being 1.38×10^{-16}. (Since the universal gas constant $R = N_A k$ was known, Planck also obtained an accurate estimate of Avogadro's number.)

1.10   Specific Heats

It was Maxwell, in 1859, who first suggested that classical physics was wrong. What he later called "the greatest difficulty yet encountered by the molecular theory" had to do with the theory of specific heats of gases.

The specific heat of a solid will in general have contributions from both electronic and vibrational degrees of freedom. Except at very high temperatures, however, the electrons are all in their ground states and make no contribution to the specific heat. Then the N atoms making up the solid may be regarded as inert vibrators, and under the approximation of harmonic vibrations the total energy for the $3N$ degrees of freedom is $U = 3NkT$. Thus $dU/dT = 3Nk$, and the specific heat per mole is

$$c_v = 3N_A k = 3R \approx 6 \text{ cal/mole–K} \quad \text{(Dulong–Petit law)}. \qquad (1.98)$$

This classical prediction is the Dulong–Petit law, named after the experimenters who observed it in 1819 for 12 metals and sulfur at room temperature. As the temperature is decreased, however, c_v is found to decrease, and $c_v \to 0$ as $T \to 0$, contradicting the classical prediction (1.98) based on the equipartition theorem.

It was found in 1840 that the specific heat of diamond is smaller than 6 cal/mole-K even at room temperature. This anomaly was first explained by Einstein in 1907. Einstein argued that Planck's equation (1.29) gives the average energy in thermal equilibrium of each (harmonic) vibrational degree of freedom, so that[7]

$$U = \frac{3Nh\nu}{e^{h\nu/kT} - 1} \qquad (1.99)$$

and

$$c_v = 3R \left(\frac{\theta}{T}\right)^2 \frac{e^{\theta/T}}{(e^{\theta/T} - 1)^2} \qquad (1.100)$$

is the specific heat per mole, where $\theta \equiv h\nu/k$ is the "Einstein temperature," the one adjustable parameter in Einstein's theory. For high temperatures $(T \gg \theta)$, equation (1.100) reduces to the Dulong–Petit law. At low temperatures, however, c_v is less than the Dulong–Petit value, and in particular $c_v \to 0$ as $T \to 0$. From a fit to experimental data Einstein deduced that $\theta \approx 1300$ K for diamond. A substance with such a large value of θ will have a small value of c_v even at room temperature.

---

[7]Einstein presented a derivation of equation (1.29) using in essence the quantum-statistical formula $U = \sum_{n=0}^{\infty} nh\nu e^{-nh\nu/kT} / \sum_{n=0}^{\infty} e^{-nh\nu/kT}$.

In 1913 Einstein and Stern, in the paper discussed in connection with the blackbody problem in Section 1.6, turned their attention to the specific heats of gases. Their work was motivated by the recent report by Eucken that the molar specific heat for H_2 at room temperature was about 5 cal/mole-K, but about 3 at $T \approx 60$ K. Einstein and Stern suggested that this behavior was a consequence of molecular rotations and *zero-point energy*.

The energy of a dumbbell rotator with moment of inertia I and rotational frequency ν is $\frac{1}{2}I(2\pi\nu)^2$. Suppose, following Einstein and Stern, that in thermal equilibrium this energy is given by the Planck equation (1.29):

$$U = \frac{1}{2}I(2\pi\nu)^2 = \frac{h\nu}{e^{h\nu/kT} - 1} \ . \tag{1.101}$$

The rotational contribution to the specific heat is then

$$c_r = N_A\frac{dU}{dT} = N_A\frac{dU}{d\nu}\frac{d\nu}{dT} = N_A\left(4\pi^2 I\nu\right)\frac{d\nu}{dT} = \left(\frac{2R}{k}\right)p\nu\frac{d\nu}{dT} \ , \tag{1.102}$$

where $p \equiv 2\pi^2 I$. From equation (1.101) it is clear that ν is a function of T; $d\nu/dT$ follows by differentiation of both sides of that equation with respect to T,

$$\frac{d\nu}{dT} = \frac{\nu}{T}\left[1 + \frac{kT}{p\nu^2 + h\nu}\right]^{-1} , \tag{1.103}$$

and it follows from (1.102) that

$$c_r = R\frac{2p\nu^2}{kT}\left[1 + \frac{kT}{p\nu^2 + h\nu}\right]^{-1} , \tag{1.104}$$

where $\nu(T)$ is found by solution of (1.101). The rotational specific heat calculated in this way for the example $p = 2.9 \times 10^{-40}$ g cm^2 considered by Einstein and Stern is shown in Figure 1.1. The predicted dependence of the specific heat on temperature is quite different from the dependence observed by Eucken, and in particular the predicted specific heats at low temperatures are much too large.

Now suppose, however, that equation (1.101) is modified to include zero-point energy:

$$U = p\nu^2 = \frac{h\nu}{e^{h\nu/kT} - 1} + \frac{1}{2}h\nu. \tag{1.105}$$

Following the same steps leading from (1.101) to (1.104), it is found that

$$c_r = R\frac{2p\nu^2}{kT}\left[1 + \frac{kT}{p\nu^2 - h^2/4p}\right]^{-1} , \tag{1.106}$$

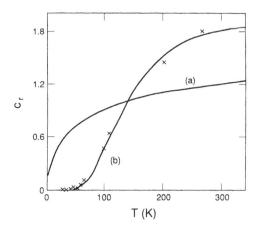

Figure 1.1: Experimental data ($\times$) of Eucken on specific heat of molecular hydrogen; specific heat computed by Einstein and Stern (a) without zero-point energy [equation (1.104)] and (b) with zero-point energy [equation (1.106)].

where $\nu(T)$ is obtained by solving (1.105) for ν in terms of T. The resulting c_r plotted in Figure 1.1 is seen to agree very well with Eucken's observations. At high temperatures c_r asymptotes to $R \approx 2$ cal/mole-K, but at low temperatures $c_r \rightarrow 0$.

Einstein and Stern thus gave a very interesting interpretation of Eucken's observation that the specific heat of H_2 decreased from 5 cal/mole-K to 3 as T decreased from 300 to 60 K: because of zero-point energy, the rotational contribution to the specific heat decreases from 2 cal/mole-K to 0 as T decreases. That is, the existence of zero-point energy causes the rotational specific heat of a gas to "freeze out." Einstein and Stern concluded that "The existence of a zero-point energy of size $\frac{1}{2}h\nu$ [is] probable."

The Einstein–Stern explanation turned out to be incorrect. The rotational energy levels of a diatomic molecule are given in quantum theory by $E_J \cong BJ(J+1)$, where B is a constant characteristic of the molecule and $J = 0, 1, 2, \ldots$ Therefore a *molecule has no zero-point rotational energy.* On the other hand, Einstein and Stern were correct in their hypothesis that the observed decrease of specific heat with temperature of H_2 was connected to molecular rotations.

According to quantum mechanics, the fact that c_v and $c_r \rightarrow 0$ as $T \rightarrow 0$ is due simply to the fact that discrete energy levels are associated with the internal degrees of freedom of a molecule. If kT is small compared with the energy separation between the lowest and first-excited energy levels, there is a high probability that only the lowest-energy state is occupied, and so the specific heat corresponding to that degree of freedom is "frozen out" in the sense that dU/dT decreases with T and approaches zero as $T \to 0$.

As a consequence of the Einstein–Stern paper, the concept of zero-point energy began to take on greater importance, especially among physical chemists. This was due in part to the growing interest at the time in low-temperature phenomena. Stern himself in 1913 used zero-point energy in a calculation of the vapor pressure of solids.

1.11   X-Ray Diffraction

An important question, prior to the first experiments, was whether x-ray diffraction would be spoiled by the thermal motions of the atoms in crystal lattices. It was first shown by Debye in 1914 that these thermal motions basically just reduce the intensity of a diffracted beam from that predicted for an idealized lattice of stationary atoms. Debye also showed that if Planck's zero-point energy were real, there should be such a reduction in intensity even as $T \to 0$. We now know that zero-point motion can indeed have a significant effect on x-ray diffraction. In this section we will briefly sketch a derivation of the so-called Debye–Waller factor that accounts for the motion of lattice atoms.

Consider the field far from a collection of identical scatterers. We assume the nth scatterer at $\mathbf{r}_n$ has strength p_n and write the total scattered field at $\mathbf{r}$ as

$$E_s(\mathbf{r}) = \sum_n \frac{p_n}{|\mathbf{r} - \mathbf{r_n}|} e^{-i\omega(t - |\mathbf{r} - \mathbf{r}_n|)/c} = e^{-i\omega t} \sum_n \frac{p_n}{|\mathbf{r} - \mathbf{r_n}|} e^{ik|\mathbf{r} - \mathbf{r}_n|} \ .$$

(1.108)

For distances large compared with the dimensions of the scattering volume we have

$$\begin{aligned} |\mathbf{r} - \mathbf{r}_n| &= [r^2 - 2\mathbf{r} \cdot \mathbf{r}_n + r_n^2]^{1/2} = r[1 - 2\mathbf{r} \cdot \mathbf{r}_n/r^2 + r_n^2/r^2]^{1/2} \\ &\cong r[1 - \mathbf{r} \cdot \mathbf{r}_n/r^2] = r - \mathbf{r} \cdot \mathbf{r}_n/r \ , \end{aligned}$$

(1.109)

so that $k|\mathbf{r} - \mathbf{r}_n| \cong kr - (kr/r)\,\mathbf{r}_n = kr - \mathbf{k} \cdot \mathbf{r}_n$ in the exponential in (1.108), where $\mathbf{k}$ is the wave vector of the (elastically) scattered wave. Thus

$$E_s(\mathbf{r}) \cong \frac{1}{r} e^{-i\omega(t - r/c)} \sum_n p_n e^{-i\mathbf{k} \cdot \mathbf{r}_n} \ .$$

(1.110)

We take the strength p_n of tho nth scatterer to be proportional to the field $E_o e^{i\mathbf{k}_o \cdot \mathbf{r}_n}$ $(k_o = k)$ incident upon it: $p_n = \alpha E_o e^{i\mathbf{k}_o \cdot \mathbf{r}_n}$ and

$$E_s(\mathbf{r}) \cong \frac{\alpha}{r} E_o e^{-i(\omega t - kr)} \sum_n e^{-i\mathbf{K} \cdot \mathbf{r}_n} \quad , \tag{1.111}$$

where $\mathbf{K} \equiv \mathbf{k} - \mathbf{k}_o$.

For a periodic lattice of scatterers the scattered field (1.111) is nonvanishing only in directions such that $\mathbf{K}$ belongs to the reciprocal lattice. For a one-dimensional lattice, for instance, this means that $Kd = 2\pi n$, where d is the lattice spacing and n is an integer. Since $K = [k^2 + k_o^2 - 2\mathbf{k}_o \cdot \mathbf{k}]^{1/2} = [2k^2 - 2k^2 \cos 2\theta]^{1/2} = 2k \sin \theta = (4\pi/\lambda) \sin \theta$, the condition that K belongs to the reciprocal lattice is just the Bragg condition, $2d \sin \theta = n\lambda$, where 2θ is the angle between the incident and scattered (diffracted) waves.

Now let us take into account the thermal motion of the atoms, replacing the preceding $\mathbf{r}_n$ by $\mathbf{r}_n + \mathbf{u}$, where $\mathbf{u}$ represents a displacement from a fixed lattice site. Then

$$\sum_n e^{-i\mathbf{K} \cdot \mathbf{r}_n} \rightarrow e^{-i\mathbf{K} \cdot \mathbf{u}} \sum_n e^{-i\mathbf{K} \cdot \mathbf{r}_n} \quad . \tag{1.112}$$

We are interested in the average of $e^{-i\mathbf{K} \cdot \mathbf{u}}$ as $\mathbf{u}$ undergoes thermal motion:

$$\langle e^{-i\mathbf{K} \cdot \mathbf{u}} \rangle = 1 - i\mathbf{K} \cdot \langle \mathbf{u} \rangle - \frac{1}{2}\langle (\mathbf{K} \cdot \mathbf{u})^2 \rangle + ... = 1 - \frac{1}{6}K^2 \langle \mathbf{u}^2 \rangle + ... \tag{1.113}$$

since $\langle \mathbf{u} \rangle = 0$. The two terms shown explicitly are the first two terms of the Taylor series for $\exp[-K^2 \langle \mathbf{u}^2 \rangle /6]$. In fact if the oscillations of $\mathbf{u}$ are assumed to be harmonic we have

$$\langle e^{-i\mathbf{K} \cdot \mathbf{u}} \rangle = e^{-K^2 \langle \mathbf{u}^2 \rangle /6} \tag{1.114}$$

and $\frac{1}{2}m\omega_o^2 \langle \mathbf{u}^2 \rangle = \frac{3}{2}kT$, where m and ω_o are the mass and frequency of the harmonic oscillations; for simplicity we assume the elastic restoring force is the same in all directions. Thus the thermal fluctuations in the atomic positions cause the diffracted beam to be reduced in intensity by the factor

$$|\langle e^{-i\mathbf{K} \cdot \mathbf{u}} \rangle|^2 \equiv e^{-2W} = e^{-K^2 kT/2m\omega_o^2}. \tag{1.115}$$

This is called the *Debye–Waller factor*. Our classical hand-waving derivation gives the correct order of magnitude for this factor.

But the classical model of lattice vibrations breaks down, of course, at low temperatures. In particular, as $T \rightarrow 0$ there is a nonvanishing $\langle \mathbf{u}^2 \rangle$ associated with zero-point energy:

$$m\omega_o^2 \langle \mathbf{u}^2 \rangle = 3(\frac{1}{2}\hbar\omega_o), \tag{1.116}$$

so that

$$c^{-2W} = e^{-\hbar K^2/2m\omega_0} \quad \text{for } T \to 0. \tag{1.117}$$

This gives the correct order of magnitude for the zero-temperature Debye–Waller factor.

1.12   Molecular Vibrations

Direct evidence for the reality of zero-point energy was provided by Mulliken in 1924. Consider the vibrational spectra of two diatomic molecules differing only by having different nuclear isotopes. The masses of these two vibrators are then different and consequently so are their vibrational frequencies. For relatively heavy molecules these differences are small but readily observable. According to quantum mechanics each molecule has vibrational energy levels given by $E_n = \hbar\omega[(n + \frac{1}{2}) - x_e(n + \frac{1}{2})^2 + y_e(n + \frac{1}{2})^3 + ...]$, where the constants $\omega, x_e, y_e, ...$ are characteristic of the particular molecule, $n = 0, 1, 2, ...$, and the zero-point contributions are included. The vibrational frequencies are given by $|E_n - E_{n'}|/\hbar$. Mulliken studied the two molecules $B^{10}O^{16}$ and $B^{11}O^{16}$. He found that a good fit to the emission spectra could be obtained only if zero-point energy were included, or in his words, "if one assumes that the true values of the vibrational quantum numbers are not n and n' but each $\frac{1}{2}$ unit greater ... It is then probable that the minimum vibrational energy of BO (and doubtless of other) molecules is $\frac{1}{2}$ quantum." It is worth noting that Mulliken reached this conclusion based on his spectroscopic data, before Heisenberg (1925) derived the zero-point energy of a harmonic oscillator from matrix mechanics.

1.13   Summary

Zero-point energy first appeared in Planck's "second theory" of blackbody radiation. The concept was quickly adopted by Einstein and Stern, who showed that it could be used to derive the Planck spectrum from largely classical considerations. They also showed that rotational zero-point energy might account for the observed decrease with temperature of the specific heat of molecular hydrogen. None of these ingenious theories turned out to be quite correct from a modern perspective.

Zero-point motion played no role in Einstein's epiphanic paper of 1917 in which he derived the Planck spectrum using his A and B coefficients. The great simplicity of Einstein's derivation, perhaps, ended speculations about the role of zero-point energy in the blackbody problem. However, we have seen that zero-point energy of the electromagnetic field has something to do with the A coefficient for spontaneous emission, although it cannot be regarded as the sole "cause" of emission. The role of the zero-point electromagnetic field in spontaneous emission and other electromagnetic processes will be discussed in much greater detail in the following chapters.

We have described how zero-point energy appeared and was used during the development of quantum theory.[8] Although interest in the concept in connection with blackbody theory declined after Einstein's 1917 paper, it was by no means abandoned. In particular, direct spectroscopic evidence for the reality of zero-point energy was provided by Mulliken in 1924, just months before it appeared so naturally in the quantum formalism established in 1925–26, and long before it was to become central to the world-view of modern physicists.

1.14   Bibliography

Debye, P., "Interferenz von Röntgenstrahlen und Wärmebewegung," *Ann. d. Phys.* **43**, 49 (1914).

Einstein, A., "Über einen die Erzeugung und Verwandlung des Lichtes betreffenden heuristischen Gesichtspunkt," *Ann. d. Phys.* **17**, 132 (1905).

Einstein, A., "Die Plancksche Theorie der Strahlung und die Theorie der spezifischen Wärme," *Ann. d. Phys.* **22**, 180 (1907).

Einstein, A., "Zur gegenwärtigen Stand des Strahlungsproblems," *Phys. Zs.* **10**, 185 (1909).

Einstein, A., "Zur Quantentheorie der Strahling," *Phys. Zs.* **18**, 121 (1917).

Einstein, A. and L. Hopf, "Über einen Satz der Wahrscheinlich-keitsrechnung und seine Anwendung in der Strahlungstheorie," *Ann. d. Phys.* **33**, 1096 (1910a).

Einstein, A. and L. Hopf, "Statistische Untersuchung der Bewegung eines Resonators in einem Strahlungsfeld," *Ann. d. Phys.* **33**, 1105 (1910b).

Einstein, A. and O. Stern, "Einige Argumente für die Annahme einer molekularen Agitation beim absoluten Nullpunkt," *Ann. d. Phys.* **40**, 551 (1913).

Einstein, A., "Zur Quantentheorie der Strahlung," *Phys. Zs.* **18**, 121 (1917).

Frisch, O. R., "Experimenteller Nachweis des Einsteinschen Strahlungs-

---

[8] Our survey is not exhaustive. We have not discussed, for instance, the necessity of zero-point energy in Stern's calculation of the vapor pressure of solids.

rückstoβes," *Z. Phys.* **86**, 42 (1933).

Klein, M. J., "Max Planck and the Beginnings of Quantum Theory," *Arch. Hist. Exact Sciences* **1**, 459 (1962).

Klein, M. J., "Thermodynamics and Quanta in Planck's Work," *Physics Today* (November 1966), 23.

Kuhn, T. S., *Black-Body Theory and the Quantum Discontinuity, 1894-1912* (Oxford University Press, New York, 1978).

Mulliken, R. S., "The Band Spectrum of Boron Monoxide," *Nature* **114**, 349 (1924).

Nernst, W., "Über einen Versuch von quantentheoretischen Betrachtungen zur Annahme stetiger Energieänderungen zurückzukehren, " *Verhandl. Deut. Phys. Gen.* **18**, 83 (1916).

Pais, A., *'Subtle is the Lord ...' The Science and the Life of Albert Einstein* (Oxford University Press, New York, 1982).

Planck, M., "Über das Gesetz der Energieverteilung im Normal- spectrum," *Ann. d. Phys.* **4**, 553 (1901).

Planck, M., "Über die Begründung des Gesetzes der schwarzen Strahlung," *Ann. d. Phys.* **37**, 642 (1912).

Chapter 2

The Electromagnetic Vacuum

In five minutes you will say that it is all so absurdly simple.
– Sherlock Holmes, "The Adventure of the Dancing Men"
Arthur Conan Doyle

2.1 Introduction

The quantum theory of the free electromagnetic field in the absence of any sources was formulated by Born, Heisenberg, and Jordan (1926) in one of the founding papers of quantum theory. The first application was made by Dirac (1927), who treated the emission and absorption of radiation. The new quantum electrodynamics (QED) predicted a fluctuating zero-point or "vacuum" field existing even in the absence of any sources. In this chapter we consider the quantization of the electromagnetic field, with particular emphasis on the vacuum state.

According to contemporary physics the universe is made up of matter fields, whose quanta are fermions (e.g., electrons and quarks), and force fields, whose quanta are bosons (e.g., photons and gluons). *All these fields have zero-point energy.* The oldest and best-known quantized force field is the electromagnetic one. It is important for us to understand the main features of the quantized electromagnetic field, not only because quantum electrodynamics is "the best theory we have," but also because it is in many ways characteristic of all quantum field theories.

2.2   The Harmonic Oscillator

A monochromatic electromagnetic field is mathematically equivalent to a harmonic oscillator of the same frequency. Before showing this we will briefly review the harmonic oscillator in quantum mechanics.

The Hamiltonian has the same form as in classical mechanics:

$$H = p^2/2m + \frac{1}{2}m\omega^2 q^2, \qquad (2.1)$$

where now q and p are quantum-mechanical operators in a Hilbert space. The Heisenberg equations of motion have the same form as the classical Hamilton equations:

$$\dot{q} = (i\hbar)^{-1}[q, H] = p/m, \qquad (2.2)$$

$$\dot{p} = (i\hbar)^{-1}[p, H] = -m\omega^2 q. \qquad (2.3)$$

These follow from the commutation rule $[q, p] \equiv qp - pq = i\hbar$. We define the (non-Hermitian) operator

$$a = \frac{1}{\sqrt{2m\hbar\omega}}(p - im\omega q) \qquad (2.4)$$

and its adjoint

$$a^\dagger = \frac{1}{\sqrt{2m\hbar\omega}}(p + im\omega q), \qquad (2.5)$$

or equivalently

$$q = i\sqrt{\frac{\hbar}{2m\omega}}(a - a^\dagger), \qquad (2.6)$$

$$p = \sqrt{\frac{m\hbar\omega}{2}}(a + a^\dagger). \qquad (2.7)$$

From $[q, p] = i\hbar$ it follows that

$$[a, a^\dagger] = 1. \qquad (2.8)$$

Equations (2.6)–(2.8) allow us to write the Hamiltonian (2.1) in the form

$$H = \frac{1}{2}\hbar\omega(aa^\dagger + a^\dagger a) = \hbar\omega(a^\dagger a + \frac{1}{2}). \qquad (2.9)$$

The energy levels of the harmonic oscillator are thus determined by the eigenvalues of the operator $N \equiv a^\dagger a$. We denote the eigenvalues and (normalized) eigenkets of N by n and $|n\rangle$, respectively:

$$N|n\rangle = n|n\rangle. \qquad (2.10)$$

Now $\langle n|N|n\rangle = \langle n|a^\dagger a|n\rangle$ is the scalar product of the vector $a|n\rangle$ with itself. It then follows from (2.10) that $n\langle n|n\rangle = n$ is real and positive.

Consider the effect on the vector $a|n\rangle$ of the operator N. Obviously $Na|n\rangle = (aN + [N,a])|n\rangle = na|n\rangle + [N,a]|n\rangle$. But (2.8) implies $[N,a] = -a$, and therefore $Na|n\rangle = (n-1)a|n\rangle$. In other words, if $|n\rangle$ is an eigenstate of N with eigenvalue n, then $a|n\rangle$ is an eigenstate of N with eigenvalue $n-1$: $a|n\rangle = C|n-1\rangle$. By taking the norm of both sides of this equation we obtain $|C|^2 = n$, and without any loss of generality we can choose the phase such that $C = \sqrt{n}$. Thus

$$a|n\rangle = \sqrt{n}|n-1\rangle. \qquad (2.11)$$

We find similarly that

$$a^\dagger|n\rangle = \sqrt{n+1}|n+1\rangle. \qquad (2.12)$$

For obvious reasons a and $a^\dagger$ are called *lowering* and *raising* operators.

We have already noted that the eigenvalues $n \geq 0$. But equation (2.11) shows that we can generate eigenstates with lower and lower eigenvalues by successive applications of the lowering operator a. Consistency then requires that $a|n\rangle = 0$ for $n < 1$, and (2.11) indicates that this is satisfied only for $n = 0$. The eigenvalues n of $N = a^\dagger a$ are therefore zero and all the positive integers. That is, the energy levels of the harmonic oscillator are given by

$$E_n = (n + \frac{1}{2})\hbar\omega, \quad n = 0, 1, 2, \dots \qquad (2.13)$$

Let us briefly connect this operator approach to that based on the Schrödinger equation in the coordinate representation. From (2.11) we have $a|0\rangle = 0$, or $(p - im\omega q)|0\rangle = 0$ for the ground state $|0\rangle$. Thus $\langle q|(p - im\omega q)|0\rangle = \langle q|p|0\rangle - im\omega q\langle q|0\rangle = 0$. Now $\langle q|0\rangle$ is the wave function $\psi_0(q)$ and $\langle q|p|0\rangle = (\hbar/i)\partial\psi_0/\partial q$, so that

$$(\frac{\hbar}{i}\frac{\partial}{\partial q} - im\omega q)\psi_0(q) = 0 \qquad (2.14)$$

or $\psi_0(q) = (m\omega/\pi\hbar)^{1/4}e^{-m\omega q^2/2\hbar}$ when normalized such that $\int_{-\infty}^{\infty} dq|\psi_0(q)|^2 = 1$. The excited-state eigenfunctions $\psi_n(q)$ may be obtained by application of $a^\dagger$ according to (2.12): $|n\rangle = (n!)^{-1/2}(a^\dagger)^n|0\rangle$ and

$$
\begin{aligned}
\psi_n(q) &= \langle q|n\rangle = (n!)^{-1/2}(2m\hbar\omega)^{-n/2}\langle q|(p + im\omega q)^n|0\rangle \\
&= [(2m\hbar\omega)^n n!]^{-1/2}(\frac{\hbar}{i}\frac{\partial}{\partial q} + im\omega q)^n \psi_0(q) \\
&= i^n(2^n n!)^{-1/2}(\frac{m\omega}{\pi\hbar})^{1/4}(\xi - \frac{\partial}{\partial\xi})^n e^{-\xi^2/2}, \qquad (2.15)
\end{aligned}
$$

where $\xi = (m\omega/\hbar)^{1/2}q$. These eigenstates are proportional to $e^{-\xi^2/2}H_n(\xi)$, where H_n is a Hermite polynomial of degree n.

Various properties of the harmonic oscillator can be worked out using either the raising and lowering operators a and $a^\dagger$ or the eigenfunctions $\psi_n(q)$. For instance, we find that $\langle n|q|n\rangle = \langle n|p|n\rangle = 0$ and

$$\langle n|q^2|n\rangle = -\frac{\hbar}{2m\omega}\langle n|(a-a^\dagger)^2|n\rangle = \frac{\hbar}{m\omega}(n+\frac{1}{2}), \qquad (2.16)$$

$$\langle n|p^2|n\rangle = m\hbar\omega(n+\frac{1}{2}), \qquad (2.17)$$

since $\langle n|m\rangle = \delta_{nm}$. (Recall that eigenkets corresponding to different eigenvalues are orthogonal in the case of a Hermitian operator like $a^\dagger a$.) Thus $\Delta q_n \Delta p_n = (n+\frac{1}{2})\hbar$, where $(\Delta q_n)^2 \equiv \langle n|q^2|n\rangle - \langle n|q|n\rangle^2$. This is consistent with the general uncertainty relation $\Delta q \Delta p \geq \hbar/2$ and shows that the ground state of the harmonic oscillator is a state of minimal uncertainty product. In other words, the ground state is a *coherent state* of the harmonic oscillator.

2.3   A Field Mode Is a Harmonic Oscillator

We will now take the most elementary route to the quantization of the electromagnetic field. The first step is to show that a field mode is equivalent to a harmonic oscillator.

The Maxwell equations for the "free" field, i.e., the field in a region where there are no sources, are

$$\nabla \cdot \mathbf{E} = 0, \qquad (2.18)$$

$$\nabla \cdot \mathbf{B} = 0, \qquad (2.19)$$

$$\nabla \times \mathbf{E} = -\frac{1}{c}\frac{\partial \mathbf{B}}{\partial t}, \qquad (2.20)$$

$$\nabla \times \mathbf{B} = \frac{1}{c}\frac{\partial \mathbf{E}}{\partial t}. \qquad (2.21)$$

We introduce the vector potential $\mathbf{A}$ by writing $\mathbf{B} = \nabla \times \mathbf{A}$. Since $\nabla \cdot (\nabla \times \mathbf{A}) = 0$, (2.19) is automatically satisfied. Equation (2.20) implies $\mathbf{E} = -(1/c)\partial\mathbf{A}/\partial t - \nabla\phi$, where ϕ is the scalar potential. From (2.21) we have

$$\nabla^2 \mathbf{A} - \frac{1}{c^2}\frac{\partial^2 \mathbf{A}}{\partial t^2} = 0 \qquad (2.22)$$

in the Coulomb gauge defined by $\nabla \cdot \mathbf{A} = 0$ and, in the absence of any sources, $\phi = 0$. Obviously (2.18) is then also satisfied. Thus we can obtain a solution of the free-space Maxwell equations by solving (2.22) for the Coulomb-gauge vector potential subject to appropriate boundary conditions.

Separation of variables gives monochromatic solutions

$$
\begin{aligned}
\mathbf{A}(\mathbf{r}, t) &= \alpha(t)\mathbf{A}_o(\mathbf{r}) + \alpha^*(t)\mathbf{A}_o^*(\mathbf{r}) \\
&= \alpha(0)e^{-i\omega t}\mathbf{A}_o(\mathbf{r}) + \alpha^*(0)e^{i\omega t}\mathbf{A}_o^*(\mathbf{r}),
\end{aligned} \tag{2.23}
$$

where $\mathbf{A}_o(\mathbf{r})$ satisfies the Helmholtz equation,

$$
\nabla^2\mathbf{A}_o(\mathbf{r}) + k^2\mathbf{A}_o(\mathbf{r}) = 0 \quad (k = \omega/c), \tag{2.24}
$$

and $\alpha(t)$ satisfies $\ddot{\alpha}(t) = -\omega^2\alpha(t)$. The electric and magnetic field vectors are given by

$$
\mathbf{E}(\mathbf{r}, t) = -\frac{1}{c}[\dot{\alpha}(t)\mathbf{A}_o(\mathbf{r}) + \dot{\alpha}^*(t)\mathbf{A}_o^*(\mathbf{r})], \tag{2.25}
$$

$$
\mathbf{B}(\mathbf{r}, t) = \alpha(t)\nabla \times \mathbf{A}_o(\mathbf{r}) + \alpha^*(t)\nabla \times \mathbf{A}_o^*(\mathbf{r}), \tag{2.26}
$$

and the electromagnetic energy is proportional to

$$
\int d^3r(\mathbf{E}^2 + \mathbf{B}^2) = \frac{1}{c^2}\dot{\alpha}(t)^2 \int d^3r\mathbf{A}_o(\mathbf{r})^2
$$

$$
+ \frac{1}{c^2}\dot{\alpha}^*(t)^2 \int d^3r\mathbf{A}_o^*(\mathbf{r})^2 + \frac{2}{c^2}|\dot{\alpha}(t)|^2 \int d^3r|\mathbf{A}_o(\mathbf{r})|^2
$$

$$
+ \alpha(t)^2 \int d^3r[\nabla \times \mathbf{A}_o(\mathbf{r})]^2 + \alpha^*(t)^2 \int d^3r[\nabla \times \mathbf{A}_o^*(\mathbf{r})]^2
$$

$$
+ 2|\alpha(t)|^2 \int d^3r|\nabla \times \mathbf{A}_o(\mathbf{r})|^2. \tag{2.27}
$$

We show in Appendix C that we may take

$$
\int d^3r[\nabla \times \mathbf{A}_o(\mathbf{r})]^2 = k^2 \int d^3r\mathbf{A}_o(\mathbf{r})^2, \tag{2.28}
$$

with similar expressions for the terms involving $[\nabla \times \mathbf{A}_o^*(\mathbf{r})]^2$ and $|\nabla \times \mathbf{A}_o(\mathbf{r})|^2$ in (2.27) We also note that $\dot{\alpha}(t)^2 = -\omega^2\alpha(t)^2$, since $\dot{\alpha}(t) = -i\omega\alpha(t)$. Then (2.27) simplifies to

$$
H_F = \frac{1}{8\pi} \int d^3r(\mathbf{E}^2 + \mathbf{B}^2) = \frac{k^2}{2\pi}|\alpha(t)|^2, \tag{2.29}
$$

where, without any loss of generality, we assume the "mode function" $\mathbf{A}_o(\mathbf{r})$ is normalized such that

$$\int d^3r |\mathbf{A}_o(\mathbf{r})|^2 = 1. \tag{2.30}$$

Define the real quantities

$$q(t) = \frac{i}{c\sqrt{4\pi}}[\alpha(t) - \alpha^*(t)], \tag{2.31}$$

$$p(t) = \frac{k}{\sqrt{4\pi}}[\alpha(t) + \alpha^*(t)], \tag{2.32}$$

in terms of which equation (2.29) is

$$H_F = \frac{1}{2}(p^2 + \omega^2 q^2). \tag{2.33}$$

The notation suggests that our field mode of frequency ω is mathematically equivalent to a harmonic oscillator of frequency ω. To prove this we must, of course, show that q and p are indeed canonically conjugate coordinate and momentum variables. But this is trivial: from the definitions (2.31) and (2.32) and $\dot\alpha = -i\omega\alpha$, we have $\dot q = p$ and $\dot p = -\omega^2 q$, which are the Hamilton equations that follow from the Hamiltonian H_F.

2.4   Quantization of a Field Mode

To describe a field mode quantum mechanically, we simply describe the equivalent harmonic oscillator quantum mechanically. Since the oscillator with Hamiltonian (2.33) has unit mass, we introduce raising and lowering operators a and $a^\dagger$ using (2.6) and (2.7) with $m = 1$. Comparing with (2.31) and (2.32), we see that this quantization procedure is equivalent to replacing the classical variable $\alpha(t)/c\sqrt{4\pi}$ by the quantum-mechanical operator $(\hbar/2\omega)^{1/2}a(t)$, or $\alpha(t)$ by $(2\pi\hbar c^2/\omega)^{1/2}a(t)$ and $\alpha^*(t)$ by $(2\pi\hbar c^2/\omega)^{1/2}a^\dagger(t)$. That is, except for trivial constants that depend upon the arbitrary normalization chosen for the mode function, $\alpha(t)$ and $\alpha^*(t)$ in the classical theory are replaced by the lowering and raising operators $a(t)$ and $a^\dagger(t)$, respectively, in quantum theory.

The classical vector potential (2.23) is thus replaced by the *operator*

$$\mathbf{A}(\mathbf{r},t) = \left(\frac{2\pi\hbar c^2}{\omega}\right)^{1/2} [a(t)\mathbf{A}_o(\mathbf{r}) + a^\dagger(t)\mathbf{A}_o^*(\mathbf{r})], \tag{2.34}$$

and the operators corresponding to the electric and magnetic fields are similarly

$$\mathbf{E}(\mathbf{r}, t) = i(2\pi\hbar\omega)^{1/2}[a(t)\mathbf{A}_{\mathrm{o}}(\mathbf{r}) - a^{\dagger}(t)\mathbf{A}_{\mathrm{o}}^{*}(\mathbf{r})], \tag{2.35}$$

$$\mathbf{B}(\mathbf{r}, t) = \left(\frac{2\pi\hbar c^2}{\omega}\right)^{1/2}[a(t)\nabla \times \mathbf{A}_{\mathrm{o}}(\mathbf{r}) + a^{\dagger}(t)\nabla \times \mathbf{A}_{\mathrm{o}}^{*}(\mathbf{r})]. \tag{2.36}$$

The Hamiltonian (2.33) for the quantized field mode is now obviously equivalent to

$$H_{\mathrm{F}} = \hbar\omega(a^{\dagger}a + \frac{1}{2}). \tag{2.37}$$

The energy eigenvalues of a field mode of frequency ω are given by equation (2.13). The integer n is the number of energy quanta or *photons* in the field mode described by the state $|n\rangle$. The vacuum state $|0\rangle$ has no photons, but it nevertheless has an energy $\frac{1}{2}\hbar\omega$. The quantum theory of radiation thus predicts the existence of a zero-point electromagnetic field. In the vacuum state, and in all stationary states $|n\rangle$, the expectation values of the electric and magnetic fields vanish:

$$\langle\mathbf{E}(\mathbf{r}, t)\rangle = \langle\mathbf{B}(\mathbf{r}, t)\rangle = 0, \tag{2.38}$$

since $\langle n|a|n\rangle = 0$. This means that the electric and magnetic field vectors fluctuate with zero mean in the state $|n\rangle$, although the field has a definite, nonfluctuating energy $(n + \frac{1}{2})\hbar\omega$.

Consider the expectation value of the square of the electric field. From (2.35) this is given by

$$\langle\mathbf{E}^2(\mathbf{r}, t)\rangle = -(2\pi\hbar\omega)[\langle a^2(t)\rangle\mathbf{A}_{\mathrm{o}}^2(\mathbf{r}) - \langle a(t)a^{\dagger}(t) + a^{\dagger}(t)a(t)\rangle$$
$$\times |\mathbf{A}_{\mathrm{o}}(\mathbf{r})|^2 + \langle a^{\dagger}(t)^2\rangle\mathbf{A}_{\mathrm{o}}^{*}(\mathbf{r})^2]. \tag{2.39}$$

In the state $|n\rangle$ we have $\langle a^2\rangle = \langle a^{\dagger 2}\rangle = 0$, $\langle aa^{\dagger} + a^{\dagger}a\rangle = \langle 2a^{\dagger}a + 1\rangle = 2n + 1$, and

$$\langle\mathbf{E}^2(\mathbf{r}, t)\rangle = (n + \frac{1}{2})4\pi\hbar\omega|\mathbf{A}_{\mathrm{o}}(\mathbf{r})|^2$$
$$= 4\pi\hbar\omega|\mathbf{A}_{\mathrm{o}}(\mathbf{r})|^2 n + 2\pi\hbar\omega|\mathbf{A}_{\mathrm{o}}(\mathbf{r})|^2$$
$$\equiv 4\pi\hbar\omega|\mathbf{A}_{\mathrm{o}}(\mathbf{r})|^2 n + \langle\mathbf{E}^2(\mathbf{r})\rangle_0. \tag{2.40}$$

From the first term on the right we can begin to understand how the quantum theory of radiation resolves the "paradox" of the wave-particle duality of light, for this term, which is a measure of the "intensity" (energy density) of the field at $\mathbf{r}$, has both wave and particle factors. The factor n is the number of photons in the field, whereas the factor $|\mathbf{A}_o(\mathbf{r})|^2$ gives the same spatial dependence for the intensity predicted by the classical wave theory. Even in the case of a single photon ($n = 1$), the classical wave theory gives the same spatial dependence as quantum theory for the intensity, but this pattern represents the relative *probability distribution* for finding the photon. The detection of a single photon does not produce the spread-out classical intensity pattern $|\mathbf{A}_o(\mathbf{r})|^2$. Instead there is relatively high probability of detecting the photon at points where $|\mathbf{A}_o(\mathbf{r})|^2$ is large, and low probability where $|\mathbf{A}_o(\mathbf{r})|^2$ is small. If $|\mathbf{A}_o(\mathbf{r})|^2 = 0$, the probability of detecting the photon at $\mathbf{r}$ is zero. The wave and particle aspects of the field are thus reconciled by this association of a particle (photon) with the classical (wave) intensity pattern. Comparing (2.25) and (2.26) to (2.35) and (2.36), we can say that the spatial pattern of the field is exactly the same as predicted classically: the quantum mechanics of the field is entirely contained, as it were, in its time dependence.

Quantities like $\langle \mathbf{E}^m(\mathbf{r}, t) \rangle$ with $m > 2$ are also easy to calculate. Suppose, for simplicity, that $\mathbf{A}_o^*(\mathbf{r}) = \mathbf{A}_o(\mathbf{r})$, so that

$$\mathbf{E}(\mathbf{r}, t) = i(2\pi\hbar\omega)^{1/2}[a(t) - a^\dagger(t)]\mathbf{A}_o(\mathbf{r}) = (4\pi\omega^2)^{1/2}q(t)\mathbf{A}_o(\mathbf{r}). \quad (2.41)$$

Then from the probability distribution $|\psi_0(q)|^2 = (\omega/\pi\hbar)^{1/2}e^{-\omega q^2/\hbar}$ for a ground-state harmonic oscillator of unit mass, we easily obtain the probability distribution

$$P[\mathbf{E}(\mathbf{r}, t)] = [2\pi\langle \mathbf{E}^2(\mathbf{r})\rangle_0]^{1/2} \exp\left[-\mathbf{E}^2(\mathbf{r}, t)/2\langle \mathbf{E}^2(\mathbf{r})\rangle_0\right] \quad (2.42)$$

for the electric field in the vacuum state $|0\rangle$. Thus $\langle \mathbf{E}^m(\mathbf{r}, t)\rangle_0 = 0$ for odd m and

$$
\begin{aligned}
\langle \mathbf{E}^m(\mathbf{r}, t)\rangle_0 &= [2\pi\langle \mathbf{E}^2(\mathbf{r})\rangle_0]^{1/2} \int_0^\infty dE\, E^m \exp\left[-E^2/2\langle \mathbf{E}^2(\mathbf{r})\rangle_0\right] \\
&= 2^{m/2}\pi^{-1/2}\Gamma\left(\frac{m+1}{2}\right)\langle \mathbf{E}^2(\mathbf{r})\rangle_0^{m/2} \quad \text{(m even)}. \quad (2.43)
\end{aligned}
$$

Similar results, with the appearance of Hermite polynomials H_n, are found for the expectation values of field powers in photon states $|n\rangle$.

What is the physical significance of these vacuum-state expectation values, and in particular of $\langle \mathbf{E}^2(\mathbf{r})\rangle_0$? One thing they indicate is that the electromagnetic vacuum is a stationary *state* of the field *with statistical fluctuations* of the electric and magnetic fields. As far as measurements are concerned, however, it is often argued that the entire universe is evidently bathed in a zero-point electromagnetic field, which can add only some constant amount to expectation values, as in equation (2.40). Physical measurements will therefore reveal only *deviations* from the vacuum state. Thus the field Hamiltonian (2.37), for example, can be replaced by

$$
\begin{aligned}
H_{\mathrm{F}} - \langle 0|H_{\mathrm{F}}|0\rangle &= \frac{1}{2}\hbar\omega(aa^\dagger + a^\dagger a) - \frac{1}{2}\hbar\omega \\
&= \frac{1}{2}\hbar\omega(2a^\dagger a + 1) - \frac{1}{2}\hbar\omega \\
&= \hbar\omega a^\dagger a
\end{aligned}
\tag{2.44}
$$

without affecting any physical predictions of the theory. The new Hamiltonian (2.44) is said to be *normally ordered* (or Wick ordered), the raising operator $a^\dagger$ appearing to the left of the lowering operator a. The normally ordered Hamiltonian is denoted $:H_{\mathrm{F}}:$, i.e.,

$$
:H_{\mathrm{F}}: = :\frac{1}{2}\hbar\omega(aa^\dagger + a^\dagger a): = \hbar\omega a^\dagger a.
\tag{2.45}
$$

In other words, within the normal ordering symbol we can commute a and $a^\dagger$. Since zero-point energy is intimately connected to the noncommutativity of a and $a^\dagger$, the normal ordering procedure eliminates any contribution from the zero-point field. This is especially reasonable in the case of the field Hamiltonian, since the zero-point term merely adds a constant energy which can be eliminated by a simple redefinition of the zero of energy. Moreover, this constant energy in the Hamiltonian obviously commutes with a and $a^\dagger$ and so cannot have any effect on the quantum dynamics described by the Heisenberg equations of motion.

So the argument goes. However, things are not quite that simple, for in general relativity the zero of energy is not arbitrary. Furthermore we shall see that it is possible to attribute measurable effects, such as the Casimir force and the Lamb shift, to *changes* in zero-point energy. And finally, as discussed in Section 2.6, the zero-point field is not eliminated by dropping its energy from the Hamiltonian.

2.5 The Field in Free Space

The generalization of the quantization procedure to a multimode field is straightforward. In this section we consider the field in free space with no physical boundaries, in which case the number of allowed modes is infinite.

Obviously the field intensity for infinite free space should be independent of position so that, from (2.40), $|\mathbf{A}_\circ(\mathbf{r})|^2$ should be independent of $\mathbf{r}$ for each mode of the field. Of course $\mathbf{A}_0(\mathbf{r})$ must still satisfy the Helmholtz equation (2.24). A mode function satisfying these conditions is obviously $\mathbf{A}_0(\mathbf{r}) = \mathbf{e_k}e^{i\mathbf{k}\cdot\mathbf{r}}$, where $\mathbf{k} \cdot \mathbf{e_k} = 0$ in order to have the transversality condition $\nabla \cdot \mathbf{A}(\mathbf{r}, t) = 0$ satisfied for the Coulomb gauge in which we are working.

We also wish to normalize our mode functions according to equation (2.30). To achieve the desired normalization we pretend that space is divided into cubes of volume $V = L^3$ and impose on the field the periodic boundary condition

$$\mathbf{A}(x + L, y + L, z + L, t) = \mathbf{A}(x, y, z, t), \qquad (2.46)$$

or equivalently

$$(k_x, k_y, k_z) = \frac{2\pi}{L}(n_x, n_y, n_z), \qquad (2.47)$$

where each n can assume any integer value. Of course this artificial periodic boundary condition will be of no physical consequences if L is very large compared with any physical dimensions of interest. It allows us to consider the field in any *one* of the imaginary cubes, and to define a mode function $\mathbf{A_k}(\mathbf{r}) = V^{-1/2}\mathbf{e_k}e^{i\mathbf{k}\cdot\mathbf{r}}$ satisfying the Helmholtz equation, transversality, and the "box normalization"

$$\int_V d^3r |\mathbf{A_k}(\mathbf{r})|^2 = 1, \qquad (2.48)$$

where $\mathbf{e_k}$ is chosen to be a unit vector.

The unit vector $\mathbf{e_k}$, which we take to be real, specifies the polarization of the field mode. The condition $\mathbf{k}\cdot\mathbf{e_k} = 0$ means there are two independent choices for $\mathbf{e_k}$, which we call $\mathbf{e_{k1}}$ and $\mathbf{e_{k2}}$, $\mathbf{e_{k1}}\cdot\mathbf{e_{k2}} = 0$ and $\mathbf{e}_{\mathbf{k1}}^2 = \mathbf{e}_{\mathbf{k2}}^2 = 1$. Thus we define the mode functions

$$\mathbf{A_{k\lambda}}(\mathbf{r}) = V^{-1/2}\mathbf{e_{k\lambda}}e^{i\mathbf{k}\cdot\mathbf{r}} \quad (\lambda = 1, 2), \qquad (2.49)$$

in terms of which the vector potential (2.34) becomes

$$\mathbf{A_{k\lambda}}(\mathbf{r}, t) = \left(\frac{2\pi\hbar c^2}{\omega_k V}\right)^{1/2} [a_{\mathbf{k\lambda}}(t)e^{i\mathbf{k}\cdot\mathbf{r}} + a_{\mathbf{k\lambda}}^{\dagger}(t)e^{-i\mathbf{k}\cdot\mathbf{r}}]\mathbf{e_{k\lambda}}, \qquad (2.50)$$

or

$$\mathbf{A_{k\lambda}}(\mathbf{r}, t) = \left(\frac{2\pi\hbar c^2}{\omega_k V}\right)^{1/2} [a_{\mathbf{k\lambda}}(0)e^{-i(\omega_k t - \mathbf{k}\cdot\mathbf{r})} + a_{\mathbf{k\lambda}}^{\dagger}(0)e^{i(\omega_k t - \mathbf{k}\cdot\mathbf{r})}]\mathbf{e_{k\lambda}}, \qquad (2.51)$$

where $\omega_k = kc$ and $a_{k\lambda}$, $a_{k\lambda}^\dagger$ are respectively the photon annihilation and creation operators for the mode with wave vector $\mathbf{k}$ and polarization λ. This gives the vector potential for a plane-wave mode of the field. The condition (2.47) shows that there is an infinite number of such modes. The linearity of Maxwell's equations allows us to write

$$\mathbf{A}(\mathbf{r},t) = \sum_{\mathbf{k}\lambda} \left(\frac{2\pi\hbar c^2}{\omega_k V}\right)^{1/2} [a_{\mathbf{k}\lambda}(t)e^{i\mathbf{k}\cdot\mathbf{r}} + a_{\mathbf{k}\lambda}^\dagger(t)e^{-i\mathbf{k}\cdot\mathbf{r}}]\mathbf{e}_{\mathbf{k}\lambda} \qquad (2.52)$$

for the total vector potential in free space.

Using the fact that

$$\int_V d^3r\, \mathbf{A}_{\mathbf{k}\lambda}(\mathbf{r}) \cdot \mathbf{A}_{\mathbf{k}'\lambda'}^*(\mathbf{r}) = \delta^3_{\mathbf{k},\mathbf{k}'}\delta_{\lambda\lambda'}\ , \qquad (2.53)$$

we find from the same sort of analysis as in the preceding section that the field Hamiltonian is

$$H_{\mathrm{F}} = \sum_{\mathbf{k}\lambda} \hbar\omega_k(a_{\mathbf{k}\lambda}^\dagger a_{\mathbf{k}\lambda} + \frac{1}{2}) \qquad (2.54)$$

for the infinity of modes in free space. This is the Hamiltonian for an infinite number of *uncoupled* harmonic oscillators. Thus the different modes of the field are independent and satisfy the commutation relations

$$[a_{\mathbf{k}\lambda}(t), a_{\mathbf{k}'\lambda'}^\dagger(t)] = \delta^3_{\mathbf{k},\mathbf{k}'}\delta_{\lambda\lambda'} \qquad (2.55)$$

and $[a_{\mathbf{k}\lambda}(t), a_{\mathbf{k}'\lambda'}(t)] = [a_{\mathbf{k}\lambda}^\dagger(t), a_{\mathbf{k}'\lambda'}^\dagger(t)] = 0$. From (2.52) it follows that

$$\mathbf{E}(\mathbf{r},t) = i\sum_{\mathbf{k}\lambda} \left(\frac{2\pi\hbar\omega_k}{V}\right)^{1/2} [a_{\mathbf{k}\lambda}(t)e^{i\mathbf{k}\cdot\mathbf{r}} - a_{\mathbf{k}\lambda}^\dagger(t)e^{-i\mathbf{k}\cdot\mathbf{r}}]\mathbf{e}_{\mathbf{k}\lambda}, \qquad (2.56)$$

$$\mathbf{B}(\mathbf{r},t) = i\sum_{\mathbf{k}\lambda} \left(\frac{2\pi\hbar c^2}{\omega_k V}\right)^{1/2} [a_{\mathbf{k}\lambda}(t)e^{i\mathbf{k}\cdot\mathbf{r}} - a_{\mathbf{k}\lambda}^\dagger(t)e^{-i\mathbf{k}\cdot\mathbf{r}}]\mathbf{k} \times \mathbf{e}_{\mathbf{k}\lambda}. \qquad (2.57)$$

It is worth noting that the free-space mode functions (2.49) form a complete set for transverse vector fields satisfying our periodic boundary condition. That is, the plane-wave modes $\mathbf{A}_{\mathbf{k}\lambda}(\mathbf{r})$ form a complete set in terms of which any mode of the field may be expanded. This is essentially just a statement of Fourier's theorem about the completeness of sines and

cosines Of course the $\mathbf{A}_{\mathbf{k}\lambda}(\mathbf{r})$ are complete only for modes satisfying the
periodic boundary condition, but in a slightly more sophisticated approach
we can work with a complete *continuum* of plane-wave mode functions in
which the $\mathbf{k}$ vectors are not restricted to the discrete spectrum (2.47) (Chapter 10). This has formal consequences such as the replacement of $\delta^3_{\mathbf{k},\mathbf{k}'}$ in
(2.53) and (2.55) by $\delta^3(\mathbf{k} - \mathbf{k}')$, but since it has no physical consequences
here, we will just stick to the periodic boundary condition.

The linear momentum of the field is given classically by $\mathbf{P} = (1/4\pi c)$
$\times \int_V d^3r(\mathbf{E} \times \mathbf{B})$. In the case of the quantized field we use (2.56) and (2.57)
in this expression and obtain, after straightforward manipulations,

$$\mathbf{P} = \sum_{\mathbf{k}\lambda} \hbar\mathbf{k}(a^{\dagger}_{\mathbf{k}\lambda}a_{\mathbf{k}\lambda} + \frac{1}{2}). \tag{2.58}$$

Obviously $[\mathbf{P}, H_{\mathbf{F}}] = 0$, so that the linear momentum of the field in the
absence of any sources is a constant of the motion. It is also obvious that
the eigenvalues of $\mathbf{P}$ are $\sum_{\mathbf{k}\lambda} \hbar\mathbf{k}(n_{\mathbf{k}\lambda} + \frac{1}{2})$, where each n is a positive integer
or zero. A stationary state of the free field is thus characterized by the set
of photon numbers $\{n_{\mathbf{k}\lambda}\}$. The state $|\{n_{\mathbf{k}\lambda}\}\rangle$ has a total photon number
$\sum_{\mathbf{k}\lambda} n_{\mathbf{k}\lambda}$, an energy

$$E = \sum_{\mathbf{k}\lambda} \hbar\omega_k(n_{\mathbf{k}\lambda} + \frac{1}{2}), \tag{2.59}$$

and a linear momentum

$$\mathbf{P} = \sum_{\mathbf{k}\lambda} \hbar\mathbf{k}(n_{\mathbf{k}\lambda} + \frac{1}{2}) \tag{2.60}$$

or

$$E = \sum_{\mathbf{k}\lambda} \hbar\omega_k n_{\mathbf{k}\lambda} \ , \tag{2.61}$$

$$\mathbf{P} = \sum_{\mathbf{k}\lambda} \hbar\mathbf{k} n_{\mathbf{k}\lambda} \tag{2.62}$$

if the zero-point energy and linear momentum associated with the vacuum
state are discarded. Note that the zero-point momentum $\sum_{\mathbf{k}\lambda} \frac{1}{2}\hbar\mathbf{k}$ in fact
vanishes since for each $\mathbf{k}$ there is an equal contribution from $-\mathbf{k}$ in the
summation.

We have thus arrived at the quantum theory of the free electromagnetic
field in which stationary states are described by photons of energy $\hbar\omega_k$ and
linear momentum $\hbar\mathbf{k}$. Since $E^2 - P^2 c^2 = \hbar^2(\omega_k^2 - k^2 c^2) = 0$ for each photon,

the photons have zero rest mass. The theory also implies that photons are bosons, i.e., that the stationary states are symmetric with respect to permutations of identical photons. To see this, note from equation (2.12) that the n-photon state $|n\rangle$ of a field mode may be written in the form

$$|n\rangle = \frac{(a^\dagger)^n}{\sqrt{n!}}|0\rangle, \tag{2.63}$$

which is obviously symmetric with respect to any permutations of the n photons. Of course the boson character of photons is just a consequence of the commutation rule (2.55), from which (2.63) follows.

The $\mathbf{k}$ vector of a photon of mode $(\mathbf{k}, \lambda)$ specifies the energy and linear momentum of the photon. The polarization index λ is connected with the intrinsic angular momentum, or spin, of the photon. To establish this connection we first note that the intrinsic angular momentum may be defined by the formula[1]

$$\mathbf{M_s} = \frac{1}{4\pi c}\int_V d^3 r (\mathbf{E} \times \mathbf{A}). \tag{2.64}$$

From (2.52) and (2.56) we obtain for $\mathbf{M_s}$ the expression

$$\mathbf{M_s} = i\hbar \sum_{\mathbf{k}\lambda} \hat{\mathbf{k}}(a_{\mathbf{k}2}^\dagger a_{\mathbf{k}1} - a_{\mathbf{k}1}^\dagger a_{\mathbf{k}2}), \tag{2.65}$$

where the unit vector $\hat{\mathbf{k}} \equiv \mathbf{k}/k = \mathbf{e}_{\mathbf{k}1} \times \mathbf{e}_{\mathbf{k}2}$. This operator does not commute with $a_{\mathbf{k}\lambda}^\dagger a_{\mathbf{k}\lambda}$, and therefore a photon number state $|n_{\mathbf{k}\lambda}\rangle$ is not an eigenstate of $\mathbf{M_s}$. To construct simultaneous eigenstates of energy, linear momentum, and intrinsic angular momentum of a photon we define the complex unit polarization vectors

$$\mathbf{e}_{\mathbf{k},+1} = -\sqrt{\frac{1}{2}}(\mathbf{e}_{\mathbf{k}1} + i\mathbf{e}_{\mathbf{k}2}), \tag{2.66}$$

$$\mathbf{e}_{\mathbf{k},-1} = \sqrt{\frac{1}{2}}(\mathbf{e}_{\mathbf{k}1} - i\mathbf{e}_{\mathbf{k}2}), \tag{2.67}$$

satisfying $\mathbf{e}_{\mathbf{k}\alpha}^* \cdot \mathbf{e}_{\mathbf{k}\alpha'} = \delta_{\alpha\alpha'}$, $\mathbf{e}_{\mathbf{k},\alpha}^* \times \mathbf{e}_{\mathbf{k},\alpha'} = i\alpha\hat{\mathbf{k}}\delta_{\alpha\alpha'}$, $\alpha = \pm 1$. It is easily seen that, whereas our original polarization vectors $\mathbf{e}_{\mathbf{k}\lambda}$ with $\lambda = 1, 2$

---

[1] See, for instance, Heitler (1966), Appendix, Section 1. It is worth noting that (2.64) is gauge–invariant, since the vector potential is transverse in the Coulomb gauge employed here, and the transverse part of the vector potential is unaffected by gauge transformations.

correspond to two orthogonal linear polarizations, the new polarization vectors $e_{\mathbf{k}\alpha}$ with $\alpha = \pm 1$ correspond to opposite circular polarizations. We define the photon annihilation operators for the circularly polarized modes $(\mathbf{k}, \alpha)$ by

$$a_{\mathbf{k},+1} = -\sqrt{\frac{1}{2}}(a_{\mathbf{k}1} - ia_{\mathbf{k}2}),\tag{2.68}$$

$$a_{\mathbf{k},-1} = \sqrt{\frac{1}{2}}(a_{\mathbf{k}1} + ia_{\mathbf{k}2}),\tag{2.69}$$

in terms of which

$$\mathbf{M_s} = \hbar \sum_{\mathbf{k}} \hat{\mathbf{k}}(a^\dagger_{\mathbf{k},+1}a_{\mathbf{k},+1} - a^\dagger_{\mathbf{k},-1}a_{\mathbf{k},-1}) = \sum_{\mathbf{k}\alpha} \alpha\hbar\hat{\mathbf{k}}a^\dagger_{\mathbf{k}\alpha}a_{\mathbf{k}\alpha}\tag{2.70}$$

and $H_F = \sum_{\mathbf{k}\alpha} \hbar\omega_k a^\dagger_{\mathbf{k}\alpha}a_{\mathbf{k}\alpha}$, $\mathbf{P} = \sum_{\mathbf{k}\alpha} \hbar\mathbf{k}a^\dagger_{\mathbf{k}\alpha}a_{\mathbf{k}\alpha}$. With circularly polarized mode functions, therefore, $\mathbf{M_s}$ commutes with H_F and has the photon number state $|n_{\mathbf{k}\alpha}\rangle$ as an eigenstate with eigenvalue $\alpha\hbar\hat{\mathbf{k}}, \alpha = \pm 1$. In other words, the component of the photon spin along the direction of propagation, the photon *helicity*, is ± 1 in units of $\hbar$, which means that a photon is a boson of spin 1. Ordinarily we will not be concerned with spin and will employ the linear polarization basis.

2.6   Necessity of the Vacuum Field

The vacuum state $|\text{vac}\rangle$ of the free field is defined as the ground state in which $n_{\mathbf{k}\lambda} = 0$ for all modes $(\mathbf{k}, \lambda)$. The vacuum state, like all stationary states of the field, is an eigenstate of the Hamiltonian but not the electric and magnetic field operators. In the vacuum state, therefore, the electric and magnetic fields do not have definite values. We can imagine them to be fluctuating about their mean values of zero, as discussed in Section 2.4 for the case of a single mode of the field.

In a process in which a photon is annihilated (absorbed), we can think of the photon as making a transition into the vacuum state. Similarly, when a photon is created (emitted), it is occasionally useful to imagine that the photon has made a transition out of the vacuum state. In the words of Dirac (1927),

> The light-quantum has the peculiarity that it apparently ceases
> to exist when it is in one of its stationary states, namely, the zero
> state, in which its momentum, and therefore also its energy, are zero.
> When a light-quantum is absorbed it can be considered to jump into

this zero state, and when one is emitted it can be considered to jump from the zero state to one in which it is physically in evidence, so that it appears to have been created. Since there is no limit to the number of light-quanta that may be created in this way, we must suppose that there are an infinite number of light-quanta in the zero state ...

We shall see later that an atom, for instance, can be considered to be "dressed" by emission and reabsorption of "virtual photons" from the vacuum.

The most glaring characteristic of the vacuum state is that its energy $\sum_{\mathbf{k}\lambda} \frac{1}{2}\hbar\omega_k$ is infinite. Let us use (2.47) to make the well-known replacement

$$\sum_{\mathbf{k}\lambda} \rightarrow \sum_{\lambda} \left(\frac{L}{2\pi}\right)^3 \int d^3k = \frac{V}{8\pi^3} \sum_{\lambda} \int d^3k. \qquad (2.71)$$

The zero-point energy density is thus

$$\frac{1}{V} \sum_{\mathbf{k}\lambda} \frac{1}{2}\hbar\omega_k = \frac{2}{8\pi^3} \int d^3k \frac{1}{2}\hbar\omega_k = \frac{4\pi}{4\pi^3} \int dk\, k^2 (\frac{1}{2}\hbar\omega_k)$$

$$= \frac{\hbar}{2\pi^2 c^3} \int d\omega\, \omega^3, \qquad (2.72)$$

or in other words the spectral energy density of the vacuum field is

$$\rho_0(\omega) = \frac{\hbar\omega^3}{2\pi^2 c^3}, \qquad (2.73)$$

which is familiar from Chapter 1. The zero-point energy density in the frequency range from ω_1 to ω_2 is therefore

$$\int_{\omega_1}^{\omega_2} d\omega \rho_0(\omega) = \frac{\hbar}{8\pi^2 c^3} (\omega_2^4 - \omega_1^4). \qquad (2.74)$$

This can be large even in relatively narrow, "low-frequency" regions of the spectrum. In the optical region from 400 nm to 700 nm, for instance, equation (2.74) yields about 220 erg/cm^3.

In Section 2.4 we noted that the zero-point energy of the field can be eliminated from the Hamiltonian by the normal ordering prescription. However, this elimination does not mean that the vacuum field has been rendered unimportant or without physical consequences! To illustrate this point we consider now a linear dipole oscillator in the vacuum.

The Hamiltonian for the oscillator plus the field with which it interacts is

$$H = \frac{1}{2m}(\mathbf{p} - \frac{e}{c}\mathbf{A})^2 + \frac{1}{2}m\omega_o^2\mathbf{x}^2 + H_\mathbf{F}. \tag{2.75}$$

Of course this has the same form as the corresponding classical Hamiltonian, and the Heisenberg equations of motion for the oscillator and the field are formally the same as their classical counterparts. For instance, the Heisenberg equations for the coordinate $\mathbf{x}$ and the canonical momentum $\mathbf{p} = m\dot{\mathbf{x}} + e\mathbf{A}/c$ of the oscillator are[2]

$$\dot{\mathbf{x}} = (i\hbar)^{-1}[\mathbf{x}, H] = \frac{1}{m}(\mathbf{p} - \frac{e}{c}\mathbf{A}), \tag{2.76}$$

$$
\begin{aligned}
\dot{\mathbf{p}} = (i\hbar)^{-1}[\mathbf{p}, H] &= -\frac{1}{2m}\nabla(\mathbf{p} - \frac{e}{c}\mathbf{A})^2 - m\omega_o^2\mathbf{x} \\
&= -\frac{1}{m}[(\mathbf{p} - \frac{e}{c}\mathbf{A}) \cdot \nabla][-\frac{e}{c}\mathbf{A}] - \frac{1}{m}(\mathbf{p} - \frac{e}{c}\mathbf{A}) \\
&\quad \times \nabla \times [-\frac{e}{c}\mathbf{A}] - m\omega_o^2\mathbf{x} \\
&= \frac{e}{c}(\dot{\mathbf{x}} \cdot \nabla)\mathbf{A} + \frac{e}{c}\dot{\mathbf{x}} \times \mathbf{B} - m\omega_o^2\mathbf{x}, \tag{2.77}
\end{aligned}
$$

or

$$
\begin{aligned}
m\ddot{\mathbf{x}} &= \dot{\mathbf{p}} - \frac{e}{c}\dot{\mathbf{A}} = -\frac{e}{c}[\dot{\mathbf{A}} - (\dot{\mathbf{x}} \cdot \nabla)\mathbf{A}] + \frac{e}{c}\dot{\mathbf{x}} \times \mathbf{B} - m\omega_o^2\mathbf{x} \\
&= e\mathbf{E} + \frac{e}{c}\dot{\mathbf{x}} \times \mathbf{B} - m\omega_o^2\mathbf{x}, \tag{2.78}
\end{aligned}
$$

since the rate of change of the vector potential in the frame of the moving charge is given by the convective derivative $\dot{\mathbf{A}} = \partial\mathbf{A}/\partial t + (\dot{\mathbf{x}} \cdot \nabla)\mathbf{A}$.[3] For nonrelativistic motion we may neglect the magnetic force and replace (2.78) by

$$\ddot{\mathbf{x}} + \omega_o^2\mathbf{x} \cong \frac{e}{m}\mathbf{E} \cong i\frac{e}{m}\sum_{\mathbf{k}\lambda}\left(\frac{2\pi\hbar\omega_k}{V}\right)^{1/2}[a_{\mathbf{k}\lambda}(t) - a_{\mathbf{k}\lambda}^\dagger(t)]\mathbf{e}_{\mathbf{k}\lambda}. \tag{2.79}$$

As in Chapter 1 we have made the electric dipole approximation in which the spatial dependence of the field is neglected. The Heisenberg equation

---

[2] The Hamiltonian for a charged particle in an electromagnetic field is reviewed in Chapter 4. In (2.77) we employ the vector generalization of the identity $[p, F(q, p)] = -i\hbar\partial F/\partial q$.

[3] This follows from the general relation $i\hbar dA_x/dt = [A_x, H] + i\hbar\partial A_x/\partial t$.

for $a_{\mathbf{k}\lambda}$ is found similarly from the Hamiltonian (2.75) to be

$$\dot{a}_{\mathbf{k}\lambda} = -i\omega_k a_{\mathbf{k}\lambda} + ie \left(\frac{2\pi}{\hbar\omega_k V}\right)^{1/2} \dot{\mathbf{x}} \cdot \mathbf{e}_{\mathbf{k}\lambda} \tag{2.80}$$

in the electric dipole approximation. In deriving these equations for $\mathbf{x}$, $\mathbf{p}$, and $a_{\mathbf{k}\lambda}$ we have used the fact that equal-time particle and field operators commute. This follows from the assumption that particle and field operators commute at some time (say, $t = 0$) when the matter–field interaction is presumed to begin, together with the fact that a Heisenberg-picture operator $A(t)$ evolves in time as $A(t) = U^{\dagger}(t)A(0)U(t)$, where $U(t)$ is the time evolution operator satisfying $i\hbar\dot{U} = HU$, $U^{\dagger}(t) = U^{-1}(t)$, $U(0) = 1$. Alternatively, we can argue that these operators must commute if we are to obtain the correct equations of motion from the Hamiltonian, just as the corresponding Poisson brackets in classical theory must vanish in order to generate the correct Hamilton equations (see also Section 1.2).

The formal solution of the field equation (2.80) is

$$a_{\mathbf{k}\lambda}(t) = a_{\mathbf{k}\lambda}(0)e^{-i\omega_k t} + ie \left(\frac{2\pi}{\hbar\omega_k V}\right)^{1/2} \int_0^t dt' \mathbf{e}_{\mathbf{k}\lambda} \cdot \dot{\mathbf{x}}(t')e^{i\omega_k(t'-t)}, \tag{2.81}$$

and therefore equation (2.79) may be written

$$\ddot{\mathbf{x}} + \omega_o^2 \mathbf{x} = \frac{e}{m}\mathbf{E}_o(t) + \frac{e}{m}\mathbf{E}_{\mathrm{RR}}(t), \tag{2.82}$$

where

$$\mathbf{E}_o(t) = i\sum_{\mathbf{k}\lambda} \left(\frac{2\pi\hbar\omega_k}{V}\right)^{1/2} [a_{\mathbf{k}\lambda}(0)e^{-i\omega_k t} - a^{\dagger}_{\mathbf{k}\lambda}(0)e^{i\omega_k t}]\mathbf{e}_{\mathbf{k}\lambda} \tag{2.83}$$

and

$$\mathbf{E}_{\mathrm{RR}}(t) = -\frac{4\pi e}{V}\sum_{\mathbf{k}\lambda} \int_0^t dt' [\mathbf{e}_{\mathbf{k}\lambda} \cdot \dot{\mathbf{x}}(t')]\mathbf{e}_{\mathbf{k}\lambda} \cos\omega_k(t' - t). \tag{2.84}$$

We show in Appendix D that we may take

$$\mathbf{E}_{\mathrm{RR}}(t) = \frac{2e}{3c^3}\dddot{\mathbf{x}} \tag{2.85}$$

for the radiation reaction field, if the mass m in (2.82) is regarded as the "observed" mass.

The total field acting on the dipole has two parts, $\mathbf{E}_o(t)$ and $\mathbf{E}_{RR}(t)$. $\mathbf{E}_o(t)$ is the free or zero-point field acting on the dipole. It is the homogeneous solution of the Maxwell equation for the field acting on the dipole, i.e., the solution, at the position of the dipole, of the wave equation $[\nabla^2 - c^{-2}\partial^2/\partial t^2]\mathbf{E} = 0$ satisfied by the field in the (source-free) vacuum. For this reason $\mathbf{E}_o(t)$ is often referred to as the *vacuum field*, although it is of course a Heisenberg-picture operator acting on whatever state of the field happens to be appropriate at $t = 0$. $\mathbf{E}_{RR}(t)$ is the source field, the field generated *by* the dipole and acting *on* the dipole.

Using (2.85) in (2.82), we obtain an equation for the Heisenberg-picture operator $\mathbf{x}(t)$ that is formally the same as the classical equation (1.41):

$$\ddot{\mathbf{x}} + \omega_o^2 \mathbf{x} - \tau\,\dddot{\mathbf{x}} = \frac{e}{m}\mathbf{E}_o(t), \qquad (2.86)$$

where again $\tau = 2e^2/3mc^3$. But here we have considered a dipole in the vacuum, without any "external" field acting on it. The role of the "external" field in equation (2.86) is played by the *vacuum* electric field acting on the dipole.

Classically, of course, a dipole in the vacuum is not acted upon by any "external" field: if there are no sources other than the dipole itself, then the only field acting on the dipole is its own radiation reaction field. In quantum theory, however, there is always an "external" field, namely, the source-free or vacuum field $\mathbf{E}_o(t)$.

According to equation (2.81) the free field is the only field in existence at $t = 0$. This defines $t = 0$ as the time at which the interaction between the dipole and the field is "switched on." The state vector of the dipole–field system at $t = 0$ is therefore of the form $|\Psi\rangle = |\text{vac}\rangle|\psi_D\rangle$, where $|\text{vac}\rangle$ is the vacuum state of the field and $|\psi_D\rangle$ is the initial state of the dipole oscillator. The expectation value of the free field is therefore at all times equal to zero: $\langle \mathbf{E}_o(t)\rangle = \langle \Psi|\mathbf{E}_o(t)|\Psi\rangle = 0$ since $a_{\mathbf{k}\lambda}(0)|\text{vac}\rangle = 0$. However, the energy density associated with the free field is infinite:

$$
\begin{aligned}
\frac{1}{4\pi}\langle \mathbf{E}_o^2(t)\rangle &= \frac{1}{4\pi}\sum_{\mathbf{k}\lambda}\sum_{\mathbf{k}'\lambda'}\left(\frac{2\pi\hbar\omega_k}{V}\right)^{1/2}\left(\frac{2\pi\hbar\omega_{k'}}{V}\right)^{1/2} \\
&\quad \times \langle a_{\mathbf{k}\lambda}(0)a_{\mathbf{k}'\lambda'}^{\dagger}(0)\rangle \\
&= \frac{1}{4\pi}\sum_{\mathbf{k}\lambda}\left(\frac{2\pi\hbar\omega_k}{V}\right) = \int_0^\infty d\omega\,\rho_o(\omega). \qquad (2.87)
\end{aligned}
$$

The important point is this: the zero-point field energy in H_F does not affect the Heisenberg equation for $a_{\mathbf{k}\lambda}$, since it is a *c*-number (i.e., an ordinary number rather than an operator) and commutes with $a_{k\lambda}$. We can therefore drop the zero-point field energy from the Hamiltonian, as is usually done. But the zero-point field re-emerges, so to speak, as the homogeneous solution of the field equation. *A charged particle in the vacuum will therefore always see a zero-point field of infinite energy density.* This is the origin of one of the infinities of quantum electrodynamics, and it cannot be eliminated by the trivial expedient of dropping the term $\sum_{k\lambda} \frac{1}{2}\hbar\omega_k$ in the field Hamiltonian.

The free field is in fact *necessary* for the formal consistency of the theory. In particular, it is necessary for the preservation of commutation relations, which is required by the unitarity of time evolution in quantum theory: $[z(t), p_z(t)] = [U^\dagger(t)z(0)U(t), U^\dagger(t)p_z(0)U(t)] = U^\dagger(t)[z(0), p_z(0)]U(t) = i\hbar U^\dagger(t)U(t) = i\hbar$. We can calculate $[z(t), p_z(t)]$ from the formal solution of the operator equation of motion (2.86). Using the fact that $[a_{k\lambda}(0), a^\dagger_{k'\lambda'}(0)] = \delta^3_{k,k'}\delta_{\lambda\lambda'}$, and that equal-time particle and field operators commute, we readily obtain

$$
\begin{aligned}
[z(t), p_z(t)] &= [z(t), m\dot{z}(t)] + [z(t), \frac{e}{c}A_z(t)] = [z(t), m\dot{z}(t)] \\
&= \left(\frac{i\hbar e^2}{2\pi^2 mc^3}\right)\left(\frac{8\pi}{3}\right)\int_0^\infty \frac{d\omega\omega^4}{(\omega^2 - \omega_o^2)^2 + \tau^2\omega^6} \quad (2.88)
\end{aligned}
$$

in the mode continuum limit (2.71). For the dipole oscillator under consideration it can sensibly be assumed that the radiative damping rate is small compared with the natural oscillation frequency, i.e., $\tau\omega_o \ll 1$. Then the integrand in (2.88) is sharply peaked at $\omega = \omega_o$, and[4]

$$
\begin{aligned}
[z(t), p_z(t)] &\cong \frac{2i\hbar e^2}{3\pi mc^3}\omega_o^3 \int_{-\infty}^\infty \frac{dx}{x^2 + \tau^2\omega_o^6} = \left(\frac{2i\hbar e^2\omega_o^3}{3\pi mc^3}\right)\left(\frac{\pi}{\tau\omega_o^3}\right) \\
&= i\hbar. \quad (2.89)
\end{aligned}
$$

We can appreciate further the necessity of the vacuum field by making the small-damping approximation directly in (2.86): $\ddot{x} \cong -\omega_o^2 x(t)$, $\dddot{x} \cong -\omega_o^2\dot{x}$, and

$$
\ddot{x} + \tau\omega_o^2\dot{x} + \omega_o^2 x \cong \frac{e}{m}E_o(t). \quad (2.90)
$$

Without the free field $E_o(t)$ in this equation the *operator* $x(t)$ would be exponentially damped, and commutators like $[z(t), p_z(t)]$ would approach zero for $t \gg (\tau\omega_o^2)^{-1}$. With the vacuum field included, however, the

---

[4] Actually (2.89) follows *exactly* from (2.88), as may be shown using the residue theorem.

commutator is $i\hbar$ at all times, as required by unitarity, and as we have just shown. A similar result is easily worked out for the case of a free particle instead of a dipole oscillator (Milonni, 1981b).

What we have here is an example of a "fluctuation–dissipation relation." Generally speaking, if a system is coupled to a "bath" that can take energy from the system in an effectively irreversible way, then the bath must also cause fluctuations. The fluctuations and the dissipation go hand in hand; we cannot have one without the other. In the present example the coupling of a dipole oscillator to the electromagnetic field has a dissipative component, in the form of radiation reaction, and a fluctuation component, in the form of the zero-point (vacuum) field; given the existence of radiation reaction, the vacuum field must also exist in order to preserve the canonical commutation rule and all it entails.

The spectral density of the vacuum field is fixed by the form of the radiation reaction field, or vice versa: because the radiation reaction field varies with the *third* derivative of $\mathbf{x}$, the spectral energy density of the vacuum field must be proportional to the *third* power of ω in order for (2.88) to hold. In the case of a dissipative force proportional to $\dot{\mathbf{x}}$, by contrast, the fluctuation force must be proportional to ω in order to maintain the canonical commutation relation (Milonni, 1981b). This relation between the form of the dissipation and the spectral density of the fluctuation is the essence of the fluctuation–dissipation theorem.[5]

The fact that the canonical commutation relation for a harmonic oscillator coupled to the vacuum field is preserved implies that the zero-point energy of the oscillator is preserved. It is easy to show that after a few damping times the zero-point motion of the oscillator is in fact sustained by the driving zero-point field (Senitzky, 1960).

The reader may well wonder whether the vacuum field is merely some sort of formal mathematical artifice of quantum electrodynamics, whether it really has any unambiguous experimental manifestations. In fact the zero-point field does appear to be quite "real," as we shall see in the following section.

2.7 The Casimir Effect

Casimir showed in 1948 that one consequence of the zero-point field is an attractive force between two uncharged, perfectly conducting parallel plates (Figure 2.1). In this section we review a standard calculation of the Casimir force, and in the following chapter we present a somewhat more physical

---

[5] H. B. Callen and T. A. Welton, *Phys. Rev.* **83**, 34 (1951).

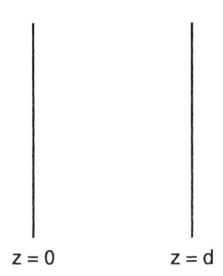

$$z = 0 \qquad\qquad z = d$$

Figure 2.1: Two conducting parallel plates experience an attractive force attributable to the zero-point electromagnetic field. This is the Casimir effect.

variation of this calculation. Various Casimir effects, and experimental evidence for them, are discussed in Chapters 7 and 8.

The physical situation shown in Figure 2.1 leads us to consider a different set of modes than the free-space plane-wave modes we have dealt with thus far. Consider first the modes appropriate to the interior of a rectangular parallelepiped of sides $L_x = L_y = L$ and L_z. For perfectly conducting walls the mode functions satisfying the boundary condition that the tangential component of the electric field vanishes on the walls are $\mathbf{A}(\mathbf{r}) = A_x(\mathbf{r})\mathbf{i} + A_y(\mathbf{r})\mathbf{j} + A_z(\mathbf{r})\mathbf{k}$, where

$$A_x(\mathbf{r}) = (8/V)^{1/2} a_x \cos(k_x x) \sin(k_y y) \sin(k_z z), \qquad (2.91)$$

$$A_y(\mathbf{r}) = (8/V)^{1/2} a_y \sin(k_x x) \cos(k_y y) \sin(k_z z), \qquad (2.92)$$

$$A_z(\mathbf{r}) = (8/V)^{1/2} a_z \sin(k_x x) \sin(k_y y) \cos(k_z z), \qquad (2.93)$$

with $a_x^2 + a_y^2 + a_z^2 = 1, V = L^2 L_z$, and

$$k_x = \frac{\ell\pi}{L}, \quad k_y = \frac{m\pi}{L}, \quad k_z = \frac{n\pi}{L_z}, \qquad (2.94)$$

with ℓ, m, and n each taking on all positive integer values and zero. In order to satisfy the transversality condition $\nabla \cdot \mathbf{A} = 0$ we also require

$$k_x A_x + k_y A_y + k_z A_z = \frac{\pi}{L}(\ell A_x + m A_y) + \frac{\pi}{L_z}(n A_z) = 0. \qquad (2.95)$$

Thus there are two independent polarizations, unless one of the integers ℓ, m, or n is zero, in which case (2.95) indicates that there is only one polarization. It is easy to check that equations (2.91)–(2.93) define transverse mode functions satisfying the Helmholtz equation (2.24) as well as the condition that the transverse components of $\mathbf{E}$ vanish on the cavity walls. Furthermore these mode functions are orthogonal and satisfy the normalization condition (2.30), i.e.,

$$\int_0^L dx \int_0^L dy \int_0^{L_z} dz [A_x^2(\mathbf{r}) + A_y^2(\mathbf{r}) + A_z^2(\mathbf{r})] = 1. \qquad (2.96)$$

Actually all we really require for the calculation of the Casimir force are the allowed frequencies defined by (2.94):

$$\omega_{\ell mn} = k_{\ell mn} c = \pi c \left[\frac{\ell^2}{L^2} + \frac{m^2}{L^2} + \frac{n^2}{L_z^2} \right]^{1/2}. \qquad (2.97)$$

The zero-point energy of the field inside the cavity is therefore

$$\sum_{\ell,m,n} {}'(2)\frac{1}{2}\hbar\omega_{\ell mn} = \sum_{\ell mn} {}' \pi \hbar c \left[\frac{\ell^2}{L^2} + \frac{m^2}{L^2} + \frac{n^2}{L_z^2} \right]^{1/2}. \qquad (2.98)$$

The factor 2 arises from the two independent polarizations of modes with $\ell, m, n \neq 0$, and the prime on the summation symbol implies that a factor $1/2$ should be inserted if one of these integers is zero, for then we have just one independent polarization, as noted earlier.

In the physical situation of interest L is so large compared with $L_z = d$ that we may replace the sums over ℓ and m in (2.98) by integrals: $\sum_{\ell mn} \rightarrow \sum_n {}'(L/\pi)^2 \int \int dk_x dk_y$ and

$$\begin{aligned} E(d) &= \sum_{\ell mn} {}'(2)\frac{1}{2}\hbar\omega_{\ell mn} \rightarrow \frac{L^2}{\pi^2}(\hbar c) \sum_n {}' \int_0^\infty dk_x \int_0^\infty dk_y \\ &\quad \times \left(k_x^2 + k_y^2 + \frac{n^2\pi^2}{d^2} \right)^{1/2}. \end{aligned} \qquad (2.99)$$

This is infinite; the zero-point energy of the vacuum is infinite in *any* finite volume.

If d were also made arbitrarily large, the sum over n could be replaced by an integral. Then the zero-point energy (2.99) would be

$$E(\infty) = \frac{L^2}{\pi^2}(\hbar c)\frac{d}{\pi} \int_0^\infty dk_x \int_0^\infty dk_y \int_0^\infty dk_z (k_x^2 + k_y^2 + k_z^2)^{1/2}. \quad (2.100)$$

which is also infinite.

The potential energy of the system when the plates are separated by a distance d is $U(d) = E(d) - E(\infty)$, the energy required to bring the plates from a large separation to the separation d:

$$
\begin{aligned}
U(d) \;=\; & \frac{L^2\hbar c}{\pi^2}\left[\sum_n{}' \int_0^\infty dk_x \int_0^\infty dk_y (k_x^2 + k_y^2 + \frac{n^2\pi^2}{d^2})^{\frac{1}{2}} \right. \\
& \left. -\frac{d}{\pi}\int_0^\infty dk_x \int_0^\infty dk_y \int_0^\infty dk_z (k_x^2 + k_y^2 + k_z^2)^{1/2}\right].
\end{aligned}
$$
$$(2.101)$$

This is the difference between two infinite quantities, but we shall now show that it is nonetheless possible to extract from it a physically meaningful, finite value.[6]

In polar coordinates u, θ in the k_x, k_y plane ($dk_x dk_y = u\,du\,d\theta$) we have

$$
\begin{aligned}
U(d) \;=\; & \frac{L^2\hbar c}{\pi^2}\left(\frac{\pi}{2}\right)\left[\sum_{n=0}{}' \int_0^\infty du\,u\left(u^2 + \frac{n^2\pi^2}{d^2}\right)^{1/2} \right. \\
& \left. -\left(\frac{d}{\pi}\right)\int_0^\infty dk_z \int_0^\infty du\,u(u^2 + k_z^2)^{1/2}\right],
\end{aligned}
$$
$$(2.102)$$

since θ ranges from 0 to $\pi/2$ for $k_x, k_y > 0$. We now introduce a cutoff function $f(k) = f([u^2 + k_z^2]^{1/2})$ such that $f(k) = 1$ for $k << k_m$ and $f(k) = 0$ for $k >> k_m$. Physically, it can be argued that $f(k)$ is necessary because the assumption of perfectly conducting walls breaks down at small wavelengths and especially for wavelengths small compared with an atomic dimension. We might then suppose that $k_m \approx 1/a_o$, where a_o is the Bohr radius. What we are assuming here is that the Casimir effect is primarily a low-frequency, nonrelativistic effect. We thus replace (2.102) by

$$
\begin{aligned}
U(d) \;=\; & \frac{L^2\hbar c}{\pi^2}(\frac{\pi}{2})\left[\sum{}' \int_0^\infty du\,u(u^2 + \frac{n^2\pi^2}{d^2})^{1/2}f([u^2 + \frac{n^2\pi^2}{d^2}]^{1/2}) \right. \\
& \left. -\left(\frac{d}{\pi}\right)\int_0^\infty dk_z \int_0^\infty du\,u(u^2 + k_z^2)^{1/2}f([u^2 + k_z^2]^{1/2})\right]
\end{aligned}
$$

---

[6] See Section 10.7 for a different approach.

$$= \frac{L^2 \hbar c}{4\pi} (\frac{\pi^3}{d^3}) \left[\sum_{n=0}^{\infty} {}' \int_0^{\infty} dx(x + n^2)^{1/2} f(\frac{\pi}{d}[x \mid n^2]^{1/2}) \right.$$

$$\left. - \int_0^{\infty} d\kappa \int_0^{\infty} dx(x + \kappa^2)^{1/2} f(\frac{\pi}{d}[x + \kappa^2]^{1/2}) \right], \qquad (2.103)$$

where we have defined the new integration variables $x = u^2 d^2/\pi^2$ and $\kappa = k_z d/\pi$. Now

$$U(d) = \left(\frac{\pi^2 \hbar c}{4d^3} \right) L^2 \left[\frac{1}{2} F(0) + \sum_{n=1}^{\infty} F(n) - \int_0^{\infty} d\kappa F(\kappa) \right], \qquad (2.104)$$

where

$$F(\kappa) \equiv \int_0^{\infty} dx(x + \kappa^2)^{1/2} f(\frac{\pi}{d}[x + \kappa^2]^{1/2}). \qquad (2.105)$$

According to the Euler–Maclaurin summation formula[7]

$$\sum_{n=1}^{\infty} F(n) - \int_0^{\infty} d\kappa F(\kappa) = -\frac{1}{2} F(0) - \frac{1}{12} F'(0) + \frac{1}{720} F'''(0) \ldots \quad (2.106)$$

for $F(\infty) \to 0$. To evaluate the nth derivative $F^{(n)}(0)$ we note that

$$F(\kappa) = \int_{\kappa^2}^{\infty} du \sqrt{u} f(\frac{\pi}{d}\sqrt{u}), \quad F'(\kappa) = -2\kappa^2 f(\frac{\pi}{d}\kappa). \qquad (2.107)$$

Then $F'(0) = 0$, $F'''(0) = -4$, and all higher derivatives $F^{(n)}(0)$ vanish if we assume that all derivatives of the cutoff function vanish at $\kappa = 0$. Thus $\sum_{n=1}^{\infty} F(n) - \int_0^{\infty} d\kappa F(\kappa) = -\frac{1}{2} F(0) - \frac{4}{720}$ and

$$U(d) = \left(\frac{\pi^2 \hbar c}{4d^3} \right) L^2 \left(\frac{-4}{720} \right) = - \left(\frac{\pi^2 \hbar c}{720 d^3} \right) L^2, \qquad (2.108)$$

which is finite and independent of the cutoff function. The attractive force per unit area between the plates is then $F(d) = -\pi^2 \hbar c/240 d^4$. This is the Casimir force, which we shall revisit in the following chapter and again in Chapters 7 and 8. The principal message of this section is that *changes* in the infinite zero-point energy of the electromagnetic vacuum can be finite and observable.

---

[7] See M. Abramowitz and I. A. Stegun, *Handbook of Mathematical Functions* (Dover Books, New York, 1971), Formula 3.6.28. For a derivation of the Euler–Maclaurin formula, see, for instance, E. T. Whittaker and G. N. Watson, *A Course of Modern Analysis*, 4th ed. (Cambridge University Press, New York, 1969), p. 127.

2.8   Field Commutators

The fundamental field commutator (2.55) holds for all times t, and regardless of whether there are any sources of radiation. From this commutator
one readily obtains commutation relations for the field vectors, such as

$$
\begin{aligned}
[F_i(\mathbf{r}_1, t), E_j(\mathbf{r}_2, t_2)] &= [B_i(\mathbf{r}_1, t), B_j(\mathbf{r}_2, t_2)] \\
&= 4\pi i\hbar c \left(\frac{\delta_{ij}}{c^2} \frac{\partial^2}{\partial t_1 \partial t_2} - \frac{\partial^2}{\partial r_{1i} \partial r_{2j}} \right) \\
&\quad \times D(|\mathbf{r}_1 - \mathbf{r}_2|, t_1 - t_2), \qquad (2.109)
\end{aligned}
$$

where

$$
\begin{aligned}
D(\mathbf{r}, t) &\equiv -\left(\frac{1}{2\pi} \right)^3 \int d^3k \frac{1}{k} e^{i\mathbf{k}\cdot\mathbf{r}} \sin\omega_k t \\
&= -\frac{1}{2\pi^2 r} \int_{-\infty}^{\infty} dk \sin kr \sin kct \\
&= \frac{1}{4\pi r}[\delta(r + ct) - \delta(r - ct)]. \qquad (2.110)
\end{aligned}
$$

These "Pauli–Jordan commutators" imply that the fields at space–time
points $(\mathbf{r}_1, t_1)$ and $(\mathbf{r}_2, t_2)$ cannot in general be simultaneously measured if
these points can be connected by a light signal, i.e., if $|\mathbf{r}_1 - \mathbf{r}_2| = \pm c(t_1 - t_2)$.
Similarly

$$
[E_i(\mathbf{r}_1, t_1), B_j(\mathbf{r}_2, t_2)] = 4\pi i\hbar c\epsilon_{ijk} \frac{\delta^2}{\delta t_1 \delta r_{2k}} D(|\mathbf{r}_1 - \mathbf{r}_2|, t_1 - t_2). \qquad (2.111)
$$

The physical significance of these commutators was discussed by Bohr and
Rosenfeld (1950): since the field of a charged particle provides information
about the motion of the particle, the uncertainty relations ($\Delta x \Delta p_x \geq \hbar/2$,
etc.) for the particles must, for the consistency of quantum theory, imply
uncertainty relations also for the field. These uncertainty relations for the
electromagnetic field are embodied in the Pauli Jordan commutators. Note
that $D(\mathbf{r}, t) = -D(\mathbf{r}, -t)$, and so $\lim_{t\to 0} D(\mathbf{r}, t) = \lim_{t\to 0}(\partial^2/\partial t^2)D(\mathbf{r}, t) =$
0. Equation (2.110) then indicates that in principle the electric and mag-
netic fields can be simultaneously measured everywhere in space at a fixed
instant of time.

Note also that these field commutators are derived for free space. The
presence of boundaries or even simple point sources will in general lead to
different commutation relations, simply because $e^{i\mathbf{k}\cdot\mathbf{r}}$ in (2.110) must be
replaced by different mode functions.[8]

---

[8] This is discussed in the papers by Milonni (1982) and Cresser (1984) cited at the
end of the chapter.

2.9   Zero-Point Spectrum Invariance

We have seen in connection with the classical Einstein–Hopf model that a dipole oscillating with frequency ω, moving with velocity v through a thermal field, experiences a frictional force $F = -Rv$, where R is proportional to $\rho(\omega) - (\omega/3)d\rho/d\omega$. The same result holds when the dipole is treated quantum mechanically or if, as shown in Appendix B, the dipole is replaced by an atom.

At $T = 0$ we have $\rho(\omega) = \rho_{o}(\omega) = \hbar\omega^3/2\pi^2c^3$ and

$$\rho_{o}(\omega) - \frac{\omega}{3}\frac{d\rho_{o}}{d\omega} = 0. \tag{2.112}$$

In other words, there is no frictional force acting on a dipole or atom moving with constant velocity in the vacuum. The zero-point spectrum proportional to ω^3, which is precisely the form required by the fluctuation-dissipation relation (Section 2.6), is thus the the *unique* spectral energy density for which there is no force. Alternatively, we can say that, since the number of modes per unit volume in free space is proportional to ω^2, the energy $\frac{1}{2}\hbar\omega$ per mode is the unique zero-point energy for which there is no force. The "uniqueness" refers, of course, to the functional dependence on ω; any zero-point energy proportional to ω, or any spectral energy density proportional to ω^3, will satisfy (2.112).

In fact it has been shown explicitly by Boyer (1969), using the Lorentz transformations for the electric and magnetic fields, that $\rho_{o}(\omega)$ is the unique Lorentz-invariant spectral energy density of the electromagnetic field. That is, the condition that $\rho_{o}(\omega)$ be the same in all inertial frames requires it to be proportional to ω^3. This conforms with our expectation that an observer moving with constant velocity in the electromagnetic vacuum cannot tell that he is moving!

2.10   The Unruh–Davies Effect

What if the observer is moving with constant (proper) *acceleration* in the vacuum? Then a remarkable thing happens: the observer perceives himself to be immersed in a thermal bath at the temperature $T = \hbar a/2\pi kc$, where a is the acceleration. This result was obtained by Unruh (1976), following a closely related result of Davies (1975). In this section we shall demonstrate this thermal effect of acceleration for the case of a scalar field, for which the calculation is simpler. The electromagnetic case is somewhat more complicated but the result is the same.

We consider a massless scalar field $\phi(\mathbf{x}, t)$ satisfying the wave equation $(\nabla^2 - c^{-2}\partial^2/\partial t^2)\phi = 0$ and having an energy density $(1/8\pi)[(\nabla\phi)^2 + c^{-2}(\partial\phi/\partial t)^2]$.[9] When quantized in free space $\phi(\mathbf{x}, t)$ has the form

$$\phi(\mathbf{x}, t) = \sum_{\mathbf{k}} \left(\frac{2\pi\hbar c^2}{\omega_k V}\right)^{1/2} [a_{\mathbf{k}}(t)e^{i\mathbf{k}\cdot\mathbf{x}} + a_{\mathbf{k}}^\dagger(t)e^{-i\mathbf{k}\cdot\mathbf{x}}], \qquad (2.113)$$

where again we assume periodic boundary conditions. Here, $a_{\mathbf{k}}(t)$ and $a_{\mathbf{k}}^\dagger(t)$ are boson annihilation and creation operators and $a_{\mathbf{k}}(t) = a_{\mathbf{k}}(0)e^{-i\omega_k t}$ for the free field, with $\omega_k = kc$. The Hamiltonian is

$$H_{\mathrm{F}} = \frac{1}{8\pi}\int d^3x\left[(\nabla\phi)^2 + \frac{1}{c^2}(\frac{\partial\phi}{\partial t})^2\right] = \sum_{\mathbf{k}}\hbar\omega_k(a_{\mathbf{k}}^\dagger a_{\mathbf{k}} + \frac{1}{2}). \qquad (2.114)$$

Everything here is much the same as in the case of the electromagnetic field, but simpler.

Consider the field correlation function $\langle\phi(0, t)\phi(0, t + \tau)\rangle$ at a point in space for a field in thermal equilibrium at temperature T. In this case $\langle a_{\mathbf{k}}^\dagger(0)a_{\mathbf{k}'}(0)\rangle = \delta_{\mathbf{k}, \mathbf{k}'}^3\,\overline{n}(\omega)$, $\overline{n}(\omega) \equiv (e^{\hbar\omega/kT} - 1)^{-1}$. These results are intuitively obvious. Basically they imply that different modes of a thermal field are uncorrelated, each mode amplitude having zero expectation value, and that a mode of frequency ω has an average number of quanta $\overline{n}(\omega)$. Thus

$$
\begin{aligned}
\langle\phi(0, t)\phi(0, t + \tau)\rangle &= \sum_{\mathbf{k}}\left(\frac{2\pi\hbar c^2}{\omega_k V}\right)\left[\langle a_{\mathbf{k}}(t)a_{\mathbf{k}}^\dagger(t + \tau)\rangle\right. \\
&\quad + \left.\langle a_{\mathbf{k}}^\dagger(t)a_{\mathbf{k}}(t + \tau)\rangle\right] \\
&\quad - \sum_{\mathbf{k}}\left(\frac{2\pi\hbar c^2}{\omega_k V}\right)\left[(\overline{n}(\omega_k) + 1)e^{i\omega_k\tau} + \overline{n}(\omega_k)e^{-i\omega_k\tau}\right] \\
&\to \frac{\hbar}{\pi c}\left[\int_0^\infty d\omega\omega e^{i\omega\tau} + 2\int_0^\infty \frac{d\omega\omega\cos\omega\tau}{e^{\hbar\omega/kT} - 1}\right]. \quad (2.115)
\end{aligned}
$$

The first integral may be evaluated as follows:

$$\int_0^\infty d\omega\omega e^{i\omega\tau} = \lim_{x\to 0}\int_0^\infty d\omega\omega e^{i\omega(\tau + ix)} = \lim_{x\to 0}\frac{1}{(x - i\tau)^2} = -\frac{1}{\tau^2}. \qquad (2.116)$$

---

[9] See Section 10.3.

The second integral follows from the general formula[10]

$$\int_0^\infty dx \frac{x^{2m+1}\cos bx}{e^x - 1} = (-1)^m \frac{\partial^{2m+1}}{\partial b^{2m+1}} \left[\frac{\pi}{2} \coth \pi b - \frac{1}{2b} \right].$$ (2.117)

Thus

$$\int_0^\infty \frac{d\omega\,\omega\,\cos\omega\tau}{e^{\hbar\omega/kT} - 1} = \frac{1}{\tau^2} - \left(\frac{\pi kT}{\hbar}\right)^2 \mathrm{csch}^2\left(\frac{\pi kT\tau}{\hbar}\right)$$ (2.118)

and

$$\begin{aligned}
\langle\phi(0,t)\phi(0,t+\tau)\rangle &= \frac{\hbar}{\pi c}\left[-\frac{1}{\tau^2} + \frac{1}{\tau^2} - \left(\frac{\pi kT}{\hbar}\right)^2 \mathrm{csch}^2\left(\frac{\pi kT\tau}{\hbar}\right) \right] \\
&= -\frac{\hbar}{\pi c}\left(\frac{\pi kT}{\hbar}\right)^2 \mathrm{csch}^2\left(\frac{\pi kT\tau}{\hbar}\right).
\end{aligned}$$ (2.119)

Let us consider also the correlation function $\langle\phi(\mathbf{y},t)\phi(\mathbf{y}+\mathbf{x},t+\tau)\rangle_0$ in the vacuum state of our scalar field. In this case $\langle a_{\mathbf{k}}(0)a_{\mathbf{k}'}(0)\rangle = \langle a_{\mathbf{k}}^\dagger(0)a_{\mathbf{k}'}(0)\rangle = 0$ and $\langle a_{\mathbf{k}}(0)a_{\mathbf{k}'}^\dagger(0)\rangle = \delta_{\mathbf{k},\mathbf{k}'}^3$, and from (2.113) we obtain

$$\begin{aligned}
\langle\phi(\mathbf{y},t)\phi(\mathbf{y}+\mathbf{x},t+\tau)\rangle_0 &= \sum_{\mathbf{k}} \left(\frac{2\pi\hbar c^2}{\omega_k V}\right) e^{-i\mathbf{k}\cdot\mathbf{x}} e^{i\omega_k\tau} \\
&\rightarrow \frac{\hbar c^2}{4\pi^2}\int_0^\infty dk\,k^2\omega^{-1}e^{i\omega\tau}\int d\Omega_{\mathbf{k}} e^{-i\mathbf{k}\cdot\mathbf{x}},
\end{aligned}$$ (2.120)

where the last integral is over all solid angles about $\mathbf{k}$:

$$\int d\Omega_{\mathbf{k}} e^{-i\mathbf{k}\cdot\mathbf{x}} = \int_0^{2\pi} d\phi \int_0^\pi d\theta \sin\theta\, e^{-ikx\cos\theta} = 4\pi\frac{\sin kx}{kx}.$$ (2.121)

Thus

$$\begin{aligned}
\langle\phi(\mathbf{y},t)\phi(\mathbf{y}+\mathbf{x},t+\tau)\rangle_0 &= \frac{\hbar c^2}{\pi}\int_0^\infty dk\,k^2\omega^{-1}e^{i\omega\tau}\frac{\sin kx}{kx} \\
&= \frac{\hbar c}{\pi}\frac{1}{x^2 - c^2\tau^2}.
\end{aligned}$$ (2.122)

[10]L. S. Gradshteyn and I. M. Rhyzhik, *Table of Integrals, Series, and Products* (Academic Press, New York, 1980), p. 494, No. 13.

We turn now to an observer undergoing uniform acceleration in the vacuum. Uniform acceleration here is defined with respect to an instantaneous inertial frame in which the observer is at rest. The *proper* acceleration a is the acceleration relative to this instantaneous inertial rest frame, and if a is constant the acceleration is said to be uniform. The acceleration dv/dt in the lab frame may be related to a using standard Lorentz transformations for acceleration:

$$\frac{dv}{dt} = a\left(1 - \frac{v^2}{c^2}\right)^{3/2}. \tag{2.123}$$

Simple integrations give $v(t) = at(1 + a^2t^2/c^2)^{-1/2}$ and $x(t) = c^2/a[(1 + a^2t^2/c^2)^{1/2} - 1]$ if we assume $v = x = 0$ when $t = 0$. Using the relation $dt = d\tau(1 - v^2/c^2)^{-1/2}$ between lab and proper time intervals, respectively, we have

$$\frac{dt}{d\tau} = \left(1 - \frac{a^2t^2}{c^2 + a^2t^2}\right)^{-1/2} = \left(1 + \frac{a^2t^2}{c^2}\right)^{1/2} \tag{2.124}$$

and

$$t(\tau) = \frac{c}{a}\sinh\frac{a\tau}{c} \tag{2.125}$$

if we define $t(\tau = 0) = 0$. We can use this result to express x and v in the lab frame in terms of the proper time τ:

$$x(\tau) = \frac{c^2}{a}[\cosh\frac{a\tau}{c} - 1], \tag{2.126}$$

$$v(\tau) = c\tanh\frac{a\tau}{c}. \tag{2.127}$$

We recall as an aside the motion of a particle of rest mass m acted upon by a constant force F. In this case the linear momentum $p = Ft = mv(1 - v^2/c^2)^{-1/2}$ and so $v = (Ft/m)[1 + (Ft/mc^2)]^{-1/2}$ and $x = (mc^2/F)([1 + (Ft/mc^2)]^{1/2} - 1)$, which are the results given previously for $a = F/m$. The world line is a hyperbola in the $x - t$ plane, with asymptote $x = ct$, and consequently this motion is often called *hyperbolic motion*. For $Ft \ll mc^2$ we have the classical parabolic motion, $x(t) = \frac{1}{2}at^2$.

The vacuum correlation function $\langle\phi(x_1, t_1)\phi(x_2, t_2)\rangle_0$ measured by our uniformly accelerated observer is given by (2.122) with $x = x_2 - x_1$ and $\tau = t_2 - t_1$, or $x = (c^2/a)[\cosh(a\tau_2/c) - \cosh(a\tau_1/c)]$ and $\tau = (c/a)[\sinh(a\tau_2/c) - \sinh(a\tau_1/c)]$. Since

$$\begin{aligned}
x^2 - c^2\tau^2 &= \frac{c^4}{a^2}[\cosh\frac{a\tau_2}{c} - \cosh\frac{a\tau_1}{c}]^2 - \frac{c^4}{a^2}[\sinh\frac{a\tau_2}{c} - \sinh\frac{a\tau_1}{c}]^2 \\
&= -\frac{c^4}{a^2}\sinh^2\frac{a(\tau_2 - \tau_1)}{2c}, \tag{2.128}
\end{aligned}$$

it follows from (2.122) that

$$\langle \phi(x_1, t_1)\phi(x_2, t_2)\rangle_0 = -\frac{\hbar a^2}{\pi c^3}\operatorname{csch}^2\frac{a(\tau_2 - \tau_1)}{2c}, \qquad (2.129)$$

which is equivalent to the *thermal-field* correlation function (2.119) with temperature

$$T = \frac{\hbar a}{2\pi kc}. \qquad (2.130)$$

The meaning of this result is that a uniformly accelerated detector in the vacuum responds as it would if it were at rest in a thermal bath at temperature $T = \hbar a/2\pi kc$. In a sense the effect of the acceleration is to "promote" zero-point quantum field fluctuations to the level of thermal fluctuations. It is hardly obvious why this should be so — it took half a century after the birth of the quantum theory of radiation for the thermal effect of uniform acceleration to be discovered.

2.11   Thermal Radiation

There are two reasons for reviewing aspects of thermal radiation in this section and the next. First, certain statistical properties of thermal radiation are similar to those of the vacuum field. Second, the quantum theory of thermal radiation provides a clearer picture of some results used in Chapter 1, particularly in connection with the role of the zero-point (vacuum) field in the blackbody problem.

The probability P_n that there are n photons in a field mode of frequency ω in thermal equilibrium at temperature T — that is, that the mode is excited to the harmonic oscillator level n — is

$$\begin{aligned}
P_n &= \frac{e^{-(n+\frac{1}{2})\hbar\omega/kT}}{\sum_{n=0}^{\infty} e^{-(n+\frac{1}{2})\hbar\omega/kT}} = \frac{e^{-n\hbar\omega/kT}}{\sum_{n=0}^{\infty} e^{-n\hbar\omega/kT}} \\
&= e^{-n\hbar\omega/kT}\left(1 - e^{-\hbar\omega/kT}\right)^{-1}.
\end{aligned} \qquad (2.131)$$

The average photon number is thus

$$\bar{n} = \sum_{n=0}^{\infty} nP_n = (e^{\hbar\omega/kT} - 1)^{-1}, \qquad (2.132)$$

and we can use this result to write P_n in terms of $\bar{n}$:

$$P_n = \frac{\bar{n}^n}{(\bar{n}+1)^{n+1}}. \qquad (2.133)$$

Since there are $(\omega^2/\pi^2 c^3)V\,d\omega$ modes of the field in the frequency interval $[\omega, \omega + d\omega]$ in a volume V large compared with c/ω, the spectral energy density is $\rho(\omega) = \hbar\omega\overline{n}(\omega)\omega^2/\pi^2 c^3$, which of course is the Planck spectrum, without the zero-point contribution.

The results (2.131)–(2.133) depend only on the frequency of the radiation, not its wave vector $\mathbf{k}$ or polarization λ. *Thermal radiation as described by the Planck spectrum is isotropic and unpolarized.*

We can use (2.133) to calculate averages of functions of n. For instance,

$$\langle n^2 \rangle = \sum_{n=0}^{\infty} n^2 P_n = 2\overline{n}^2 + \overline{n}, \qquad (2.134)$$

and so

$$\langle \Delta n^2 \rangle \equiv \langle n^2 \rangle - \langle n \rangle^2 = 2\overline{n}^2 + \overline{n} - \overline{n}^2 = \overline{n}^2 + \overline{n}, \qquad (2.135)$$

which is a well known consequence of Bose–Einstein statistics.[11]

Since $\rho(\omega) = \hbar\omega^3\overline{n}(\omega)/\pi^2 c^3$, we can write (2.135) in the form

$$\begin{aligned}
\langle \Delta n(\omega)^2 \rangle &= \frac{\pi^2 c^3}{\hbar\omega^3}\left[\frac{\pi^2 c^3}{\hbar\omega^3}\rho^2(\omega) + \rho(\omega)\right] \\
&= \frac{\pi^2 c^3}{\hbar^2\omega^4}\left[\hbar\omega\rho(\omega) + \frac{\pi^2 c^3}{\omega^2}\rho^2(\omega)\right]. \qquad (2.136)
\end{aligned}$$

The variance in the energy of the thermal field is thus

$$\begin{aligned}
\langle \Delta E^2 \rangle &= \sum_{\mathbf{k}\lambda} \hbar^2\omega_k^2 \langle \Delta n(\omega_k)^2 \rangle \rightarrow \frac{V}{8\pi^3}(2)\int d^3k\, \hbar^2\omega^2 \langle \Delta n(\omega)^2 \rangle \\
&= \frac{V}{4\pi^3 c^3}(4\pi)\int d\omega\,\omega^2(\hbar^2\omega^2)\langle \Delta n(\omega)^2 \rangle \equiv \int \langle \Delta E_\omega^2 \rangle, \\
& \hspace{9cm} (2.137)
\end{aligned}$$

where

$$\langle \Delta E_\omega^2 \rangle = \frac{V}{\pi^2 c^3}\hbar^2\omega^4 \langle \Delta n(\omega)^2 \rangle d\omega = [\hbar\omega\rho(\omega) + \frac{\pi^2 c^3}{\omega^2}\rho^2(\omega)]V\,d\omega. \qquad (2.138)$$

This is the Einstein fluctuation formula (1.63).

And so Einstein's fluctuation formula can be regarded as a precursor of the result (2.135) of Bose–Einstein statistics. From the discussion in

---

[11] See, for instance, L. D. Landau and E. M. Lifshitz, *Statistical Physics* (Addison–Wesley, Reading, Mass., 1969), p. 355.

Section 1 7 we can associate $\bar{n}^2$ in (2.135) with wave fluctuations, and $\bar{n}$ with particle fluctuations.

In Section 1.7 we also inferred that the particle fluctuation term could be attributed to the zero-point energy of the field. To appreciate this from the perspective of the quantum theory of the field, note that

$$
\begin{aligned}
\langle \Delta n^2 \rangle &= \langle n^2 \rangle - \langle n \rangle^2 = \langle a^\dagger a a^\dagger a \rangle - \langle a^\dagger a \rangle^2 \\
&= \langle a^\dagger (a^\dagger a + 1) a \rangle - \langle a^\dagger a \rangle^2 \\
&= \langle a^\dagger a^\dagger a a \rangle + \langle a^\dagger a \rangle - \langle a^\dagger a \rangle^2 \\
&= \langle a^\dagger a^\dagger a a \rangle + \bar{n} - \bar{n}^2,
\end{aligned}
\tag{2.139}
$$

where a is the photon annihilation operator for the field mode under consideration. Now a mode of a thermal field is described by the density matrix

$$
\rho = \sum_n P_n |n\rangle\langle n| ,
\tag{2.140}
$$

and therefore

$$
\langle a^\dagger a^\dagger a a \rangle = \sum_n \langle n | a^\dagger a^\dagger a a | n \rangle.
\tag{2.141}
$$

But $aa|n\rangle = \sqrt{n} a |n-1\rangle = \sqrt{n(n-1)} |n-2\rangle$, so that

$$
\langle a^\dagger a^\dagger a a \rangle = \sum_{n=0}^{\infty} n(n-1) P_n = \sum_{n=0}^{\infty} n(n-1) \frac{\bar{n}^n}{(\bar{n}+1)^{n+1}} = 2\bar{n}^2
\tag{2.142}
$$

for a thermal field. Then (2.139) reproduces (2.135). But note that the particle term $\bar{n}$ in this formula arises from the second term in the last line of (2.139), i.e., *from the fact that the commutator* $[a, a^\dagger] = 1$. Note furthermore that this same commutator gives rise to the zero-point energy of a harmonic oscillator such as a field mode, as is clear from equation (2.9). The conclusion is obvious: *the particle term in the Einstein fluctuation formula, or equivalently (2.135), is closely linked to the existence of zero-point energy.*

The "wave" fluctuation term $\bar{n}^2$ in the variance $\langle \Delta n^2 \rangle$ for thermal radiation arises from the factor of 2 in $\langle a^\dagger a^\dagger a a \rangle = 2\bar{n}^2$. This important factor is the origin of Brown–Twiss correlations,[12] also known as *photon bunching*. Suppose we take a spectrally filtered beam of thermal radiation and

---

[12] R. Hanbury Brown and R. Q. Twiss, *Nature* **127**, 27 (1956); *Proc. Roy. Soc. Lond.* **A242**, 300 (1957). For a discussion of the Brown–Twiss effect see, for instance, Knight and Allen (1983); Loudon (1983); Milonni (1984).

employ a detection scheme in which photons are counted by two-photon absorption rather than ordinary one-photon absorption. That is, a photo-electron is produced by the simultaneous absorption of two photons. As shown in Appendix E, such a detector responds to the *normally ordered* field correlation function $\langle a^\dagger a^\dagger aa \rangle$ if we have a single field mode. The fact that this quantity exceeds n^2 indicates that the photons have a statistical tendency to arrive in pairs. Such photon bunching of thermal radiation was first measured by Brown and Twiss in the 1950s.

It is worth noting that this "photon bunching" may be understood in purely classical terms, based on the Einstein–Hopf model of a thermal field as a superposition of waves with independent random phases (Section 1.5). Comparing equations (1.67) and (1.68), we note that $\langle \mathbf{E}^4(\mathbf{r},t) \rangle = 2\langle \mathbf{E}^2(\mathbf{r},t) \rangle^2$, or $\overline{I^2} = 2\overline{I}^2$. Thus there are positive intensity correlations or, in photon language, a tendency for photons to arrive in pairs.

It should be emphasized that photon bunching is not a universal property of light. An ideal laser, for example, gives $\langle a^\dagger a^\dagger aa \rangle = \overline{n}^2$, indicating that the photon arrivals are uncorrelated. In other words, an ideal laser has no wave fluctuations: $\langle \Delta n^2 \rangle = \overline{n}$. It is the closest we can get to the idealized, nonfluctuating classical wave of light.

A thermal field, like the vacuum field, is described by Gaussian statistics. Consider for simplicity a single mode of the field, for which the electric field operator is given by equation (2.35). The characteristic function of a single component of this field, which is defined as

$$C[E(\mathbf{r},t), \xi] = \langle e^{i\xi E(\mathbf{r},t)} \rangle, \qquad (2.143)$$

gives the probability distribution $P[E(\mathbf{r},t)]$ via a Fourier transform:

$$P[E(\mathbf{r},t)] = \frac{1}{2\pi} \int_{-\infty}^{\infty} d\xi\, e^{-i\xi E} C[E, \xi]. \qquad (2.144)$$

Using (2.35), we have

$$C[E(\mathbf{r},t), \xi] = \langle e^{i\xi(\alpha a + \alpha^* a^\dagger)} \rangle, \qquad (2.145)$$

where $\alpha \equiv i(2\pi\hbar\omega)^{1/2} A_o(\mathbf{r})$. For a thermal field, according to Bloch's theorem for a harmonic oscillator in thermal equilibrium,[13]

$$\langle e^{i\xi(\alpha a + \alpha^* a^\dagger)} \rangle = e^{-\xi^2 |\alpha|^2 (\overline{n} + \frac{1}{2})}, \qquad (2.146)$$

---

[13] F. Bloch, Z. Phys. **74**, 295 (1932). See also W. H. Louisell, *Radiation and Noise in Quantum Electronics* (McGraw-Hill, New York, 1964), p. 244.

so that, from (2.144),

$$P[E(\mathbf{r},t)] = \frac{1}{\sqrt{2\pi\mu}}e^{-E^2(\mathbf{r},t)/2\mu}, \qquad (2.147)$$

with

$$\mu = 2|\alpha|^2(\overline{n}+\frac{1}{2}) = 4\pi\hbar\omega|A_o(\mathbf{r})|^2(\overline{n}+\frac{1}{2}). \qquad (2.148)$$

Comparing (2.147) with the distribution (2.42) for the case of the vacuum field, we see that both the vacuum and thermal fields are distributed according to a Gaussian probability distribution. The vacuum distribution is just the $T \to 0$ limit of the thermal distribution. These results are easy to generalize to the multimode case.

2.12   Thermal Equilibrium

We now turn our attention once more to the Einstein–Hopf model of thermal equilibrium between radiation and matter, this time treating both the radiation and the dipole oscillators quantum mechanically (Milonni, 1981a). This will allow us, among other things, to better understand the Einstein–Stern derivation of the Planck spectrum discussed in Section 1.6.

The impulse imparted to a dipole oscillator in the quantum-mechanical version of the Einstein–Hopf model is given by equation (1.48), but now $z(t)$ and $\partial E_z(t)/\partial x$ are quantum-mechanical operators. For the dipole oscillator we introduce lowering and raising operators, σ and $\sigma^\dagger$, respectively, as in Section 2.2: $z = i(\hbar/2m\omega_o)^{1/2}(\sigma - \sigma^\dagger), [\sigma,\sigma^\dagger] = 1$. For $\partial E_z/\partial x$ we have, from (2.56),

$$\frac{\partial E_z}{\partial x} = -\sum_{\mathbf{k}\lambda}\left(\frac{2\pi\hbar\omega_k}{V}\right)^{1/2}k_x e_{\mathbf{k}\lambda z}[a_{\mathbf{k}\lambda}(t) + a_{\mathbf{k}\lambda}^\dagger(t)] \qquad (2.149)$$

at the position $\mathbf{r} = 0$ of one of the dipole oscillators. Thus

$$
\begin{aligned}
\Delta &= e\int_0^\tau dt z(t)\frac{\partial E_z(t)}{\partial x} \\
&= -ie\left(\frac{\hbar}{2m\omega_o}\right)^{1/2}\sum_{\mathbf{k}\lambda}\left(\frac{2\pi\hbar\omega_k}{V}\right)^{1/2}k_x e_{\mathbf{k}\lambda z} \\
&\quad \times \int_0^\tau dt[\sigma(t) - \sigma^\dagger(t)][a_{\mathbf{k}\lambda}(t) + a_{\mathbf{k}\lambda}^\dagger(t)].
\end{aligned}
\qquad (2.150)
$$

We shall assume that the dipole-field coupling is sufficiently weak over the time interval $[0, \tau]$ that $\sigma(t)$ and $a_{\mathbf{k}\lambda}(t)$ follow approximately their free evolution in (2.150): $\sigma(t) \cong \sigma(0)e^{-i\omega_o t} \equiv \sigma e^{-i\omega_o t}$ and $a_{\mathbf{k}\lambda}(t) \cong a_{\mathbf{k}\lambda}(0)e^{-i\omega_k t} \equiv a_{\mathbf{k}\lambda}e^{-i\omega_k t}$:

$$
\begin{aligned}
\Delta \;\cong\; & -ie\left(\frac{\hbar}{2m\omega_o}\right)^{1/2}\sum_{\mathbf{k}\lambda}\left(\frac{2\pi\hbar\omega_k}{V}\right)^{1/2}k_x e_{\mathbf{k}\lambda z}\left[\sigma a_{\mathbf{k}\lambda}\int_0^\tau dt\, e^{-i(\omega_k+\omega_o)t}\right.\\
& \left.+\sigma a_{\mathbf{k}\lambda}^\dagger\int_0^\tau dt\, e^{i(\omega_k-\omega_o)t} - \text{h.c.}\right]\\
=\;& ie\left(\frac{\hbar}{2m\omega_o}\right)^{1/2}\sum_{\mathbf{k}\lambda}\left(\frac{2\pi\hbar\omega_k}{V}\right)^{1/2}k_x e_{\mathbf{k}\lambda z}\\
& \times\left[\sigma a_{\mathbf{k}\lambda}e^{-i(\omega_k+\omega_o)\tau/2}\frac{\sin\frac{1}{2}(\omega_k+\omega_o)\tau}{\frac{1}{2}(\omega_k+\omega_o)}\right.\\
& \left.-\sigma a_{\mathbf{k}\lambda}^\dagger e^{i(\omega_k-\omega_o)\tau/2}\frac{\sin\frac{1}{2}(\omega_k-\omega_o)\tau}{\frac{1}{2}(\omega_k-\omega_o)} - \text{h.c.}\right]. \quad (2.151)
\end{aligned}
$$

Terms involving $\omega_k+\omega_o$ do not in the end contribute to $\langle\Delta^2\rangle$, just as such "energy nonconserving" terms do not contribute to transition probabilities in standard second-order perturbation theory. Thus

$$
\begin{aligned}
\Delta \;\cong\; & -ie\left(\frac{\hbar}{2m\omega_o}\right)^{1/2}\sum_{\mathbf{k}\lambda}\left(\frac{2\pi\hbar\omega_k}{V}\right)^{1/2}k_x e_{\mathbf{k}\lambda z}\\
& \times[\sigma a_{\mathbf{k}\lambda}^\dagger e^{i(\omega_k-\omega_o)\tau/2} - \text{h.c.}]\frac{\sin\frac{1}{2}(\omega_k-\omega_o)\tau}{\frac{1}{2}(\omega_k-\omega_o)} \quad (2.152)
\end{aligned}
$$

and

$$
\begin{aligned}
\langle\Delta^2\rangle \;\cong\; & e^2\left(\frac{2\hbar}{m\omega_o}\right)\sum_{\mathbf{k}\lambda}\left(\frac{2\pi\hbar\omega_k}{V}\right)k_x^2 e_{\mathbf{k}\lambda z}^2\frac{\sin^2\frac{1}{2}(\omega_k-\omega_o)\tau}{(\omega_k-\omega_o)^2}\\
& \times\left[\langle\sigma\sigma^\dagger\rangle\langle a_{\mathbf{k}\lambda}^\dagger a_{\mathbf{k}\lambda}\rangle + \langle\sigma^\dagger\sigma\rangle\langle a_{\mathbf{k}\lambda}a_{\mathbf{k}\lambda}^\dagger\rangle\right]. \quad (2.153)
\end{aligned}
$$

Let us write $\langle a_{\mathbf{k}\lambda}^\dagger a_{\mathbf{k}\lambda}\rangle = \bar{n}(\omega)$ and $\langle a_{\mathbf{k}\lambda}a_{\mathbf{k}\lambda}^\dagger\rangle = \langle a_{\mathbf{k}\lambda}^\dagger a_{\mathbf{k}\lambda}\rangle+1 = \bar{n}(\omega)+1$, and proceed to the mode continuum limit $V\to\infty$ in (2.153):

$$
\langle\Delta^2\rangle \;\cong\; e^2\left(\frac{2\hbar}{m\omega_o}\right)\left(\frac{V}{8\pi^3}\right)\int d^3k\left(\frac{2\pi\hbar\omega}{V}\right)k_x^2\sum_\lambda e_{\mathbf{k}\lambda z}^2\frac{\sin^2\frac{1}{2}(\omega-\omega_o)\tau}{(\omega-\omega_o)^2}
$$

$$\times \left[\langle \sigma\sigma^\dagger\rangle \overline{n}(\omega) + \langle \sigma^\dagger\sigma\rangle(\overline{n}(\omega) + 1) \right]$$

$$= \frac{e^2\hbar^2}{2\pi^2 m\omega_o} \int d^3k\,\omega k_x^2 \left(1 - \frac{k_z^2}{k^2}\right) \frac{\sin^2 \frac{1}{2}(\omega - \omega_o)\tau}{(\omega - \omega_o)^2}$$

$$\times \left[\langle \sigma\sigma^\dagger\rangle \overline{n}(\omega) + \langle \sigma^\dagger\sigma\rangle(\overline{n}(\omega) + 1) \right]. \tag{2.154}$$

In writing this expression we have used the identity $\hat{\mathbf{z}} = (\hat{\mathbf{k}}\cdot\hat{\mathbf{z}})\hat{\mathbf{k}} + \sum_\lambda e_{\mathbf{k}\lambda z}\mathbf{e}_{\mathbf{k}\lambda}$ for any unit vector $\hat{\mathbf{z}}$, and therefore $\sum_\lambda e_{\mathbf{k}\lambda z}^2 = 1 - (\hat{\mathbf{k}} \cdot \hat{\mathbf{z}})^2 = 1 - k_z^2/k^2$. Now

$$\int d^3k\,k_x^2 \left(1 - \frac{k_z^2}{k^2}\right) = \int dk\,k^2 \int d\Omega_\mathbf{k}\,k_x^2 \left(\frac{k_x^2 + k_y^2}{k^2}\right)$$

$$= \int dk \int d\Omega_\mathbf{k}\,k_x^2(k_x^2 + k_y^2)$$

$$= \int dk\,k^4 \int_0^{2\pi} d\phi \int_0^\pi d\theta \sin\theta(\sin\theta\cos\phi)^2 \sin^2\theta$$

$$= \frac{16\pi}{15} \int dk\,k^4 = \frac{16\pi}{15c^5} \int d\omega\,\omega^4 \tag{2.155}$$

and so

$$\langle \Delta^2\rangle = \frac{8e^2\hbar^2}{15\pi m\omega_o c^5} \int_0^\infty d\omega\,\omega^5 \frac{\sin^2 \frac{1}{2}(\omega - \omega_o)\tau}{(\omega - \omega_o)^2}$$

$$\times \left[\langle \sigma\sigma^\dagger\rangle \overline{n}(\omega) + \langle \sigma^\dagger\sigma\rangle(\overline{n}(\omega) + 1) \right]$$

$$\cong \frac{8e^2\hbar^2\omega_o^4}{15\pi mc^5} \left[\langle \sigma\sigma^\dagger\rangle \overline{n}(\omega_o) + \langle \sigma^\dagger\sigma\rangle(\overline{n}(\omega_o) + 1) \right]$$

$$\times \int_0^\infty d\omega \frac{\sin^2 \frac{1}{2}(\omega - \omega_o)\tau}{(\omega - \omega_o)^2}$$

$$\cong \frac{4e^2\hbar^2\omega_o^4}{15mc^5} \left[\langle \sigma\sigma^\dagger\rangle \overline{n}(\omega_o) + \langle \sigma^\dagger\sigma\rangle(\overline{n}(\omega_o) + 1) \right] \tau. \tag{2.156}$$

As in the classical Einstein–Hopf model the condition for thermal equilibrium is $\tau^{-1}\langle\Delta^2\rangle = 2RkT$ [Equation (1.47)]. From (2.156) and (1.43) this condition is

$$\rho(\omega_o) - \frac{\omega_o}{3}\frac{d\rho}{d\omega_o} = \frac{\hbar^2\omega_o^4}{6\pi^2 c^3 kT} \left[\langle \sigma\sigma^\dagger\rangle \overline{n}(\omega_o) + \langle \sigma^\dagger\sigma\rangle(\overline{n}(\omega_o) + 1) \right]. \tag{2.157}$$

Now the dipole is just another harmonic oscillator, and so in thermal equilibrium $\langle \sigma \sigma^\dagger \rangle = \langle \sigma^\dagger \sigma \rangle + 1 = \bar{n}(\omega_o) + 1$, whence

$$\langle \sigma \sigma^\dagger \rangle \bar{n}(\omega_o) + \langle \sigma^\dagger \sigma \rangle [\bar{n}(\omega_o) + 1] = 2\bar{n}(\omega_o)[\bar{n}(\omega_o) + 1]$$
$$= 2 \left(\frac{\pi^2 e^3}{\hbar \omega_o^3} \right)^2 [p^2(\omega_o) + \frac{\hbar \omega_o^3}{\pi^2 c^3} \rho(\omega_o)],$$

$$(2.158)$$

where we have again employed the relation $\bar{n}(\omega_o) = (\pi^2 c^3/\hbar \omega_o^3)\rho(\omega_o)$. Then equation (2.157) yields exactly the Einstein–Stern equation (1.55), whose solution is the Planck spectrum.

It is hardly surprising that the quantum theory of the Einstein–Hopf model produces the Planck spectrum for the spectral energy density of radiation at thermal equilibrium. What is of interest is to see just what about the quantum theory leads to the Planck spectrum rather than the Rayleigh–Jeans spectrum of the classical Einstein–Hopf model. To this end we use the identities $\sigma \sigma^\dagger = \sigma^\dagger \sigma + 1$ and $a_{\mathbf{k}\lambda} a_{\mathbf{k}\lambda}^\dagger = a_{\mathbf{k}\lambda}^\dagger a_{\mathbf{k}\lambda} + 1$ to write

$$\langle \sigma \sigma^\dagger \rangle \langle a_{\mathbf{k}\lambda}^\dagger a_{\mathbf{k}\lambda} \rangle + \langle \sigma^\dagger \sigma \rangle \langle a_{\mathbf{k}\lambda} a_{\mathbf{k}\lambda}^\dagger \rangle = 2 \left[\langle \sigma^\dagger \sigma \rangle \langle a_{\mathbf{k}\lambda}^\dagger a_{\mathbf{k}\lambda} \rangle + \frac{1}{2} \langle \sigma^\dagger \sigma \rangle \right.$$
$$\left. + \frac{1}{2} \langle a_{\mathbf{k}\lambda}^\dagger a_{\mathbf{k}\lambda} \rangle \right]. \qquad (2.159)$$

Without the term $\frac{1}{2}\langle \sigma^\dagger \sigma \rangle + \frac{1}{2}\langle a_{\mathbf{k}\lambda}^\dagger a_{\mathbf{k}\lambda} \rangle$ in this expression we are led to the Rayleigh–Jeans spectrum. In other words, the Planck spectrum is a consequence of the quantum-mechanical commutation rules $[\sigma, \sigma^\dagger] = [a_{\mathbf{k}\lambda}, a_{\mathbf{k}\lambda}^\dagger] = 1$.

For a more physical interpretation of the role of quantum mechanics, let us note that in the final expression for $\tau^{-1}\langle \Delta^2 \rangle$ only the field modes at $\omega_k = \omega_o$ contribute. These modes impart a mean-square momentum transfer proportional to

$$(\hbar \omega_o)^2 \left[\langle \sigma^\dagger \sigma \rangle \bar{n}(\omega_o) + \frac{1}{2} \langle \sigma^\dagger \sigma \rangle + \frac{1}{2} \bar{n}(\omega_o) \right] = \langle H_{\mathrm{osc}} \rangle \langle H_{\mathrm{F}} \rangle$$
$$+ H_{\mathrm{F}}^{\mathrm{zp}} \langle H_{\mathrm{osc}} \rangle + H_{\mathrm{osc}}^{\mathrm{zp}} \langle H_{\mathrm{F}} \rangle, \qquad (2.160)$$

where $H_{\mathrm{osc}} = \hbar \omega_o \langle \sigma^\dagger \sigma \rangle$ and $H_{\mathrm{F}} = \hbar \omega_o a^\dagger a$ are the Hamiltonian operators for the dipole oscillator and a resonant field mode, respectively, excluding zero-point energies, and $H_{\mathrm{osc}}^{\mathrm{zp}}$ and $H_{\mathrm{F}}^{\mathrm{zp}}$ are the corresponding zero-point energies ($= \frac{1}{2}\hbar \omega_o$). Were it not for the zero-point energies in (2.160), we would obtain the classical Rayleigh- Jeans spectrum instead of the Planck spectrum. These terms give rise to the "particle" term proportional to $\rho(\omega_o)$ in (2.158).

All this is consistent with the classical discussion in Section 1.6. Since

$$H_{\mathrm{F}}^{\mathrm{zp}}\langle H_{\mathrm{osc}}\rangle + H_{\mathrm{osc}}^{\mathrm{zp}}\langle H_{\mathrm{F}}\rangle = 2H_{\mathrm{osc}}^{\mathrm{zp}}\langle H_{\mathrm{F}}\rangle = \hbar\omega_o\langle H_{\mathrm{F}}\rangle, \qquad (2.161)$$

a classical theory of the Einstein–Hopf model that includes a zero-point energy $\hbar\omega_o$ for a material oscillator, but not for any field oscillator, leads to the same equation (2.157) of quantum theory, and therefore gives the Planck spectrum. This was the approach of Einstein and Stern.

Alternatively, we can include in the classical theory a zero-point energy $\frac{1}{2}\hbar\omega_o$ for both the material oscillator and a field mode of frequency ω_o, and this too leads to the Planck spectrum, as discussed in Section 1.6. This approach is closer to the (quantum-mechanical) truth. But in such a classical approach we must follow the ad hoc procedure of dropping a contribution $\frac{1}{2}\hbar\omega_o\rho_0(\omega_o)$, which arises from the product $H_{\mathrm{osc}}^{\mathrm{zp}}H_{\mathrm{F}}^{\mathrm{zp}}$. There was really no justification of this *Ansatz* in Section 1.6 other than the fact that it gave the Planck spectrum.

In the quantum theory just presented, the terms $H_{\mathrm{F}}^{\mathrm{zp}}\langle H_{\mathrm{osc}}\rangle$ and $H_{\mathrm{osc}}^{\mathrm{zp}} \times \langle H_{\mathrm{F}}\rangle$ leading to the Planck spectrum arise "automatically" from the zero-point energies of the dipole and field oscillators or, more formally, from the commutation properties of the dipole and field operators. But there is no term $H_{\mathrm{osc}}^{\mathrm{zp}}H_{\mathrm{F}}^{\mathrm{zp}}$ that had to be dropped ad hoc in the classical approach to the Planck spectrum presented in Section 1.6. In other words, *the quantum theory of the Einstein–Hopf model apparently does not allow for any effect of the interaction between a ground-state dipole oscillator and the vacuum field.*

We must be careful here about what we mean by the "effect" of the vacuum field on a ground- state dipole oscillator. The dipole coordinate obeys the Heisenberg equation of motion (2.86), and we have shown that the vacuum field is necessary for the preservation of the canonical commutation relations for the dipole coordinate and momentum operators, regardless of the state of the dipole. In this sense the vacuum field certainly has a formal "effect." Physically, however, a dipole oscillator in its ground state shows no obvious effect of its interaction with the vacuum field: a ground-state oscillator in the vacuum remains forever in its ground state. Whereas an excited dipole oscillator can undergo spontaneous emission attributable in part to the vacuum field, there is no such thing as "spontaneous absorption" by a ground-state oscillator in vacuum. We shall see that in the ground state of an atom spontaneous absorption is precluded by an exact cancellation of vacuum field fluctuations by fluctuations in the atom.

2.13 Summary

In the quantum theory of the electromagnetic field, classical wave amplitudes α, α^* are replaced by operators $a, a^\dagger$ satisfying $[a, a^\dagger] = 1$. The quantity $|\alpha|^2$ appearing in the classical expression for the energy of a field mode [cf. equation (2.29)] is replaced in quantum theory by the photon number operator $a^\dagger a$. The fact that $[a, a^\dagger a] \neq 0$ implies that quantum theory does not allow states of the radiation field for which the photon number and a field amplitude can be precisely defined, i.e., we cannot have simultaneous eigenstates of $a^\dagger a$ and a. The reconciliation of wave and particle attributes of the field is accomplished via the association of a probability *amplitude* with a classical mode pattern, as discussed in Section 2.4. The calculation of field modes is an entirely classical problem, while the quantum properties of the field are carried by the mode "amplitudes" a and $a^\dagger$ associated with these classical modes.

The zero-point energy of the field arises formally from the noncommutativity of a and $a^\dagger$. This is true for *any* harmonic oscillator: the zero-point energy $\frac{1}{2}\hbar\omega$ appears when we write the Hamiltonian $H = p^2/2m + \frac{1}{2}m\omega^2 q^2 = \frac{1}{2}\hbar\omega(aa^\dagger + a^\dagger a)$ as $\hbar\omega(a^\dagger a + \frac{1}{2})$ [Equation (2.9)].

This zero-point energy can be dropped from the Hamiltonian by redefining the zero of energy, or by arguing that it is a c-number and therefore has no effect on Heisenberg equations of motion. However, when we do this and solve the Heisenberg equation for a field operator, we must include the vacuum field, which is the homogeneous part of the solution for the field operator. In fact we showed in Section 2.6 that the vacuum field is essential for the preservation of commutators and the formal consistency of the theory. When we calculate the field energy we obtain not only a contribution from any sources which may be present, but also a contribution from the vacuum field. The latter is of course the zero-point field energy. In other words, the zero-point field energy "reappears" even though we may have deleted it from the Hamiltonian.

As we saw in the first chapter, the concept of zero-point energy arose before the development of the quantum formalism. However, in quantum theory zero-point energy rests upon a much firmer foundation than was possible classically. This is illustrated by a comparison of the Einstein–Stern theory of blackbody radiation with the quantum theory presented in Section 2.12.

Observable phenomena like the Casimir effect strongly suggest that the vacuum electromagnetic field and its zero-point energy are real physical entities and not mere artifices of the quantum formalism. In the following chapter we shall turn to other things that similarly suggest the physical reality of the fluctuating vacuum electromagnetic field.

Finally, the Maxwell equations (2.18)–(2.21) are satisfied by the electric and magnetic field *operators* in the quantum theory of the field. Maxwell was lucky: his equations turned out to be Lorentz–invariant and gauge–invariant, and to retain the same form in quantum theory. But whereas in classical physics one makes the "natural" assumption that $\mathbf{E} = \mathbf{B} = 0$ in the absence of any sources, this cannot be done in quantum theory. Such an assumption is not only inconsistent with quantum theory; it would also appear to contradict experimental facts such as the Casimir force, the Lamb shift, and other effects to which we turn in the following chapter.

2.14   Bibliography

There are many excellent books on the quantum theory of radiation and quantum electrodynamics. The books by Power and Loudon are particularly readable and succinct introductions to the nonrelativistic theory. Power's book was one of the first to address the zero-point field in any sort of detail. Heitler's book is an old standard and still contains much useful material, while the treatise by Itzykson and Zuber offers a more modern and advanced perspective. Boyer's review of the role of zero-point field energy in long-range forces is also recommended.

Bohr, N. and L. Rosenfeld, "Field and Charge Measurements in Quantum Electrodynamics," *Phys. Rev.* **78**, 794 (1950).

Born, M., W. Heisenberg, and P. Jordan, "Zur Quantenmechanik II," *Z. Phys.* **35**, 557 (1926).

Boyer, T. H., "Derivation of the Blackbody Radiation Spectrum Without Quantum Assumptions," *Phys. Rev.* **182**, 1374 (1969).

Boyer, T. H., "Quantum Zero-Point Energy and Long-Range Forces," *Ann. Phys.* (New York) **56**, 474 (1970).

Casimir, H. B. G., "On the Attraction between Two Perfectly Conducting Plates," *Proc. K. Ned. Akad. Wet.* **51**, 793 (1948).

Cresser, J. D., "Electric Field Commutation Relation in the Presence of a Dipole Atom," *Phys. Rev.* **A29**, 1984 (1984).

Davies, P. C. W., "Scalar Particle Production in Schwarzschild and Rindler Metrics," *J. Phys.* **A8**, 609 (1975).

Dirac, P. A. M., "The Quantum Theory of the Emission and Absorption of Radiation," *Proc. Roy. Soc. Lond.* **A114**, 243 (1927).

Heitler, W., *The Quantum Theory of Radiation*, 3rd ed. (Oxford

University Press, London, 1966).

Itzykson, C. and J.-B. Zuber, *Quantum Field Theory* (McGraw–Hill, New York, 1980).

Knight, P. L. and L. Allen, *Concepts of Quantum Optics* (Pergamon Press, Oxford, 1983).

Loudon, R., *The Quantum Theory of Light*, 2nd ed. (Oxford University Press, London, 1983).

Milonni, P. W., "Quantum Mechanics of the Einstein–Hopf Model," *Am. J. Phys.* **49**, 177 (1981a).

Milonni, P. W., "Radiation Reaction and the Nonrelativistic Theory of the Electron," *Phys. Lett.* **82A**, 225 (1981b).

Milonni, P. W., "Casimir Forces Without the Vacuum Radiation Field," *Phys. Rev.* **A25**, 1315 (1982).

Milonni, P. W., "Wave-Particle Duality of Light: A Current Perspective," in *The Wave-Particle Dualism*, ed. S. Diner, D. Fargue, G. Lochak, and F. Selleri (Reidel, Dordrecht, 1984).

Milonni, P. W. and M.-L. Shih, "Casimir Forces," *Contemp. Phys.* **33**, 313 (1993).

Power, E. A., *Introductory Quantum Electrodynamics* (Longmans, London, 1964).

Senitzky, I. R., "Dissipation in Quantum Mechanics: The Harmonic Oscillator," *Phys. Rev.* **119**, 670 (1960).

Unruh, W. G., "Notes on Black-Hole Evaporation," *Phys. Rev.* **D14**, 870 (1976).

Chapter 3

Some QED Vacuum Effects

> [My father] said, "I understand that they say that light is emitted from an atom when it goes from one state to another, from an excited state to a state of lower energy."
>
> I said, "That's right."
>
> "And light is a kind of particle, a photon, I think they call it."
>
> "Yes."
>
> "So if the photon comes out of the atom when it goes from the excited to the lower state, the photon must have been in the atom in the excited state."
>
> I said, "Well, no."
>
> He said, "Well, how do you look at it so you can think of a particle photon coming out without it having been there in the excited state?"
>
> I thought a few minutes, and I said, "I'm sorry; I don't know. I can't explain it to you."
>
> — Richard P. Feynman, *The Physics Teacher* (September 1969).

3.1 Introduction

We noted in the preceding chapter that Dirac's theory of emission and absorption (1927) was the first application of the quantum theory of radiation. The importance of Dirac's theory of spontaneous emission has been emphasized by Weinberg (1977):

> ... This problem was of crucial importance, because the process of spontaneous emission of radiation is one in which "particles" are

actually created. Before the event, the system consists of an excited atom, whereas after the event, it consists of an atom in a state of lower energy, plus one photon. If quantum mechanics could not deal with processes of creation and destruction, it could not be an all-embracing physical theory ... Dirac's successful theory of spontaneous emission of radiation confirmed the universal character of quantum mechanics.

Dirac argued that his theory "must presumably give the effect of radiation reaction on the emitting system." Spontaneous emission was also interpreted in terms of radiation reaction in the theory of Landau (1927) and, before the development of the quantum formalism, by van Vleck (1924). However, contemporary physicists, when asked to give a physical explanation for the occurrence of spontaneous emission, generally invoke the vacuum electromagnetic field. This view was popularized by Weisskopf (1935) and later by Welton (1948), who argued that spontaneous emission "can be thought of as forced emission taking place under the action of the fluctuating field." In the following chapters we show that these two interpretations — based on radiation reaction or vacuum field fluctuations — are in fact closely related in the quantum theory of radiation. We show furthermore that various other effects can be interpreted equally well in terms of radiation reaction or vacuum field fluctuations.

What are these "vacuum fluctuation effects"? The first example that is usually cited is the Lamb shift, or sometimes the Casimir force between conducting plates. In this chapter we consider these and other manifestations of the vacuum electromagnetic field. Our aim is not to present detailed calculations, but to emphasize the *physics* of the vacuum field. For this reason we adhere strictly to the formalism of the quantized field only when it is absolutely necessary.

3.2   Spontaneous Emission

Spontaneous emission is ultimately responsible for most of the light around us. For a thermal source the ratio of the spontaneous and stimulated emission rates for radiation of frequency ω_o is (Section 1.8)

$$\frac{A_{21}}{B_{21}\rho(\omega_o)} = e^{\hbar\omega_o/kT} - 1. \tag{3.1}$$

The sun may be regarded for our purposes as a blackbody radiator at the temperature $T = 6000$ K. At this temperature the ratio (3.1) is about 400 at the wavelength $\lambda = 400$ nm, and about 30 at $\lambda = 700$ nm. Most of the visible output from the sun, therefore, is due to spontaneous rather than stimulated emission.

As we shall see in the following chapter, spontaneous emission can be correctly described only when the radiation field is quantized; if the field is not treated quantum mechanically we obtain predictions in conflict with experiment. Nevertheless some aspects of spontaneous emission are adequately described without the full machinery of quantum electrodynamics. In this section we use some simplistic arguments to derive the A coefficient for spontaneous emission and to provide some preliminary evidence for the interplay between radiation reaction and vacuum field fluctuations.

Consider the rate R_{21} of stimulated emission in a broadband field of spectral energy density $\rho(\omega_o)$. According to the discussion in Section 1.8, this rate is $B_{21}\rho(\omega_o)$ for an atomic transition from level 2 to level 1 with transition frequency $\omega_o = (E_2 - E_1)/\hbar$. The Einstein B coefficient for stimulated emission is given by the standard formula

$$B_{21} = \frac{4\pi^2 d^2}{3\hbar^2} , \tag{3.2}$$

where $\mathbf{d}$ is the electric dipole matrix element for the transition $2 \to 1$ and $d \equiv |\mathbf{d}|$. Thus

$$R_{21} = \frac{4\pi^2 d^2}{3\hbar^2}\rho(\omega_o). \tag{3.3}$$

This result follows also from the classical formula (1.7) when we replace e^2/m by $(e^2/m)f$, with f the transition oscillator strength defined as $2md^2\omega_o/e^2\hbar$. (See the remark at the end of Appendix A.) That is,

$$\frac{\dot{W}_A}{\hbar\omega_o} = \frac{2\pi^2 e^2}{3m\hbar\omega_o}\rho(\omega_o) \to \frac{2\pi^2 e^2}{3m\hbar\omega_o}f\rho(\omega_o) = R_{21}. \tag{3.4}$$

According to equation (3.3) the vacuum electromagnetic field should induce an atom in the excited level 2 to make a downward transition to level 1 at the rate (transition probability per unit time)

$$R_{\mathrm{VF}} = \left(\frac{4\pi^2 d^2}{3\hbar^2}\right)\left(\frac{\hbar\omega_o^3}{2\pi^2 c^3}\right) = \frac{2d^2\omega_o^3}{3\hbar c^3} = \frac{1}{2}A_{21}, \tag{3.5}$$

where A_{21} is the Einstein A coefficient for spontaneous emission (Chapter 4).

We have thus arrived at the same result found in Section 1.9: the vacuum field induces transitions at a rate equal to *half* the spontaneous emission rate. Evidently spontaneous emission *cannot* simply "be thought of as forced emission taking place under the action of the fluctuating field."

Consider now the effect of radiation reaction. As shown in Appendix A, the radiation reaction field is responsible for the rate $2e^2a^2/3c^3$ at which an oscillating charge loses energy to the electromagnetic field. For oscillation at frequency ω_o with amplitude x_o this rate is

$$\frac{d\epsilon}{dt} = \frac{2e^2\omega_o^4 x_o^2}{3c^3}\cos^2\omega_o t \rightarrow \frac{e^2\omega_o^4 x_o^2}{3c^3} \quad (3.6)$$

when we average over a cycle of oscillation. Since $\epsilon = m\omega_o^2 x_o^2$ is the energy of oscillation, we have

$$\frac{d\epsilon}{dt} = \left(\frac{e^2\omega_o^2}{3mc^3}\right)\epsilon, \quad (3.7)$$

and so $e^2\omega_o^2/3mc^3$ is the rate of emission attributable to radiation reaction. The replacement of e^2/m by $e^2 f/m$ as previously gives

$$R_{RR} = \left(\frac{e^2\omega_o^2}{3mc^3}\right)\left(\frac{2md^2\omega_o}{e^2\hbar}\right) = \frac{2d^2\omega_o^3}{3\hbar c^3} = \frac{1}{2}A_{21} \quad (3.8)$$

for the emission rate due to radiation reaction.

On the basis of this simplistic semiclassical analysis, therefore, we have arrived at the conclusion that $R_{RR} = R_{VF} = \frac{1}{2}A_{21}$ and

$$A_{21} = R_{VF} + R_{RR}. \quad (3.9)$$

In other words, both the vacuum field and radiation reaction induce transitions at the rate $\frac{1}{2}A_{21}$, and the two together give the Einstein A coefficient for spontaneous emission.

As noted earlier, modern physicists generally think of spontaneous emission as a consequence of the vacuum field. Weisskopf (1981), for instance, writes that "spontaneous emission appears as a forced emission caused by the zero-point oscillations of the electromagnetic field." The fact that the vacuum field gives only half the correct A coefficient in this simplified picture does not seem to be widely appreciated, although it has been emphasized by several authors (Ginzburg, 1983; Milonni, 1984), and in his well-known textbook Schiff (1968) indirectly acknowledges it:

> ... From a formal point of view, we can say that the spontaneous emission probability is equal to the probability of emission that would be induced by the presence of one quantum in each state of the radiation field. Now ... the smallest possible energy of the field corresponds to the presence of one-half quantum per state. This suggests that we regard the spontaneous emission as being induced by the zero-point oscillations of the electromagnetic field; note, however, that these oscillations are twice as effective in producing emissive transitions as are real photons and are of course incapable of producing absorptive transitions.

These arguments indicate that the "missing one-half" comes from radiation reaction. Of course these arguments are semiclassical and oversimplified; the main purpose of the following chapter is to refine these arguments using the quantum theory of radiation. We will also show why the vacuum field is "of course incapable of producing absorptive transitions": in the lower state of an atomic transition the effects of the vacuum field and radiation reaction cancel, so that the "spontaneous absorption" rate is

$$A_{12} = R_{\mathrm{VF}} - R_{\mathrm{RR}} = \frac{1}{2}A_{21} - \frac{1}{2}A_{21} = 0. \tag{3.10}$$

3.3   Atomic Stability

The fact that an accelerating charge loses energy by radiating implies, according to classical ideas, that an electron should spiral into the nucleus and that atoms should not be stable. The balancing of the effects of radiation reaction and the vacuum field implied by (3.10), however, suggests that the stability of atoms might be attributable to the influence on the atom of the vacuum field. We now give a simplistic argument in support of this idea.

Using equation (3.4) we write

$$\dot{W}_{\mathrm{A}} = \frac{2\pi^2 e^2}{3m} f\rho(\omega_o) = \frac{e^2 f\hbar\omega_o^3}{3mc^3} \tag{3.11}$$

for the rate at which an atom absorbs energy from the vacuum field. But according to (3.6) there is also a loss of energy at the rate

$$\dot{W}_{\mathrm{EM}} = \frac{e^2 f\omega_o^4 x_o^2}{3c^3} \tag{3.12}$$

due to radiation, where we have again made the replacement $e^2/m \rightarrow e^2 f/m$. Equating (3.11) and (3.12), we obtain

$$mx_o^2\omega_o = \hbar, \tag{3.13}$$

which will be recognized as the Bohr quantization condition for the ground state of a one-electron atom.

This "derivation" of the Bohr quantization condition obviously should not be taken very seriously. It suggests only how Bohr's quantization condition, at least for $n = 1$, might have been interpreted by physicists in 1913. We now know that the vacuum field is in fact formally necessary for the stability of atoms in quantum theory: as we saw in Section 2.6, radiation reaction will cause canonical commutators like $[x, p_x]$ to decay to zero unless the fluctuating vacuum field is included, in which case commutators are consistently preserved.

3.4   The Lamb Shift

The solution of the Schrödinger equation for the hydrogen atom gives energy levels depending only on the principal quantum number n. In the solution of the Dirac equation (Chapter 9) the spin–orbit coupling partially lifts this degeneracy, but states with the same n and the same total angular momentum quantum number j, such as $2s_{1/2}$ and $2p_{1/2}$, remain degenerate:

$$ E = mc^2 \left[1 + \left(\frac{\alpha}{n - (j + \frac{1}{2}) + \sqrt{(j + \frac{1}{2})^2 - \alpha^2}} \right)^2 \right]^{-1/2} , \qquad (3.14) $$

where $\alpha = e^2/\hbar c \cong 1/137$ is the fine structure constant, $n = 1, 2, 3, ...\infty$, and $j + \frac{1}{2} \leq n$. Experiments in the 1930s indicated that the $2s_{1/2}$ and $2p_{1/2}$ energies might actually differ, but the data were not sufficiently accurate to draw any definite conclusions, and other experiments appeared to confirm the prediction of degeneracy. In 1947, however, Lamb and Retherford performed experiments showing convincingly that the $2s_{1/2}$ level lies about 1000 MHz, or 0.030 cm^{-1}, above the $2p_{1/2}$ level. Shortly thereafter they reported a more accurate value near 1060 MHz. This tiny energy difference is called the Lamb shift.

According to the energy level formula (3.14) predicted by the Dirac equation, the energy difference between the $2p_{3/2}$ and $2p_{1/2}, 2s_{1/2}$ levels is $\cong \alpha^4 mc^2/32$, corresponding to a frequency of about 11,000 MHz or a wavelength of about 2.7 cm. A simplified energy level diagram for the $n = 2$ states of hydrogen, including the Lamb shift, is shown in Figure 3.1.

The fact that the $2s_{1/2} - 2p_{3/2}$ (and $2s_{1/2} - 2p_{1/2}$) transition wavelength lies in the microwave region allowed Lamb and Retherford to utilize advances in microwave technology made during World War II. The basic idea of their experiment is as follows. First a beam of H atoms is produced by thermal dissociation of H_2 in an oven. The atomic beam is then bombarded with an electron beam that collisionally excites about 10^{-8} of the atoms into the $2s_{1/2}$ state. This state is metastable, since (one-photon) spontaneous emission to the $1s_{1/2}$ ground state is forbidden ($\Delta l = 0$). The radiative lifetime of the $2s_{1/2}$ state is thus very large ($\cong 1/7$ sec) and is due to two-photon spontaneous emission to the ground state. The $2s_{1/2}$ atoms are detected by the fact that they cause emission of electrons when they are incident on a metal target. Excited atoms incident on the metal thus produce an electric current, while ground-state atoms do not. Now the application of a field at the $2s_{1/2} - 2p_{3/2}$ (or $2s_{1/2} - 2p_{1/2}$) transition frequency induces transitions to a p state which quickly decays to the ground

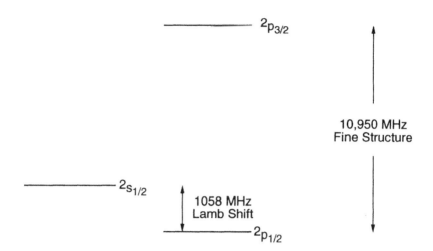

Figure 3.1: Energy level diagram for the $n = 2$ states of the hydrogen atom.

state by spontaneous emission. Application of a field of the appropriate frequency thus reduces the electric current at the detector of excited atoms. (The microwave field can be held fixed while the transition frequency is Zeeman-shifted with a magnetic field.) In this way Lamb and Retherford (1947) determined that the $2s_{1/2} - 2p_{1/2}$ level shift was about 1000 MHz. In 1952 they reported a value of 1058.27 ± 1.0 MHz based on more refined measurements. The Lamb–Retherford experiments and analysis were remarkably accurate. Since then various other experimental techniques have been employed, and the currently accepted value for the $2s_{1/2} - 2p_{1/2}$ shift in hydrogen is about 1057.85 MHz.[1]

The Lamb shift and its explanation marked the beginning of modern quantum electrodynamics. In the words of Dirac (1989), "No progress was made for 20 years. Then a development came, initiated by Lamb's discovery and explanation of the Lamb shift, which fundamentally changed the character of theoretical physics. It involved setting up rules for discarding ... infinities ... "

The reason the Dirac theory leading to (3.14) fails to account for the Lamb shift is that it ignores the coupling of the atomic electron to the vacuum electromagnetic field. Actually the Lamb shift turns out to be a predominantly nonrelativistic effect, and can be understood in part by modifying the Schrödinger theory of the hydrogen atom to include the cou-

---

[1] For a review see Drake (1982).

pling to the vacuum field. We therefore consider the nonrelativistic theory
with Hamiltonian (Chapter 4)

$$H = H_A + H_F - \frac{e}{mc}\mathbf{A}\cdot\mathbf{p} + \frac{e^2}{2mc^2}\mathbf{A}^2, \qquad (3.15)$$

where H_A is the Hamiltonian operator for the atomic electron and, as in the
preceding chapter, $\mathbf{A}$ is the vector potential, H_F is the field Hamiltonian,
and we make the electric dipole approximation of neglecting any spatial
variation of $\mathbf{A}$.

If the field is treated according to standard classical electromagnetic
theory, the vector potential $\mathbf{A} = 0$ in the vacuum and so there is no field
to perturb the atomic energy levels. This is not the case when the field
is quantized; standard second-order perturbation theory gives the following expression for the shift in the atomic level n due to the interaction
$-(e/mc)\mathbf{A}\cdot\mathbf{p}$:

$$\Delta E_n = \sum_m \sum_{\mathbf{k}\lambda} \frac{|\langle m, 1_{\mathbf{k}\lambda}|h_{\mathbf{k}\lambda}|n, \mathrm{vac}\rangle|^2}{E_n - E_m - \hbar\omega_k}, \qquad (3.16)$$

$$h_{\mathbf{k}\lambda} = -\frac{e}{mc}\left(\frac{2\pi\hbar c^2}{\omega_k V}\right)^{1/2} a^\dagger_{\mathbf{k}\lambda}(\mathbf{e}_{\mathbf{k}\lambda}\cdot\mathbf{p}), \qquad (3.17)$$

where we follow the notation of Chapter 2 for the field in free space. Hence,
ΔE_n is the energy shift of the state $|n, \mathrm{vac}\rangle$ in which the atom is in stationary state n and the field is in its vacuum state of no photons. The
intermediate state $|m, 1_{\mathbf{k}\lambda}\rangle$ corresponds to the atom in state m and one
photon in mode $(\mathbf{k}, \lambda)$. This intermediate state has energy $E_m + \hbar\omega_k$ which
appears in the denominator in (3.16). Only one-photon intermediate states
appear because $\mathbf{A}$ can only connect the vacuum state to such states , and
furthermore only $a^\dagger_{\mathbf{k}\lambda}$ in $\mathbf{A}$ contributes to the matrix element in (3.16)
because $a_{\mathbf{k}\lambda}|\mathrm{vac}\rangle = 0$. The expression (3.16) for ΔE_n is derived in the
following chapter in both the Schrödinger and Heisenberg pictures.

Since

$$\langle m, 1_{\mathbf{k}\lambda}|h_{\mathbf{k}\lambda}|n, \mathrm{vac}\rangle = -\frac{e}{mc}\left(\frac{2\pi\hbar c^2}{\omega_k V}\right)^{1/2}\mathbf{p}_{mn}\cdot\mathbf{e}_{\mathbf{k}\lambda}, \qquad (3.18)$$

we can write

$$\Delta E_n = \frac{2\pi e^2}{m^2}\frac{1}{V}\sum_m \sum_{\mathbf{k}\lambda}\frac{1}{\omega_k}\frac{|\mathbf{p}_{mn}\cdot\mathbf{e}_{\mathbf{k}\lambda}|^2}{\omega_{nm} - \omega_k}, \qquad (3.19)$$

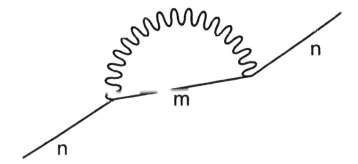

Figure 3.2: Diagrammatic representation of (3.21) in terms of emission and absorption of virtual photons.

where $\hbar\omega_{nm} = E_n - E_m$. Using (2.71) and other simple manipulations, furthermore, we have

$$
\begin{aligned}
\Delta E_n &= \frac{2e^2}{3\pi m^2 c^3} \sum_m |\mathbf{p}_{mn}|^2 \int_0^\infty \frac{d\omega\,\omega}{\omega_{nm} - \omega} \\
&= \frac{2\alpha}{3\pi} \left(\frac{1}{mc}\right)^2 \sum_m |\mathbf{p}_{mn}|^2 \int_0^\infty \frac{dE\,E}{E_n - E_m - E} .
\end{aligned}
\tag{3.20}
$$

In these expressions the integrals are to be understood in terms of the Cauchy principal part.

At the risk of laboring the obvious, we emphasize that ΔE_n arises from the vacuum field. Writing

$$
\begin{aligned}
|\langle m, 1_{\mathbf{k}\lambda}|a_{\mathbf{k}\lambda}^\dagger(\mathbf{p} \cdot \mathbf{e}_{\mathbf{k}\lambda})|n, \text{vac}\rangle|^2 &= \langle n, \text{vac}|a_{\mathbf{k}\lambda}(\mathbf{p} \cdot \mathbf{e}_{\mathbf{k}\lambda})|m, 1_{\mathbf{k}\lambda}\rangle \\
&\times \langle m, 1_{\mathbf{k}\lambda}|a_{\mathbf{k}\lambda}^\dagger(\mathbf{p} \cdot \mathbf{e}_{\mathbf{k}\lambda})|n, \text{vac}\rangle,
\end{aligned}
\tag{3.21}
$$

we are led to interpret ΔE_n in terms of an emission process $n \to m + \gamma$ followed by the absorption process $m + \gamma \to n$, where γ denotes a photon. This emission and absorption of "virtual photons" is indicated diagrammatically in Figure 3.2.

We have ignored the contribution of the interaction $(e^2/2mc^2)\mathbf{A}^2$ to ΔE_n. Since this term does not involve atomic operators, it contributes the same energy to every state $|n, \text{vac}\rangle$,

$$
\Delta E_\circ = \langle n, \text{vac}|\frac{e^2}{2mc^2}\mathbf{A}^2|n, \text{vac}\rangle
$$

$$= \frac{e^2}{2mc^2} \sum_{\mathbf{k}\lambda} \sum_{\mathbf{k'}\lambda'} \left(\frac{2\pi\hbar c^2}{\omega_k V} \right)^{1/2} \left(\frac{2\pi\hbar c^2}{\omega'_k V} \right)^{1/2}$$

$$\times \langle \mathrm{vac}|a_{\mathbf{k'}\lambda'} a^{\dagger}_{\mathbf{k}\lambda}|\mathrm{vac}\rangle \mathbf{e}_{\mathbf{k'}\lambda'} \cdot \mathbf{e}_{\mathbf{k}\lambda}$$

$$= \frac{e^2}{2mc^2} \sum_{\mathbf{k}\lambda} \left(\frac{2\pi\hbar c^2}{\omega_k V} \right) = \frac{e^2 \hbar}{\pi mc^3} \int_0^\infty d\omega\, \omega, \qquad (3.22)$$

and therefore it does not affect observable *frequency* shifts and may be ignored. This term is indicated diagrammatically in Figure 3.3.

Both ΔE_n and ΔE_o are seen to be infinite. This is especially problematic for ΔE_n, since this presumably corresponds to the Lamb shift, which experiment shows to be not infinite but *small*. It was the resolution of this dilemma that "fundamentally changed the character of theoretical physics" (Dirac, 1989). The first person to calculate a finite value for the Lamb shift was Bethe (1947), and we now turn our attention to his calculation.

3.5   Bethe's Mass Renormalization

The energy of a *free* electron due to its coupling to the field may be obtained from (3.20) by taking the limit in which all the transition frequencies $\omega_{nm} \rightarrow 0$. Thus

$$\Delta E_n^{\mathrm{free}} = -\frac{2\alpha}{3\pi} \left(\frac{1}{mc} \right)^2 \sum_m |\mathbf{p}_{mn}|^2 \int_0^\infty dE \qquad (3.23)$$

is the expectation value in state n of the operator corresponding to the energy of a free electron due to its coupling to the field. (As noted earlier,

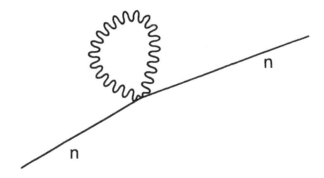

Figure 3.3: Diagrammatic representation of (3.22).

the contribution ΔE_o is the same for every state, free or bound, and may be neglected for our purposes.)

Now it is reasonable that the observed level shift for an atom in state n should be $\Delta E_n - \Delta E_n^{\text{free}}$, the difference between the shift in the electron energy when it is bound and when it is free:

$$
\begin{aligned}
\Delta E_n^{\text{obs}} \quad &- \quad \Delta E_n - \Delta E_n^{\text{free}} \\
&= \frac{2\alpha}{3\pi}\left(\frac{1}{mc}\right)^2 \sum_m |\mathbf{p}_{mn}|^2 \left[\int_0^\infty \frac{dE\,E}{E_n - E_m - E} + \int_0^\infty dE\right] \\
&= \frac{2\alpha}{3\pi}\left(\frac{1}{mc}\right)^2 \sum_m |\mathbf{p}_{mn}|^2 (E_n - E_m)\int_0^\infty \frac{dE}{E_n - E_m - E}\,.
\end{aligned}
\tag{3.24}
$$

This expression is still infinite but, unlike (3.20), the divergence is "only" logarithmic. The subtraction of ΔE_n^{free} from ΔE_n has thus reduced the divergence from linear to logarithmic. This subtraction was done by Bethe (1947). He correctly suggested that in a relativistic theory, where ΔE_n and ΔE_n^{free} themselves turn out to diverge only logarithmically, the subtraction of ΔE_n^{free} would produce a finite value for ΔE_n^{obs}.

Bethe assumed that the main part of the Lamb shift was due to the interaction of the electron with vacuum field modes of frequency small enough to justify a nonrelativistic approach. In this case it is reasonable to cut off the upper limit of integration in (3.24) by some $E_{\max}$, which Bethe took to be mc^2. Then

$$
\begin{aligned}
\Delta E_n^{\text{obs}} \quad &\rightarrow \quad \frac{2\alpha}{3\pi}\left(\frac{1}{mc}\right)^2 \sum_m |\mathbf{p}_{mn}|^2 (E_n - E_m)\int_0^{mc^2} \frac{dE}{E_n - E_m - E} \\
&\cong \frac{2\alpha}{3\pi}\left(\frac{1}{mc}\right)^2 \sum_m |\mathbf{p}_{mn}|^2 (E_m - E_n)\log\frac{mc^2}{|E_m - E_n|}
\end{aligned}
\tag{3.25}
$$

for $mc^2 \gg |E_n - E_m|$. Since the argument of the logarithm is accordingly very large, Bethe replaced the logarithm by an average value, independent of m, as a first approximation:

$$
\Delta E_n^{\text{obs}} \cong \frac{2\alpha}{3\pi}\left(\frac{1}{mc}\right)^2 \log\frac{mc^2}{|E_m - E_n|_{\text{avg}}}\sum_m |\mathbf{p}_{mn}|^2(E_m - E_n).
\tag{3.26}
$$

Now

$$
\sum_m |\mathbf{p}_{mn}|^2(E_m - E_n) = \sum_m \langle n|\mathbf{p}|m\rangle \cdot \langle m|\mathbf{p}|n\rangle (E_m - E_n)
$$

$$= \sum_m {}' \langle n|[\mathbf{p}, H_A]|m\rangle \cdot \langle m|\mathbf{p}|n\rangle$$

$$= -i\hbar\langle n|\nabla V \cdot \mathbf{p}|n\rangle = -\frac{1}{2}i\hbar\langle n|[\nabla V, \mathbf{p}]|n\rangle$$

$$= \frac{1}{2}\hbar^2\langle n|\nabla^2 V|n\rangle = \frac{1}{2}\hbar^2 \int d^3 r|\psi_n(\mathbf{r})|^2\nabla^2 V(\mathbf{r}),$$

$$(3.27)$$

where V is the binding potential. ($H_A = p^2/2m + V$) For the Coulomb potential $V = -Ze^2/r$, we have $\nabla^2 V = 4\pi Ze^2\delta^3(\mathbf{r})$ and

$$\sum_m |\mathbf{p}_{mn}|^2(E_m - E_n) = 2\pi\hbar^2 e^2 Z|\psi_n(0)|^2, \qquad (3.28)$$

$$\Delta E_n^{\mathrm{obs}} \cong \frac{4\alpha Z}{3}\left(\frac{e\hbar}{mc}\right)^2 |\psi_n(0)|^2 \log\frac{mc^2}{|E_m - E_n|_{\mathrm{avg}}}. \qquad (3.29)$$

This expression already exhibits an important element of truth: the Lamb shift should be largest for s states, for which $|\psi_n(0)|^2 \neq 0$. For an s state with principal quantum number n, $|\psi_n(0)|^2 = (Z/na_o)^3/\pi$ and

$$\Delta E_{ns}^{\mathrm{obs}} \cong \frac{8\alpha^3 Z^4}{3\pi n^3}R_\infty \log\frac{mc^2}{|E_m - E_n|_{\mathrm{avg}}}, \qquad (3.30)$$

where R_∞ is the Rydberg unit of energy ($e^2/2a_o \cong 13.6$ eV) for infinite nuclear mass. Using a numerical estimate of 17.8 R_∞ for the average excitation energy $|E_n - E_m|_{\mathrm{avg}}$ defined by[2]

$$\log|E_m - E_n|_{\mathrm{avg}} = \frac{\sum_m |\mathbf{p}_{mn}|^2(E_m - E_n)\log|E_m - E_n|}{\sum_m |\mathbf{p}_{mn}|^2(E_m - E_n)}, \qquad (3.31)$$

Bethe obtained for the $2s$ state of hydrogen a level shift in excellent agreement with experiment:

$$\Delta E_{2s}^{\mathrm{obs}} \cong 1040 \text{ MHz}. \qquad (3.32)$$

---

[2] The sums in this expression include continuum states. In fact the continuum states make a larger contribution to the average excitation energy than discrete states. This explains why the average excitation energy turns out to be so large compared with bound–bound transition frequencies. See H. A. Bethe and E. E. Salpeter, *Quantum Mechanics of One- and Two-Electron Atoms* (Springer-Verlag, Berlin, 1957), pp. 318–320.

The crux of Bethe's calculation is the "mass renormalization" implied by the subtraction of ΔE_n^{free} from ΔE_n in (3.24). Recall from Appendix D that a charged particle has an "electromagnetic mass" δm due to its own radiation reaction field. The nonrelativistic calculation in Appendix D gives

$$\delta m = \frac{4\alpha}{3\pi c^2} \int_0^\infty dE. \tag{3.33}$$

As discussed in Chapter 5, the observed electron mass m is $m_o + \delta m$, where m_o is the "bare mass," i.e., the contribution to the electron mass that is not associated with radiation reaction. Although (3.33) is infinite, and δm is also infinite when calculated relativistically, we might suppose that in some future, more refined theory, δm will be finite. If we assume furthermore that $\delta m/m$ in reality is small, then the kinetic energy of the electron is

$$\frac{\mathbf{p}^2}{2m} = \frac{\mathbf{p}^2}{2(m_o + \delta m)} \cong \frac{\mathbf{p}^2}{2m_o} - \frac{\delta m}{2m^2}\mathbf{p}^2. \tag{3.34}$$

Now the basic idea behind mass renormalization is this: when we write $\mathbf{p}^2/2m$ for the electron kinetic energy in the Schrödinger equation, m is the *observed mass* ($\cong 9.1 \times 10^{-28}$ g), which includes δm. But when we "turn on" the coupling of the electron to the field in our calculations, the radiation reaction on the electron adds δm to its mass. Since we have implicitly already accounted for δm in writing the electron mass as m in the Schrödinger (or Dirac) equation with no coupling to the field, we must be careful to avoid "double counting" δm. In particular, we should *subtract* the "additional" contribution, $-(\delta m/2m^2)\mathbf{p}^2$, that we incur after coupling the electron to the field. That is, we must subtract the *self-energy*

$$
\begin{aligned}
-\frac{\delta m}{2m^2}\langle n|\mathbf{p}^2|n\rangle &= -\frac{4\alpha}{3\pi c^2}\left(\frac{1}{2m^2}\right)\int_0^\infty dE\langle n|\mathbf{p}^2|n\rangle \\
&= -\frac{2\alpha}{3\pi}\left(\frac{1}{mc}\right)^2 \sum_m |\mathbf{p}_{mn}|^2 \int_0^\infty dE \\
&= \Delta E_n^{\text{free}}
\end{aligned}
\tag{3.35}
$$

from the calculated shift in E_n arising from the coupling of the electron to the radiation field. This is exactly what Bethe did in order to reduce the order of divergence of ΔE_n.

The idea behind renormalization is attributable to Kramers (Dresden, 1987) and also to Weisskopf (1936). Bethe applied it to the Lamb shift immediately after a conference on Shelter Island, where the Lamb–Retherford experiments and the theoretical difficulties with infinities in electrodynamics were discussed (Bethe, 1989):

> ... I thought that it ought to be possible to get Lamb's result by applying the idea of Kramers. So on the train from Shelter Island ... I wrote down some ... equations ... and found out that the effect on the 2s-state or any state of hydrogen would involve the logarithm of the energy ... Stupidly or boldly, I just assumed that the higher energy was mc^2, and with this assumption, I got about the right answer. Of course, I was afraid that I might have made a mistake by a factor of 2 ... after all one cannot remember factors of 2 on a train. So the next morning, as early as I could, I looked for Heitler's book in the General Electric library, and found that I had not made a mistake. Indeed I got a result of about a thousand megacycles which was about the right answer.

In his Nobel lecture (1966) Feynman called Bethe's estimate "the most important discovery in the history of quantum electrodynamics."

Renormalizability, like Lorentz invariance and gauge invariance, is presently believed to be required of any fundamental theory of physics. However, dissatisfaction with renormalization has been expressed at various times by many physicists, including Dirac (1978), who felt that "This is just not sensible mathematics. Sensible mathematics involves neglecting a quantity when it turns out to be small — not neglecting it just because it is infinitely great and you do not want it!"

On the other hand, it can be argued that mass renormalization, for instance, would be necessary to avoid double counting in calculations *even if the electromagnetic mass δm turned out to be finite.* It can also be argued that δm in a more refined theory would turn out to be small, and that mass differences between the particles π^+ and π^0, or K^+ and K^0, etc., are "almost certainly electromagnetic in origin" (Feynman, 1961).

Discussion of these matters further would take us too far from our present subject. We shall return to the numerical value of the Lamb shift later in connection with vacuum polarization, but in the next few sections we wish to develop a more physical understanding of the dominant (nonrelativistic) contribution to the Lamb shift calculated by Bethe.

3.6   Welton's Interpretation

Welton (1948) interpreted the Lamb shift as follows. The vacuum field causes the position of the electron to fluctuate. The fluctuation $\Delta \mathbf{r}$ is determined by $m\Delta\ddot{\mathbf{r}} = e\mathbf{E}_o$, where $\mathbf{E}_o$ is the zero-point electric field. If we make a Fourier decomposition of both $\mathbf{E}_o$ and $\Delta\mathbf{r}$, then

$$\Delta\mathbf{r}_\omega = -\frac{e}{m\omega^2}\mathbf{E}_{o,\omega} \tag{3.36}$$

gives the component of $\Delta \mathbf{r}$ at the frequency ω. Thus

$$\langle (\Delta \mathbf{r}_\omega)^2 \rangle = \frac{e^2}{m^2 \omega^4} \langle \mathbf{E}_{o,\omega}^2 \rangle, \qquad (3.37)$$

where the expectation values are over the vacuum state of the field. Since $(\omega^2 / \pi^2 c^3) d\omega$ is the number of field modes per unit volume in the frequency interval $[\omega, \omega + d\omega]$, and each mode has a zero-point energy $\frac{1}{2} \hbar \omega$, we have

$$\frac{1}{2} \hbar \omega \left(\frac{\omega^2}{\pi^2 c^3} \right) d\omega = \frac{1}{8\pi} [\langle \mathbf{E}_{o,\omega}^2 \rangle + \langle \mathbf{B}_{o,\omega}^2 \rangle] d\omega = \frac{1}{4\pi} \langle \mathbf{E}_{o,\omega}^2 \rangle d\omega \qquad (3.38)$$

and

$$\begin{aligned}
\langle (\Delta \mathbf{r})^2 \rangle &= \int_0^\infty d\omega \left(\frac{e^2}{m^2 \omega^4} \right) \langle \mathbf{E}_{o,\omega}^2 \rangle = \frac{2\alpha}{\pi} \left(\frac{\hbar}{mc} \right)^2 \int_0^\infty \frac{d\omega}{\omega} \\
&= \frac{2\alpha}{\pi} \left(\frac{\hbar}{mc} \right)^2 \int_0^\infty \frac{dE}{E},
\end{aligned} \qquad (3.39)$$

where again α is the fine structure constant and $\hbar/mc$ is the electron Compton radius divided by 2π.

Now the fluctuation in $\mathbf{r}$ causes the potential energy $V(\mathbf{r})$ to fluctuate,

$$V(\mathbf{r} + \Delta \mathbf{r}) = V(\mathbf{r}) + \Delta \mathbf{r} \cdot \nabla V(\mathbf{r}) + \frac{1}{6} (\Delta \mathbf{r})^2 \nabla^2 V(\mathbf{r}) + \dots \qquad (3.40)$$

for a spherically symmetric potential, so that an electron in state n should experience an energy shift with leading term

$$\begin{aligned}
\Delta E_n' &= \frac{1}{6} \langle (\Delta \mathbf{r})^2 \rangle \langle n | \nabla^2 V(\mathbf{r}) | n \rangle = \frac{1}{6} \langle (\Delta \mathbf{r})^2 \rangle 4\pi Z e^2 |\psi_n(0)|^2 \\
&= \frac{8\alpha^3 Z^4}{3\pi n^3} R_\infty \int_0^\infty \frac{dE}{E}
\end{aligned} \qquad (3.41)$$

for an s state with principal quantum number n. This is infinite, but if we replace the upper limit of integration by mc^2 in this nonrelativistic model, and the lower limit by Bethe's average excitation energy, then we recover exactly Bethe's expression (3.30) for the Lamb shift.

Note that $E_n' = 0$ for a free electron ($\nabla^2 V = 0$), so that there is no need here to subtract away a free-electron contribution in order to obtain an observable shift. That is, no mass renormalization is necessary in Welton's heuristic approach to the Lamb shift.

The steps leading to (3.41) are the essence of Welton's interpretation of the Lamb shift, and this interpretation is mentioned in many textbooks in advanced quantum mechanics. Welton's argument seems to leave little doubt about the "reality" of the vacuum fluctuations of the electromagnetic field.

3.7   A Feynman Interpretation of the Lamb Shift

Thus far we have described two ways of thinking about the role of the vacuum field in the Lamb shift: we can explain the level shift as the result of emission into and re-absorption from the vacuum of virtual photons (Figure 3.2), or as the result of the fluctuations in the position of an electron due to the fluctuations in the vacuum electric field. In this section and the next we descibe two variations on this theme.

The first is a simplification of an argument due originally to Feynman (1961) (Power 1966). Consider a dilute gas of N atoms per unit volume in a large box of volume V. Since the allowed wavelengths are fixed by the dimensions of the box, the effect of the refractive index of the gas is to change the frequencies ω_k to $\omega_k/n(\omega_k)$, where $n(\omega_k)$ is the index at ω_k. The change in the zero-point field energy due to the presence of the atoms is therefore

$$\Delta E = \sum_{\mathbf{k}\lambda} \frac{1}{2}\frac{\hbar\omega_k}{n(\omega_k)} - \sum_{\mathbf{k}\lambda} \frac{1}{2}\hbar\omega_k \cong -\sum_{\mathbf{k}\lambda}[n(\omega_k)-1]\frac{1}{2}\hbar\omega_k \qquad (3.42)$$

for $n(\omega_k) \cong 1$. Now $n(\omega_k)$ is given, for a dilute gas of atoms in level n, by[3]

$$n(\omega_k) \cong 1 + \frac{4\pi N}{3\hbar}\sum_m \frac{\omega_{mn}|\mathbf{d}_{mn}|^2}{\omega_{mn}^2 - \omega_k^2}, \qquad (3.43)$$

where $\mathbf{d}_{mn}$ is the $m \leftrightarrow n$ transition dipole moment. Thus

$$\begin{aligned}\Delta E_n &\cong -\frac{2\pi N}{3}\sum_{\mathbf{k}\lambda}\omega_k\sum_m \frac{\omega_{mn}|\mathbf{d}_{mn}|^2}{\omega_{mn}^2 - \omega_k^2} \\ &\rightarrow -\frac{2NV}{3\pi c^3}\sum_m \omega_{mn}|\mathbf{d}_{mn}|^2 \int_0^\infty \frac{d\omega\,\omega^3}{\omega_{mn}^2 - \omega^2}.\end{aligned} \qquad (3.44)$$

To obtain an observable shift in level n we subtract from this expression the change in zero-point energy due to N *free* electrons per unit volume in the box. This is obtained by ignoring ω_{mn}^2 compared with ω_k^2 in (3.44), i.e., by taking a limit of effectively continuous electron energies:

$$\Delta E_o = \frac{2NV}{3\pi c^3}\sum_m \omega_{mn}|\mathbf{d}_{mn}|^2 \int_0^\infty d\omega\,\omega. \qquad (3.45)$$

---

[3] See, for instance, P. W. Milonni and J. H. Eberly, *Lasers* (Wiley, New York, 1988), Chapter 7.

The Thomas–Reiche–Kuhn sum rule,[4] $\sum_m \omega_{mn}|d_{mn}|^2 = 3\hbar e^2/2m$, allows us to write this free-electron energy as

$$\Delta E_o = (NV)\frac{e^2\hbar}{\pi mc^3}\int_0^\infty d\omega\omega \; . \tag{3.46}$$

which is just the vacuum expectation value of the energy $(e^2/2mc^2)\mathbf{A}^2$ for NV electrons. [See equation (3.22).] The observable shift in level n should therefore be

$$
\begin{aligned}
\Delta E_n^{\text{obs}} &= -\frac{2NV}{3\pi c^3}\sum_m \omega_{mn}|d_{mn}|^2 \int_0^\infty d\omega[\frac{\omega^3}{\omega_{mn}^2 - \omega^2} + \omega] \\
&= -\frac{2NV}{3\pi c^3}\sum_m \omega_{mn}^3|d_{mn}|^2 \int_0^\infty \frac{d\omega\omega}{\omega_{mn}^2 - \omega^2} \\
&\to \frac{2NV}{3\pi c^3}\sum_m \omega_{mn}^3|d_{mn}|^2 \log\frac{mc^2}{|E_m - E_n|} \tag{3.47}
\end{aligned}
$$

when we introduce a high-frequency cutoff $mc^2/\hbar$.

Finally we recall that[5] $|\mathbf{p}_{mn}|^2 = m^2\omega_{mn}^2|\mathbf{x}_{mn}|^2 = (m^2\omega_{mn}^2/e^2)|d_{mn}|^2$ and write (3.47) in the form

$$\Delta E_n^{\text{obs}} = (NV)\frac{2\alpha}{3\pi}\left(\frac{1}{mc}\right)^2 \sum_m (E_m - E_n)|\mathbf{p}_{mn}|^2 \log\frac{mc^2}{|E_m - E_n|} \; , \tag{3.48}$$

which is exactly Bethe's expresssion (3.25) obtained after mass renormalization when we take $NV = 1$, i.e., when we let our original box contain one atom. Note that, as in Welton's argument and for basically the same reason, no mass renormalization is required in this approach.

3.8   The Lamb Shift as a Stark Shift

There is yet another interpretation of the "Bethe log." Consider the energy $W = -\frac{1}{2}\mathbf{d}\cdot\mathbf{E}$ associated with a dipole moment $\mathbf{d}$ induced by an electric field $\mathbf{E}$.[6] Writing $\mathbf{d}_\omega = \alpha(\omega)\mathbf{E}_\omega$ for the Fourier component of the dipole moment induced by the Fourier component $\mathbf{E}_\omega$ of the field, where $\alpha(\omega)$ is

---

[4] See, for instance, J. J. Sakurai, *Advanced Quantum Mechanics* (Addison–Wesley, Reading, Mass., 1970), p. 74.

[5] Ibid., p. 42.

[6] The factor $1/2$ is due to the fact that the dipole moment is induced rather than permanent. See, for instance, J. D. Jackson, *Classical Electrodynamics*, 2nd ed. (Wiley, New York, 1975), p. 161.

the polarizability, we have $W = -\frac{1}{2}\alpha(\omega)E_\omega^2$ and, if there is a continuous distribution of field frequencies,

$$W = -\frac{1}{2}\int \alpha(\omega)[4\pi\rho(\omega)d\omega], \tag{3.49}$$

where $\rho(\omega)$ is the spectral energy density of the field, $\mathbf{E}_\omega^2 = 4\pi\rho(\omega)d\omega$. For an atom in level n, therefore, we expect a level shift

$$\Delta E_n = -2\pi \int_0^\infty d\omega\, \alpha_n(\omega)\rho(\omega) \tag{3.50}$$

due to an applied field of spectral energy density $\rho(\omega)$, where $\alpha_n(\omega)$ is the polarizability for level n and is given by the Kramers–Heisenberg formula,[7]

$$\alpha_n(\omega) = \frac{2}{3\hbar}\sum_m \frac{\omega_{mn}|\mathbf{d}_{mn}|^2}{\omega_{mn}^2 - \omega^2}\ . \tag{3.51}$$

For a monochromatic field, equation (3.50) reduces to the standard formula for the second-order Stark shift produced by an external field.

For an atom in the vacuum we use the spectral energy density $\rho_o(\omega) = \hbar\omega^3/2\pi^2c^3$ of the zero-point field and obtain the level shift

$$\Delta E_n = -\frac{2}{3\pi c^3}\sum_m \omega_{mn}|\mathbf{d}_{mn}|^2 \int_0^\infty \frac{d\omega\,\omega^3}{\omega_{mn}^2 - \omega^2}\ . \tag{3.52}$$

This is identical to equation (3.44) in the case of one atom ($NV = 1$). Therefore we can regard the Lamb shift as a Stark shift produced by the vacuum electromagnetic field.

The equivalence of this interpretation of the Lamb shift to that given in the preceding section follows from the relation

$$n(\omega) = 1 + 2\pi N\alpha(\omega) \tag{3.53}$$

between the refractive index and the polarizability for a gas with $n(\omega) \cong 1$. These interpretations of the Lamb shift can be "dressed up" by relating $\alpha(\omega)$ to the real part of the forward scattering amplitude $f(\omega)$ [$\propto \omega^2\alpha(\omega)$] for a photon of frequency ω, but this is hardly necessary to bring out the point that the Bethe log may be attributed to the coupling of the atom to the vacuum radiation field.

---

[7]See, for instance, A. S. Davydov, *Quantum Mechanics* (Pergamon Press, Oxford, 1965), pp. 316–321.

3.9 Retardation

The Bethe log arising from vacuum field fluctuations accounts for all but a few percent of the $2s_{1/2} - 2p_{1/2}$ Lamb shift in hydrogen. The Lamb shift provides one of the most delicate tests of QED, and various other effects contributing to it, such as vacuum polarization, finite nuclear mass, etc., must be accounted for in any detailed comparison with experiment. However, for a basic understanding of why there is a Lamb shift and for an estimate of its magnitude in hydrogen, it is sufficient to concentrate on the Bethe log. For this reason we have discussed various physical interpretations of this contribution, all of them involving the vacuum electromagnetic field.

It is obviously of interest to determine the effect of higher order corrections to the Bethe log. One correction is to go beyond the dipole approximation in which the factors $e^{\pm i \mathbf{k} \cdot \mathbf{x}}$ in the field are dropped. This results in the replacement of (3.19) by

$$\Delta E_n = \frac{2\pi c^2}{m^2} \frac{1}{V} \sum_{\mathbf{k}\lambda} \frac{1}{\omega_k} \sum_m \frac{\langle n|\mathbf{p} \cdot \mathbf{e}_{\mathbf{k}\lambda} e^{i\mathbf{k}\cdot\mathbf{x}}|m\rangle\langle m|e^{-i\mathbf{k}\cdot\mathbf{x}}\mathbf{p} \cdot \mathbf{e}_{\mathbf{k}\lambda}|n\rangle}{\omega_{nm} - \omega_k} \; ,$$

$$(3.54)$$

which, unlike (3.19), is logarithmically divergent *without mass renormalization* (Au and Feinberg, 1974). We can see this by writing

$$\sum_m \frac{\mathbf{p} \cdot \mathbf{e}_{\mathbf{k}\lambda} e^{i\mathbf{k}\cdot\mathbf{x}}|m\rangle\langle m|e^{-i\mathbf{k}\cdot\mathbf{x}}\mathbf{p} \cdot \mathbf{e}_{\mathbf{k}\lambda}}{\omega_{nm} - \omega_k} = \hbar \sum_m \mathbf{p} \cdot \mathbf{e}_{\mathbf{k}\lambda} e^{i\mathbf{k}\cdot\mathbf{x}} G_n |m\rangle\langle m|$$

$$\times \, e^{-i\mathbf{k}\cdot\mathbf{x}}\mathbf{p} \cdot \mathbf{e}_{\mathbf{k}\lambda} \qquad (3.55)$$

in (3.54), where the operator $G_n \equiv (E_n - \mathbf{p}^2/2m - V - \hbar\omega_k)^{-1}$, with $(\mathbf{p}^2/2m + V)|m\rangle = E_m|m\rangle$. Then the general identity

$$e^{i\mathbf{k}\cdot\mathbf{x}} F(\mathbf{p}) e^{-i\mathbf{k}\cdot\mathbf{x}} = F(\mathbf{p} - \hbar\mathbf{k}) \qquad (3.56)$$

gives

$$e^{i\mathbf{k}\cdot\mathbf{x}} G_n e^{-i\mathbf{k}\cdot x} = [E_n - \frac{1}{2m}(\mathbf{p} - \hbar\mathbf{k})^2 - V - \hbar\omega_k]^{-1} \; . \qquad (3.57)$$

The effect of the retardation, i.e., of not making the dipole approximation, is then to replace ω by ω^2 for large frequencies in the denominator of the integrand in (3.20). This leads to logarithmic rather than linear divergence.

Physically, the effect of retardation is to give *recoil* of the electron in photon emission and absorption. The replacement of $\mathbf{p}$ by $\mathbf{p} - \hbar\mathbf{k}$ in the

intermediate electron momenta then reflects the conservation of linear mo
mentum in the interaction between electrons and photons.

We can aproach the free-particle self-energy with retardation included
by taking $E_n = \mathbf{p}^2/2m$ and $V = 0$ in (3.57), and ignoring the term $\mathbf{k} \cdot \mathbf{p}$ in
order to obtain from (3.56) a contribution proportional to $\mathbf{p}^2$ for small $\mathbf{p}$.
Then (3.54) gives

$$\Delta E_n^{\text{free}} = -\frac{2\pi e^2}{m^2} \frac{1}{V} \sum_{\mathbf{k}\lambda} \frac{1}{\omega_k} \frac{\langle n|(\mathbf{p} \cdot \mathbf{e}_{\mathbf{k}\lambda})^2|n\rangle}{\omega_k + \hbar\omega_k^2/2mc^2} = -\frac{\delta m}{2m^2}\langle n|\mathbf{p}^2|n\rangle, \quad (3.58)$$

where

$$\delta m = \frac{4e^2}{3\pi c^3} \int_0^\infty \frac{d\omega\omega}{\omega + \hbar\omega^2/2mc^2} , \quad (3.59)$$

which is logarithmically divergent as opposed to the linear divergence of
the result (3.33) obtained without retardation.

Using a momentum representation of G_n (Schwinger, 1964; see also
Lieber, 1968), Au and Feinberg (1974) have numerically computed the
mass-renormalized shifts $\Delta E_n - (\delta m/2m^2)\langle n|\mathbf{p}^2|n\rangle$, with ΔE_n and δm
given by (3.54) and (3.59), respectively, for the $2s$ and $2p$ levels of hydrogen
with retardation. Each term diverges logarithmically, and the difference is
finite without any high-frequency cutoff. They obtained $\Delta E_{2s} = 931.1$
MHz and $\Delta E_{2s} - \Delta E_{2p} = 950.3$ MHz. (When the $\mathbf{k} \cdot \mathbf{p}$ term was kept
in the denominator of (3.58) in the evaluation of ΔE_n^{free}, they obtained
$\Delta E_{2s} = 1330$ MHz and $\Delta E_{2s} - \Delta E_{2p} = 996.6$ MHz.) The nonrelativistic
computation of the Lamb shift with retardation included, therefore, gives a
finite value, but this value is significantly different from the experimentally
observed Lamb shift (see also Grotch, 1981).

The nonrelativistic theory with retardation as just described, of course,
involves photon frequencies at which nonrelativistic theory breaks down. In
this sense the theory is inconsistent. However, it does show that retardation
is important in the numerical value of the Lamb shift, and furthermore
it provides some insight into why the nonrelativistic, nonretarded Bethe
log with a high-frequency cutoff works so well (Au and Feinberg, 1974):
the dominant contribution to the Bethe log comes from frequencies that
are too small for retardation to be important, while, for high frequencies,
where Bethe introduced a cutoff, there is an *effective* cutoff resulting from
retardation.

It is worth noting that the $\Delta E_{2s} - \Delta E_{2p}$ separation is convergent in the
nonrelativistic theory with retardation even without mass renormalization.

3.10   Another Look at the Casimir Force

In Section 2.7 we obtained the Casimir force between two conducting plates in the conventional way, by calculating the difference between the zero-point field energies for finite and infinite plate separations. Having interpreted the Lamb shift in different ways based on vacuum field fluctuations, we now turn to an alternative interpretation of the Casimir force.

The idea here is that the virtual photons of the vacuum carry linear momentum $\frac{1}{2}\hbar\mathbf{k}$; recall equation (2.60). Then the reflections off the plates of the zero-point field outside the plates act to push the plates together, while reflections of the field confined between the plates push them apart. Loosely speaking, there are more field modes outside the plates than inside, since only certain discrete frequencies are allowed between the plates. The net effect of the zero-point radiation pressure is then to push the plates together. We shall now show that the force calculated in this way is exactly the Casimir force (Milonni, Cook, and Goggin, 1988).

Consider the radiation pressure exerted by a plane wave incident normally on a plate. This pressure is twice the energy u per unit volume of the incident field (Section 1.2). If the wave has an angle of incidence θ, however, the radiation pressure is reduced to $P = F/A = 2u\cos^2\theta$. There are two factors of $\cos\theta$ here because (1) the normal component of the linear momentum imparted to the plate is proportional to $\cos\theta$, and (2) the element of area A is increased by $(\cos\theta)^{-1}$ compared with the case of normal incidence.

Between the plates the modes formed by reflections off the plates obviously act to push the plates apart. A mode of frequency ω contributes a pressure

$$P = 2(\frac{1}{2})(\frac{1}{2}\hbar\omega)V^{-1}\cos^2\theta = \frac{\hbar\omega}{2V}\frac{k_z^2}{k^2}\ , \tag{3.60}$$

where, as usual, $k = \omega/c$ and V is a quantization volume. A factor $1/2$ has been inserted because the zero-point energy of each mode is divided equally between waves propagating toward or away from each plate. For large plates, k_x and k_y take on a continuum of values, whereas $k_z = n\pi/d$, where n is a positive integer. Adding the contributions from all modes of the space between the plates, we have the total outward pressure

$$P_{\text{out}} = \frac{\hbar c}{\pi^2 d}\sum_{n=1}^{\infty}\int_0^{\infty}dk_x\int_0^{\infty}dk_y\frac{(n\pi/d)^2}{[k_x^2 + k_y^2 + (n\pi/d)^2]^{1/2}} \tag{3.61}$$

on each plate. In writing this expression a factor of 2 has been included to allow for the two independent polarizations. The replacement of sums by integrals over k_x and k_y brings in a factor $(L/\pi)^2 = V/\pi^2 d$, as in Section 2.7.

The field outside the "resonator" formed by the plates has a continuum of allowed frequencies. These modes obviously act to push the plates together by reflections off the plates. The total inward pressure exerted by these modes may be obtained from (3.61) by replacing $\sum_n$ by $(d/\pi)\int dk_z$:

$$P_{\text{in}} = \frac{\hbar c}{\pi^3} \int_0^\infty dk_x \int_0^\infty dk_y \int_0^\infty dk_z \frac{k_z^2}{(k_x^2 + k_y^2 + k_z^2)^{1/2}} . \qquad (3.62)$$

Both P_{out} and P_{in} are infinite, but it is only their difference that is physically meaningful. After some simple algebra we can write this difference as

$$P_{\text{out}} - P_{\text{in}} = \frac{\pi^2 \hbar c}{4d^4} \left[\sum_{n=1}^\infty n^2 \int_0^\infty \frac{dx}{(x+n^2)^{1/2}} - \int_0^\infty du\, u^2 \int_0^\infty \frac{dx}{(x+u^2)^{1/2}} \right] .$$
$$(3.63)$$

Application of the Euler–Maclaurin summation formula as in Section 2.7 then leads to the Casimir result

$$P_{\text{out}} - P_{\text{in}} = -\frac{\pi^2 \hbar c}{240 d^4} \qquad (3.64)$$

for the force per unit area between the plates.

We can therefore regard the Casimir force as a consequence of the radiation pressure associated with the zero-point energy $\frac{1}{2}\hbar\omega$ per mode of the field. This interpretation is directly connected to the conventional one (Section 2.7) through the Maxwell stress tensor for the quantized field (Milonni et al., 1988).[8]

3.11   Van der Waals Forces

In order to account for observed deviations from the ideal gas law, in 1873 J. D. van der Waals proposed the equation of state

$$(P + \frac{a}{V^2})(V - b) = RT \qquad (3.65)$$

for 1 mole of a gas at temperature T. P and V are as usual the pressure and volume, R is the universal gas constant, and a and b are now called the *van der Waals constants*, obtained by fitting (3.65) to experimental data.

---

[8] See also A. E. González, *Physica* **131**A, 228 (1985).

Van der Waals interpreted the constant b as the volume excluded by two atoms. If the atoms were imagined to be spheres of radius r_o, then $b = 16\pi r_o^3/3$. The constant a was associated with an attractive force between two atoms. Van der Waals later suggested an interaction potential of the form $V(r) = -Ar^{-1}e^{-Br}$, where A and B are constants.

Much later Keesom obtained the potential $V(r) = -p_1^2 p_2^2/3kTr^6$ for two polar molecules, i.e., molecules with permanent dipole moments. Here p_1 and p_2 are the dipole moments of the two molecules and $V(r)$ is obtained as a consequence of molecular rotations. There is an attractive force because attractive orientations are statistically favored over repulsive ones.

Debye and others recognized that more general attractive forces must exist between molecules, since gases of nonpolar molecules have nonvanishing values of the van der Waals constant a. Moreover a temperature-independent potential was needed. Debye noted that many molecules have a permanent quadrupole moment, which can induce a dipole moment in a second molecule, and the resulting dipole–quadrupole force is temperature independent. Such an "induction force" occurs also if the first molecule has a permanent dipole moment. However, neither case is sufficiently general to account for the van der Waals equation of state.

London (1930) employed fourth-order quantum-mechanical perturbation theory to derive the interaction potential[9]

$$V(r) = -\frac{3\hbar\omega_o\alpha^2}{4r^6} \tag{3.66}$$

between two identical atoms (or molecules) with transition frequency ω_o between the ground and first excited levels, with α the static (zero-frequency) polarizability. London's result, which was considered a major accomplishment of the new quantum mechanics, showed that there is a general force of attraction between two molecules even if neither has a permanent moment; it is necessary only that a dipole moment can be *induced* in each molecule, i.e., that each molecule is polarizable ($\alpha \neq 0$). And London's result, unlike Keesom's, is temperature independent.

Since it involves the polarizability, which in turn is related to the refractive index and dispersion [cf. (3.53)], London's force is often called the *dispersion force.* Dispersion forces, together with the orientation and induction forces of Keesom and Debye, are now regarded as three general types of van der Waals forces. In this section we will consider the origin of the dispersion force between two neutral polarizable particles, and show that this type of van der Waals force may be attributed to zero-point energy.

---

[9]Before the work of London, S. C. Wang [*Phys. Zs.* **28**, 663 (1927)] presented somewhat indirect quantum-mechanical arguments for an r^{-6} interaction between two hydrogen atoms.

In fact London originally proposed such an interpretation. The essence of the argument is as follows. Consider two identical dipole oscillators of frequency ω_o coupled through their near fields. For this system we write the equations of motion

$$\ddot{x}_1 + \omega_o^2 x_1 = K x_2, \tag{3.67}$$

$$\ddot{x}_2 + \omega_o^2 x_2 = K x_1, \tag{3.68}$$

with $K = qe^2/mr^3$, q being the dipole–dipole orientation factor, $3(\hat{\mu}_1 \cdot s)(\hat{\mu}_2 \cdot s) - \hat{\mu}_1 \cdot \hat{\mu}_2$, where $\hat{\mu}_1, \hat{\mu}_2$ are unit vectors in the directions of the dipole moments and s is a unit vector pointing from one dipole to the other. Two points about these equations are worth noting. First, since atoms that remain with high probability in their ground states are accurately represented for many purposes as harmonic oscillators, these equations provide a reasonable qualitative description of the coupling between two ground-state atoms.[10] Second, we are not assuming permanent dipole moments; equations (3.67) and (3.68) can be thought of as operator equations, with the expectation values $\langle x_1 \rangle = \langle x_2 \rangle = 0$ implying a vanishing permanent dipole moment. In fact the only thing of interest for the present discussion is that the normal mode frequencies of this coupled oscillator system are given by

$$\omega_\pm = (\omega_o^2 \pm K)^{1/2} \ . \tag{3.69}$$

The quantum-mechanical ground-state energy of the system is

$$E = \frac{1}{2}\hbar(\omega_+ + \omega_-) \cong \hbar\omega_o - \frac{\hbar K^2}{8\omega_o^3} \tag{3.70}$$

to lowest order in K/ω_o^2. This implies an interaction energy

$$V(r) = -\frac{\hbar}{8\omega_o^3} \left(\frac{qe^2}{mr^3} \right)^2 = -\frac{q^2\hbar\omega_o\alpha^2}{8r^6} \ , \tag{3.71}$$

where $\alpha = e^2/m\omega_o^2$ is the classical static polarizability. Now if we use the fact that a quantum-mechanical evaluation of q^2 gives an average value of 2, and multiply by 3 to account for the three-dimensionality of the atoms, then (3.71) yields $V(r) = -3\hbar\omega_o\alpha^2/4r^6$, which is London's result (3.66).[11]

---

[10] See, for instance, M. Cray, M.-L. Shih, and P. W. Milonni, *Am. J. Phys.* **50**, 1016 (1982).

[11] If the two induced dipoles are parallel to each other and perpendicular to the axis joining them, $q^2 = 1$, whereas if they are parallel and aligned along the axis, $q^2 = 4$. The orientationally averaged value of q^2 is then $(2/3)(1) + (1/3)(4) = 2$. See also P. W. Milonni and P. L. Knight, *Phys. Rev.* A10, 1096 (1974); A11, 1090 (1975).

We saw in Section 2.6 that the vacuum field is required to maintain the commutation relation, and therefore the zero-point energy, of a dipole oscillator. This suggests that the r^{-6} van der Waals interaction might be attributed physically to the fluctuating vacuum electromagnetic field. That this is so may be seen from the following argument.

As in Section 3.8 we begin with the formula for the Stark shift of an atomic energy level, but now without the assumption that the field is isotropic:

$$W_A = -\frac{1}{2} \sum_{k\lambda} \alpha_A(\omega_k) E_{k\lambda}^2(x_A, t) \tag{3.72}$$

for an atom A located at x_A with polarizability $\alpha_A(\omega)$. The total field in mode (k, λ) acting on A is assumed to be the zero-point field plus the field at A produced by a second atom B:

$$E_{k\lambda}(x_A, t) = E_{o,k\lambda}(x_A, t) + E_{B,k\lambda}(x_A, t). \tag{3.73}$$

Then the part of W_A due to the interaction between the two atoms is

$$
\begin{aligned}
W_{AB} = & -\frac{1}{2} \sum_{k\lambda} \alpha_A(\omega_k) [E_{o,k\lambda}(x_A, t) \cdot E_{B,k\lambda}(x_A, t) \\
& + E_{B,k\lambda}(x_A, t) \cdot E_{o,k\lambda}(x_A, t)].
\end{aligned}
\tag{3.74}
$$

This is the only part of W_A that will involve the distance r between the atoms. We have seen in Section 3.8 that $E_{o,k\lambda}^2(x_A, t)$, for instance, will contribute to the Lamb shift in atom A.

Now actually the right side of (3.74) should be a vacuum expectation value involving field operators. Let us write the operator $E_{o,k\lambda}$ as $E_{o,k\lambda}^{(+)} + E_{o,k\lambda}^{(-)}$, where

$$E_{o,k\lambda}^{(+)}(x_A, t) = i\left(\frac{2\pi\hbar\omega_k}{V}\right)^{1/2} a_{k\lambda}(0) e^{-i\omega_k t} e^{ik \cdot x_A} e_{k\lambda} \tag{3.75}$$

and

$$E_{o,k\lambda}^{(-)}(x_A, t) = -i\left(\frac{2\pi\hbar\omega_k}{V}\right)^{1/2} a_{k\lambda}^\dagger(0) e^{i\omega_k t} e^{-ik \cdot x_A} e_{k\lambda} \tag{3.76}$$

are called the positive- and negative-frequency parts, respectively, of $E_{o,k\lambda}$. Since $E_{o,k\lambda}^{(+)}(x_A, t)|vac\rangle = \langle vac|E_{o,k\lambda}^{(-)}(x_A, t) = 0$, (3.74) is equivalent to

$$W_{AB} = -\frac{1}{2} \sum_{k\lambda} \alpha_A(\omega_k) \left[\langle E_{o,k\lambda}^{(+)}(x_A, t) \cdot E_{B,k\lambda}(x_A, t) \rangle \right.$$

$$+ \langle \mathbf{E}_{B,\mathbf{k}\lambda}(\mathbf{x}_A, t) \cdot \mathbf{E}^{(-)}_{o,\mathbf{k}\lambda}(\mathbf{x}_A, t) \rangle \Big] \tag{3.77}$$

when a vacuum expectation value is taken.

The field from atom B has the same form as the classical field of an electric dipole $\mathbf{p}_B = \hat{\mu}_B p_B(t)$, where as before $\hat{\mu}_B$ is the unit vector in the direction of $\mathbf{p}_B$:

$$\mathbf{E}_B(\mathbf{x}_A, t) = -[\hat{\mu}_B - (\hat{\mu}_B \cdot \mathbf{s})\mathbf{s}] \frac{1}{c^2 r} \ddot{p}_B(t - \frac{r}{c}) + [3(\hat{\mu}_B \cdot \mathbf{s})\mathbf{s} - \hat{\mu}_B]$$

$$\times [\frac{1}{r^3} p_B(t - \frac{r}{c}) + \frac{1}{cr^2} \dot{p}_B(t - \frac{r}{c})]. \tag{3.78}$$

Here $\mathbf{s}$ is the unit vector pointing from atom B to atom A and now $\mathbf{E}_B$ and p_B are quantum-mechanical operators.[12] The dipole moment $\mathbf{p}_B$ has zero expectation value for an atom in a stationary state (the ground state in the situation of interest here), but this does not of course mean that the dipole moment of atom B is identically zero. Rather, this dipole moment *fluctuates* about zero mean due to the influence of the vacuum field at $\mathbf{x}_B$:

$$\mathbf{p}_B(t) = \sum_{\mathbf{k}\lambda} \alpha_B(\omega_k) \left[\mathbf{E}^{(+)}_{o,\mathbf{k}\lambda}(\mathbf{x}_B, t) + \mathbf{E}^{(-)}_{o,\mathbf{k}\lambda}(\mathbf{x}_B, t) \right]. \tag{3.79}$$

Since $\mathbf{E}^{(+)}_{o,\mathbf{k}\lambda} |vac\rangle = 0$, only the negative-frequency (creation) part of $\mathbf{E}_{B,\mathbf{k}\lambda}(\mathbf{x}_B, t)$, determined by $\mathbf{E}^{(-)}_{o,\mathbf{k}\lambda}(\mathbf{x}_B, t)$, will contribute to the first term in brackets in (3.77). This is easily read off from (3.78) and (3.79):

$$\mathbf{E}^{(-)}_{B,\mathbf{k}\lambda}(\mathbf{x}_A, t) = k^3 \alpha_B(\omega_k) e^{-ikr} \left\{ [\mathbf{e}_{\mathbf{k}\lambda} - (\mathbf{e}_{\mathbf{k}\lambda} \cdot \mathbf{s})\mathbf{s}] \frac{1}{kr} \right.$$

$$+ [3(\mathbf{e}_{\mathbf{k}\lambda} \cdot \mathbf{s})\mathbf{s} - \mathbf{e}_{\mathbf{k}\lambda}]$$

$$\left. [\frac{1}{(kr)^3} + \frac{i}{(kr)^2}] \right\} E^{(-)}_{o,\mathbf{k}\lambda}(\mathbf{x}_B, t), \tag{3.80}$$

where $E^{(-)}_{o,\mathbf{k}\lambda}(\mathbf{x}_B, t) \equiv \mathbf{e}_{\mathbf{k}\lambda} \cdot \mathbf{E}^{(-)}_{o,\mathbf{k}\lambda}(\mathbf{x}_B, t)$. Similarly, only $\mathbf{E}^{(+)}_{B,\mathbf{k}\lambda}(\mathbf{x}_A, t)$ will contribute to the second term in brackets in (3.77). Then, using the fact that

$$\langle vac | a_{\mathbf{k}\lambda}(0) a^{\dagger}_{\mathbf{k}\lambda}(0) | vac \rangle = 1, \tag{3.81}$$

we obtain from (3.77) the expression

$$W_{AB} \equiv V(r) = -\frac{2\pi\hbar}{V} \text{Re} \sum_{\mathbf{k}\lambda} k^3 \omega_k \alpha_A(\omega_k) \alpha_B(\omega_k) e^{-ikr} e^{i\mathbf{k}\cdot\mathbf{r}}$$

---

[12] See Chapter 4 for a discussion of the correspondence between classical and quantum solutions of the Maxwell equations.

$$\times \left\{ [1 - (\mathbf{e}_{\mathbf{k}\lambda} \cdot \mathbf{s})^2]\frac{1}{kr} + [3(\mathbf{e}_{\mathbf{k}\lambda} \cdot \mathbf{s})^2 - 1][\frac{1}{(kr)^3} + \frac{i}{(kr)^2}] \right\} . \tag{3.82}$$

Note that this expression is symmetric in A and B, as it should be.

Now as usual $\sum_{\mathbf{k}\lambda} \rightarrow (V/8\pi^3) \int dk k^2 \sum_\lambda \int d\Omega_{\mathbf{k}}$, and the sum over polarizations plus the integration over solid angles about $\mathbf{k}$ are easily carried out, using the identity $\sum_\lambda (\mathbf{e}_{\mathbf{k}\lambda} \cdot \mathbf{s})^2 = 1 - (\hat{\mathbf{k}} \cdot \mathbf{s})^2$, with $\hat{\mathbf{k}} \equiv \mathbf{k}/k$:

$$\sum_\lambda \int d\Omega_{\mathbf{k}} e^{i\mathbf{k} \cdot \mathbf{r}} \left\{ [1 - (\mathbf{e}_{\mathbf{k}\lambda} \cdot \mathbf{s})^2]\frac{1}{kr} + [3(\mathbf{e}_{\mathbf{k}\lambda} \cdot \mathbf{s})^2 - 1][\frac{1}{(kr)^3} + \frac{i}{(kr)^2}] \right\}$$

$$= \int d\Omega_{\mathbf{k}} e^{i\mathbf{k} \cdot \mathbf{s} r} \left\{ [1 + (\hat{\mathbf{k}} \cdot \mathbf{s})^2]\frac{1}{kr} + [1 - 3(\hat{\mathbf{k}} \cdot \mathbf{s})^2][\frac{1}{(kr)^3} + \frac{i}{(kr)^2}] \right\} . \tag{3.83}$$

This integral is easily performed by choosing the polar (z) axis to lie along s:

$$\frac{1}{8\pi} \int_0^{2\pi} d\phi \int_0^\pi d\theta \sin\theta e^{ix\cos\theta}[(1 + \cos^2\theta)\frac{1}{x} + (1 - 3\cos^2\theta)$$

$$\times(\frac{1}{x^3} + \frac{i}{x^2})] = \left(\frac{\sin x}{x^2} - \frac{i\sin x}{x^3} + \frac{\cos x}{x^3} - \frac{2\sin x}{x^4} - \frac{3i\cos x}{x^4} \right.$$

$$\left. + \frac{3i\sin x}{x^5} - \frac{3\cos x}{x^5} + \frac{3\sin x}{x^6} \right) , \tag{3.84}$$

with $x \equiv kr$. Then (3.82) gives

$$V(r) = -\frac{\hbar}{\pi c^6} \int_0^\infty d\omega \omega^6 \alpha_A(\omega)\alpha_B(\omega) G\left(\frac{\omega r}{c}\right) , \tag{3.85}$$

$$G(x) \equiv \frac{\sin 2x}{x^2} + \frac{2\cos 2x}{x^3} - \frac{5\sin 2x}{x^4} - \frac{6\cos 2x}{x^5} + \frac{3\sin 2x}{x^6} . \tag{3.86}$$

For small r the dominant contribution to $V(r)$ comes from the last term in (3.86):

$$V(r) \cong -\frac{\hbar}{\pi c^6}\frac{3c^6}{r^6} \int_0^\infty d\omega \alpha_A(\omega)\alpha_B(\omega) \sin\frac{2\omega r}{c}$$

$$= -\frac{3\hbar}{\pi r^6} \int_0^\infty du \alpha_A(iu)\alpha_B(iu)e^{-2ur/c}$$

$$= -\frac{3\hbar}{\pi r^6}\left(\frac{2}{3\hbar}\right)^2 \sum_m \sum_p \omega_{mn}\omega_{pn}|\mathbf{d}_{mn}|^2|\mathbf{d}_{pn}|^2$$

$$\times \int_0^\infty \frac{du e^{-2ur/c}}{(u^2 + \omega_{mn}^2)(u^2 + \omega_{pn}^2)} , \tag{3.87}$$

where we have used equation (3.51) for the polarizability of an atom in state n and assumed for simplicity that the two atoms are identical. The change in the path of integration implied by the second line of (3.87), where we replace an integral along the real axis by an integral along the imaginary axis plus a (vanishing) contribution along a large quarter-circle, assumes that we do not need to concern ourselves with poles of $\alpha(\omega)$. This is in fact the case because, at a resonance frequency ω_{mn}, the real part of the polarizability is found to vanish when the $m \leftrightarrow n$ transition linewidth is accounted for.[13] If $r \to 0$, or more precisely if $r << c/|\omega_{mn}|$ for all transitions $m \leftrightarrow n$, we may replace $e^{-2ur/c}$ by 1 in (3.87), and this gives the r^{-6} form of the van der Waals potential derived by London. In fact if we assume furthermore that one particular transition $m \leftrightarrow n$ makes a dominant contribution to (3.87), then

$$
V(r) \cong -\frac{3\hbar}{\pi r^6}\left(\frac{2}{3\hbar}\right)^2 \omega_o^2|\mathbf{d}|^4 \int_0^\infty \frac{du}{(u^2 + \omega_o^2)^2} = -\frac{3\hbar\omega_o\alpha^2}{4r^6} , \qquad (3.88)
$$

where ω_o and $\mathbf{d}$ are respectively the transition frequency and dipole matrix element of this transition and $\alpha = (2/3\hbar)|\mathbf{d}|^2/\omega_o$ is the static ($\omega = 0$) polarizability of a ground-state atom in the two-state approximation in which the one transition is assumed to be dominant. This result is exactly that derived by London.

However, the r^{-6} van der Waals potential does not apply in the "retarded" regime of large interatomic separations. In a study of the stability of certain (lyophobic) colloidal systems, Verwey and Overbeek (1948) found that the interatomic potential must fall off faster than r^{-6} at large distances in order for theory and experiment to be consistent. They suggested that at large atomic separations — that is, at separations large compared with atomic transition wavelengths — the London theory must be modified to account for retardation. Such a modification was worked out by Casimir and Polder in 1948. They derived an expression equivalent to (3.85) and showed that, for large r, $V(r) \propto r^{-7}$. The simplest way to obtain the Casimir–Polder result is to argue that for distances large enough for retardation to be important, (3.85) may effectively be replaced by[14]

$$
V(r) \cong -\frac{\hbar}{\pi c^6}\alpha_A\alpha_B \int_0^\infty d\omega\,\omega^6 G\left(\frac{\omega r}{c}\right) , \qquad (3.89)
$$

---

[13] See, for instance, P. W. Milonni and J. H. Eberly, *Lasers*, Chapter 3.

[14] This may be justified quantitatively by making the same change in the path of integration in (3.85) as in (3.87). Then it can be seen that the zero-frequency polarizability makes the dominant contribution for $|\omega_{mn}|r/c >> 1$, i.e., when retardation is important. This condition is roughly equivalent to $r >> 137a_o$, where a_o is the Bohr radius.

where α_A, α_B are the static polarizabilities. The integral may be evaluated by introducing a cutoff function $e^{-\lambda \omega r/c}$, $\lambda > 0$, and taking the limit $\lambda \to 0$ after integrating. For instance,

$$\int_0^\infty d\omega \omega^6 \frac{\sin 2\omega r/c}{(\omega r/c)^2} = \frac{c^7}{r^7} \int_0^\infty du\, u^4 \sin 2u$$
$$\to \frac{c^7}{r^7} \lim_{\lambda \to 0} \int_0^\infty du\, u^4 e^{-\lambda u} \sin 2u = \frac{48 c^7}{r^7}\,.$$

(3.90)

We find in this limit, for large r,

$$V(r) \cong -\frac{23\hbar c}{4\pi r^7} \alpha_A \alpha_B,$$

(3.91)

which is the Casimir–Polder result.

In either the retarded (Casimir–Polder) or nonretarded (London) limit the van der Waals interaction may be regarded as a consequence of the fluctuating vacuum electromagnetic field. Our derivation leading to (3.82) shows that the van der Waals interaction results from the fact that

$$\langle vac| E_{o,k\lambda}^{(+)}(\mathbf{x}_A, t)_i E_{o,k\lambda}^{(-)}(\mathbf{x}_B, t)_j |vac\rangle \neq 0.$$

(3.92)

In other words, the van der Waals interaction results from *correlations* of the vacuum field over distances on the order of $|\mathbf{x}_A - \mathbf{x}_B| = r$. In more physical terms, the vacuum field induces fluctuating dipole moments in the two atoms, and the dipole–dipole interaction of these zero-mean but correlated moments is the van der Waals interaction.

At the conclusion of their paper Casimir and Polder argued that the simple form of (3.91) might allow it to be derived "by more elementary considerations" than the perturbation-theoretic approach they employed and that "This would be desirable since it would also give a more physical background to our result, a result which in our opinion is rather remarkable. So far we have not been able to find such a simple argument." Not long thereafter Casimir (1949) gave a derivation based on the fluctuating zero-point field, and the derivation given in this section follows closely the spirit of his insightful analysis.[15] The following section is also based on Casimir's work.

---

[15] See also Boyer (1972,1980) and Renne (1971).

3.12   Force on an Atom near a Conducting Wall

Casimir and Polder (1948) also considered a simpler problem in their study of long-range, retarded interactions; namely, the interaction between an atom and a perfectly conducting wall. For short distances d of the atom from the wall the attractive potential $V(d)$ may be obtained from the dipole–dipoleole interaction of the atom with its image in the wall, and varies as d^{-3}. For large d, however, $V(d)$ falls off as d^{-4}; as in the case of the van der Waals force, the effect of retardation is to weaken the interaction by a factor $\propto d^{-1}$.

Consider first an atom located at the point $\mathbf{R} \equiv (L/2, L/2, d)$ inside the rectangular parallelepiped described by the mode functions (2.91)–(2.93). The energy (3.72) in this case is

$$-\frac{\alpha}{2}\sum_{\mathbf{k}\lambda}(2\pi\hbar\omega)|\mathbf{A}_{\mathbf{k}\lambda}(\mathbf{R})|^2 =$$

$$
\begin{aligned}
- \;&\frac{4\alpha}{V}\sum_{\mathbf{k}\lambda}(2\pi\hbar\omega)[e_{\mathbf{k}\lambda x}^2\cos^2\tfrac{1}{2}k_xL\sin^2\tfrac{1}{2}k_yL\sin^2 k_zd \\
+ \;&e_{\mathbf{k}\lambda y}^2\sin^2\tfrac{1}{2}k_xL\cos^2\tfrac{1}{2}k_yL\sin^2 k_zd \\
+ \;&e_{\mathbf{k}\lambda z}^2\sin^2\tfrac{1}{2}k_xL\sin^2\tfrac{1}{2}k_yL\cos^2 k_zd] \\
\to \;- \;&\left(\frac{2\pi\hbar\alpha}{V}\right)\sum_{\mathbf{k}\lambda}\omega_k[(e_{\mathbf{k}\lambda x}^2+e_{\mathbf{k}\lambda y}^2)\sin^2 k_zd + e_{\mathbf{k}\lambda z}^2]\cos^2 k_zd,
\end{aligned}
$$

$$(3.93)$$

where, as in the preceding section, we have replaced the polarizability $\alpha(\omega)$ by the static polarizability $\alpha(\omega) \equiv \alpha$, arguing that only the value of $\alpha(\omega)$ at $\omega = 0$ contributes at large distances d of the atom from one of the (conducting) walls of the parallelepiped. In the last expression we have also replaced $\sin^2\tfrac{1}{2}k_yL$, $\cos^2\tfrac{1}{2}k_yL$, etc. by their average value, $1/2$. We now define the potential $V(d)$ describing the interaction of the atom with a conducting *wall* as the difference between (3.93) for d finite and for $d \to \infty$. In the latter limit we replace $\sin^2 k_zd$, $\cos^2 k_zd$ by $1/2$. Thus

$$V(d) \;=\; -\left(\frac{2\pi\hbar\alpha}{V}\right)\sum_{\mathbf{k}\lambda}\omega_k[e_{\mathbf{k}\lambda x}^2+e_{\mathbf{k}\lambda y}^2-e_{\mathbf{k}\lambda z}^2][\sin^2 k_zd - \tfrac{1}{2}]$$

$$= \left(\frac{\pi\hbar\alpha}{V}\right) \sum_{\mathbf{k}} \omega_k \cos 2k_z d \sum_{\lambda} [e^2_{\mathbf{k}\lambda x} + e^2_{\mathbf{k}\lambda y} - e^2_{\mathbf{k}\lambda z}]$$

$$= \left(\frac{\pi\hbar\alpha}{V}\right) \sum_{\mathbf{k}} \omega_k \cos 2k_z d (2k_z^2/k^2)$$

$$= \left(\frac{2\pi\hbar\alpha}{V}\right) \frac{V}{8\pi^3} \int d^3k \omega \frac{k_z^2}{k^2} \cos 2k_z d$$

$$= \left(\frac{\alpha\hbar c}{2\pi}\right) \int_0^\infty dk k^3 \int_0^\pi d\theta \sin\theta \cos^2\theta \cos(2kd\cos\theta)$$

$$= \left(\frac{\alpha\hbar c}{\pi}\right) \int_0^\infty dk k^3 \left(\frac{\sin 2kd}{2kd} + \frac{2\cos 2kd}{4k^2 d^2} - \frac{2\sin 2kd}{8k^3 d^3}\right).$$

$$(3.94)$$

Evaluating the integral using the procedure exemplified by (3.90), we obtain the Casimir–Polder result

$$V(r) = -\frac{3\alpha\hbar c}{8\pi d^4}. \qquad (3.95)$$

In Chapter 8 we discuss experimental evidence for the Casimir–Polder force.

3.13   The Magnetic Moment of the Electron

In order to explain the spectra of atoms in magnetic fields, Uhlenbeck and Goudsmit (1926) postulated that the electron has an intrinsic (spin) angular momentum $\hbar/2$ and a magnetic dipole moment $e\hbar/2mc \equiv \mu_o$, the Bohr magneton. Both properties of the electron were later found by Dirac (1928) to be consequences of relativistically invariant quantum mechanics.

Recall that a curent loop enclosing a plane area A has a magnetic dipole moment $\mu = IA/c$, where I is the current. For a charge e moving in a circular orbit of radius r, $\mu = (\pi r^2)(e\nu)/c - (e/2mc)L$, where ν and L are, respectively, the orbital frequency and angular momentum. Therefore the gyromagnetic ratio $\mu/L = e/2mc$. For the electron magnetic dipole moment and spin angular momentum, however, $\mu_s/L_s = \mu_o/(\hbar/2) = 2(e/2mc)$. That is, the Landé g-factor for electron spin is 2, as predicted by the Dirac theory without coupling of the electron to the radiation field.

As in the case of the Lamb shift, radiative corrections give small departures from this prediction. Just prior to the first accurate measurements by Kusch et al. (see Kusch and Foley, 1948), Schwinger (1948) calculated for the "anomaly" $(g-2)/2$ the value $\alpha/2\pi \cong .00116$; the experimentalists reported a value $.00119 \pm .00005$.

And like the Lamb shift, the anomalous moment of the electron provides one of the most sensitive tests of QED. Recent experiments by Dehmelt et al.[16] give a value of $(g-2)/2$ more accurate than all previous measurements by a factor of nearly 1000:

$$\frac{g-2}{2} = .001159652188(4) \ . \tag{3.96}$$

A QED calculation up to fourth order in the fine–structure constant α yields (Kinoshita, 1989)

$$\frac{g-2}{2} = .001159652192(74) \ . \tag{3.97}$$

Here the theoretical "error" is due mainly to the uncertainty in the fine-structure constant. Such a comparison of theory and experiment explains the cliché that QED is "the best theory we have!"

As in our discussion of the Lamb shift, we will focus our attention here on the nonrelativistic theory of the anomalous moment of the electron. state

From the interaction Hamiltonian $-(e\hbar/2mc)\boldsymbol{\sigma} \cdot \mathbf{B}$ describing the coupling of electron spin to a magnetic field $\mathbf{B}$, and the commutation relations for the Pauli spin—$1/2$ operators $\sigma_x, \sigma_y, \sigma_z$, we obtain the Heisenberg equation of motion

$$\frac{d\boldsymbol{\sigma}}{dt} = \frac{e}{mc}\mathbf{B} \times \boldsymbol{\sigma} \tag{3.98}$$

for $\boldsymbol{\sigma} = (\sigma_x, \sigma_y, \sigma_z)$. If $\mathbf{B} = 0, \boldsymbol{\sigma}$ is constant in time. However, in QED $\mathbf{B}$ is an operator which, like $\mathbf{E}$, has zero expectation value but nonvanishing variance in the vacuum state of the field. Thus, for an electron in free space,

$$
\begin{aligned}
\frac{d\boldsymbol{\sigma}}{dt} &= \frac{ie}{mc}\sum_{\mathbf{k}\lambda}\left(\frac{2\pi\hbar c^2}{\omega_k V}\right)^{1/2}[a_{\mathbf{k}\lambda}(0)e^{-i\omega_k t}e^{i\mathbf{k}\cdot\mathbf{r}} - a_{\mathbf{k}\lambda}^{\dagger}(0)e^{i\omega_k t}e^{-i\mathbf{k}\cdot\mathbf{r}}] \\
&\quad \times (\mathbf{k} \times \mathbf{e}_{\mathbf{k}\lambda}) \times \boldsymbol{\sigma}(t) \ ,
\end{aligned} \tag{3.99}
$$

where we have used equation (2.57). In writing this equation we are ignoring the part of the magnetic field that depends on $\boldsymbol{\sigma}$. That is, we are including the effect of the vacuum $\mathbf{B}$ field, but not the radiation reaction $\mathbf{B}$ field.

In the lowest order of approximation we use $\boldsymbol{\sigma}(t) \cong \boldsymbol{\sigma}(0)$, the zero-coupling solution, on the right side of (3.99). Then

$$\boldsymbol{\sigma}(t) \cong \boldsymbol{\sigma}(0) - \frac{e}{mc}\sum_{\mathbf{k}\lambda}\left(\frac{2\pi\hbar c^2}{\omega_k V}\right)^{1/2}\frac{1}{\omega_k}[a_{\mathbf{k}\lambda}(0)e^{-i\omega_k t}e^{i\mathbf{k}\cdot\mathbf{r}}$$

---

[16] See Dehmelt (1990) and references therein, and Chapter 6, where the experiments are briefly described.

$$+ a_{\mathbf{k}\lambda}^{\dagger}(0)e^{i\omega_k t}{}_0{}^{-i\mathbf{k}\cdot\mathbf{r}}](\mathbf{k} \times \mathbf{c}_{\mathbf{k}\lambda}) \times \boldsymbol{\sigma}(0) , \tag{3.100}$$

where in the second term we ignore a contribution from the artificial switch-on at $t = 0$ of the interaction. Using the vacuum expectation values $\langle a_{\mathbf{k}\lambda}(0)a_{\mathbf{k}'\lambda'}(0)\rangle = \langle a_{\mathbf{k}\lambda}^{\dagger}(0)a_{\mathbf{k}'\lambda'}^{\dagger}(0)\rangle = 0$ and $\langle a_{\mathbf{k}\lambda}a_{\mathbf{k}'\lambda'}^{\dagger}(0)\rangle = \delta_{\mathbf{k},\mathbf{k}'}^{3}\delta_{\lambda\lambda'}$, we obtain

$$\langle\Delta\sigma^2\rangle \cong \langle\left(\frac{e}{mc}\right)^2 \sum_{\mathbf{k}} \left(\frac{2\pi\hbar c^2}{\omega_k V}\right) \frac{1}{\omega_k^2} \sum_{\lambda}[(\mathbf{k} \times \mathbf{e}_{\mathbf{k}\lambda}) \times \boldsymbol{\sigma}(0)]^2\rangle , \tag{3.101}$$

with $\Delta\boldsymbol{\sigma} \equiv \boldsymbol{\sigma}(t) - \boldsymbol{\sigma}(0)$. Now

$$\begin{aligned}
\sum_{\lambda}[(\mathbf{k} \times \mathbf{e}_{\mathbf{k}\lambda}) \times \boldsymbol{\sigma}(0)]^2 &= \sum_{\lambda}[\mathbf{k}(\boldsymbol{\sigma}(0) \cdot \mathbf{e}_{\mathbf{k}\lambda}) - \mathbf{e}_{\mathbf{k}\lambda}(\mathbf{k} \cdot \boldsymbol{\sigma}(0))]^2 \\
&= k^2\sum_{\lambda}(\boldsymbol{\sigma}(0) \cdot \mathbf{e}_{\mathbf{k}\lambda})^2 + (\mathbf{k} \cdot \boldsymbol{\sigma}(0))^2 \sum_{\lambda}(1) \\
&= k^2\boldsymbol{\sigma}^2(0) - (\mathbf{k} \cdot \boldsymbol{\sigma}(0))^2 + 2(\mathbf{k} \cdot \boldsymbol{\sigma}(0))^2 \\
&= k^2\boldsymbol{\sigma}^2(0) + (\mathbf{k} \cdot \boldsymbol{\sigma}(0))^2 , \tag{3.102}
\end{aligned}$$

and the integration over all solid angles about $\mathbf{k}$ of this expression is

$$\int d\Omega_{\mathbf{k}}[k^2 + (\mathbf{k}\cdot\boldsymbol{\sigma}(0))^2] = 4\pi k^2\boldsymbol{\sigma}^2(0) + \frac{4\pi}{3}k^2\boldsymbol{\sigma}^2(0) = \frac{16\pi}{3}k^2\boldsymbol{\sigma}^2(0) . \tag{3.103}$$

The replacement $\sum_{\mathbf{k}} \to (V/8\pi^3) \int dk k^2 \int d\Omega_{\mathbf{k}}$ in (3.101) then yields

$$\frac{\langle\Delta\sigma^2\rangle}{\langle\sigma^2(0)\rangle} \cong \frac{4\alpha}{3\pi} \left(\frac{\hbar}{mc}\right)^2 \int dk k \to \frac{2\alpha}{3\pi} \left(\frac{\hbar K}{mc}\right)^2 \tag{3.104}$$

when, in this nonrelativistic approach, we introduce an upper limit K in the integration over $k = \omega/c$ in order to avoid a divergence.

In the absence of any coupling to the vacuum magnetic field, the electron spin has a fixed direction. Following Welton (1948), we define the mean-square fluctuation angle

$$\langle\Delta\theta^2\rangle \equiv \frac{\langle\Delta\sigma^2\rangle}{\langle\sigma^2(0)\rangle} = \frac{2\alpha}{3\pi} \left(\frac{\hbar K}{mc}\right)^2 \tag{3.105}$$

and consider the expectation value

$$-\frac{e\hbar}{2mc}\langle\boldsymbol{\sigma}\rangle \cdot \mathbf{B}_{ext} = -\frac{e\hbar}{2mc}|\langle\boldsymbol{\sigma}\rangle|B_{ext}\cos\theta \tag{3.106}$$

of the potential energy of the spin in an *external* magnetic field, where θ is the angle between $\boldsymbol{\sigma}$ and the direction of $\mathbf{B}_{\text{ext}}$. The effect of the fluctuating vacuum $\mathbf{B}$ field is to replace $\cos\theta$ by

$$\langle\cos(\theta+\Delta\theta)\rangle \cong \cos\theta\left[1-\frac{1}{2}\langle\Delta\theta^2\rangle\right] = \cos\theta\left[1-\frac{\alpha}{3\pi}\left(\frac{\hbar K}{mc}\right)^2\right], \quad (3.107)$$

so that

$$-\frac{e\hbar}{2mc}\langle\boldsymbol{\sigma}\rangle\cdot\mathbf{B}_{\text{ext}} \rightarrow -\frac{e\hbar}{2mc}|\langle\boldsymbol{\sigma}\rangle|B_{\text{ext}}\left[1-\frac{\alpha}{3\pi}\left(\frac{\hbar K}{mc}\right)^2\right], \quad (3.108)$$

or, in effect,

$$\frac{e\hbar}{2mc} \rightarrow \frac{e\hbar}{2mc}\left[1-\frac{\alpha}{3\pi}\left(\frac{\hbar K}{mc}\right)^2\right]. \quad (3.109)$$

This implies

$$\frac{g-2}{2} = -\frac{\alpha}{3\pi}\left(\frac{\hbar K}{mc}\right)^2 \quad (3.110)$$

to first order in α.

The problem with this result is that it has the wrong sign: experiment shows that $(g-2)/2$ is *positive*. We can rectify this situation starting from the observation that radiation reaction has been ignored. Although it turns out that radiation reaction does not affect the potential energy $-(e\hbar/2mc)\boldsymbol{\sigma}\cdot\mathbf{B}$ to first order in α, it does contribute to the electron mass at this order [Equation (3.33)]. Since we have left out radiation reaction in the calculation leading to (3.109), the mass in that expression must actually be the *bare* mass m_o. What is measured experimentally, of course, involves the observed mass $m = m_o + \delta m$. Therefore we should express (3.109) in terms of the observed mass. This is accomplished by the replacement

$$\frac{e\hbar}{2m_oc}\left[1-\frac{\alpha}{3\pi}\left(\frac{\hbar K}{mc}\right)^2\right] \rightarrow \frac{e\hbar}{2mc}\left(\frac{m_o+\delta m}{m_o}\right)\left[1-\frac{\alpha}{3\pi}\left(\frac{\hbar K}{mc}\right)^2\right]$$

$$\cong \frac{e\hbar}{2mc}\left[1-\frac{\alpha}{3\pi}\left(\frac{\hbar K}{mc}\right)^2\right]\left[1+\frac{\delta m}{m}\right]$$

$$= \frac{e\hbar}{2mc}\left[1-\frac{\alpha}{3\pi}\left(\frac{\hbar K}{mc}\right)^2\right]\left[1+\frac{4\alpha}{3\pi mc^2}(\hbar Kc)\right]$$

$$\cong \frac{e\hbar}{2mc}\left[1+\frac{4\alpha}{3\pi}\left(\frac{\hbar K}{mc}\right)-\frac{\alpha}{3\pi}\left(\frac{\hbar K}{mc}\right)^2\right],$$

$$(3.111)$$

$$+ a^\dagger_{\mathbf{k}\lambda}(0)e^{i\omega_k t}e^{-i\mathbf{k}\cdot\mathbf{r}}](\mathbf{k} \times \mathbf{e}_{\mathbf{k}\lambda}) \times \boldsymbol{\sigma}(0) , \tag{3.100}$$

where in the second term we ignore a contribution from the artificial switch-on at $t = 0$ of the interaction. Using the vacuum expectation values $\langle u_{\mathbf{k}\lambda}(0)u_{\mathbf{k}'\lambda'}(0)\rangle - \langle u^\dagger_{\mathbf{k}\lambda}(0)u_{\mathbf{k}'\lambda'}(0)\rangle = 0$ and $\langle a_{\mathbf{k}\lambda}a^\dagger_{\mathbf{k}'\lambda'}(0)\rangle = \delta^0_{\mathbf{k},\mathbf{k}'}\delta_{\lambda\lambda'}$, we obtain

$$\langle\Delta\boldsymbol{\sigma}^2\rangle \cong \langle\left(\frac{e}{mc}\right)^2 \sum_{\mathbf{k}} \left(\frac{2\pi\hbar c^2}{\omega_k V}\right) \frac{1}{\omega_k^2} \sum_\lambda [(\mathbf{k} \times \mathbf{e}_{\mathbf{k}\lambda}) \times \boldsymbol{\sigma}(0)]^2\rangle , \tag{3.101}$$

with $\Delta\boldsymbol{\sigma} \equiv \boldsymbol{\sigma}(t) - \boldsymbol{\sigma}(0)$. Now

$$
\begin{aligned}
\sum_\lambda [(\mathbf{k} \times \mathbf{e}_{\mathbf{k}\lambda}) \times \boldsymbol{\sigma}(0)]^2 &= \sum_\lambda [\mathbf{k}(\boldsymbol{\sigma}(0) \cdot \mathbf{e}_{\mathbf{k}\lambda}) - \mathbf{e}_{\mathbf{k}\lambda}(\mathbf{k} \cdot \boldsymbol{\sigma}(0))]^2 \\
&= k^2 \sum_\lambda (\boldsymbol{\sigma}(0) \cdot \mathbf{e}_{\mathbf{k}\lambda})^2 + (\mathbf{k} \cdot \boldsymbol{\sigma}(0))^2 \sum_\lambda (1) \\
&= k^2\boldsymbol{\sigma}^2(0) - (\mathbf{k} \cdot \boldsymbol{\sigma}(0))^2 + 2(\mathbf{k} \cdot \boldsymbol{\sigma}(0))^2 \\
&= k^2\boldsymbol{\sigma}^2(0) + (\mathbf{k} \cdot \boldsymbol{\sigma}(0))^2 , \tag{3.102}
\end{aligned}
$$

and the integration over all solid angles about $\mathbf{k}$ of this expression is

$$\int d\Omega_{\mathbf{k}}[k^2 + (\mathbf{k}\cdot\boldsymbol{\sigma}(0))^2] = 4\pi k^2\boldsymbol{\sigma}^2(0) + \frac{4\pi}{3}k^2\boldsymbol{\sigma}^2(0) = \frac{16\pi}{3}k^2\boldsymbol{\sigma}^2(0) . \tag{3.103}$$

The replacement $\sum_{\mathbf{k}} \to (V/8\pi^3)\int dk\,k^2 \int d\Omega_{\mathbf{k}}$ in (3.101) then yields

$$\frac{\langle\Delta\boldsymbol{\sigma}^2\rangle}{\langle\boldsymbol{\sigma}^2(0)\rangle} \cong \frac{4\alpha}{3\pi}\left(\frac{\hbar}{mc}\right)^2 \int dk\,k \to \frac{2\alpha}{3\pi}\left(\frac{\hbar K}{mc}\right)^2 \tag{3.104}$$

when, in this nonrelativistic approach, we introduce an upper limit K in the integration over $k = \omega/c$ in order to avoid a divergence.

In the absence of any coupling to the vacuum magnetic field, the electron spin has a fixed direction. Following Welton (1948), we define the mean-square fluctuation angle

$$\langle\Delta\theta^2\rangle \equiv \frac{\langle\Delta\boldsymbol{\sigma}^2\rangle}{\langle\boldsymbol{\sigma}^2(0)\rangle} = \frac{2\alpha}{3\pi}\left(\frac{\hbar K}{mc}\right)^2 \tag{3.105}$$

and consider the expectation value

$$-\frac{e\hbar}{2mc}\langle\boldsymbol{\sigma}\rangle \cdot \mathbf{B}_{\text{ext}} = -\frac{e\hbar}{2mc}|\langle\boldsymbol{\sigma}\rangle|B_{\text{ext}}\cos\theta \tag{3.106}$$

of the potential energy of the spin in an *external* magnetic field, where θ is the angle between σ and the direction of $\mathbf{B}_{\text{ext}}$. The effect of the fluctuating vacuum $\mathbf{B}$ field is to replace $\cos\theta$ by

$$\langle\cos(\theta + \Delta\theta)\rangle \cong \cos\theta \left[1 - \frac{1}{2}\langle\Delta\theta^2\rangle\right] = \cos\theta \left[1 - \frac{\alpha}{3\pi}\left(\frac{\hbar K}{mc}\right)^2\right], \quad (3.107)$$

so that

$$-\frac{e\hbar}{2mc}\langle\sigma\rangle \cdot \mathbf{B}_{\text{ext}} \rightarrow -\frac{e\hbar}{2mc}|\langle\sigma\rangle|B_{\text{ext}}\left[1 - \frac{\alpha}{3\pi}\left(\frac{\hbar K}{mc}\right)^2\right], \quad (3.108)$$

or, in effect,

$$\frac{e\hbar}{2mc} \rightarrow \frac{e\hbar}{2mc}\left[1 - \frac{\alpha}{3\pi}\left(\frac{\hbar K}{mc}\right)^2\right]. \quad (3.109)$$

This implies

$$\frac{g - 2}{2} = -\frac{\alpha}{3\pi}\left(\frac{\hbar K}{mc}\right)^2 \quad (3.110)$$

to first order in α.

The problem with this result is that it has the wrong sign: experiment shows that $(g - 2)/2$ is *positive*. We can rectify this situation starting from the observation that radiation reaction has been ignored. Although it turns out that radiation reaction does not affect the potential energy $-(e\hbar/2mc)\sigma \cdot \mathbf{B}$ to first order in α, it does contribute to the electron mass at this order [Equation (3.33)]. Since we have left out radiation reaction in the calculation leading to (3.109), the mass in that expression must actually be the *bare* mass m_o. What is measured experimentally, of course, involves the observed mass $m = m_o + \delta m$. Therefore we should express (3.109) in terms of the observed mass. This is accomplished by the replacement

$$\frac{e\hbar}{2m_oc}\left[1 - \frac{\alpha}{3\pi}\left(\frac{\hbar K}{mc}\right)^2\right] \rightarrow \frac{e\hbar}{2mc}\left(\frac{m_o + \delta m}{m_o}\right)\left[1 - \frac{\alpha}{3\pi}\left(\frac{\hbar K}{mc}\right)^2\right]$$

$$\cong \frac{e\hbar}{2mc}\left[1 - \frac{\alpha}{3\pi}\left(\frac{\hbar K}{mc}\right)^2\right]\left[1 + \frac{\delta m}{m}\right]$$

$$= \frac{e\hbar}{2mc}\left[1 - \frac{\alpha}{3\pi}\left(\frac{\hbar K}{mc}\right)^2\right]\left[1 + \frac{4\alpha}{3\pi mc^2}(\hbar Kc)\right]$$

$$\cong \frac{e\hbar}{2mc}\left[1 + \frac{4\alpha}{3\pi}\left(\frac{\hbar K}{mc}\right) - \frac{\alpha}{3\pi}\left(\frac{\hbar K}{mc}\right)^2\right],$$

$$(3.111)$$

where we have used (3.33) with a cutoff $E = \hbar K c$ in the upper limit of integration. This implies

$$\frac{g-2}{2} \cong \frac{4\alpha}{3\pi} \left(\frac{\hbar K}{mc} \right) - \frac{\alpha}{3\pi} \left(\frac{\hbar K}{mc} \right)^2 \qquad (3.112)$$

which is positive for any cutoff $K < 4mc/\hbar$. Thus the predicted anomaly in the nonrelativistic theory is positive for all cutoffs for which the nonrelativistic theory is sensible. The choice $K = 0.42mc/\hbar$ yields $(g-2)/2 = \alpha/2\pi$, which is the relativistic QED result to first order in α (Grotch and Kazes, 1977).

We conclude therefore that, as in spontaneous emission, both vacuum field fluctuations and radiation reaction are important for the anomalous magnetic moment of the electron (Grotch and Kazes, 1977; Dupont-Roc, Fabre, and Cohen–Tannoudji, 1978).

However, the reader is warned not to take these calculations too seriously, for the result $(g-2)/2 = \alpha/2\pi$ could be obtained by retaining only the first (radiation reaction) term in (3.112) and choosing $K = 3mc/8\hbar$. It should also be noted that the solution $K \cong 0.42mc/\hbar$ of (3.112) with $(g-2)/2 = \alpha/2\pi$ is not unique.

3.14 Summary

We have shown in this chapter how some basic QED effects may be understood physically as consequences of the fluctuating vacuum electromagnetic field. These effects include such commonplace phenomena as spontaneous emission and van der Waals forces and also the Lamb shift and the anomalous moment of the electron, which provide the most important tests of QED. Consideration of these vacuum effects leads us to the concept of renormalization as a means of obtaining finite results from otherwise infinite quantitites. Vacuum fluctuations and renormalization are two of the most important features of modern physics.

It is hoped that this chapter has convinced (or reminded) the reader that the vacuum — or the electromagnetic vacuum, at least — is a quantum state with observable physical consequences.

These physical explanations of various QED vacuum effects have considerable esthetic appeal and seem to offer compelling evidence for the "reality" of vacuum field fluctuations. And yet the vacuum field fluctuations are not the only physical basis for understanding these phenomena. There is another basis — source fields — upon which we can construct physical interpretations of QED vacuum effects. This point is pursued further in the following chapters.

3.15   Bibliography

Au, C.-K. and G. Feinberg, "Effects of Retardation on Electromagnetic Self–Energy of Atomic States," *Phys. Rev.* A9, 1794 (1974).

Bethe, H. A., "The Electromagnetic Shift of Energy Levels," *Phys. Rev.* 72, 241 (1947).

Bethe, H. A., in *From a Life of Physics*, ed. A. Salam et al. (World Scientific, Singapore, 1989).

Boyer, T. H., "Quantum Zero-Point Energy and Long-Range Forces," *Ann. Phys.* (New York), 56, 474 (1970).

Boyer, T. H., "Asymptotic Retarded van der Waals Forces Derived from Classical Electrodynamics with Classical Electromagnetic Zero-Point Radiation," *Phys. Rev.* A5, 1799 (1972).

Boyer, T. H., "A Brief Survey of Stochastic Electrodynamics," in *Foundations of Radiation Theory and Quantum Electrodynamics*, ed. A.O. Barut (Plenum Press, New York, 1980).

Casimir, H. B. G. and D. Polder, "The Influence of Retardation on the London–van der Waals Forces," *Phys. Rev.* 73, 360 (1948).

Casimir, H. B. G., "Sur les Forces van der Waals–London," *J. Chim. Phys.* 46, 407 (1949).

Dehmelt, H., "Experiments with an Isolated Subatomic Particle at Rest," *Rev. Mod. Phys.* 62, 525 (1990).

Dirac, P. A. M., "The Quantum Theory of the Emission and Absorption of Radiation," *Proc. Roy. Soc. Lond.* A114, 243 (1927).

Dirac, P. A. M., "The Quantum Theory of the Electron," *Proc. Roy. Soc. Lond.* A117, 610 (1928).

Dirac, P. A. M., in *Directions in Physics*, ed. H. Hora and J. R. Shepanski (Wiley, New York, 1978), p. 36.

Dirac, P. A. M., in *From a Life of Physics*, ed. A. Salam et al. (World Scientific, Singapore, 1989).

Drake, G. W. F., "Quantum Electrodynamic Effects in Few-Electron Atomic Systems," *Adv. At. Mol. Phys.* 18, 399 (1982).

Dresden, M., *H. A. Kramers: Between Tradition and Revolution* (Springer-Verlag, New York, 1987).

Dupont-Roc, J., C. Fabre, and C. Cohen-Tannoudji, "Physical Interpretaions for Radiative Corrections," *J. Phys.* B11, 563 (1978).

Feynman, R. P., "The Present Status of Quantum Electrodynamics," in *The Quantum Theory of Fields*, ed. R. Stoops (Wiley Interscience, New York, 1961).

Feynman, R. P., "The Development of the Space-Time View of Quantum Electrodynamics," Nobel lecture reprinted in *Physics Today* (August

1966), 31.

Ginzburg, V. L., "The Nature of Spontaneous Radiation," *Sov. Phys. Usp.* **26**, 713 (1983)].

Grotch, H., "Lamb Shift in Nonrelativistic Quantum Electrodynamics," *Am. J. Phys.* **49**, 48 (1981).

Grotch, H. and E. Kazes, "Nonrelativistic Quantum Mechanics and the Anomalous Part of the Electron g-Factor," *Am. J. Phys.* **45**, 618 (1977).

Kinoshita, T., "Electron $g-2$ and High Precision Determination of α," in *The Hydrogen Atom*, ed. G. F. Bassani, M. Inguscio, and T. W. Hänsch (Springer-Verlag, Berlin, 1989).

Kusch, P. and H. M. Foley, "The Magnetic Moment of the Electron," *Phys. Rev.* **74**, 250 (1948).

Lamb, W. E., Jr. and R. C. Retherford, "Fine Structure of the Hydrogen Atom by a Microwave Method," *Phys. Rev.* **72**, 241 (1947).

Lamb, W. E., Jr. and R. C. Retherford, "Fine Structure of the Hydrogen Atom. Part I," *Phys. Rev.* **79**, 549 (1950); Part II, *Phys. Rev.* **81**, 222 (1951); Part III, *Phys. Rev.* **85**, 259 (1952); Part IV, *Phys. Rev.* **86**, 1014 (1952).

Landau, L. D., "Das Dämfungsproblem in der Wellenmechanik," *Z. Phys.* **45**, 430 (1927).

Lieber, M., "O(4) Symmetry of the Hydrogen Atom and the Lamb Shift," *Phys. Rev.* **174**, 2037 (1968).

London, F., "Zur Theorie und Systematik der Molekularkräfte," *Z. Phys.* **63**, 245 (1930).

London, F., "The General Theory of Molecular Forces," *Trans. Faraday Soc.* **33**, 8 (1937).

Lundeen, S. R. and F. M. Pipkin, "Measurement of the Lamb Shift in Hydrogen, n = 2," *Phys. Rev. Lett.* **46**, 232 (1981).

Milonni, P. W., "Why Spontaneous Emission?," *Am. J. Phys.* **52**, 340 (1984).

Milonni, P. W., R. J. Cook, and M. E. Goggin, "Radiation Pressure from the Vacuum: Physical Interpretation of the Casimir Force," *Phys. Rev.* **A38**, 1621 (1988).

Milonni, P. W. and M.-L. Shih, "Casimir Forces," *Contemp. Phys.* **33**, 313 (1993).

Mohr, P. J., "Self-Energy of the n = 2 States in a Strong Coulomb Field," *Phys. Rev.* **A26**, 2338 (1982).

Power, E. A., "Zero-Point Energy and the Lamb Shift," *Am. J. Phys.* **34**, 516 (1966).

Power, E. A., "The Application of Quantum Electrodynamics to Molecular

Forces," *Phys. Bull.* **19**, 369 (1968).

Puthoff, H. E., "Ground State of Hydrogen as a Zero-Point-Fluctuation-Determined State," *Phys. Rev.* **A35**, 3266 (1987).

Renne, M. J., "Retarded van der Waals Interaction in a System of Harmonic Oscillators," *Physica* **D53**, 193 (1971).

Schiff, L. I., *Quantum Mechanics*, 3rd ed. (McGraw–Hill, New York, 1968).

Schwinger, J., "On Gauge Invariance and Vacuum Polarization," *Phys. Rev.* **82**, 664 (1951).

Schwinger, J., "Coulomb Green's Function," *J. Math. Phys.* **5**, 1606 (1964).

Uhlenbeck, G. E. and S. Goudsmit, "Spinning Electrons and the Structure of Spectra," *Nature* **117**, 264 (1926).

Van Vleck, J. H., "The Absorption of Radiation by Multiply Periodic Orbits, and Its Relation to the Correspondence Principle and the Rayleigh–Jeans Law," *Phys. Rev.* **24**, 330 (1924).

Verwey, E. J. W. and J. T. G. Overbeek, *Theory of the Stability of Lyophobic Colloids* (Elsevier, Amsterdam, 1948).

Weinberg, S., "The Search for Unity: Notes for a History of Quantum Field Theory," *Daedalus* **106**, 17 (1977).

Weisskopf, V. F., "Probleme der neueren Quantentheorie des Elektrons," *Naturwissenschaften* **23**, 631 (1935).

Weisskopf. V. F., "Über die Elektrodynamik des Vakuums auf Grund der Quantentheorie des Elektrons," *Kon. Dan. Vid. Sel. Mat.-Fys. Medd.* **14**, 3 (1936).

Weisskopf, V. F., "The Development of Field Theory in the Last 50 Years," *Physics Today* (November 1981), 69.

Welton, T. A., "Some Observable Effects of the Quantum-Mechanical Fluctuations of the Electromagnetic Field," *Phys. Rev.* **74**, 1157 (1948).

Chapter 4

Nonrelativistic Theory of Atoms in a Vacuum

> When you follow two separate chains of thought, Watson,
> you will find some point of intersection which should approxi-
> mate the truth.
> – Sherlock Holmes, "The Disappearance of Lady Frances Carfax"
> Arthur Conan Doyle

4.1 Introduction

In the preceding chapter we discussed various effects of the vacuum electromagnetic field on atoms, but thus far we have not formulated the theory
of the atom–field interaction in any systematic way. One purpose of this
chapter is to formulate the nonrelativistic QED theory of atom–field interactions, beginning with the Coulomb-gauge Hamiltonian for the system
consisting of an electron and the electromagnetic field. In so doing we shall
see how various QED vacuum effects can be thought of alternatively in
terms of either vacuum field fluctuations or radiation reaction. The different interpretations have to do with the way certain *commuting* operators
are ordered.

4.2 The Hamiltonian

According to classical mechanics a system of particles is characterized by
a Lagrangian function $L(q, \dot{q}, t)$ of the coordinates, velocities, and time.

This function governs the time evolution of the system through the principle of least action: the evolution from t_1 to t_2 is such that the action $\int_{t_1}^{t_2} L(q, \dot{q}, t)dt$ is minimized (or, more precisely, is an extremum). The fact that L does not depend on $\ddot{q}$ or higher derivatives of the coordinates reflects the empirical law that the time evolution of the system is uniquely determined by the coordinates and velocities at a given time. The principle of least action implies

$$\frac{d}{dt}\left(\frac{\partial L}{\partial \dot{q}_k}\right) - \frac{\partial L}{\partial q_k} = 0 \qquad (4.1)$$

for each coordinate q_k.

For a point particle of mass m and charge e in an electromagnetic field with scalar and vector potentials ϕ and $\mathbf{A}$, a suitable Lagrangian is

$$L = \frac{1}{2}m\mathbf{v}^2 - e\phi + \frac{e}{c}\mathbf{A} \cdot \mathbf{v} . \qquad (4.2)$$

That is, we obtain from (4.1) and this Lagrangian the correct (nonrelativistic) equation of motion

$$m\ddot{\mathbf{x}} = e\mathbf{E} + \frac{e}{c}\mathbf{v} \times \mathbf{B}, \qquad (4.3)$$

where $\mathbf{E} = -\nabla\phi - c^{-1}\partial\mathbf{A}/\partial t, \mathbf{B} = \nabla \times \mathbf{A}, \mathbf{v} = \dot{\mathbf{x}}$.

From (4.1),

$$\begin{aligned}
\frac{dL}{dt} &= \sum_k \left[\frac{\partial L}{\partial q_k}\frac{dq_k}{dt} + \frac{\partial L}{\partial \dot{q}_k}\frac{d\dot{q}_k}{dt}\right] = \sum_k \left[\frac{d}{dt}\left(\frac{\partial L}{\partial \dot{q}_k}\right)\frac{dq_k}{dt} + \frac{\partial L}{\partial \dot{q}_k}\frac{d}{dt}\left(\frac{dq_k}{dt}\right)\right] \\
&= \frac{d}{dt}\sum_k \dot{q}_k \frac{\partial L}{\partial \dot{q}_k} \qquad (4.4)
\end{aligned}$$

or $dH/dt = 0$, where the Hamiltonian

$$H \equiv \sum_k \dot{q}_k \frac{\partial L}{\partial \dot{q}_k} - L = \sum_k p_k \dot{q}_k - L \qquad (4.5)$$

and $p_k \equiv \partial L/\partial \dot{q}_k$ is the momentum conjugate to q_k. For the example of a particle in a field we obtain from (4.2) the canonical momentum $\mathbf{p} = m\mathbf{v} + e\mathbf{A}/c$ and the Hamiltonian

$$\begin{aligned}
H &= \mathbf{p} \cdot \mathbf{v} - \frac{1}{2}m\mathbf{v}^2 + e\phi - \frac{e}{c}\mathbf{A} \cdot \mathbf{v} = \frac{1}{2}m\mathbf{v}^2 + e\phi \\
&= \frac{1}{2m}(\mathbf{p} - \frac{e}{c}\mathbf{A})^2 + e\phi. \qquad (4.6)
\end{aligned}$$

The Hamiltonian is to be expressed as a function of the coordinates q_k and the momenta p_k, as in the second line of (4.6). As such, it governs the time evolution of the coordinates and momenta via the Hamilton equations of motion $\dot{q}_k = \partial H / \partial p_k, \dot{p}_k = -\partial H / \partial q_k$. Thus, for the Hamiltonian (4.6),

$$\dot{\mathbf{x}} = \frac{1}{m}(\mathbf{p} - \frac{e}{c}\mathbf{A}), \qquad (4.7)$$

$$\dot{\mathbf{p}} = \frac{e}{c}[\mathbf{v} \times \mathbf{B} + (\mathbf{v} \cdot \nabla)\mathbf{A}] - e\nabla\phi, \qquad (4.8)$$

or

$$
\begin{aligned}
m\ddot{\mathbf{x}} &= \dot{\mathbf{p}} - \frac{e}{c}\left[\frac{\partial \mathbf{A}}{\partial t} + (\mathbf{v} \cdot \nabla)\mathbf{A}\right] = -\frac{e}{c}\frac{\partial \mathbf{A}}{\partial t} - e\nabla\phi + \frac{e}{c}\mathbf{v} \times \mathbf{B} \\
&= e\mathbf{E} + \frac{e}{c}\mathbf{v} \times \mathbf{B}. \qquad (4.9)
\end{aligned}
$$

where we have used the fact that $d\mathbf{A}/dt = \partial\mathbf{A}/\partial t + (\mathbf{v} \cdot \nabla)\mathbf{A}$ (recall Section 2.6).

The canonical momentum $\mathbf{p} = m\mathbf{v} + e\mathbf{A}/c$ for a charged particle in a field is the "kinetic" momentum $m\mathbf{v}$ plus the momentum $e\mathbf{A}/c$. To better appreciate this famous result, note that if $\phi = 0$ and a spatially uniform vector potential is suddenly switched on, then according to (4.9) $m\mathbf{v}$ changes to $m\mathbf{v} - e\mathbf{A}/c$, while $\mathbf{p} = m\mathbf{v} + e\mathbf{A}/c$ is unchanged. In this example $\mathbf{x}$ is a cyclic coordinate (H is independent of $\mathbf{x}$) and so its conjugate momentum is conserved. This conserved momentum is $\mathbf{p}$, not $m\mathbf{v}$. The particle's kinetic energy remains $\frac{1}{2}mv^2$, of course, and this becomes $(\mathbf{p} - e\mathbf{A}/c)^2/2m$ when expressed in terms of the canonical momentum.

Knowing the Newton equation of motion (4.9), we usually have no practical need in classical electrodynamics for the Hamiltonian. The Hamiltonian takes on fundamental importance, however, when we go over into quantum theory and work with the Schrödinger equation, $i\hbar\partial\psi/\partial t = H\psi$, to calculate transition probabilities, energy levels, etc. When the field is quantized, and not simply a classically prescribed function of space and time, we must also include field variables in the specification of the state vector $|\psi\rangle$ for the combined particle–field system. In particular, we must include the contribution of the field to the total Hamiltonian.

The Hamiltonian for the classical electromagnetic field has already been employed in Chapter 2 [Equation (2.29)]. The Hamiltonian for the system of a charged particle plus the electromagnetic field is obtained simply by adding the field Hamiltonian to $\frac{1}{2}mv^2 = (\mathbf{p} - e\mathbf{A}/c)^2/2m$:

$$H = \frac{1}{2m}(\mathbf{p} - \frac{e}{c}\mathbf{A})^2 + \frac{1}{8\pi}\int d^3r(\mathbf{E}^2 + \mathbf{B}^2). \qquad (4.10)$$

Note that the term $e\phi$ appearing in (4.6) seems to be missing here. We now turn to this point.

Recall first the Helmholtz theorem (Appendix F) stating that any vector field $\mathbf{E}$ may be divided into transverse and longitudinal parts: $\mathbf{E} = \mathbf{E}^\perp + \mathbf{E}^\parallel$, where $\nabla \cdot \mathbf{E}^\perp = 0$ and $\nabla \times \mathbf{E}^\parallel = 0$. In the Coulomb gauge this division is obvious: $\mathbf{E}^\perp = -c^{-1}\partial \mathbf{A}/\partial t$, $\mathbf{E}^\parallel = -\nabla\phi$, since $\nabla \cdot \mathbf{A} = 0$. Then

$$\int d^3 r \mathbf{E}^2 = \int d^3 r (\mathbf{E}^{\perp\,2} + \mathbf{E}^{\parallel\,2}) = \int d^3 r \mathbf{E}^{\perp\,2} + \int d^3 r (\nabla\phi)^2 , \quad (4.11)$$

since $\int d^3 r \mathbf{E}^\perp \cdot \mathbf{E}^\parallel = 0$, and

$$\int d^3 r (\nabla\phi)^2 = \int d^3 r \nabla \cdot (\phi\nabla\phi) - \int d^3 r \phi\nabla^2\phi = 4\pi \int d^3 r \rho\phi, \quad (4.12)$$

since $\nabla^2\phi = -4\pi\rho$ in the Coulomb gauge. Thus, for a system of N point charges,

$$H = \sum_{i=1}^{N} \frac{1}{2m_i}(\mathbf{p}_i - \frac{e_i}{c}\mathbf{A}_i)^2 + \frac{1}{2}\int d^3 r \rho\phi + \frac{1}{8\pi}\int d^3 r (\mathbf{E}^{\perp\,2} + \mathbf{B}^2), \quad (4.13)$$

where $\mathbf{A}_i$ is the vector potential at the position $\mathbf{r}_i$ of the ith particle and

$$\rho = \sum_{i=1}^{N} e_i \delta^3(\mathbf{r} - \mathbf{r}_i) \quad (4.14)$$

is the charge density. Furthermore

$$\phi(\mathbf{r}, t) = \int \frac{d^3 r' \rho(\mathbf{r}', t)}{|\mathbf{r} - \mathbf{r}'|} \quad (4.15)$$

in the Coulomb gauge, and so

$$\begin{aligned}
\frac{1}{2}\int d^3 r \rho(\mathbf{r}, t)\phi(\mathbf{r}, t) &= \frac{1}{2}\int d^3 r \int d^3 r' \frac{\rho(\mathbf{r}, t)\rho(\mathbf{r}', t)}{|\mathbf{r} - \mathbf{r}'|} \\
&= \sum_{i>j} \frac{e_i e_j}{|\mathbf{r}_i - \mathbf{r}_j|} \quad (4.16)
\end{aligned}$$

if we drop the infinite Coulomb interaction of each particle with itself. Therefore

$$H = \sum_{i=1}^{N} \frac{1}{2m_i}(\mathbf{p}_i - \frac{e_i}{c}\mathbf{A}_i)^2 + \sum_{i>j} \frac{e_i e_j}{|\mathbf{r}_i - \mathbf{r}_j|} + \frac{1}{8\pi}\int d^3 r (\mathbf{E}^{\perp\,2} + \mathbf{B}^2). \quad (4.17)$$

Suppose, for instance, that all but one of the particles have such large masses that they are approximately fixed in position. Then the Hamiltonian describing the dynamics of the system may be approximated by

$$H - \frac{1}{2m}(\mathbf{p} - \frac{e}{c}\mathbf{A})^2 + e\phi + \frac{1}{8\pi} \int d^3r(\mathbf{E}^{\perp 2} + \mathbf{B}^2), \qquad (4.18)$$

where e and m are the charge and mass of the one particle that is free to move, and ϕ is the scalar potential at this particle due to the other $N - 1$ fixed charges. This Hamiltonian is just (4.6) plus the Hamiltonian associated with the *transverse* electromagnetic field. Unlike (4.6), this Hamiltonian determines the time evolution for both the particle *and* the field.

4.3   Dipole Approximation

Consider now a single *bound* electron with binding potential energy $V(\mathbf{x}) = e\phi(\mathbf{x})$. Suppose that the distances over which the bound electron can move in this potential are small compared with the wavelength of any field with which the electron undergoes a significant interaction. Then it is convenient to make the *electric dipole approximation* in which spatial variations of $\mathbf{A}$ are ignored in the interaction $-(e/mc)\mathbf{A} \cdot \mathbf{p} + (e^2/2mc^2)\mathbf{A}^2$ appearing in the Hamiltonian (4.18):

$$H \rightarrow \frac{\mathbf{p}^2}{2m} + V(\mathbf{x}) - \frac{e}{mc}\mathbf{A} \cdot \mathbf{p} + \frac{e^2}{2mc^2}\mathbf{A}^2 + \frac{1}{8\pi} \int d^3r(\mathbf{E}^{\perp 2} + \mathbf{B}^2), \quad (4.19)$$

where now $\mathbf{A}$ is the vector potential evaluated at a fixed position, e.g., at the center of the region over which the electron is free to move classically. For an electron in the hydrogen atom, the dipole approximation is accurate for wavelengths large compared with the Bohr radius. In this case we may take $\mathbf{A}$ to be the vector potential at the nucleus.

The classical equations of motion following from (4.19) are

$$\dot{\mathbf{x}} = \frac{1}{m}\mathbf{p} - \frac{e}{mc}\mathbf{A}, \qquad (4.20)$$

$$\dot{\mathbf{p}} = -\nabla V(\mathbf{x}), \qquad (4.21)$$

or

$$m\ddot{\mathbf{x}} = -\nabla V(\mathbf{x}) + e\mathbf{E}^\perp, \qquad (4.22)$$

where $\mathbf{E}$ is evaluated at $\mathbf{x} = 0$. This is the type of equation used to describe a bound electron in the preceding chapters.

In this approximation the electron and nucleus in an atom form a point electric dipole. Note that there is no $\mathbf{v} \times \mathbf{B}$ force in the electric dipole approximation.

Recall that the principle of least action allows us to add any time derivative $(d/dt)S(q,t)$ to the Lagrangian without affecting equations of motion. If in the electric dipole approximation we use $S = -(e/c)\mathbf{A} \cdot \mathbf{x}$, then the Lagrangian (4.2) is changed to

$$
\begin{aligned}
L' &= L - \frac{e}{c}\dot{\mathbf{A}} \cdot \mathbf{x} - \frac{e}{c}\mathbf{A} \cdot \mathbf{v} = \frac{1}{2}mv^2 - e\phi - \frac{e}{c}\dot{\mathbf{A}} \cdot \mathbf{x} \\
&= \frac{1}{2}mv^2 - e\phi + e\mathbf{x} \cdot \mathbf{E}^{\perp} ,
\end{aligned} \tag{4.23}
$$

and the Hamiltonian (4.19) is transformed to the equivalent Hamiltonian

$$
H' = \frac{\mathbf{p}^2}{2m} + V(\mathbf{x}) - e\mathbf{x} \cdot \mathbf{E}^{\perp} + \frac{1}{8\pi} \int d^3 r (\mathbf{E}^{\perp 2} + \mathbf{B}^2), \tag{4.24}
$$

where now $\mathbf{p} = mv$ is the momentum conjugate to $\mathbf{x}$. This Hamiltonian leads trivially to the same equation (4.22) as does H.

4.4   Quantization

The quantum-mechanical theory of the system consisting of an electron and the electromagnetic field begins with the replacement of the classical Hamiltonian with the quantum-mechanical Hamiltonian. That is, we replace the classical variables $\mathbf{p}, \mathbf{x}, \mathbf{A}, \mathbf{E}^{\perp}$, and $\mathbf{B}$ by the corresponding quantum-mechanical operators. A system of charged particles plus the field is described similarly by the quantized version of the Hamiltonian (4.17). Note that in the Coulomb gauge in which we are working *it is sufficient to quantize only the transverse electromagnetic field*. The longitudinal field energy is replaced by instantaneous Coulomb interactions among particles, the second term in (4.17).

We will work with the quantized version of the electric dipole Hamiltonian (4.19) for a single bound electron:

$$
\begin{aligned}
H &= H_{\text{Atom}} + H_{\text{Field}} - \frac{e}{mc} \sum_{\mathbf{k}\lambda} \left(\frac{2\pi\hbar c^2}{\omega_k V}\right)^{1/2} [a_{\mathbf{k}\lambda} + a^{\dagger}_{\mathbf{k}\lambda}]\mathbf{p} \cdot \mathbf{e}_{\mathbf{k}\lambda} \\
&\quad + \frac{e^2}{2mc^2} \sum_{\mathbf{k}\lambda} \sum_{\mathbf{k}'\lambda'} \left(\frac{2\pi\hbar c^2}{V}\right) \left(\frac{1}{\omega_k \omega'_k}\right)^{1/2} [a_{\mathbf{k}\lambda} + a^{\dagger}_{\mathbf{k}\lambda}] \\
&\quad \times [a_{\mathbf{k}'\lambda'} + a^{\dagger}_{\mathbf{k}'\lambda'}]\mathbf{e}_{\mathbf{k}\lambda} \cdot \mathbf{e}_{\mathbf{k}'\lambda'} ,
\end{aligned} \tag{4.25}
$$

where $H_{\text{Atom}} = \mathbf{p}^2/2m + V(\mathbf{x})$ is the Hamiltonian describing the unperturbed bound electron (the "atom"),

$$H_{\text{Field}} = \sum_{\mathbf{k}\lambda} \hbar\omega_k (a_{\mathbf{k}\lambda}^\dagger a_{\mathbf{k}\lambda} + \frac{1}{2}) \qquad (4.26)$$

is the Hamiltonian for the transverse electromagnetic field, and $a_{\mathbf{k}\lambda}, a_{\mathbf{k}\lambda}^\dagger$ are the annihilation and creation operators for the field mode $(\mathbf{k}, \lambda)$. In (4.25) we have used equation (2.52) for the quantized vector potential. This Hamiltonian describes the atom–field system in free space, i.e., in the absence of any matter except for the atom. It differs from the classical Hamiltonian (4.19) in that the atomic and field variables are now *operators* acting in a Hilbert space.

Transformation of Hamiltonian

The transformation that takes us classically from the form (4.19) of the dipole Hamiltonian to the form (4.24) is effected quantum mechanically by the unitary operator (Power and Zienau, 1959)

$$U = e^{-iS/\hbar} = e^{i e\mathbf{x}\cdot\mathbf{A}/\hbar c} \qquad (4.27)$$

in the Schrödinger picture (where $\mathbf{x}$ and $\mathbf{A}$ are time independent). Thus, writing $|\psi\rangle = U|\phi\rangle$, we have

$$i\hbar\frac{\partial}{\partial t}|\psi\rangle = i\hbar U\frac{\partial}{\partial t}|\phi\rangle = H|\psi\rangle = HU|\phi\rangle, \qquad (4.28)$$

or

$$i\hbar\frac{\partial}{\partial t}|\phi\rangle = U^\dagger H U|\phi\rangle = H'|\phi\rangle. \qquad (4.29)$$

This is the Schrödinger equation for the transformed state vector $|\phi\rangle$. The transformed Hamiltonian is

$$H' = \frac{1}{2m}U^\dagger(\mathbf{p} - \frac{e}{c}\mathbf{A})^2 U + V(\mathbf{x}) + U^\dagger H_{\text{Field}} U. \qquad (4.30)$$

From the general operator identity

$$e^A B e^{-A} = B + [A, B] + \frac{1}{2!}[A, [A, B]] + \dots \qquad (4.31)$$

it follows that $U^\dagger \mathbf{p} U = \mathbf{p} + (e/c)\mathbf{A}$ and

$$H' = \frac{\mathbf{p}^2}{2m} + V(\mathbf{x}) + \frac{1}{8\pi}\int d^3 r \mathbf{B}^2(\mathbf{r}) + \frac{1}{8\pi}\int d^3 r [U^\dagger(\mathbf{x})\mathbf{E}^\perp(\mathbf{r})U(\mathbf{x})]^2. \quad (4.32)$$

To evaluate the last term we note first that

$$[A_i(\mathbf{r}), E_j^\perp(\mathbf{r}')] = -4\pi i\hbar c \delta_{ij}^\perp(\mathbf{r} - \mathbf{r}'), \qquad (4.33)$$

where $\delta_{ij}^\perp$ is the transverse delta function tensor defined by (Appendix F)

$$
\begin{aligned}
\delta_{ij}^\perp(\mathbf{r}) &= \left(\frac{1}{2\pi}\right)^3 \int d^3k \left(\delta_{ij} - \frac{k_i k_j}{k^2}\right) e^{i\mathbf{k}\cdot\mathbf{r}} \\
&= \frac{2}{3}\delta_{ij}\delta^3(\mathbf{r}) - \frac{1}{4\pi r^3}\left(\delta_{ij} - \frac{3r_i r_j}{r^2}\right). \qquad (4.34)
\end{aligned}
$$

$\delta_{ij}^\perp(\mathbf{r})$ has the property

$$F_i^\perp(\mathbf{r}) = \int d^3r' \delta_{ij}^\perp(\mathbf{r} - \mathbf{r}') F_j(\mathbf{r}'), \qquad (4.35)$$

i.e., it gives by integration the transverse part of a vector field $\mathbf{F}(\mathbf{r})$. We can similarly define a longitudinal delta function:

$$
\begin{aligned}
\delta_{ij}^\parallel(\mathbf{r}) &= \left(\frac{1}{2\pi}\right)^3 \int d^3k \frac{k_i k_j}{k^2} e^{i\mathbf{k}\cdot\mathbf{r}} \\
&= \frac{1}{3}\delta_{ij}(\mathbf{r}) + \frac{1}{4\pi r^3}\left(\delta_{ij} - \frac{3r_i r_j}{r^2}\right), \qquad (4.36)
\end{aligned}
$$

such that

$$F_i^\parallel(\mathbf{r}) = \int d^3r' \delta_{ij}^\parallel(\mathbf{r} - \mathbf{r}') F_j(\mathbf{r}'). \qquad (4.37)$$

The commutator (4.33) follows easily from the same sort of manipulations as in Section 2.8.

Next we use (4.31) and (4.33) to obtain

$$U^\dagger(\mathbf{x})E_j^\perp(\mathbf{r})U(\mathbf{x}) = E_j^\perp(\mathbf{r}) - 4\pi e x_i \delta_{ij}^\perp(\mathbf{r}) \qquad (4.38)$$

and

$$
\int d^3r[U^\dagger(\mathbf{x})\mathbf{E}^\perp(\mathbf{r})U(\mathbf{x})]^2 = \int d^3r \mathbf{E}^{\perp\,2}(\mathbf{r}) - 8\pi e\mathbf{x}\cdot\mathbf{E}^\perp(0)
$$
$$
+ 16\pi^2 \int d^3r \mathbf{P}^\perp(\mathbf{r})^2, \qquad (4.39)
$$

where $\mathbf{P}(\mathbf{r}) \equiv e\mathbf{x}\delta^3(\mathbf{r})$ is the polarization density associated with the bound electron. Thus the transformed electric dipole Hamiltonian (4.32) is

$$H' = \frac{\mathbf{p}^2}{2m} + V(\mathbf{x}) - e\mathbf{x}\cdot\mathbf{E}^\perp + \frac{1}{8\pi}\int d^3r(\mathbf{E}^{\perp\,2} + \mathbf{B}^2) + 2\pi \int d^3r \mathbf{P}^\perp(\mathbf{r})^2. \quad (4.40)$$

This Hamiltonian is equivalent to the "$\mathbf{A} \cdot \mathbf{p}$" form (4.19) [or (4.25)]: calculated physical quantities will be the same regardless of whether we use (4.19) or (4.40) as our Hamiltonian. Formally this is trivial, of course, provided we remember to transform state vectors in the fashion $|\psi\rangle = U|\phi\rangle$.[1] We will use the traditional form (4.19) of the nonrelativistic electric dipole Hamiltonian.

4.5 Heisenberg Equations

It is well known that the Heisenberg picture often facilitates physical interpretation, and this is certainly the case in the nonrelativistic theory of atom–field interactions. In this section we write the Heisenberg equations of motion for $\mathbf{x}, \mathbf{p}$, and $a_{\mathbf{k}\lambda}$.

The Heisenberg equations for the electron coordinate and momentum are easily found from (4.25) to be

$$\dot{\mathbf{x}} = (i\hbar)^{-1}[\mathbf{x}, H] = \frac{1}{m}(\mathbf{p} - \frac{e}{c}\mathbf{A}), \qquad (4.41)$$

$$\dot{\mathbf{p}} = (i\hbar)^{-1}[\mathbf{p}, H] = -\nabla V(\mathbf{x}). \qquad (4.42)$$

These are formally the same as the classical equations (4.20) and (4.21). In particular,

$$m\ddot{\mathbf{x}} = -\nabla V(\mathbf{x}) + e\mathbf{E}, \qquad (4.43)$$

which is formally the same as the Newton equation of motion for an electron in a potential $V(\mathbf{x})$ and an electric field $\mathbf{E}$.[2]

The Heisenberg equation for $a_{\mathbf{k}\lambda}$ is also easily obtained:

$$\begin{aligned}
\dot{a}_{\mathbf{k}\lambda} &= (i\hbar)^{-1}[a_{\mathbf{k}\lambda}, H] \\
&= -i\omega_k a_{\mathbf{k}\lambda} - (i\hbar)^{-1}\frac{e}{mc}\left(\frac{2\pi\hbar c^2}{\omega_k V}\right)^{1/2} \mathbf{p} \cdot \mathbf{e}_{\mathbf{k}\lambda} \\
&\quad + (i\hbar)^{-1}\frac{e^2}{mc^2}\left(\frac{2\pi\hbar c^2}{\omega_k V}\right)^{1/2} \sum_{\mathbf{k}'\lambda'}\left(\frac{2\pi\hbar c^2}{\omega_{k'} V}\right)^{1/2} \mathbf{e}_{\mathbf{k}\lambda} \cdot \mathbf{e}_{\mathbf{k}'\lambda'}
\end{aligned}$$

---

[1] Note that (4.40) is written in terms of the old (untransformed) canonical variables. A different Hamiltonian having the same form as (4.40) can be obtained by writing the original Hamiltonian (4.19) in terms of new canonical variables. In this case it is not necessary to transform state vectors. See J. R. Ackerhalt and P. W. Milonni, J. Opt. Soc. Am. B1, 116 (1984). It is worth noting that, in either case, the vector $\mathbf{E}$ that appears in the interaction term is actually the electric displacement vector $\mathbf{D}$. We follow notational convention and do not distinguish typographically between $\mathbf{E}$ and $\mathbf{D}$.

[2] We typically write $\mathbf{E}$ instead of $\mathbf{E}^{\perp}$ when it is clear from the context that we are dealing with the transverse electric field.

$$= -i\omega_k a_{\mathbf{k}\lambda} + \frac{i}{\hbar}\left(\frac{c}{mc}\right)\left(\frac{2\pi\hbar c^2}{\omega_k V}\right)^{1/2}(\mathbf{p} - \frac{c}{c}\mathbf{A})\cdot\mathbf{e}_{\mathbf{k}\lambda}$$

$$= -i\omega_k a_{\mathbf{k}\lambda} + i\left(\frac{2\pi e^2}{\hbar\omega_k V}\right)^{1/2}\dot{\mathbf{x}}\cdot\mathbf{e}_{\mathbf{k}\lambda}. \tag{4.44}$$

This equation has already been used in Chapter 2 [Equation (2.80)]. We showed there that it implies the electric field operator

$$\mathbf{E}(t) = \mathbf{E}_o(t) + \mathbf{E}_{RR}(t), \tag{4.45}$$

where $\mathbf{E}_{RR}(t)$ is the radiation reaction field and $\mathbf{E}_o(t)$ is the source-free field. Thus

$$m\ddot{\mathbf{x}} = -\nabla V(\mathbf{x}) + \frac{2e^2}{3c^3}\dddot{\mathbf{x}} + e\mathbf{E}_o(t), \tag{4.46}$$

where m is the renormalized, observed mass. Equation (2.86) is the special case, with $\nabla V(\mathbf{x}) = (1/2)m\omega_o^2\mathbf{x}^2$, of this equation.

Taking expectation values over an initial state $|\psi\rangle = |\psi_A\rangle|\text{vac}\rangle$, where $|\psi_A\rangle$ is an atomic (electron) state and $|\text{vac}\rangle$ is the vacuum state of the field, we obtain

$$m\langle\ddot{\mathbf{x}}\rangle = -\langle\nabla V(\mathbf{x})\rangle + \frac{2e^2}{3c^3}\langle\dddot{\mathbf{x}}\rangle, \tag{4.47}$$

since $\langle\text{vac}|\mathbf{E}_o(t)|\text{vac}\rangle = 0$. This has the form of the classical equation of motion for the electron in the absence of any *external* field:

$$m\ddot{\mathbf{x}}_{cl} = -\nabla V(\mathbf{x}_{cl}) + \frac{2e^2}{3c^3}\dddot{\mathbf{x}}_{cl}. \tag{4.48}$$

Note, however, that in general $\langle\nabla V(\mathbf{x})\rangle \neq \nabla V(\langle\mathbf{x}\rangle)$, and so $\langle\mathbf{x}\rangle$ does not in general follow the classical path. For the very special case of a harmonic oscillator with $\nabla V(\mathbf{x}) = m\omega_o^2\mathbf{x}$, the quantum-mechanical expectation value $\langle\mathbf{x}(t)\rangle$ does follow the classical path with $\mathbf{x}_{cl}(0) = \langle\mathbf{x}(0)\rangle$ and $\dot{\mathbf{x}}_{cl}(0) = \langle\dot{\mathbf{x}}(0)\rangle$. Of course this equality of $\langle\mathbf{x}(t)\rangle$ and $\mathbf{x}_{cl}(t)$ does not mean that the classical and quantum mechanics of the harmonic oscillator are the same! For instance, $\langle\mathbf{x}^2(t)\rangle \neq \mathbf{x}_{cl}^2(t)$. Thus, although the equations of motion (4.41) and (4.42) give expectation values having the same form as the classical equations of motion (Ehrenfest's theorem), *they do not in themselves constitute any classical limit.*

There is an obvious but profound difference between the quantum-mechanical operator equation (4.46) and the classical equation (4.48). Classically, an electron in the vacuum sees only its radiation reaction field, whereas in QED the free-field operator $\mathbf{E}_o(t)$ is always present. We showed in Section 2.6 that without $\mathbf{E}_o(t)$ the whole quantum theory of a charged particle in vacuum becomes inconsistent. Commutation relations are broken, operators decay to zero, and an atomic electron would suffer the fate of its distant classical cousin and spiral into the nucleus.

4.6   Classical–Quantum Correspondence

The transverse electric field operator at point $\mathbf{r}$ and time t in free space is defined by (2.56). For an atom at $\mathbf{r} = 0, a_{\mathbf{k}\lambda}(t)$ is given by (2.81), which is the solution of (4.44). The first term on the right side of (2.81) leads to the source-free electric field (2.83). The second term gives the part of the field due to the dipole source:

$$
\begin{aligned}
\mathbf{E}_s^{\perp}(\mathbf{r},t) &= i\sum_{\mathbf{k}\lambda}\left(\frac{2\pi\hbar\omega_k}{V}\right)^{1/2}\left[ie\left(\frac{2\pi}{\hbar\omega_k V}\right)^{1/2}\right.\\
&\qquad \times \left.\int_0^t dt'\mathbf{e}_{\mathbf{k}\lambda}\cdot\dot{\mathbf{x}}(t')e^{i\omega_k(t'-t)}\right]\mathbf{e}_{\mathbf{k}\lambda}e^{i\mathbf{k}\cdot\mathbf{r}} + \text{h.c.}\\
&= -\frac{2\pi e}{V}\sum_{\mathbf{k}\lambda}\int_0^t dt'[\mathbf{e}_{\mathbf{k}\lambda}\cdot\dot{\mathbf{x}}(t')e^{i\omega_k(t'-t)}]\mathbf{e}_{\mathbf{k}\lambda}e^{i\mathbf{k}\cdot\mathbf{r}} + \text{h.c.}\\
&\rightarrow -\frac{2\pi e}{V}\frac{V}{8\pi^3}\int d^3k e^{i\mathbf{k}\cdot\mathbf{r}}\int_0^t dt' e^{i\omega_k(t'-t)}\\
&\qquad \times \sum_\lambda[\mathbf{e}_{\mathbf{k}\lambda}\cdot\dot{\mathbf{x}}(t')]\mathbf{e}_{\mathbf{k}\lambda} + \text{h.c.}\\
&= -\frac{e}{4\pi^2}\int d^3k[\hat{\mu} - (\hat{\mu}\cdot\hat{k})\hat{k}]e^{i\mathbf{k}\cdot\mathbf{r}}\int_0^t dt'\dot{x}(t')e^{i\omega_k(t'-t)} + \text{h.c.},
\end{aligned}
$$

$$(4.49)$$

where $\hat{\mu}$ and $\hat{k}$ are unit vectors in the directions of $\dot{\mathbf{x}}(t')$ and $\mathbf{k}$, respectively.
   Now

$$
\begin{aligned}
\int d^3k\hat{\mu}e^{i\mathbf{k}\cdot\mathbf{r}} &= \hat{\mu}\int dk k^2\int d\Omega_{\mathbf{k}}e^{i\mathbf{k}\cdot\mathbf{r}}\\
&= \hat{\mu}\int dk k^2\int_0^{2\pi}d\phi\int_0^\pi d\theta\sin\theta e^{ikr\cos\theta}\\
&= \hat{\mu}\int dk k^2\left(4\pi\frac{\sin kr}{kr}\right),
\end{aligned}
$$

$$(4.50)$$

and

$$\int d^3k(\hat{\mu} \cdot \hat{k})\hat{k}e^{i\mathbf{k}\cdot\mathbf{r}} = \int dk\, k^2 \int d\Omega_{\mathbf{k}} \frac{1}{k^2}[k_x\mu_x + k_y\mu_y + k_z\mu_z]$$
$$\times [k_x\hat{x} + k_y\hat{y} + k_z\hat{z}]e^{ik_z r}, \qquad (4.51)$$

where $d\Omega_{\mathbf{k}}$ represents an element of solid angle in $\mathbf{k}$ space, and for the purpose of performing the integral over all solid angles about $\mathbf{k}$, we have chosen $\mathbf{r}$ to define the z direction. Retaining only the nonvanishing contributions to (4.51), we have

$$\int d^3k(\hat{\mu} \cdot \hat{k})\hat{k}e^{i\mathbf{k}\cdot\mathbf{r}} = \int dk\, k^2 \int_0^{2\pi} d\phi \int_0^{\pi} d\theta \sin\theta[\mu_x\hat{x}\sin^2\theta\cos^2\phi$$
$$+ \mu_y\hat{y}\sin^2\theta\sin^2\phi + \mu_z\hat{z}\cos^2\theta]e^{ikr\cos\theta}$$
$$= 2\pi \int dk\, k^2[(\mu_x\hat{x} + \mu_y\hat{y})\left(\frac{2\sin kr}{k^3 r^3} - \frac{2\cos kr}{k^2 r^2}\right)$$
$$+ 2\mu_z\hat{z}\left(\frac{\sin kr}{kr} - \frac{2\sin kr}{k^3 r^3} + \frac{2\cos kr}{k^2 r^2}\right)]. \qquad (4.52)$$

The combination of (4.49), (4.50), and (4.52) yields

$$\mathbf{E}_s^{\perp}(\mathbf{r}, t) = -\frac{e}{\pi}\int_0^t dt'\, \dot{x}(t') \int_0^{\infty} dk\, k^2 e^{ikc(t'-t)}\left([\hat{\mu} - (\hat{\mu} \cdot \hat{r})\hat{r}]\frac{\sin kr}{kr}\right.$$
$$\left. -[\hat{\mu} - 3(\hat{\mu} \cdot \hat{r})\hat{r}][\frac{\sin kr}{k^3 r^3} - \frac{\cos kr}{k^2 r^2}]\right) + \text{h.c.}$$
$$= -\frac{2e}{\pi}\int_0^t dt'\, \dot{x}(t')\left(\frac{1}{r}[\hat{\mu} - (\hat{\mu} \cdot \hat{r})\hat{r}]\int_0^{\infty} dk\, k\sin kr\cos kc(t' - t)\right.$$
$$- [\hat{\mu} - 3(\hat{\mu} \cdot \hat{r})\hat{r}][\frac{1}{r^3}\int_0^{\infty} dk\frac{1}{k}\sin kr\cos kc(t' - t)$$
$$\left. -\frac{1}{r^2}\int_0^{\infty} dk\cos kr\cos kc(t' - t)]\right). \qquad (4.53)$$

Since

$$\int_0^{\infty} dk\, k\sin kr\cos kc(t' - t) = -\frac{\pi}{2c^2}\frac{\partial}{\partial t'}[\delta(t' - t + \frac{r}{c}) - \delta(t' - t - \frac{r}{c})], \quad (4.54)$$

$$\int_0^{\infty} dk\cos kr\cos kc(t' - t) = \frac{\pi}{2c}[\delta(t' - t + \frac{r}{c}) + \delta(t' - t - \frac{r}{c})], \quad (4.55)$$

it follows that

$$\mathbf{E}_s^{\perp}(\mathbf{r}, t) = -\frac{e}{c^2 r}[\hat{\mu} - (\mu \cdot \hat{r})\hat{r}]\ddot{x}(t - \frac{r}{c}) - \frac{e}{cr^2}[\hat{\mu} - 3(\hat{\mu} \cdot \hat{r})\hat{r}]\dot{x}(t - \frac{r}{c})$$

$$+ \frac{2e}{\pi r^3}[\hat{\mu} - 3(\hat{\mu} \cdot \hat{r})\hat{r}] \int_0^t dt' \dot{x}(t') F(t' - t), \qquad (4.56)$$

where

$$F(t' - t) \equiv \int_0^\infty dk \frac{1}{k} \sin kr \cos kc(t' - t) . \qquad (4.57)$$

The last term in (4.56) may be evaluated by partial integration:

$$\int_0^t dt' \dot{x}(t') F(t' - t) = -\frac{\pi}{2} x(t - \frac{r}{c}) + \frac{\pi}{2} x(t) , \qquad (4.58)$$

whereupon

$$\begin{aligned}
\mathbf{E}_s^\perp(\mathbf{r}, t) &= -\frac{e}{c^2 r}[\hat{\mu} - (\hat{\mu} \cdot \hat{r})\hat{r}]\ddot{x}(t - \frac{r}{c}) \\
&\quad - \frac{e}{cr^2}[\hat{\mu} - 3(\hat{\mu} \cdot \hat{r})\hat{r}]\dot{x}(t - \frac{r}{c}) - \frac{e}{r^3}[\hat{\mu} - 3\hat{\mu} \cdot \hat{r})\hat{r}]x(t - \frac{r}{c}) \\
&\quad + \frac{e}{r^3}[\hat{\mu} - 3(\hat{\mu} \cdot \hat{r})\hat{r}]x(t) .
\end{aligned} \qquad (4.59)$$

The last term in this expression for the transverse part of the electric field operator due to an electric dipole $e\hat{\mu}x(t)$ at $\mathbf{r} = 0$ is *unretarded*. There is nothing wrong with this unretarded contribution to the transverse field: only the complete, transverse plus longitudinal field must be purely retarded. The longitudinal part of $\mathbf{E}_s(\mathbf{r}, t)$ is easily obtained from (4.15) and has the same form as the classical electrostatic dipole field:

$$\mathbf{E}_s^\parallel(\mathbf{r}, t) = -\nabla\phi(\mathbf{r}, t) = \frac{e}{r^3}[3(\hat{\mu} \cdot \hat{r})\hat{r} - \hat{\mu}]x(t). \qquad (4.60)$$

The complete electric field operator associated with an electric dipole $e\hat{\mu}x(t)$ is therefore

$$\begin{aligned}
\mathbf{E}_s(\mathbf{r}, t) &= \mathbf{E}_s^\perp(\mathbf{r}, t) + \mathbf{E}_s^\parallel(\mathbf{r}, t) \\
&= -\frac{e}{c^2 r}[\hat{\mu} - (\hat{\mu} \cdot \hat{r})\hat{r}]\ddot{x}(t - \frac{r}{c}) - \frac{e}{cr^2}[\hat{\mu} - 3(\hat{\mu} \cdot \hat{r})\hat{r}]\dot{x}(t - \frac{r}{c}) \\
&\quad - \frac{e}{r^3}[\hat{\mu} - 3(\hat{\mu} \cdot \hat{r})\hat{r}]x(t - \frac{r}{c}) .
\end{aligned} \qquad (4.61)$$

This field is purely retarded, and has the same form as the *classical* electric field of an electric dipole. This formal correspondence between the classical and quantum dipole fields was employed in Section 3.11 in connection with the van der Waals interaction.

The formal correspondence between classical electromagnetic fields and QED Heisenberg-picture field operators holds for all multipole orders. In fact *the Maxwell equations themselves have the same form in classical and quantum electromagnetism.* The crucial difference is that in the quantum theory the fields are operators in a Hilbert space.

One consequence of this difference is that in QED the fields must have zero-point energy and fluctuations even in the absence of any sources of radiation. This is not to say that we cannot have source-free fields in classical electromagnetism. Rather, the absence of zero-point fields in standard classical theory lies in the *assumption* that there are no fields in the absence of any sources.

We can go beyond standard classical theory and postulate the existence of zero-point electric and magnetic fields in the absence of any sources. The resulting electromagnetic theory differs profoundly from QED, but it is able to account for some vacuum-electrodynamical effects within a fully classical framework. During the past few decades this classical theory of "stochastic electrodynamics" has been a fairly active area of research, and we describe its principal features in Chapter 8.

4.7   Two-State Model for an Atom

The Heisenberg operator equation of motion (4.46) for a bound electron is not very convenient for calculations pertaining to atoms in fields. This is because it does not exhibit explicit information about the unperturbed energy levels of the electron in the potential $V(\mathbf{x})$. To obtain Heisenberg equations that explicitly account for the bound energy levels, we return to the Hamiltonian (4.25) and write

$$
\begin{aligned}
H_{\text{Atom}} &= \left(\sum_i |i\rangle\langle i|\right) H_{\text{Atom}} \left(\sum_j |j\rangle\langle j|\right) \\
&= \sum_{i,j} |i\rangle\langle j|\langle i|H_{\text{Atom}}|j\rangle = \sum_{i,j} |i\rangle\langle j|E_j\langle i|j\rangle \\
&= \sum_i E_i |i\rangle\langle i| \equiv \sum_i E_i \sigma_{ii} \ .
\end{aligned}
\tag{4.62}
$$

Here $|i\rangle$ denotes an eigenstate of H_{Atom}, and the set $\{|i\rangle\}$ is complete, so that $\sum_i |i\rangle\langle i| = 1$.[3]

---

[3] Completeness of the set $\{|i\rangle\}$ is guaranteed by the Hermiticity of H_{Atom} and the fact that the energy levels E_i are bounded from below but not from above. See T. D. Lee, *Particle Physics and Introduction to Field Theory* (Harwood Academic Publishers, Chur, Switzerland, 1981), p. 12. If the Hilbert space spanned by the eigenvectors of H_{Atom} were finite, completeness would of course be trivial.

Similarly

$$\mathbf{p} = \sum_{i,j} |i\rangle\langle i|\mathbf{p}|j\rangle|j\rangle = \sum_{i,j} \mathbf{p}_{ij}|i\rangle\langle j| \equiv \sum_{i,j} \mathbf{p}_{ij}\sigma_{ij} \ . \tag{4.63}$$

Then the Hamiltonian (4.25) is

$$
\begin{aligned}
H \ = \ & \sum_i E_i\sigma_{ii} + \sum_{\mathbf{k}\lambda} \hbar\omega_k[a^\dagger_{\mathbf{k}\lambda}a_{\mathbf{k}\lambda} + \tfrac{1}{2}] - \hbar\sum_{i,j}\sum_{\mathbf{k}\lambda} C_{\mathbf{k}\lambda ij}[a_{\mathbf{k}\lambda} + a^\dagger_{\mathbf{k}\lambda}]\sigma_{ij} \\
& + \frac{e^2}{2mc^2}\sum_{\mathbf{k}\lambda}\sum_{\mathbf{k}'\lambda'}\left(\frac{2\pi\hbar c^2}{V}\right)\left(\frac{1}{\omega_k\omega'_k}\right)^{1/2}[a_{\mathbf{k}\lambda} + a^\dagger_{\mathbf{k}\lambda}] \\
& \times [a_{\mathbf{k}'\lambda'} + a^\dagger_{\mathbf{k}'\lambda'}]\mathbf{e}_{\mathbf{k}\lambda}\cdot\mathbf{e}_{\mathbf{k}'\lambda'} \ ,
\end{aligned}
\tag{4.64}
$$

where

$$C_{\mathbf{k}\lambda ij} \equiv \frac{e}{m}\left(\frac{2\pi}{\hbar\omega_k V}\right)^{1/2}\mathbf{p}_{ij}\cdot\mathbf{e}_{\mathbf{k}\lambda} = ie\left(\frac{2\pi}{\hbar\omega_k V}\right)^{1/2}\omega_{ij}\mathbf{x}_{ij}\cdot\mathbf{e}_{\mathbf{k}\lambda} \ . \tag{4.65}$$

A considerable simplification is realized if we restrict ourselves to the two-state model of an atom. In this model the Hilbert space of the atom is artificially truncated to the two states $|1\rangle$ and $|2\rangle$ with unperturbed energy eigenvalues E_1 and E_2, $E_2 - E_1 \equiv \hbar\omega_o > 0$. Then, since $\sigma_{11} + \sigma_{22} = 1$,

$$
\begin{aligned}
\sum_i E_i\sigma_{ii} \ = \ & E_1\sigma_{11} + E_2\sigma_{22} \\
\\
= \ & \frac{1}{2}E_1(\sigma_{11} + 1 - \sigma_{22}) + \frac{1}{2}E_2(\sigma_{22} + 1 - \sigma_{11}) \\
= \ & \frac{1}{2}(E_2 - E_1)(\sigma_{22} - \sigma_{11}) + \frac{1}{2}(E_1 + E_2) \\
= \ & \frac{1}{2}\hbar\omega_o\sigma_z + \frac{1}{2}(E_1 + E_2) \ ,
\end{aligned}
\tag{4.66}
$$

where

$$\sigma_z \equiv \sigma_{22} - \sigma_{11} \ . \tag{4.67}$$

Also

$$
\begin{aligned}
\sum_{i,j} C_{\mathbf{k}\lambda ij}\sigma_{ij} \ = \ & C_{\mathbf{k}\lambda 12}\sigma_{12} + C_{\mathbf{k}\lambda 21}\sigma_{21} \\
\\
= \ & -C_{\mathbf{k}\lambda 21}(\sigma_{12} - \sigma_{21}) = -C_{\mathbf{k}\lambda}\sigma_y \ ,
\end{aligned}
\tag{4.68}
$$

where

$$C_{\mathbf{k}\lambda} \equiv -iC_{\mathbf{k}\lambda 21} = e\omega_o\left(\frac{2\pi}{\hbar\omega_k V}\right)^{1/2}\mathbf{x}_{12}\cdot\mathbf{e}_{\mathbf{k}\lambda} \tag{4.69}$$

and
$$\sigma_y \equiv i(\sigma_{12} - \sigma_{21}) \ . \tag{4.70}$$

In the second line of (4.68) we are assuming $x_{12} = x_{21}$ for simplicity and without any loss of generality whatsoever for our purposes. (This equality can always be satisfied with an appropriate choice of the phases of the wave functions ψ_1 and ψ_2.) We also define

$$\sigma_x = \sigma_{12} + \sigma_{21} \ , \tag{4.71}$$

and it is easily shown that σ_x, σ_y, and σ_z satisfy the spin–1/2 Pauli algebra: $[\sigma_x, \sigma_y] = 2i\sigma_z, [\sigma_y, \sigma_z] = 2i\sigma_x, [\sigma_z, \sigma_x] = 2i\sigma_y$, and $\sigma_x^2 = \sigma_x, \sigma_y^2 = \sigma_y, \sigma_z^2 = \sigma_z$. In terms of these two-state operators,

$$H = \frac{1}{2}\hbar\omega_o\sigma_z + \sum_{\mathbf{k}\lambda} \hbar\omega_k a_{\mathbf{k}\lambda}^\dagger a_{\mathbf{k}\lambda} + \hbar \sum_{\mathbf{k}\lambda} C_{\mathbf{k}\lambda}[a_{\mathbf{k}\lambda} + a_{\mathbf{k}\lambda}^\dagger]\sigma_y + \frac{e^2}{2mc^2}\mathbf{A}^2 \ , \tag{4.72}$$

where now we have dropped the additive constants $\frac{1}{2}(E_1+E_2)$ and $\sum_{\mathbf{k}\lambda} \frac{1}{2}\hbar\omega_k$ from the Hamiltonian. It is also convenient for some purposes to write (4.72) as

$$H = \frac{1}{2}\hbar\omega_o\sigma_z + \sum_{\mathbf{k}\lambda} \hbar\omega_k a_{\mathbf{k}\lambda}^\dagger a_{\mathbf{k}\lambda} + i\hbar \sum_{\mathbf{k}\lambda} C_{\mathbf{k}\lambda}[a_{\mathbf{k}\lambda} + a_{\mathbf{k}\lambda}^\dagger][\sigma - \sigma^\dagger] + \frac{e^2}{2mc^2}\mathbf{A}^2 \ , \tag{4.73}$$

where $\sigma \equiv \sigma_{12}, \sigma^\dagger = \sigma_{21}$. Note that

$$\sigma|1\rangle = \sigma_{12}|1\rangle = |1\rangle\langle 2|1\rangle = 0, \tag{4.74}$$

and similarly

$$\sigma^\dagger|1\rangle = |2\rangle, \quad \sigma|2\rangle = |1\rangle, \quad \sigma^\dagger|2\rangle = 0. \tag{4.75}$$

Because of these properties, σ and $\sigma^\dagger$ are called atomic lowering and raising operators, respectively. It is easily shown from the definitions of these operators that

$$[\sigma, \sigma^\dagger] = -\sigma_z, \quad [\sigma, \sigma_z] = 2\sigma. \tag{4.76}$$

The Hamiltonian, of course, may be used in either the Schrödinger or Heisenberg picture, or any other picture. We will work in the Heisenberg picture with the equations of motion

$$i\hbar\dot\sigma_{ij} = [\sigma_{ij}, H], \tag{4.77}$$

$$i\hbar\dot a_{\mathbf{k}\lambda} = [a_{\mathbf{k}\lambda}, H], \tag{4.78}$$

for the atomic and field operators. Now

$$[\sigma_{ij}, \sigma_{kl}] = [|i\rangle\langle j|, |k\rangle\langle l|] = |i\rangle\langle j|k\rangle\langle l| - |k\rangle\langle l|i\rangle\langle j| = \delta_{jk}\sigma_{il} - \delta_{li}\sigma_{kj} \quad (4.79)$$

and, of course, $[a_{\mathbf{k}\lambda}, a^\dagger_{\mathbf{k}',\lambda'}] = \delta^3_{\mathbf{k},\mathbf{k}'}\delta_{\lambda,\lambda'}$. These commutation relations, together with the commutativity of equal-time atomic and field operators (i.e., $[\sigma_{ij}(t), a_{\mathbf{k}\lambda}(t)] = 0$, etc.), lead via (4.77) and (4.78) to the explicit equations of motion for the operators σ_{ij} and $a_{\mathbf{k}\lambda}$.

From the Hamiltonian (4.73) and the commutation relations we obtain the Heisenberg equations of motion

$$\dot{\sigma} = -i\omega_o\sigma + \sum_{\mathbf{k}\lambda} C_{\mathbf{k}\lambda}[a_{\mathbf{k}\lambda} + a^\dagger_{\mathbf{k}\lambda}]\sigma_z, \quad (4.80)$$

$$\dot{\sigma}_z = -2\sum_{\mathbf{k}\lambda} C_{\mathbf{k}\lambda}[a_{\mathbf{k}\lambda} + a^\dagger_{\mathbf{k}\lambda}][\sigma + \sigma^\dagger], \quad (4.81)$$

$$\dot{a}_{\mathbf{k}\lambda} = -i\omega_k a_{\mathbf{k}\lambda} + C_{\mathbf{k}\lambda}[\sigma - \sigma^\dagger]. \quad (4.82)$$

In writing (4.82) we have ignored the effect of the $\mathbf{A}^2$ term. This point is discussed in Section 4.9.

4.8   Operator Orderings

Since equal-time atomic and field operators commute, we can write the Heisenberg equations (4.80) and (4.81) in different but equivalent ways. For instance, we can use the *normal ordering* in which photon annihilation operators $a_{\mathbf{k}\lambda}$ appear at the right and creation operators $a^\dagger_{\mathbf{k}\lambda}$ appear at the left in operator products:

$$\dot{\sigma} = -i\omega_o\sigma + \sum_{\mathbf{k}\lambda} C_{\mathbf{k}\lambda}[\sigma_z a_{\mathbf{k}\lambda} + a^\dagger_{\mathbf{k}\lambda}\sigma_z], \quad (4.83)$$

$$\dot{\sigma}_z = -2\sum_{\mathbf{k}\lambda} C_{\mathbf{k}\lambda}[(\sigma + \sigma^\dagger)a_{\mathbf{k}\lambda} + a^\dagger_{\mathbf{k}\lambda}(\sigma + \sigma^\dagger)]. \quad (4.84)$$

Or we can antinormally order the field operators:

$$\dot{\sigma} = -i\omega_o\sigma + \sum_{\mathbf{k}\lambda} C_{\mathbf{k}\lambda}[a_{\mathbf{k}\lambda}\sigma_z + \sigma_z a^\dagger_{\mathbf{k}\lambda}], \quad (4.85)$$

$$\dot{\sigma}_z = -2\sum_{\mathbf{k}\lambda} C_{\mathbf{k}\lambda}[a_{\mathbf{k}\lambda}(\sigma + \sigma^\dagger) + (\sigma + \sigma^\dagger)a^\dagger_{\mathbf{k}\lambda}]. \quad (4.86)$$

Another ordering of interest is the symmetric one:

$$\dot{\sigma} = -i\omega_o\sigma + \frac{1}{2}\sum_{\mathbf{k}\lambda}C_{\mathbf{k}\lambda}[\sigma_z(a_{\mathbf{k}\lambda} + a_{\mathbf{k}\lambda}^\dagger) + (a_{\mathbf{k}\lambda} + a_{\mathbf{k}\lambda}^\dagger)\sigma_z], \quad (4.87)$$

$$\dot{\sigma}_z = -\sum_{\mathbf{k}\lambda}C_{\mathbf{k}\lambda}[(\sigma + \sigma^\dagger)(a_{\mathbf{k}\lambda} + a_{\mathbf{k}\lambda}^\dagger) + (a_{\mathbf{k}\lambda} + a_{\mathbf{k}\lambda}^\dagger)(\sigma + \sigma^\dagger)].$$

$$(4.88)$$

To get a sense of the implications of different orderings, consider the formal solution of (4.82):[4]

$$a_{\mathbf{k}\lambda}(t) = a_{\mathbf{k}\lambda}(0)e^{-i\omega_k t} + C_{\mathbf{k}\lambda}\int_0^t dt'[\sigma(t') - \sigma^\dagger(t')]e^{i\omega_k(t'-t)}$$

$$\equiv a_{\mathbf{k}\lambda}(0)e^{-i\omega_k t} + a_{\mathbf{k}\lambda,s}(t). \quad (4.89)$$

The first term is the source-free part of $a_{\mathbf{k}\lambda}(t)$, whereas the second term is the part due to the source, which in this case is a two-state "atom." Now if we use (4.89) in (4.83), and take expectation values over an initial atom–field state $|\text{vac}\rangle|\psi_A\rangle$, where $|\psi_A\rangle$ is an arbitrary state of our two-state atom, then

$$\langle\dot{\sigma}(t)\rangle = -i\omega_o\langle\sigma(t)\rangle + \sum_{\mathbf{k}\lambda}C_{\mathbf{k}\lambda}[\langle\sigma_z(t)a_{\mathbf{k}\lambda,s}(t)\rangle + \langle a_{\mathbf{k}\lambda,s}^\dagger(t)\sigma_z(t)\rangle]. \quad (4.90)$$

Since $a_{\mathbf{k}\lambda}(0)|\text{vac}\rangle = 0$, the source-free part of $a_{\mathbf{k}\lambda}(t)$ does not appear explicitly in this equation. If, however, we use (4.89) in the antinormally ordered equation (4.85), then

$$\langle\dot{\sigma}(t)\rangle = -i\omega_o\langle\sigma(t)\rangle + \sum_{\mathbf{k}\lambda}C_{\mathbf{k}\lambda}[\langle a_{\mathbf{k}\lambda}(0)\sigma_z(t)\rangle e^{-i\omega_k t} + \langle a_{\mathbf{k}\lambda,s}(t)\sigma_z(t)\rangle$$

$$+ \langle\sigma_z(t)a_{\mathbf{k}\lambda}^\dagger(0)\rangle e^{i\omega_k t} + \langle\sigma_z(t)a_{\mathbf{k}\lambda,s}^\dagger(t)\rangle]. \quad (4.91)$$

In this case, since $\langle\text{vac}|a_{\mathbf{k}\lambda}(0)$ and $a_{\mathbf{k}\lambda}^\dagger(0)|\text{vac}\rangle$ do not vanish, the source-free part of the field *does* make an explicit contribution. In physical terms, the vacuum field appears to make an explicit contribution in (4.91) but not in (4.90), where only the *source* field appears. Since equations (4.90) and (4.91) are equivalent, this suggests that there are different but equivalent interpretations in terms of vacuum and source fields. These interpretations are discussed in Sections 4.10–4.13.

---

[4]Note that in writing equation (4.89) we assume that the atom–field interaction is "switched on" at $t = 0$. Prior to this time the atomic and field operators act in different Hilbert spaces and therefore commute. Unitarity guarantees that they commute at all times.

4.9   Spontaneous Emission and the Lamb Shift

It follows from (4.89) that

$$\sum_{\mathbf{k}\lambda} C_{\mathbf{k}\lambda} a_{\mathbf{k}\lambda,s}(t) = \sum_{\mathbf{k}\lambda} C_{\mathbf{k}\lambda}^2 \int_0^t dt'[\sigma(t') - \sigma^\dagger(t')]e^{i\omega_k(t'-t)} \qquad (4.92)$$

Let us assume that the atom–field coupling is sufficiently weak that we can sensibly take

$$\sigma(t') \cong \sigma(t)e^{-i\omega_o(t'-t)} , \qquad (4.93)$$

$$\sigma^\dagger(t') \cong \sigma^\dagger(t)e^{i\omega_o(t'-t)} \qquad (4.94)$$

in the integrand of (4.92). This is called a *Markovian approximation* because it replaces the field operator (4.92) by an operator that depends on the atomic variables at the same time t, without any "memory" of these variables at earlier times:

$$\sum_{\mathbf{k}\lambda} C_{\mathbf{k}\lambda} a_{\mathbf{k}\lambda,s}(t) \cong \sigma(t)\sum_{\mathbf{k}\lambda} C_{\mathbf{k}\lambda}^2 \int_0^t dt' e^{i(\omega_k-\omega_o)(t'-t)}$$

$$- \sigma^\dagger(t)\sum_{\mathbf{k}\lambda} C_{\mathbf{k}\lambda}^2 \int_0^t dt' e^{i(\omega_k+\omega_o)(t'-t)} . \qquad (4.95)$$

In the mode continuum limit,

$$\sum_{\mathbf{k}\lambda} C_{\mathbf{k}\lambda}^2 \int_0^t dt' e^{i(\omega_k-\omega_o)(t'-t)} \longrightarrow \frac{e^2\omega_o^2}{4\pi^2\hbar c^3}\int_0^\infty d\omega\omega \int d\Omega_{\mathbf{k}} \sum_\lambda |\mathbf{x}_{12}\cdot\mathbf{e}_{\mathbf{k}\lambda}|^2$$

$$\times \int_0^t dt' e^{i(\omega-\omega_o)(t'-t)}$$

$$= \frac{2e^2\omega_o^2|\mathbf{x}_{12}|^2}{3\pi\hbar c^3}\int_0^\infty d\omega\omega \int_0^t dt' e^{i(\omega-\omega_o)(t'-t)} . \qquad (4.96)$$

The integral over time,

$$\int_0^t e^{i(\omega-\omega_o)(t'-t)} = -i\left[\frac{1-\cos(\omega-\omega_o)t}{\omega-\omega_o}\right] + \frac{\sin(\omega-\omega_o)t}{\omega-\omega_o} , \qquad (4.97)$$

arises frequently in QED and is treated as follows. The bracketed factor in the first term vanishes if $\omega - \omega_o = 0$, but is effectively $(\omega-\omega_o)^{-1}$ otherwise because of the rapid oscillations of $\cos\omega_o t$, $\sin\omega_o t$ for $\omega_o t \gg 1$. Similarly the second term effectively vanishes unless $\omega - \omega_o = 0$, in which case it becomes t. Thus, when (4.97) appears in an integral over ω, we make the replacement

$$\int_0^t dt' e^{i(\omega-\omega_o)(t'-t)} \rightarrow -iP\left(\frac{1}{\omega - \omega_o}\right) + \pi\delta(\omega - \omega_o) \qquad (4.98)$$

for sufficiently large times. Here as usual P and δ denote the Cauchy principal part and the delta function, respectively. In similar fashion

$$\int_0^t dt' e^{i(\omega+\omega_o)(t'-t)} \rightarrow -i\left(\frac{1}{\omega + \omega_o}\right) \qquad (4.99)$$

and therefore

$$\sum_{\mathbf{k}\lambda} C_{\mathbf{k}\lambda} a_{\mathbf{k}\lambda,s}(t) \cong (\beta - i\Delta_2)\sigma(t) + i\Delta_1\sigma^\dagger(t), \qquad (4.100)$$

where

$$\beta \equiv \frac{2\omega_o^3|\mathbf{d}_{12}|^2}{3\hbar c^3}, \qquad (4.101)$$

$$\Delta_2 \equiv \frac{2\omega_o^2|\mathbf{d}_{12}|^2}{3\pi\hbar c^3} P \int_0^\infty \frac{d\omega\omega}{\omega - \omega_o}, \qquad (4.102)$$

$$\Delta_1 \equiv \frac{2\omega_o^2|\mathbf{d}_{12}|^2}{3\pi\hbar c^3} \int_0^\infty \frac{d\omega\omega}{\omega + \omega_o}, \qquad (4.103)$$

and $\mathbf{d}_{12} = e\mathbf{x}_{12}$ is the electric dipole transition matrix element.

We now use the approximation (4.100) in the equation (4.90) for the expectation value $\langle\sigma(t)\rangle$ when a normal ordering is chosen:

$$\begin{aligned}
\langle\dot{\sigma}(t)\rangle &\cong -i\omega_o\langle\sigma(t)\rangle + [(\beta - i\Delta_2)\langle\sigma_z(t)\sigma(t)\rangle + i\Delta_1\langle\sigma_z(t)\sigma^\dagger(t)\rangle \\
&\quad + (\beta + i\Delta_2)\langle\sigma^\dagger(t)\sigma_z(t)\rangle - i\Delta_1\langle\sigma(t)\sigma_z(t)\rangle] \\
&= -i\omega_o\langle\sigma(t)\rangle - [\beta - i(\Delta_2 - \Delta_1)]\langle\sigma(t)\rangle \\
&\quad - [\beta + i(\Delta_2 - \Delta_1)]\langle\sigma^\dagger(t)\rangle \\
&= -i[\omega_o - (\Delta_2 - \Delta_1)]\langle\sigma(t)\rangle - \beta\langle\sigma(t)\rangle \\
&\quad - [\beta + i(\Delta_2 - \Delta_1)]\langle\sigma^\dagger(t)\rangle, \qquad (4.104)
\end{aligned}$$

where we have employed the equal-time identities $\sigma_z\sigma = -\sigma, \sigma\sigma_z = \sigma$, and their Hermitian conjugates.

Now according to (4.93) and (4.94), the time evolution of $\langle\sigma(t)\rangle$ is primarily an oscillation $\sim e^{-i\omega_o t}$ and that of $\langle\sigma^\dagger(t)\rangle$ an oscillation $\sim e^{i\omega_o t}$.

The effect of $\langle \sigma^\dagger(t) \rangle$ on $\langle \sigma(t) \rangle$ in equation (4.104) may then be assumed to be negligible. In this so-called rotating-wave approximation (RWA) we write

$$\langle \dot{\sigma}(t) \rangle \cong -i[\omega_o - (\Delta_2 - \Delta_1)]\langle \sigma(t) \rangle - \beta \langle \sigma(t) \rangle. \tag{4.105}$$

Note that

$$\beta = \frac{1}{2} A_{21} , \tag{4.106}$$

where A_{21} is the Einstein A coefficient giving the rate of spontaneous emission for the $2 \rightarrow 1$ transition. The reason that β is *half* the A coefficient becomes obvious if we consider the atomic state $c_1|1\rangle + c_2|2\rangle$, for which $\langle \sigma(0) \rangle = c_1^* c_2$. Since the upper-state probability $|c_2|^2$ decays at the rate $A = 2\beta$ and the lower state does not decay in our two-state model, we can infer that $\langle \sigma(t) \rangle$ should decay at the rate β.

It is also obvious that $-(\Delta_2 - \Delta_1)$ represents a shift of the transition frequency ω_o. Note that, since $m^2 \omega_o^2 |\mathbf{d}_{12}|^2 = e^2 |\mathbf{p}_{12}|^2$,[5]

$$-\hbar \Delta_2 = \frac{2\alpha}{3\pi} \left(\frac{1}{mc} \right)^2 |\mathbf{p}_{12}|^2 \int_0^\infty \frac{dE\,E}{E_2 - E_1 - E} , \tag{4.107}$$

$$-\hbar \Delta_1 = \frac{2\alpha}{3\pi} \left(\frac{1}{mc} \right)^2 |\mathbf{p}_{12}|^2 \int_0^\infty \frac{dE\,E}{E_1 - E_2 - E} \tag{4.108}$$

for our two-state atom. The generalization to a real atom with energy levels $\{E_n\}$ is

$$- \hbar \Delta_n \equiv \Delta E_n = \frac{2\alpha}{3\pi} \left(\frac{1}{mc} \right)^2 \sum_m |\mathbf{p}_{mn}|^2 \int_0^\infty \frac{dE\,E}{E_n - E_m - E} , \tag{4.109}$$

which is the expression (3.20) for the shift in the energy E_n due to the interaction of the atom with the vacuum electromagnetic field. Mass renormalization and a high-frequency cutoff in the nonrelativistic theory lead to an accurate estimate of the Lamb shift in hydrogen, as discussed in Chapter 3.

From equation (4.84) we obtain similarly the equation

$$\langle \dot{\sigma}_z(t) \rangle = -2 \sum_{\mathbf{k}\lambda} C_{\mathbf{k}\lambda}[\langle \sigma(t) a_{\mathbf{k}\lambda,s}(t) \rangle + \langle a_{\mathbf{k}\lambda,s}^\dagger(t) \sigma(t) \rangle$$
$$+ \langle \sigma^\dagger(t) a_{\mathbf{k}\lambda,s}(t) \rangle + \langle a_{\mathbf{k}\lambda,s}^\dagger(t) \sigma^\dagger(t) \rangle], \tag{4.110}$$

---

[5] Following a notational convention, we do not explicitly indicate the fact that the integrals for the radiative shifts should be understood in the sense of the Cauchy principal part.

when the normal ordering is used and vacuum expectation values are taken The Markovian approximation (4.100) then gives

$$\langle \dot{\sigma}_z(t) \rangle = -4\beta \langle \sigma^\dagger(t)\sigma(t) \rangle = -2\beta[1 + \langle \sigma_z(t) \rangle], \qquad (4.111)$$

when we use the fact that σ^2 is identically zero at all times. Thus

$$\langle \sigma_z(t) \rangle = -1 + [1 + \sigma_z(0) \rangle]e^{-2\beta t}. \qquad (4.112)$$

If $\langle \sigma_z(0) \rangle = |c_2(0)|^2 - |c_1(0)|^2 = 1$, i.e., if the two-state atom is initially in the upper state, then there is exponential decay to the lower state (with $\langle \sigma_z \rangle = -1$) at the rate 2β. If $\langle \sigma_z(0) \rangle = -1$, then $\langle \sigma_z(t) \rangle = -1$, i.e., if the atom is initially in the lower state of the transition, it remains for all time in that state.

The Markovian approximation therefore leads to the correct Einstein A coefficient and the same radiative level shift given by second-order perturbation theory. It also predicts that the spontaneous decay of the excited-state probability is purely exponential. As discussed in Section 4.14, these results are the same as those obtained by "standard" Weisskopf–Wigner theory.

It is also worth noting that exponential decay here is an *approximation*. Indeed it is known that a system whose energy levels are bounded from below cannot exhibit purely exponential decay. However, the deviation from exponential decay for a radiating atom turns out to be very small, as discussed in Section 4.14.

It should be recalled that we have neglected the $\mathbf{A}^2$ term in the Hamiltonian. As noted in Section 3.4, this term does not contribute to observable level shifts in the dipole approximation, and it is also easy to show that it does not affect the calculation of the A coefficient. The $\mathbf{A}^2$ term can affect the calculation of the ("natural") lineshape of spontaneous emission,[6] but for our purposes this effect of the $\mathbf{A}^2$ term may be ignored.

4.10 Normal Ordering and Source Field

In the calculations of the preceding section we employed a normal ordering in which the source-free part of the electric field operator made no contribution to vacuum expectation values. That is, the vacuum field made no

---

[6] P. W. Milonni, R. J. Cook, and J. R. Ackerhalt, *Phys. Rev.* A40, 3764 (1989). If the $\mathbf{x} \cdot \mathbf{E}$ form of the interaction is used, $\mathbf{A}^2$ does not appear explicitly but is included "automatically" in the calculated level shifts. The expectation value of $(e^2/2mc^2)\mathbf{A}^2$ can therefore be subtracted from the level shifts calculated with the $\mathbf{x} \cdot \mathbf{E}$ interaction. In the calculation of frequency shifts, of course, the $\mathbf{A}^2$ contribution to the levels shifts will cancel out anyway. See Milonni (1976).

explicit contribution to the radiative decay or level shifts; these "radiative corrections" were due entirely to the *source* or radiation reaction field.

This is remarkable in view of the simple and seemingly natural ways in which the level shift, in particular, was interpreted in terms of the vacuum field in Chapter 3. In spite of such appealing derivations we have now apparently found that the vacuum field has no effect on the level shifts! This is because we have employed the normal ordering of field operators, where the vacuum field contributes nothing to vacuum expectation values because $a_{\mathbf{k}\lambda}(0)|\text{vac}\rangle = \langle\text{vac}|a_{\mathbf{k}\lambda}^{\dagger}(0) = 0$.

Two caveats are in order here. First, in our calculation the atom does not really "see" the radiation reaction field as such: the vector potential associated with the radiation reaction field is

$$\mathbf{A}_{\text{RR}}(t) = \sum_{\mathbf{k}\lambda}\left(\frac{2\pi\hbar c^2}{\omega_k V}\right)^{1/2}[a_{\mathbf{k}\lambda,s}(t) + a_{\mathbf{k}\lambda,s}^{\dagger}(t)], \qquad (4.113)$$

whereas in (4.90), for instance, $a_{\mathbf{k}\lambda,s}(t)$ and $a_{\mathbf{k}\lambda,s}^{\dagger}(t)$ appear in the combination $\sigma_z(t)a_{\mathbf{k}\lambda,s}(t) + a_{\mathbf{k}\lambda,s}^{\dagger}(t)\sigma_z(t)$ rather than $[a_{\mathbf{k}\lambda,s}(t) + a_{\mathbf{k}\lambda,s}^{\dagger}(t)]\sigma_z(t)$ or $\sigma_z(t)[a_{\mathbf{k}\lambda,s}(t) + a_{\mathbf{k}\lambda,s}^{\dagger}(t)]$. This difference arises as a consequence of our normal ordering. In this approach it is not the vector potential that ultimately appears in vacuum expectation values, but its photon annihilation and creation parts, and these do not appear in the symmetric combination that we identify with the vector potential operator.

Second, the radiation reaction electric field[7] involving a third derivative with respect to time [equation (4.46)] is not obtained in the Markovian approximation we have employed. This approximation proceeded from the assumption that the time evolution of $\sigma(t)$ is primarily a sinusoidal oscillation at frequency ω_o. If we use this approximation in (4.46), for instance, then

$$m\ddot{\mathbf{x}} \cong -\nabla V(\mathbf{x}) - \frac{2e^2\omega_o^2}{3c^3}\dot{\mathbf{x}} + e\mathbf{E}_o(t). \qquad (4.114)$$

This implies that radiation reaction damps $\dot{\mathbf{x}}$ at the rate $2e^2\omega_o^2/3mc^3$. If we replace e^2 by e^2f, where $f = 2m\omega_o|\mathbf{d}_{12}|^2/\hbar e^2$ is the oscillator strength of the $2 \rightarrow 1$ transition, then this dipole damping rate becomes the Einstein A coefficient (see Appendix A).

---

[7] For the quantum-mechanical calculation, as well as the classical, we must retain the $\mathbf{A}^2$ term in the Hamiltonian to obtain the correct radiation reaction field. The $\mathbf{A}^2$ term leads to expressions involving $\mathbf{p} + (e/c)\mathbf{A} = m\dot{\mathbf{x}}$, whereas only $\mathbf{p}$ appears when $\mathbf{A}^2$ is ignored. For the purpose of getting the correct result without the $\mathbf{A}^2$ term we can simply (but not rigorously) replace $\mathbf{p}$ by $m\dot{\mathbf{x}}$ in the calculation of $\mathbf{E}_{\text{RR}}(t)$.

Although these things should be borne in mind when we attribute spontaneous emission and the level shift to radiation reaction, we are certainly justified in saying that only the *source* part of the field is responsible for these effects when they are calculated with a normal ordering of field operators. We emphasize again how different this interpretation is compared with those based on vacuum field fluctuations in Chapter 3. However, the latter interpretations also contain an element of truth, as we shall now show.

4.11　Nonnormal Ordering and Vacuum

We have emphasized that the vacuum field is absolutely necessary in the quantum theory of radiation, if only to preserve commutation relations and the formal consistency of the theory (recall Section 2.6). To see an explicit role for the vacuum field in the theory of spontaneous emission and the Lamb shift, all we need to do is perform a calculation similar to that in Section 4.9, using a *nonnormal* ordering of field operators.

Suppose we use the antinormal ordering, in which $\langle \dot{\sigma}(t) \rangle$, for instance, is given by equation (4.91). The Markovian approximation (4.100) for the source part of the field operator, $\sum_{\mathbf{k}\lambda} C_{\mathbf{k}\lambda} a_{\mathbf{k}\lambda,s}(t)$, gives

$$
\begin{aligned}
\langle \dot{\sigma}(t) \rangle \cong & -i\omega_o \langle \sigma(t) \rangle + [\beta - i(\Delta_2 - \Delta_1)]\langle \sigma(t) \rangle \\
& + \sum_{\mathbf{k}\lambda} C_{\mathbf{k}\lambda}[\langle a_{\mathbf{k}\lambda}(0)\sigma_z(t) \rangle e^{-i\omega_k t} + \langle \sigma_z(t) a_{\mathbf{k}\lambda}^\dagger(0) \rangle e^{i\omega_k t}],
\end{aligned}
$$

$$(4.115)$$

when we make the RWA and use the identities $\sigma(t)\sigma_z(t) = \sigma(t)$ and $\sigma_z(t)\sigma(t) = -\sigma(t)$. The first two terms on the right side of this equation are similar to the right side of (4.105), but the terms associated with the level shift and spontaneous decay in (4.105) have the opposite signs in (4.115). Obviously this difference must be made up for by the vacuum field terms involving $a_{\mathbf{k}\lambda}(0)$ and $a_{\mathbf{k}\lambda}^\dagger(0)$ in (4.115).

To evaluate the vacuum field terms we use the formal solution of equation (4.86),

$$
\begin{aligned}
\sigma_z(t) = & \sigma_z(0) - 2\sum_{\mathbf{k}\lambda} C_{\mathbf{k}\lambda} \left[\int_0^t dt' a_{\mathbf{k}\lambda}(t')[\sigma(t') + \sigma^\dagger(t')] \right. \\
& \left. + \int_0^t dt'[\sigma(t') + \sigma^\dagger(t')]a_{\mathbf{k}\lambda}^\dagger(t') \right],
\end{aligned}
$$

$$(4.116)$$

in (4.115). Thus

$$
\sum_{\mathbf{k}\lambda} C_{\mathbf{k}\lambda} \langle a_{\mathbf{k}\lambda}(0)\sigma_z(t)\rangle e^{-i\omega_k t} \cong -2\sum_{\mathbf{k}\lambda} C_{\mathbf{k}\lambda}^2 \left[\int_0^t dt' \langle a_{\mathbf{k}\lambda}(0)[\sigma(t') + \sigma^\dagger(t')] \right.
$$

$$
\left. \times a_{\mathbf{k}\lambda}(t')\rangle e^{-i\omega_k t} + \int_0^t dt' \langle a_{\mathbf{k}\lambda}(0)a_{\mathbf{k}\lambda}^\dagger(t')[\sigma(t') + \sigma^\dagger(t')]\rangle \right] e^{-i\omega_k t}.
$$

$$(4.117)$$

Here we have used the fact that $\langle a_{\mathbf{k}\lambda}(0)\sigma_z(0)\rangle = \langle \sigma_z(0)a_{\mathbf{k}\lambda}(0)\rangle = 0$, as well as the equal-time commutation relations $a_{\mathbf{k}\lambda}(t')\sigma^\dagger(t') = \sigma^\dagger(t')a_{\mathbf{k}\lambda}(t')$, $\sigma(t')a_{\mathbf{k}\lambda}(t') = a_{\mathbf{k}\lambda}(t')\sigma(t')$. Furthermore, to remain to the same order of approximation as in the calculation of Section 4.9, we replace $a_{\mathbf{k}\lambda}(t')$ by $a_{\mathbf{k}\lambda}(0)e^{-i\omega_k t'}$:

$$
\sum_{\mathbf{k}\lambda} C_{\mathbf{k}\lambda} \langle a_{\mathbf{k}\lambda}(0)\sigma_z(t)\rangle e^{-i\omega_k t} \cong
$$

$$
-2\sum_{\mathbf{k}\lambda} C_{\mathbf{k}\lambda}^2 \int_0^t dt' \langle a_{\mathbf{k}\lambda}(0)a_{\mathbf{k}\lambda}^\dagger(0)[\sigma(t') + \sigma^\dagger(t')]\rangle e^{i\omega_k(t'-t)}. \quad (4.118)
$$

We have used here the fact that the vacuum expectation value $\langle a_{\mathbf{k}\lambda}(0)\sigma^\dagger(t') \times a_{\mathbf{k}\lambda}(0)\rangle$ is 0. And since

$$
\langle \text{vac}| a_{\mathbf{k}\lambda}(0)a_{\mathbf{k}\lambda}^\dagger(0) = \langle \text{vac}|[a_{\mathbf{k}\lambda}^\dagger(0)a_{\mathbf{k}\lambda}(0) + 1] = \langle \text{vac}|(0+1), \quad (4.119)
$$

$$
\begin{aligned}
\sum_{\mathbf{k}\lambda} C_{\mathbf{k}\lambda} \langle a_{\mathbf{k}\lambda}(0)\sigma_z(t)\rangle e^{-i\omega_k t} &\cong -2\sum_{\mathbf{k}\lambda} C_{\mathbf{k}\lambda}^2 \left[\int_0^t dt' \langle \sigma(t')\rangle e^{i\omega_k(t'-t)} \right. \\
&\qquad \left. + \int_0^t dt' \langle \sigma^\dagger(t')\rangle e^{i\omega_k(t'-t)} \right] \\
&\cong -2\langle \sigma(t)\rangle \sum_{\mathbf{k}\lambda} C_{\mathbf{k}\lambda}^2 \int_0^t dt' e^{i(\omega_k - \omega_o)(t'-t)} \\
&\qquad - 2\langle \sigma^\dagger(t)\rangle \sum_{\mathbf{k}\lambda} C_{\mathbf{k}\lambda}^2 \int_0^t dt' e^{i(\omega_k + \omega_o)(t'-t)} \\
&\cong -2(\beta - i\Delta_2)\langle \sigma(t)\rangle + 2i\Delta_1 \langle \sigma^\dagger(t)\rangle
\end{aligned}
$$

$$(4.120)$$

in the Markovian approximation. It follows by complex conjugation that

$$\sum_{\mathbf{k}\lambda} C_{\mathbf{k}\lambda}\langle \sigma_z(t)a_{\mathbf{k}\lambda}^\dagger(0)\rangle e^{i\omega_k t} \cong -2(\beta+i\Delta_2)\langle\sigma^\dagger(t)\rangle - 2i\Delta_1\langle\sigma(t)\rangle \quad (4.121)$$

and finally, using (4.120) and (4.121) in (4.115),

$$\langle\dot\sigma(t)\rangle \cong -i[\omega_o - (\Delta_2 - \Delta_1)]\langle\sigma(t)\rangle - \beta\langle\sigma(t)\rangle \qquad (4.122)$$

in the RWA. We have thus recovered (4.105) starting with an *antin*ormal ordering of field operators.

Looking back over this calculation, we can see that the source field contributes $(\Delta_2 - \Delta_1)$ to the radiative frequency shift, whereas the vacuum field contributes $-2(\Delta_2 - \Delta_1)$, giving the complete shift $-(\Delta_2 - \Delta_1)$ of Section 4.9. We can similarly describe the origin of the decay rate β in this calculation.

In other words, when a normal ordering is used, the entire contribution to the level shift $-\hbar\Delta_n$ comes from the source field, whereas when an antinormal ordering is used, $-2\hbar\Delta_n$ comes from the vacuum field and another $\hbar\Delta_n$ from the source field. Although we have shown this explicitly only in the two-state model for an atom, the same conclusions apply for a real atom (Milonni and Smith, 1975; Milonni, 1976).

A calculation of $\langle\dot\sigma_z(t)\rangle$, using antinormal ordering and the Markovian approximation, gives contributions $-4\beta\langle\sigma_z(t)\rangle$ from the vacuum field and $-2\beta[1 - \langle\sigma_z(t)\rangle]$ from the source, so that

$$\begin{aligned}\langle\dot\sigma_z(t)\rangle &\cong -4\beta\langle\sigma_z(t)\rangle - 2\beta[1 - \langle\sigma_z(t)\rangle] \\ &\cong -2\beta[1 + \langle\sigma_z(t)\rangle],\end{aligned} \qquad (4.123)$$

as in Section 4.9. As in the case of the level shift, therefore, the decay of the population difference $\langle\sigma_z(t)\rangle$ has contributions attributable both to the source field and the vacuum field when the field operators are antinormally ordered. These are important results for the physical interpretation of spontaneous emission and the Lamb shift, but before discussing them further it will be useful to consider one other ordering.

4.12 Symmetric Ordering

The symmetric ordering defined in Section 4.8 is special in that the operators $a_{\mathbf{k}\lambda}$ and $a_{\mathbf{k}\lambda}^\dagger$ appear in the combination that defines the vector potential. Equation (4.87), for instance, may be written as

$$\dot\sigma = -i\omega_o\sigma + \frac{\omega_o}{2\hbar c}\mathbf{d}_{12}\cdot[\sigma_z\mathbf{A} + \mathbf{A}\sigma_z]. \qquad (4.124)$$

Since

$$\frac{\omega_o}{2\hbar c}\mathbf{d}_{12}\cdot\mathbf{A}(t) = \frac{1}{2}\sum_{\mathbf{k}\lambda}C_{\mathbf{k}\lambda}[a_{\mathbf{k}\lambda}(0)e^{-i\omega_k t} + a_{\mathbf{k}\lambda,s}(t)] + \text{h.c.}$$

$$\cong \frac{1}{2}\sum_{\mathbf{k}\lambda}C_{\mathbf{k}\lambda}u_{\mathbf{k}\lambda}(0)e^{-i\omega_k t} + \frac{1}{2}(\beta - i\Delta_2)v(t)$$

$$+ i\Delta_1\sigma^\dagger(t) + \text{h.c.} \qquad (4.125)$$

in the Markovian approximation (4.100), we have

$$\langle\dot{\sigma}(t)\rangle \cong -i\omega_o\langle\sigma(t)\rangle + \frac{1}{2}[\beta - i(\Delta_2 + \Delta_1)]\langle\sigma_z(t)\sigma(t) + \sigma(t)\sigma_z(t)\rangle$$

$$+ \frac{1}{2}[\beta + i(\Delta_2 + \Delta_1)]\langle\sigma_z(t)\sigma^\dagger(t) + \sigma^\dagger(t)\sigma_z(t)\rangle$$

$$+ \frac{1}{2}\sum_{\mathbf{k}\lambda}C_{\mathbf{k}\lambda}[\langle\sigma_z(t)a_{\mathbf{k}\lambda}^\dagger(0)\rangle e^{i\omega_k t} + \langle a_{\mathbf{k}\lambda}(0)\sigma_z(t)\rangle]e^{-i\omega_k t}$$

$$(4.126)$$

when a vacuum expectation value is taken. Since the two-state operators satisfy $\sigma_z(t)\sigma(t) + \sigma(t)\sigma_z(t) = 0$, furthermore, (4.126) simplifies to

$$\langle\dot{\sigma}(t)\rangle \cong -i\omega_o\langle\sigma(t)\rangle + \frac{1}{2}\sum_{\mathbf{k}\lambda}C_{\mathbf{k}\lambda}[\langle\sigma_z(t)a_{\mathbf{k}\lambda}^\dagger(0)\rangle e^{i\omega_k t} + \langle a_{\mathbf{k}\lambda}(0)\sigma_z(t)\rangle]e^{-i\omega_k t}.$$

$$(4.127)$$

Without continuing the calculation further we can already draw an important conclusion: when the field operators are symmetrically ordered there is no contribution from the *source* field to the level shifts. The shifts come entirely from the interaction of the atom with the vacuum field. The vacuum contribution to (4.127) can be read from (4.120) and (4.121), with the familiar (RWA) result

$$\langle\dot{\sigma}(t)\rangle \cong -i[\omega_o - i(\Delta_2 - \Delta_1)]\langle\sigma(t)\rangle - \beta\langle\sigma(t)\rangle. \qquad (4.128)$$

Thus, whereas a normal ordering leads us to interpret the level shifts in terms of the source field alone, and an antinormal ordering leads us to interpret them in terms of both source and vacuum fields, a symmetric ordering suggests an interpretation in terms of *only the vacuum field* (Milonni, Ackerhalt, and Smith, 1973; Milonni and Smith, 1975).

The situation is quite different when we calculate $\langle\dot{\sigma}_z(t)\rangle$ with a symmetric ordering of field operators. In this case we obtain a vacuum field contribution $-2\beta\langle\sigma_r(t)\rangle$ and a source field contribution -2β. In fact there is no ordering scheme in which the decay of $\langle\sigma_z(t)\rangle$, and therefore the loss of atomic energy due to spontaneous emission, can be attributed entirely to the vacuum field (Milonni et al., 1973; Milonni and Smith, 1975).

Cohen-Tannoudji and others have advocated symmetric ordering as a means of removing the "ambiguity" between source and vacuum field effects. They arrive at the symmetric ordering as the preferred one by imposing an additional condition that source and vacuum field contributions be separately Hermitian (Dalibard, Dupont-Roc, and Cohen–Tannoudji, 1982, 1984; Cohen-Tannoudji, 1984). This condition facilitates physical interpretation along classical lines in the sense that the field annihilation and creation operators appear in the symmetric combination defining the vector potential. Of course this too is a matter of taste, since nothing in the quantum formalism requires the source and vacuum contributions to be *separately* Hermitian. In matters of taste there can be no disputes.

4.13   Remarks

The freedom to order atomic and field operators in any way we like allows us to interpret spontaneous emission and radiative shifts in different ways and reconciles what for many years were perceived as two quite distinct physical interpretations.[8]

Spontaneous emission is one of the most ubiquitous of natural phenomena, and in the Introduction to Chapter 3 we stressed the importance of the theory of spontaneous emission in the development of quantum mechanics. We noted that Landau and van Vleck, among others, interpreted spontaneous emission as a consequence of radiation reaction, as did Dirac, who wrote in his 1927 paper that his theory "must presumably give the effect of radiation reaction on the [atom]." Similarly, Slater wrote that "The part of the field originating from the given atom is supposed to induce a probability that the atom lose energy spontaneously, while radiation from external sources is regarded as inducing additional probabilities that it gain or lose energy, much as Einstein has suggested ...".[9]

Interpretations in terms of the vacuum field arose in connection with efforts to understand radiative corrections and to remove divergences in QED calculations. We noted in Chapter 3 that such an interpretation

---

[8] This way of reconciling the radiation reaction and vacuum field interpretations was found independently by P. W. Milonni, I. R. Senitzky, and W. A. Smith. (Senitzky, 1973; Milonni et al., 1973)

[9] J. C. Slater, *Nature* **113**, 307 (1924).

was advanced by Weisskopf and Welton, among many others, in this connection, and that it has persisted more or less intact up to more recent times. Thus Welton, in his 1948 paper interpreting the Lamb shift, wrote that "spontaneous emission can be thought of as forced emission taking place under the action of the fluctuating field," and Weisskopf, in a 1981 overview of quantum field theory, said that "spontaneous emission appears as a forced oscillation caused by the zero-point oscillations of the electromagnetic field."

We can now see that these interpretations were oversimplified. It is true that we can describe the Lamb shift, for instance, in terms of radiation reaction alone. It is also true that we can describe it solely in terms of the vacuum field, as in the derivations by Welton and Feynman, or the derivation that treats the Lamb shift as a Stark shift (Chapter 3). These interpretations emerge when the field operators are normally or symmetrically ordered, respectively. But we can use other orderings to interpret the Lamb shift as partly due to radiation reaction and partly to the vacuum field.

The situation is similar with regard to the loss of atomic energy due to spontaneous emission. Here, however, there is no ordering that allows us to attribute the process solely to the vacuum field. This points not only to an oversimplification in the glib statement that "spontaneous emission is stimulated by the vacuum field," but also to an *error*: as we noted in Section 3.2, a calculation based on this commonplace yields only *half* the spontaneous emission rate. We are now in a position to understand why this is so, and also why there is no "spontaneous absorption" of zero-point radiation.

The ordering that brings us closest to a classical explanation in terms of the electric field or vector potential is the symmetric one, where $a_{\mathbf{k}\lambda}(t)$ and its Hermitian conjugate appear in the combination corresponding to the electric field or vector potential.[10] We noted in the preceding section that in this ordering

$$\langle \dot{\sigma}_z(t) \rangle = \langle \dot{\sigma}_z(t) \rangle_{\mathrm{RR}} + \langle \dot{\sigma}_z(t) \rangle_{\mathrm{VF}} , \qquad (4.129)$$

where the radiation reaction and vacuum field contributions are respectively

$$\langle \dot{\sigma}_z(t) \rangle_{\mathrm{RR}} = -2\beta \qquad (4.130)$$

and

$$\langle \dot{\sigma}_z(t) \rangle_{\mathrm{VF}} = -2\beta \langle \sigma_z(t) \rangle. \qquad (4.131)$$

---

[10] Compare, for instance, equations (4.83) and (4.124).

Now $\langle \sigma_z(t) \rangle = p_2(t) - p_1(t) = 2p_2(t) - 1$, where $p_1(t)$ and $p_2(t)$ are respectively the lower- and upper-state probabilities. Thus

$$\dot{p}_2(t)_{RR} = -\beta = -\frac{1}{2}A_{21} \qquad (4.132)$$

and

$$\dot{p}_2(t)_{VF} = -\beta \langle \sigma_z(t) \rangle = -\frac{1}{2}A_{21}[2p_2(t) - 1]. \qquad (4.133)$$

Equation (4.133) confirms our conclusion in Section 3.2 that, if we take the vacuum-field picture of spontaneous emission seriously and perform a calculation based on it, we calculate only half the spontaneous emission rate A_{21} for an initially excited atom. The other half comes from radiation reaction, which is not part of the usual vacuum-field interpretation of spontaneous emission.

Moreover, the conventional interpretation in terms of the vacuum field alone does not even pose the question as to why there is no spontaneous *absorption* of zero-point radiation. Indeed, equation (4.133) implies

$$\dot{p}_2(0)_{VF} = \frac{1}{2}A_{21} \qquad (4.134)$$

for an atom initially in the lower state of the transition $[p_2(0) = 0]$. This would predict spontaneous absorption were it not for the fact that the radiation reaction contribution (4.132) adds to (4.134) to give $\dot{p}_2(t) = 0$ for a ground-state atom. In other words, *the dissipative effect of radiation reaction precludes spontaneous absorption of energy from the vacuum field.*

The balancing of the effects of radiation reaction and the vacuum field on a ground-state atom is consistent with our heuristic discussion of atomic stability in Section 3.3, where we in fact *assumed* such a balance. The same balance holds when the atom is replaced by a harmonic oscillator, and explains the need to omit the term $H_{osc}^{zp} H_F^{zp}$ in our discussion of thermal equilibrium in Section 2.12.

4.14   Weisskopf–Wigner Theory

In the Schrödinger picture the operators in the Hamiltonian (4.73) are time-independent, and the time evolution of the atom–field system is described by the Schrödinger equation, $i\hbar(\partial/\partial t)|\psi(t)\rangle = H|\psi(t)\rangle$. The initial state of interest is $|\psi(0)\rangle = |\phi\rangle \equiv |2\rangle|\text{vac}\rangle$, the state in which the atom is in the upper state and there are no photons in the field. This is not a stationary state of the atom–field system, for there are nonvanishing matrix elements of H connecting this state to other atom–field states. One such state is $|\phi_{k\lambda}\rangle = |1\rangle|1_{k\lambda}\rangle$, the state in which the atom is in the lower state and there is one photon in some mode (k, λ) of the field.

If we truncate our Hilbert space to the states $|\phi\rangle$ and $|\phi_{k\lambda}\rangle$, and write

$$|\psi(t)\rangle = b(t)|\phi\rangle + \sum_{k\lambda} b_{k\lambda}(t)|\phi_{k\lambda}\rangle, \qquad (4.135)$$

then the Schrödinger equation reduces to coupled differential equations for the amplitudes $b(t)$ and $b_{k\lambda}(t)$:

$$i\hbar\dot{b}(t) = \frac{1}{2}\hbar\omega_o b(t) - i\hbar\sum_{k\lambda} C_{k\lambda} b_{k\lambda}(t), \qquad (4.136)$$

$$i\hbar\dot{b}_{k\lambda}(t) = (\hbar\omega_k - \frac{1}{2}\hbar\omega_o)b_{k\lambda}(t) + i\hbar C_{k\lambda} b(t). \qquad (4.137)$$

We are again ignoring in (4.73) the $\mathbf{A}^2$ term, which has the effect of coupling the states $|1\rangle|1_{k\lambda}\rangle$ to states $|1\rangle|1_{k'\lambda'}\rangle$ and $|1\rangle|1_{k\lambda}, 1_{k'\lambda'}, 1_{k''\lambda''}\rangle$. As noted earlier, $\mathbf{A}^2$ does not affect the decay rate or observable level shifts in the dipole approximation, and therefore will not be of concern here.

Even without $\mathbf{A}^2$, however, the Hamiltonian (4.73) requires a larger set of basis states then is implied by (4.135): the interaction term

$$H_{\mathrm{INT}} = i\hbar \sum_{k\lambda} C_{k\lambda}[a_{k\lambda} + a_{k\lambda}^\dagger][\sigma - \sigma^\dagger] \qquad (4.138)$$

couples $|\phi_{k\lambda}\rangle$ not only to $|\phi\rangle$, but also to states $|2\rangle|1_{k\lambda}, 1_{k'\lambda'}\rangle$, which in turn couple to other states to further enlarge the set of basis states needed to represent $|\psi(t)\rangle$. If we restrict ourselves only to "energy-conserving" processes and make the replacement

$$H_{\mathrm{INT}} \rightarrow i\hbar \sum_{k\lambda} C_{k\lambda}[a_{k\lambda}^\dagger \sigma - \sigma^\dagger a_{k\lambda}], \qquad (4.139)$$

however, then the expansion (4.135) encompasses all the so-called essential states required to describe an evolution of $|\psi(t)\rangle$ from the initial state $|\phi\rangle$.

We shall first work within the essential-states approximation, where the time-dependent Schrödinger equation reduces to the coupled amplitude equations (4.136) and (4.137). It is convenient to write these equations in terms of amplitudes $c(t)$ and $c_{k\lambda}(t)$ defined by writing

$$b(t) = c(t)e^{-i\omega_o t/2}, \qquad (4.140)$$

$$b_{k\lambda}(t) = c_{k\lambda}(t)e^{-i\omega_o t/2}, \qquad (4.141)$$

from which

$$\dot{c}(t) = -\sum_{\mathbf{k}\lambda} C_{\mathbf{k}\lambda} c_{\mathbf{k}\lambda}(t), \tag{4.142}$$

$$\dot{c}_{\mathbf{k}\lambda}(t) = -i(\omega_k - \omega_o)c_{\mathbf{k}\lambda}(t) + C_{\mathbf{k}\lambda}c(t). \tag{4.143}$$

Thus

$$\dot{c}(t) = -\sum_{\mathbf{k}\lambda} C_{\mathbf{k}\lambda}^2 \int_0^t dt' c(t') e^{i(\omega_k - \omega_o)(t'-t)} \tag{4.144}$$

and

$$\dot{c}(t) \cong -c(t) \sum_{\mathbf{k}\lambda} C_{\mathbf{k}\lambda}^2 \int_0^t dt' e^{i(\omega_k - \omega_o)(t'-t)} \tag{4.145}$$

in the Markovian approximation. Then (4.96) and (4.98) give

$$\dot{c}(t) \cong -(\beta - i\Delta_2)c(t), \tag{4.146}$$

or

$$c(t) \cong e^{-\beta t} e^{i\Delta_2 t} = e^{-A_{21}t/2} e^{-i(-\hbar\Delta_2)t/\hbar} \tag{4.147}$$

for $c(0) = 1$. This implies that the upper-state probability decays exponentially ($|b(t)|^2 = e^{-A_{21}t}$) and that there is a shift $-\hbar\Delta_2$ from the unperturbed energy of the state $|2\rangle|vac\rangle$.

The essential-states and Markovian approximations in the Schrödinger picture therefore lead to the same decay rate and upper-level shift obtained within the rotating-wave and Markovian approximations in the Heisenberg picture. Note, however, that the lower-level shift, $-\hbar\Delta_1$, has not appeared in our Schrödinger-picture calculation. This is a consequence of the essential-states approximation used in connection with our assumed initial state.

Suppose we assume the initial state $|\phi'\rangle = |1\rangle|vac\rangle$ instead of $|2\rangle|vac\rangle$. The interaction (4.138) couples this state to $|\phi'_{\mathbf{k}\lambda}\rangle = |2\rangle|1_{\mathbf{k}\lambda}\rangle$ via the "energy-nonconserving" process associated with $\sigma^\dagger a_{\mathbf{k}\lambda}^\dagger$, and if we truncate our Hilbert space by writing

$$\begin{aligned}
|\psi(t)\rangle &= b'(t)|\phi'\rangle + \sum_{\mathbf{k}\lambda} b'_{\mathbf{k}\lambda}(t)|\phi'_{\mathbf{k}\lambda}\rangle \\
&= c'(t)e^{i\omega_o t/2}|\phi'\rangle + \sum_{\mathbf{k}\lambda} c'_{\mathbf{k}\lambda}(t)e^{i\omega_o t/2}|\phi'_{\mathbf{k}\lambda}\rangle, \tag{4.148}
\end{aligned}$$

we obtain

$$\dot{c}'(t) = -\sum_{\mathbf{k}\lambda} C_{\mathbf{k}\lambda} c'_{\mathbf{k}\lambda}(t), \qquad (4.149)$$

$$\dot{c}'_{\mathbf{k}\lambda}(t) = -i(\omega_k + \omega_o)c'_{\mathbf{k}\lambda}(t) + C_{\mathbf{k}\lambda} c'_{\mathbf{k}\lambda}(t), \qquad (4.150)$$

as our Schrödinger equation. The Markovian approximation then yields $c'(t) = e^{i\Delta_1 t}$ for $c'(0) = 1$, implying the level shift $-\hbar\Delta_1$ for the state $|1\rangle|\text{vac}\rangle$. To obtain this level shift we have had to include states coupled to the initial state $|\phi'\rangle$ by "energy-nonconserving" processes. Such processes were not included in the derivation leading to (4.147), and this is the reason for the absence of any shift of the unperturbed energy of the state $|\phi_{\mathbf{k}\lambda}\rangle$ in that derivation.

Of course the transitions $|1\rangle|\text{vac}\rangle \rightarrow |2\rangle|1_{\mathbf{k}\lambda}\rangle$ responsible for the energy shift in the state $|1\rangle|\text{vac}\rangle$ are virtual transitions, not real ones: for times long compared with $\hbar/\Delta E = \omega_o^{-1}$, $|b'(t)|^2 = 1$ and $|b'_{\mathbf{k}\lambda}(t)|^2 = 0$. [This "long-time" assumption is in fact used in (4.99).] The virtual transitions appear simply because none of the states in the expansion (4.148) of $|\psi(t)\rangle$ are eigenstates of the atom–field system, and therefore their unperturbed energies are not eigenvalues of the atom–field Hamiltonian. Some sort of coupling ("virtual transitions") between these states is required in order to get nonvanishing corrections to the unperturbed energies.

The Schrödinger-picture theory of spontaneous emission outlined previously is similar in spirit to that presented by Weisskopf and Wigner in 1930. They were concerned with the long-time dynamics of spontaneous emission, rather than simply the upper-state decay rate, in order to obtain the "natural" lineshape of a radiative transition. This lineshape may be obtained in terms of the probabilities for finding the emitted photon at different frequences ω_k. From (4.137),

$$
\begin{aligned}
|b_{\mathbf{k}\lambda}(t)|^2 &= C_{\mathbf{k}\lambda}^2 \Big| \int_0^t dt' b(t') e^{i(\omega_k - \frac{1}{2}\omega_o)(t'-t)} \Big|^2 \\
&= C_{\mathbf{k}\lambda}^2 \Big| \int_0^t dt' c(t') e^{i(\omega_k - \omega_o)(t'-t)} \Big|^2 \\
&\cong C_{\mathbf{k}\lambda}^2 \Big| \int_0^t dt' e^{i(\omega_k - \omega_o + i\beta)t'} \Big|^2 \\
&\cong \frac{C_{\mathbf{k}\lambda}^2}{(\omega_k - \omega_o')^2 + \beta^2}, \qquad (4.151)
\end{aligned}
$$

where in the third line we have used the approximation (4.147), denoting by ω_o' the Lamb-shifted frequency, and in the last line we have used $\beta t \gg 1$.

This result implies that the spectral lineshape of the spontaneously emitted radiation is approximately Lorentzian near the atomic transition frequency, with linewidth (half-width at half-maximum) β.

The principal approximation of Weisskopf and Wigner was that the decay of the upper-state probability amplitude is exponential; they assumed exponential decay and then argued that it is an excellent approximation. For this reason the approximations used to arrive at exponential decay are often referred to collectively as the *Weisskopf–Wigner approximation.* In our approach the Weisskopf–Wigner approximation is equivalent to the Markovian approximation plus the rotating-wave or esential-states approximation, depending on whether the problem is formulated in the Heisenberg or Schrödinger picture.

As noted in Section 4.9, a quantum-mechanical system whose energy spectrum is bounded from below cannot undergo purely exponential decay from an excited state to a state of lower energy.[11] It is not too difficult to obtain corrections to exponential decay by going beyond the Markovian approximation. In the Heisenberg picture, for instance, one finds, instead of the exponential decay implied by (4.105), that[12]

$$\langle \sigma(t) \rangle \cong \langle \sigma(0) \rangle \left[e^{-i\omega_o t} e^{-\beta t} + \left(\frac{\beta}{2\pi\omega_o} \right) \left(\frac{1}{\omega_o t} \right)^2 \right] \qquad (4.152)$$

for $\beta t \gtrsim 1$. Since $\beta << \omega_o$, the correction to exponential decay represented by the second term in brackets is significant only for times $t >> \beta^{-1}$. The correction to purely exponential decay is thus very small – too small to have been observed in experiments thus far.

4.15   Neoclassical Theory

The impetus to consider more carefully the physical interpretation of spontaneous emission and the Lamb shift in the early 1970s came from Jaynes's "neoclassical theory" (Crisp and Jaynes, 1969; Stroud and Jaynes, 1970; Jaynes 1973, 1978). Neoclassical theory was based on the recognition that rather few phenomena in nonrelativistic radiation theory actually require field quantization for their explanation, and its purpose was to explore just how far one could get without field quantization and possibly to point the way to alternatives to QED. Its primary focus was the theory of spontaneous emission and the Lamb shift.

---

[11] See L. A. Khalfin, *Sov. Phys. JETP* **6**, 1053 (1958); L. Fonda and G. C. Ghirardi, *Nuovo Cim.* **A10**, 850 (1972).
[12] See, for instance, P. L. Knight and P. W. Milonni, *Phys. Lett.* **A56**, 275 (1976).

In neoclassical theory all electromagnetic fields are treated classically. Thus the Hilbert space for an atom in the vacuum is the space spanned by a complete set of atomic states. There is no vacuum field, and even the radiation reaction field of an atom is regarded as a classical, c-number field.

The quickest way to arrive at the neoclassical equations for a two-state atom is to go back to (4.104) and note that the operators $\sigma(t)$ and $\sigma^\dagger(t)$ in the products $\sigma_z(t)\sigma(t)$ and $\sigma_z(t)\sigma^\dagger(t)$ stem from the atom's radiation reaction field. In neoclassical theory this field is treated classically, and $\sigma(t)$ and $\sigma^\dagger(t)$ are therefore replaced by the c-numbers $\langle\sigma(t)\rangle$ and $\langle\sigma^\dagger(t)\rangle$, respectively. In other words, in neoclassical theory the expectation value $\langle\sigma_z(t)\sigma(t)\rangle$ is replaced by $\langle\sigma_z(t)\rangle\langle\sigma(t)\rangle$, and similarly all such expectation values of operator products are factored into a product of two expectation values. Thus, in the rotating-wave approximation, (4.104) is replaced by

$$\langle\dot\sigma(t)\rangle = -i\omega_o\langle\sigma(t)\rangle + [\beta - i(\Delta_2 + \Delta_1)]\langle\sigma_z(t)\rangle\langle\sigma(t)\rangle \qquad (4.153)$$

in neoclassical theory. Writing $\langle\sigma(t)\rangle = u(t)e^{-i\omega_o t}$, and assuming $u(t)$ and $\langle\sigma_z(t)\rangle$ are slowly varying compared with $e^{-i\omega_o t}$, we obtain

$$\dot x(t) = \beta x(t)z(t) - \gamma y(t)z(t), \qquad (4.154)$$

$$\dot y(t) = \beta y(t)z(t) + \gamma x(t)z(t), \qquad (4.155)$$

where $\frac{1}{2}x$ and $-\frac{1}{2}y$ are the real and imaginary parts, respectively, of u, and $\gamma \equiv \Delta_2 + \Delta_1, z(t) \equiv \langle\sigma_z(t)\rangle$. The neoclassical equation for $z(t)$ is found similarly by replacing $\langle\sigma^\dagger(t)\sigma(t)\rangle$ by $\langle\sigma^\dagger(t)\rangle\langle\sigma(t)\rangle = \frac{1}{4}[x^2(t) + y^2(t)]$ in (4.111):

$$\dot z(t) = -\beta[1 - z^2(t)]. \qquad (4.156)$$

We have used the facts that $x^2(t) + y^2(t) + z^2(t)$ is a constant and that this constant is 1 since that is its value for an atom in the upper or lower state ($x = y = 0, z = \pm 1$). Equations (4.154)–(4.156) are the neoclassical equations for a two-state atom in vacuum (Jaynes and Cummings, 1963; Stroud and Jaynes, 1970).

The most obvious difference between the neoclassical equations and the corresponding QED equations (4.105) and (4.111) is that the neoclassical equations are nonlinear. Furthermore, for an initially excited atom ($z(0) = +1$), neoclassical theory predicts that $\dot z(t) = 0$, so that the atom remains in the upper state. However, it is trivial to show that this initial state is unstable against small perturbations, and so it is not immediately obvious that neoclassical theory is in conflict with experiments on the spontaneous decay of excited atomic states. The general solution of equation (4.156) is

$$z(t) = -\tanh \beta(t - t_o), \qquad (4.157)$$

where t_e specifies the initial value $\gamma(0)$

Regarding the neoclassical frequency shifts, note first of all that

$$
\begin{aligned}
\gamma = \Delta_2 + \Delta_1 \quad &= \quad \frac{2\omega_o^2 |\mathbf{d}_{12}|^2}{3\pi\hbar c^3} \int_0^\infty d\omega \left(\frac{\omega}{\omega - \omega_o} + \frac{\omega}{\omega + \omega_o} \right) \\
&\rightarrow \quad \frac{4\omega_o^2 |\mathbf{d}_{12}|^2 \Omega}{3\pi\hbar c^3} \ ,
\end{aligned}
\tag{4.158}
$$

where Ω is a high-frequency cutoff, is the *sum* rather than the difference of the upper- and lower-state energy shifts divided by $\hbar$. It is the difference, of course, that appears in the QED theory [Equation (4.105)]. Furthermore, owing to the nonlinearity of equations (4.154) and (4.155), the frequency shift of the radiation emitted by the atom is predicted to be time dependent in neoclassical theory (Stroud and Jaynes, 1970). There appears to be no experimental evidence for such "chirped" level shifts.[13]

Such predictions of neoclassical theory are quite different from the QED predictions. But the most compelling evidence against neoclassical theory comes from the photon polarization correlations measured in three-level atomic cascades, the same type of experiments that have confirmed the predictions of quantum theory vis-à-vis those of local hidden variable theories (Clauser, 1972; Clauser and Shimony, 1978). The failure of neoclassical theory here is basically a failure of classical radiation theory in general to properly describe photon polarization (see also Milonni 1976, 1984).

Although neoclassical theory does not appear to suggest any viable alternative to QED, the controversy surrounding it in the 1970s served a very useful purpose. In particular, it led to a widespread appreciation of how successful semiclassical radiation theories can be and how the necessity of nonrelativistic QED can be fully supported only by resorting to rather subtle effects and experiments. And neoclassical theory led the way to a QED source–field approach to spontaneous emission and the Lamb shift, based mainly on the fact that the expectation values of operator products involving the atom's reaction field could *not* be factored as in our derivation of the neoclassical equations (Ackerhalt, Knight, and Eberly, 1973). This is turn raised old questions about the role of the vacuum field, and resulted in the reconciliation of the source and vacuum interpretations discussed in Sections 4.10–4.13.

---

[13] It might also be noted that what is regarded as an electromagnetic mass term and removed by mass renormalization in QED is taken to be physically observable in neoclassical theory. See Milonni (1976).

4.16   Discussion

We have shown how some old ideas about the physical origin of spontaneous emission and the Lamb shift can be formulated in the Heisenberg picture. For clarity we have dealt with a two-state "atom," but everything we have deduced about the nearly interchangeable roles of radiation reaction and the vacuum field in "causing" spontaneous emission and the Lamb shift applies also to a real atom (Milonni and Smith, 1975; Milonni, 1976).

There is one feature that is not brought out in the two-state model, and it can be understood most easily in terms of radiation reaction: the evolution of $\langle\sigma_{ij}(t)\rangle$ for a real atom is influenced not only by the radiation reaction field associated with the dipole moment for the $i \leftrightarrow j$ transition, but also by dipole moments of other transitions. This leads to the appearance in the theory of generalized decay rates of the form

$$\beta_{lmnp} = \frac{2\omega_{pn}^3}{3\hbar c^3} \mathbf{d}_{lm} \cdot \mathbf{d}_{np} \qquad (4.159)$$

and similarly generalized frequency shifts (see Milonni and Smith, 1975; Milonni, 1976). Such terms are usually negligible, and indeed the effect of one transition on another is typically ignored altogether in an approximation akin to the RWA. For some purposes, however, these generalized damping and frequency shift terms should be retained.[14] Since they have no specific bearing on radiation reaction or the vacuum field and can be described in terms of either, we shall not discuss them further here.

It can be argued that the physical equivalence of radiation-reaction and vacuum-field interpretations of level shifts is already clear from Figure 3.2, which represents a level shift in terms of the emission and absorption of virtual photons. This is obviously a vacuum field process, in that the field is in the vacuum state "before" the emission of the virtual photon and "after" its absorption. As discussed in Sections 3.6–3.8, we can in fact derive the level shift by taking the viewpoint that the fluctuating vacuum field acts as an external perturbation of the atom. On the other hand we can regard the emission and absorption of virtual photons in Figure 3.2 as defining a self-interaction. This viewpoint, too, can be supported by deriving the level shift using the atom's radiation reaction field as the perturbation. The interchangeability of these two viewpoints may be explicitly demonstrated

---

[14] R. J. Cook, *Phys. Rev.* A29, 1583 (1984), and references therein. The generalized damping and frequency shift terms were apparently first discussed in the modern literature by Milonni and Smith (1975), although such terms appeared in an old paper by Landau that is more noteworthy as an early application of the density-matrix formalism. See *Collected Papers of L. D. Landau*, ed. D. ter Haar (Gordon and Breach, New York, 1965), pp. 8–18.

using different but equivalent orderings of field operators, as we have shown in this chapter.

One can take a stance "against interpretation," arguing that the calculation of the spontaneous emission rate or the level shift can be carried through without ever having to worry about physical interpretation. But, if "QED is the best theory we have," and if in fact the vacuum "holds the key to a full understanding of the forces of nature,"[15] then certainly every attempt should be made to identify the physical mechanisms underlying QED vacuum effects.

Indeed, as discussed in Sections 3.2 and 4.13, physicists have for many years sought a "cause" for spontaneous emission. We have emphasized that the prevailing view of spontaneous emission as being induced by the vacuum field is, however, not quite correct; it is not possible to describe the loss of atomic energy to radiation without invoking the source field. The source field provides a "missing 1/2" and prevents an atom from undergoing "spontaneous excitation" by absorption of energy from the vacuum.

4.17   Bibliography

Ackerhalt, J. R., P. L. Knight, and J. H. Eberly, "Radiation Reaction and Radiative Frequency Shifts," *Phys. Rev. Lett.* **30**, 456 (1973).

Clauser, J. F., "Experimental Limitations to the Validity of Semiclassical Radiation Theories," *Phys. Rev.* **A6**, 49 (1972).

Clauser, J. F. and A. Shimony, "Bell's Theorem: Experimental Tests and Implications," *Rep. Prog. Phys.* **41**, 1881 (1978).

Cohen-Tannoudji, C., "Simple Physical Pictures for Radiative Processes: Vacuum Polarization versus Radiation Reaction," in *Quantum Electrodynamics and Quantum Optics*, ed. A.O. Barut (Plenum Press, New York, 1984).

Crisp, M. D. and E. T. Jaynes, "Radiative Effects in Semiclassical Theory," *Phys. Rev.* **179**, 1253 (1969).

Dalibard, J., J. Dupont-Roc, and C. Cohen-Tannoudji, "Vacuum Fluctuations and Radiation Reaction: Identification of Their Respective Contributions," *J. Physique* **43**, 1617 (1982).

Dalibard, J., J. Dupont-Roc, and C. Cohen-Tannoudji, "Dynamics of a Small System Coupled to a Reservoir: Reservoir Fluctuations and Self-Reaction," *J. Physique* **45**, 637 (1984).

Jaynes, E. T. and F. W. Cummings, "Comparison of Quantum and Semiclassical Radiation Theories with Applications to the Beam Maser,"

---

[15]P. C. W. Davies, *Superforce* (Simon and Schuster, New York, 1985), p. 104.

Proc. IEEE **51**, 89 (1963).

Jaynes, E. T., "Survey of the Present Status of Neoclassical Radiation Theory," in *Coherence and Quantum Optics III*, by L. Mandel and E. Wolf (Plenum Press, New York, 1973).

Jaynes, E. T., "Electrodynamics Today," in *Coherence and Quantum Optics IV*, ed. L. Mandel and E. Wolf (Plenum Press, New York, 1978).

Louisell, W. H., *Radiation and Noise in Quantum Electronics* (McGraw–Hill, New York, 1964).

Milonni, P. W., "Semiclassical and Quantum-Electrodynamical Approaches in Nonrelativistic Radiation Theory," *Phys. Rep.* **25**, 1 (1976).

Milonni, P. W., "Wave-Particle Duality of Light: A Current Perspective," in *The Wave-Particle Dualism*, ed. S. Diner, D. Fargue, G. Lochak, and F. Selleri (Reidel, Dordrecht, 1984).

Milonni, P. W., J. R. Ackerhalt, and W. A. Smith, "Interpretation of Radiative Corrections in Spontaneous Emission," *Phys. Rev. Lett.* **31**, 958 (1973).

Milonni, P. W. and W. A. Smith, "Radiation Reaction and Vacuum Fluctua tions in Spontaneous Emission," *Phys. Rev.* **A11**, 814 (1975).

Power. E. A. and S. Zienau, "Coulomb Gauge in Non-Relativistic Quantum Electrodynamics and the Shape of Spectral Lines," *Phil. Trans. Roy. Soc.* **A251**, 427 (1959).

Sakurai, J. J., *Advanced Quantum Mechanics* (Addison–Wesley, Reading, Mass., 1976).

Senitzky, I. R., "Radiation-Reaction and Vacuum–Field Effects in Heisenberg–Picture Picture Quantum Electrodynamics," *Phys. Rev. Lett.* **31**, 955 (1973).

Stroud, C. R. and E. T. Jaynes, "Long-Term Solutions in Semiclassical Radiation Theory," *Phys. Rev.* **A1**, 106 (1970).

Weisskopf, V. and E. Wigner, "Berechnung der natürlichen Linienbreite auf Grund der Diracshen Lichttheorie," *Z. Phys.* **63**, 54 (1930).

Weisskopf, V., "The Development of Field Theory in the Last 50 Years," *Physics Today* **34** (November 1981), 69.

Welton, T. A., "Some Observable Effects of the Quantum-Mechanical Fluctuations of the Electromagnetic Field," *Phys. Rev.* **74**, 1157 (1948).

Chapter 5

Interlude: Radiation Reaction

> ... we have allowed what is perhaps a silly thing, the possibility of the 'point' electron acting on itself.
> — Richard P. Feynman (1964)

5.1 Introduction

In QED a charged particle in vacuum is acted upon by its own radiation reaction field as well as the vacuum field. In standard classical electrodynamics, however, there is only the radiation reaction field to act on a single particle in the vacuum. The possibility of ascribing spontaneous emission and the Lamb shift to radiation reaction thus provides a more classical interpretation than that based on the fluctuating vacuum field. Mass renormalization, for instance, may be viewed as a natural extension of the classical theory, where the electromagnetic mass of a point charge is likewise found to be infinite.

As noted in Section 2.6, the vacuum and radiation reaction fields have a fluctuation–dissipation connection, and both are required for the consistency of QED. The idea of a fluctuating vacuum field existing in the absence of any sources may seem peculiar from a conventional classical standpoint, but when it comes to radiation reaction the classical theory itself has some curious features. In this chapter we briefly review some aspects of the theory of radiation reaction. Sections 5.2–5.4 are deliberately sketchy, since much of the material is covered in standard texts or reviews (Erber, 1961;

Plass, 1961; Jackson, 1975; Klepikov, 1985; Grandy, 1991), and we focus primarily on the physical ideas rather than details of the formalism.

5.2   The Abraham–Lorentz Equation

The electric field of a point charge e at the position of the point charge is (Appendix D)

$$\mathbf{E}_{\mathrm{RR}} = \frac{2e}{3c^3}\,\dddot{\mathbf{x}} - \frac{\delta m}{e}\ddot{\mathbf{x}}, \tag{5.1}$$

where δm is the electromagnetic mass, which is discussed in Section 3.5 and in more detail later. The equation of motion for a point charge in the vacuum is therefore

$$m_o\ddot{\mathbf{x}} = \tau m\,\dddot{\mathbf{x}} - \delta m\ddot{\mathbf{x}} + e\mathbf{E}_o(t), \tag{5.2}$$

where $\tau = 2e^2/3mc^3$, m_o is the "bare mass," and $\mathbf{E}_o(t)$ is the source-free, vacuum electric field operator. Thus we have

$$m\langle\ddot{\mathbf{x}}\rangle = \tau m\langle\dddot{\mathbf{x}}\rangle \tag{5.3}$$

for the vacuum expectation value of $\mathbf{x}$, where $m = m_o + \delta m$ is the observed particle mass. This equation has the same form as the classical equation for a point charge in the vacuum. If a force $\mathbf{F}_{\mathrm{ext}}$ is also applied to the particle, then the classical equation of motion is

$$m(\ddot{\mathbf{x}} - \tau\,\dddot{\mathbf{x}}) = \mathbf{F}_{\mathrm{ext}}\,. \tag{5.4}$$

We will for now restrict ourselves to the classical nonrelativistic theory of a point charge. The equation (5.4) of this theory is called the *Abraham–Lorentz equation*.

The classical theory of radiation reaction leading to the Abraham–Lorentz equation has the peculiarity that the third derivative of $\mathbf{x}$ appears. This means that the particle motion is determined by its position, velocity, *and* acceleration at time $t = 0$. This is in marked contrast to the usual situation in classical theory, where the initial positions and velocities alone determine particle trajectories. If $\mathbf{F}_{\mathrm{ext}} = 0$, for instance, (5.4) implies that the acceleration $\mathbf{a}(t) = \ddot{\mathbf{x}}(t)$ grows exponentially:

$$\mathbf{a}(t) = \mathbf{a}(0)e^{t/\tau}\,. \tag{5.5}$$

This is a so-called runaway solution of the Abraham–Lorentz equation. The runaway problem persists when $\mathbf{F}_{\mathrm{ext}} \neq 0$:

$$\mathbf{a}(t) = [\mathbf{a}(0) - \frac{1}{m\tau}\int_0^t dt'\,\mathbf{F}_{\mathrm{ext}}(t')e^{-t'/\tau}]e^{t/\tau}\,. \tag{5.6}$$

Here, $\mathbf{a}(t)$ eventually increases as $e^{t/\tau}$ *unless* the initial acceleration $\mathbf{a}(0)$ has the particular value

$$\mathbf{a}(0) = \frac{1}{m\tau} \int_0^\infty dt' \mathbf{F}_{\text{ext}}(t') e^{-t'/\tau} , \qquad (5.7)$$

in which case the solution of the Abraham–Lorentz equation is

$$\mathbf{a}(t) = \frac{1}{m\tau} \int_t^\infty dt' \mathbf{F}_{\text{ext}}(t') e^{(t-t')/\tau} = \frac{1}{m\tau} \int_0^\infty dt' \mathbf{F}_{\text{ext}}(t' + t) e^{-t'/\tau} . \qquad (5.8)$$

Dirac (1938) proposed the particular initial acceleration (5.7) as a way of avoiding runaway solutions of the Abraham–Lorentz equation. The solution (5.8) is physically very peculiar: the acceleration $\mathbf{a}(t)$ depends on the force $\mathbf{F}_{\text{ext}}(t + t')$ at times greater than t. That is, the particle *anticipates* future variations of the applied force. This disturbing feature of classical radiation reaction without runaways is called *preacceleration*.

Although preacceleration does not seem to make good physical sense, it is not as a practical matter an observable phenomenon. The time τ ($\cong$ 6.3×10^{-24} sec) is so short that for realistic variations of the applied force,

$$\mathbf{a}(0) \cong \frac{1}{m\tau} \mathbf{F}_{\text{ext}}(0) \int_0^\infty dt' e^{-t'/\tau} = \frac{1}{m} \mathbf{F}_{\text{ext}}(0) \qquad (5.9)$$

and

$$\mathbf{a}(t) \cong \frac{1}{m} \mathbf{F}_{\text{ext}}(t), \qquad (5.10)$$

which is the acceleration predicted when radiation reaction is ignored. Here, τ is on the order of the time it takes for light to propagate a distance equal to the classical electron radius ($r_o = e^2/mc^2 \cong 2.82 \times 10^{-13}$ cm); more generally it is roughly on the order of the time it takes for light to propagate across an elementary particle. Therefore we cannot expect the "acausal" preacceleration in the solution (5.8) of the Abraham–Lorentz equation to have any *observable* consequences.

And more important, classical theory itself breaks down at distances and times down to r_o and τ. If we could turn on a force $\mathbf{F}_{\text{ext}}$ in a time on the order of τ, we would have an uncertainty in energy on the order of $\Delta E \approx \hbar/\tau \approx \hbar mc^3/e^2 = \alpha^{-1}(mc^2) \cong 137mc^2$. Thus the energy uncertainty associated with such a small time scale exceeds the electron's rest energy, and distinctly quantum effects cannot be ignored. A force with such rapid time variations would have frequency components large enough to produce electron–positron pairs, and the very idea of a single classical electron becomes irrelevant (Chapter 9).

It is interesting to note that Dirac (1938) suggested that preacceleration might be associated with a failure of the theory when applied to the interior of the electron, that perhaps "it is possible for a signal to be transmitted faster than light through the interior of an electron. The finite size of the electron now reappears in a new sense, the interior of the electron being a region of failure ... of some of the elementary properties of space–time." However, Wheeler and Feynman (1945) considered a macroscopic situation where charges are packed together with separations $\approx r_o$ and found that the time scale of preacceleration is substantially *decreased* compared with that in the case of a single charge. They concluded from this result that "it is misleading to attribute the phenomenon of preacceleration to an abnormal velocity of light or to a failure of the usual conceptions of space–time in the immediate neighborhood of a charged particle."

5.3   Electromagnetic Mass

Another difficulty with the classical theory of radiation reaction is that the electromagnetic mass δm turns out to be infinite. According to the calculation in Appendix D,

$$\delta m = \frac{4e^2}{3\pi c^3} \int_0^\infty d\omega. \tag{5.11}$$

Now the nonrelativistic theory used in obtaining this result breaks down for frequencies approaching $mc^2/\hbar$. It might therefore be argued that we should cut off the upper limit of integration in (5.11) with some frequency $\omega_{\max} \approx mc^2/\hbar$, as in the nonrelativistic theory of the Lamb shift in Chapter 3. However, this hardly solves the divergence problem, since the fully relativistic theory also yields a divergent electromagnetic mass (Chapters 11 and 12).

One of the hopes of theorists at the beginning of the twentieth century was that the mass of the electron would turn out to be entirely electromagnetic. If we introduce a high-frequency cutoff $\omega_{\max}$ in (5.11) and set δm equal to the observed electron mass m, we find $\omega_{\max} = 3\pi mc^3/4e^2$, or

$$\lambda_{\min} = \frac{2\pi c}{\omega_{\max}} = \frac{8e^2}{3mc^2} = \frac{8}{3}r_o \tag{5.12}$$

for the smallest wavelength of radiation interacting with the electron after the cutoff.

The nonrelativistic equation of motion (5.4) was first derived by Lorentz in the 1890s. He assumed that the electron is a rigid body with finite dimensions. In his theory (5.4) was obtained as an approximation in which structure-dependent terms were dropped. Since (5.4) was regarded only as an approximation, Lorentz was not concerned with runaway solutions.

The distinctive feature of Lorentz's calculation is a series expansion in powers of the retardation time across the assumed *finite* extent of the charge c. The result, in terms of the velocity $\mathbf{v}$ of the center of mass and charge, is (Appendix D)

$$\mathbf{F}_{\mathrm{RR}} = -\frac{2e^2}{3c^3} \sum_{n=0}^{\infty} \frac{A_n}{n!} \left(\frac{a}{c}\right)^{n-1} \frac{d^{n+1}\mathbf{v}}{dt^{n+1}} \tag{5.13}$$

in the approximation of dropping terms of second and higher order in v/c. Here

$$A_n \equiv \frac{(-1)^n}{e^2} \int d^3x' \int d^3x \left[\frac{|\mathbf{x} - \mathbf{x}'|}{a}\right]^{n-1} \rho(\mathbf{x})\rho(\mathbf{x}') , \tag{5.14}$$

$\rho(\mathbf{x})$ is the (spherically symmetric) charge density, and a is a length characterizing the extent of the charge distribution. The coefficient

$$A_1 = -\frac{1}{e^2} \int d^3x' \int d^3x \rho(\mathbf{x})\rho(\mathbf{x}') = -1, \tag{5.15}$$

and

$$\frac{e^2}{2a} A_0 = \frac{1}{2} \int d^3x' \int d^3x \frac{\rho(\mathbf{x})\rho(\mathbf{x}')}{|\mathbf{x} - \mathbf{x}'|} \equiv U \tag{5.16}$$

is the electrostatic energy of the charge distribution. Thus

$$\mathbf{F}_{\mathrm{RR}} = -\frac{4}{3} \frac{U}{c^2} \frac{d\mathbf{v}}{dt} + \frac{2e^2}{3c^3} \frac{d^2\mathbf{v}}{dt^2} - \frac{2e^2}{3c^3} \left(\frac{a}{2c}\right) A_2 \frac{d^3\mathbf{v}}{dt^3} + \dots. \tag{5.17}$$

If we assume that the charge e is uniformly distributed over the surface of a sphere of radius a, for instance, then

$$U = \frac{1}{8\pi} \int_a^{\infty} \left(\frac{e}{r^2}\right)^2 4\pi r^2 dr = \frac{e^2}{2a} \tag{5.18}$$

and $a \approx r_o$ if we wish to have $U \approx mc^2$, i.e., if the mass is to be totally electromagnetic in origin.

The factor $(4/3)U/c^2$ in (5.17) is evidently the electromagnetic mass. Now if U were the total self-energy, special relativity would require $U = mc^2$. For this reason the factor of 4/3 has historically been a source of considerable discussion. One point of view holds that the very stability of

the electron requires there to be nonelectromagnetic, attractive forces acting on it, the so-called Poincaré stresses. Then the mass cannot be totally
electromagnetic, and special relativity would require only that the *total* self-
energy and mass are related by $U = mc^2$.[1] It is worth noting, nevertheless,
that one can define a purely electromagnetic self-energy-momentum four-
vector having the correct Lorentz transformation properties (Kwal, 1949;
Rohrlich, 1960; Jackson, 1975).

The second term in the series (5.17) is the Abraham–Lorentz force,
$m\tau\ddot{x}$. This term is independent of the assumed distribution of charge. The
third and higher order terms in (5.17) all vanish in the limit $a \to 0$ of a
point charge. The electromagnetic mass term, however, varies as a^{-1}, and
diverges in the limit of a point charge.

In QED the electron is viewed as a point particle. There is no experimental evidence of any internal electron structure. However, quantum
fluctuations act in effect to spread out the point electron (Chapter 11). As
recalled in the preceding section, classical considerations collapse at very
small scales of time and distance. These scales are of order $\Delta t \sim \hbar/mc^2$
and $\Delta d \sim \hbar/mc \equiv \lambda_c$, the electron Compton wavelength divided by 2π.
And λ_c is much larger than the classical electron radius:

$$\frac{\lambda_c}{r_o} = \frac{\hbar/mc}{e^2/mc^2} = \frac{\hbar c}{e^2} = \alpha^{-1} \cong 137. \tag{5.19}$$

Therefore, if we make a classical model of an electron in such a way that the
mass is totally electromagnetic, so that the electron radius $\approx r_o$, we are attempting a description on a distance scale much finer than quantum theory
deems sensible. Any attempt to localize the electron within such a short
distance would involve such a large uncertainty in energy that electron–
positron pairs can be created, and again the notion of a single classical
electron loses relevance. Pais has remarked, perhaps with some exaggera-
tion, that[2] "The investigations of the self-energy problem of the electron
by men like Abraham, Lorentz, and Poincaré have long since ceased to be
relevant. All that has remained from those early times is that we still do
not understand the problem."

The infinite electromagnetic mass of the point electron is often invoked
to reconcile the existence of runaway solutions with the conservation of
energy: the indefinite increase of the electron kinetic energy in a runaway

---

[1] It should be borne in mind, of course, that the theory just outlined is nonrelativistic.
See the interesting historical remarks near the end of Section 28-3 of Feynman, Leighton,
and Sands (1964).

[2] A. Pais, *'Subtle is the Lord ... '* *The Science and the Life of Albert Einstein* (Oxford
University Press, New York, 1982), p. 155.

mode comes at the expense of the *infinite* self-energy of the electron. However, all that is really necessary to reconcile runaway solutions with conservation of energy is the fact that the nonelectromagnetic, bare mass m_o is negative. Indeed the principle of energy conservation applied to the free electron involves the bare mass; the total energy is the (positive) energy of the field plus the (negative) kinetic energy associated with the bare mass. An accelerated motion of the electron implies a continually decreasing kinetic energy associated with the bare mass, and therefore a continually increasing radiated energy, as occurs in the case of a runaway mode (Coleman, 1961). The fact that the bare mass is negative is obviously peculiar, but it prevents any contradiction between conservation of energy and the existence of runaway solutions of the Abraham–Lorentz equation. We shall see later that runaways occur in classical theory *only* if the bare mass is negative.

Runaway solutions, and the divergence of δm in the point-charge limit, also plague the classical relativistic theory of radiation reaction. The divergence of δm is sometimes regarded as a failure of classical electrodynamics (Feynman et al., 1964). However, the electromagnetic self-energy of an electron also diverges in QED; the divergence there is "swept under the rug" by mass renormalization.[3] Mass renormalization in classical electrodynamics makes it no worse off in this regard than QED.

One problem that seems peculiar to the classical theory is that a non-runaway solution of the relativistic Lorentz–Dirac equation for radiation reaction is not necessarily unique (Baylis and Huschilt, 1976).

There have been attempts to modify the classical Maxwell theory by the adoption of boundary conditions different from the "logical" one in which only retarded fields are allowed. Dirac (1938), before the invention of renormalization, proposed to eliminate the divergent self-energy by using not the usual retarded fields but rather half the difference of retarded and advanced fields. The *advanced* reaction force may be obtained from the retarded force

$$\mathbf{F}_{RR}^{\text{ret}} = -\frac{4}{3}\frac{U}{c^2}\frac{d^2\mathbf{x}}{dt^2} + \frac{2e^2}{3c^3}\frac{d^3\mathbf{x}}{dt^3} - \frac{2e^2}{3c^3}\left(\frac{a}{2c}\right)A_2\frac{d^4\mathbf{x}}{dt^4} + \dots \qquad (5.20)$$

by the replacement $t \rightarrow -t$:

$$\mathbf{F}_{RR}^{\text{adv}} = -\frac{4}{3}\frac{U}{c^2}\frac{d^2\mathbf{x}}{dt^2} - \frac{2e^2}{3c^3}\frac{d^3\mathbf{x}}{dt^3} - \frac{2e^2}{3c^3}\left(\frac{a}{2c}\right)A_2\frac{d^4\mathbf{x}}{dt^4} + \dots , \qquad (5.21)$$

---

[3] In relativistic QED the divergence of δm is weaker than in classical theory — logarithmic rather than linear, see Chapters 11 and 12.

so that

$$\frac{1}{2}[\mathbf{F}_{\text{RR}}^{\text{ret}} - \mathbf{F}_{\text{RR}}^{\text{adv}}] = \frac{2e^2}{3c^3}\frac{d^3\mathbf{x}}{dt^3} \quad (a \to 0). \tag{5.22}$$

This boundary condition thus preserves the $\dddot{\mathbf{x}}$ term required to damp the particle motion in a way consistent with the loss of energy to radiation (Appendix A), while eliminating the (divergent) self-energy. In this approach *there is no electromagnetic mass*.

The "absorber" theory of Wheeler and Feynman (1945) offered a physical basis for Dirac's somewhat arbitrary choice of boundary conditions. In this classical theory there is fundamentally no radiation reaction at all: point charges interact only with *other* point charges, not with themselves. The interaction between charges is assumed to occur via half the retarded plus half the advanced fields. The radiative damping term proportional to $\dddot{\mathbf{x}}$ appears as a consequence of the *absorption* of the outgoing field of a charge by the rest of the universe. In particular, in this theory a charge that begins to accelerate is acted upon by a force from charges that will in the future absorb its outgoing radiation. Wheeler and Feynman went to considerable pains to show with specific models that this theory is a physically consistent one. As in Dirac's theory, there is no electromagnetic mass and no need for mass renormalization.

It was originally hoped that a quantized version of the Wheeler–Feynman theory would also be free of infinities. Unfortunately, the theory has eluded canonical quantization because it cannot be formulated as an initial-value problem. Furthermore it is presently believed that electromagnetic mass, which does not appear in the Dirac or Wheeler–Feynman theories, is necessary to explain mass differences such as that between the neutral and charged pions.

"But we still do not know what causes the electron to weigh."[4]

5.4   Does a Uniformly Accelerated Charge Radiate?

This perennial question arises from the observation that the rate at which an electron with acceleration a radiates is given (nonrelativistically) by the Larmor formula, $2e^2a^2/3c^3$, while the radiation reaction force in (5.4) is $2e^2\dot{a}/3c^3$. If a is constant, therefore, the electron apparently radiates without any radiation reaction force to account for its energy loss. That is, if a uniformly accelerated electron does indeed radiate, there seems to be a

---

[4] A. Pais, *"Subtle Is the Lord ...,"* p. 159.

conflict with the conservation of energy. One might be led to the conclusion that a uniformly accelerated charge does *not* radiate.[5]

The argument against radiation based on the presumed violation of energy conservation is simplistic and incorrect. Multiplication of both sides of (5.4) by $\mathbf{v} = \dot{\mathbf{x}}$ yields

$$
\begin{aligned}
\mathbf{F}_{\text{ext}} \cdot \mathbf{v} &= \frac{d}{dt}(\frac{1}{2}mv^2) - m\tau\mathbf{v} \cdot \ddot{\mathbf{v}} \\
&= \frac{d}{dt}(\frac{1}{2}mv^2) + \frac{2e^2}{3c^3}\mathbf{a}^2 - \frac{2e^2}{3c^3}\frac{d}{dt}(\mathbf{v} \cdot \mathbf{a}).
\end{aligned} \tag{5.23}
$$

Now the simple "derivation" of the radiation reaction force in Appendix A, for instance, assumes that the motion is periodic, or in any case such that the last term on the right in (5.23) may be ignored. That is, it is assumed that

$$
\mathbf{a}^2 = \frac{d}{dt}(\mathbf{v} \cdot \mathbf{a}) - \mathbf{v} \cdot \ddot{\mathbf{v}} \tag{5.24}
$$

may be replaced by $-\mathbf{v} \cdot \ddot{\mathbf{v}} = -\dot{\mathbf{x}} \cdot \dddot{\mathbf{x}}$. In general, however, and in particular for uniformly accelerated (hyperbolic) motion, the term $d(\mathbf{v} \cdot \mathbf{a})/dt$ is not zero, and

$$
\mathbf{F}_{\text{RR}} \cdot \mathbf{v} = \frac{2e^2}{3c^3}\mathbf{v} \cdot \dot{\mathbf{a}} = \frac{2e^2}{3c^3}\frac{d}{dt}(\mathbf{v} \cdot \mathbf{a}) - \frac{2e^2}{3c^3}\mathbf{a}^2 . \tag{5.25}
$$

Thus, for constant acceleration $\mathbf{a}$, $\mathbf{F}_{\text{RR}} = 0$ and

$$
\frac{2e^2}{3c^3}\mathbf{a}^2 = \frac{2e^2}{3c^3}\frac{d}{dt}(\mathbf{v} \cdot \mathbf{a}), \tag{5.26}
$$

so that the radiated power is not zero in spite of the fact that the radiation reaction force is zero.

The "Schott energy" $(2e^2/3c^3)(\mathbf{v} \cdot \mathbf{a})$ has been interpreted either as part of the internal energy of the charge (Fulton and Rohrlich, 1960) or as part of the energy of the field in the immediate vicinity of the particle (Coleman, 1961). It arises from the interference of the "velocity" and "acceleration" fields of a point charge.

It is now generally accepted that a uniformly accelerated point charge does radiate and that the radiation does not contradict the fact that the radiation reaction force vanishes during uniform acceleration. It has also been shown that the fact of radiation does not contradict the Principle of Equivalence in the general theory of relativity: an observer falling with a charge in a uniform gravitational field will detect no radiation, but only an electrostatic Coulomb field (Fulton and Rohrlich, 1960; Coleman, 1961).

---

[5] This conclusion was reached for different reasons by Pauli, von Laue, and others. See Fulton and Rohrlich (1960) for some of the history of work on this question.

Among the quantum effects ignored in the (classical) electron theory considered up to this point is the possibility, noted earlier, of creating electron–positron pairs out of the vacuum. There is another quantum feature that has been ignored in connection with a uniformly accelerated charge, namely the Unruh–Davies effect, the fact that the vacuum for a uniformly accelerated observer appears as a thermal equilibrium field of temperature $T = \hbar a/2\pi kc$ (Section 2.10). The fact that the vacuum appears in the instantaneous rest frame of a uniformly accelerated charge to be a thermal bath implies that there is no net transfer of energy and momentum with the perceived field; this is consistent with the facts that $\mathbf{F}_{RR} = 0$ and that the field produced by the charge is Coulombic (nonradiative) in its instantaneous rest frame (Sciama, Candelas, and Deutsch, 1981).

The thermal field perceived by the uniformly accelerated charge imparts momentum fluctuations to the particle, just as in the Einstein–Hopf model discussed in Chapter 1. If the particle does have an exactly constant acceleration, therefore, the external force responsible for this acceleration must itself fluctuate in such a way as to compensate for the fluctuations in the thermal bath. If, on the other hand, the external force were strictly constant, the fluctuations of the thermal field would produce nonuniform acceleration of the particle and so lead to a damping of its motion; the combination of this damping and the fluctuations would be expected to lead, as in the Einstein–Hopf model, to a steady-state momentum distribution appropriate to the temperature T (Sciama et al., 1981).

5.5   Remarks

Consider an electron in a spatially uniform electric field. The nonrelativistic equation of motion, including radiation reaction, is

$$\ddot{\mathbf{x}} - \tau\,\dddot{\mathbf{x}} = \frac{e}{m}\mathbf{E}_o(t). \tag{5.27}$$

Suppose we regard the radiation reaction force as a small perturbation, so that $\ddot{\mathbf{x}} \cong (e/m)\mathbf{E}_o(t)$ in the first approximation and

$$\ddot{\mathbf{x}} \;\cong\; \frac{e}{m}\mathbf{E}_o(t) + \frac{1}{m}\mathbf{f}_{RR}\ , \tag{5.28}$$

$$\mathbf{f}_{RR} \;=\; \tau e\dot{\mathbf{E}}_o(t), \tag{5.29}$$

in the next order of approximation. Now we ask under what conditions $\mathbf{f}_{RR}$ is in fact small compared with the applied force $e\mathbf{E}_o$.

Suppose $\mathbf{E}_o(t)$ is sinusoidal, so that $\mathbf{E}_o(t) = \mathbf{A}e^{-i\omega t}$ and $\mathbf{f}_{RR} = -i\omega\tau e$ $\times \mathbf{A}e^{-i\omega t}$. Then

$$\left| \frac{\mathbf{f}_{RR}}{e\mathbf{E}_o} \right| = \omega\tau \approx \frac{\omega r_o}{c} \cong \left(\frac{1}{137} \right) \frac{\omega\lambda_c}{c} \; . \tag{5.30}$$

But classical theory is inapplicable when ω is so large that $\omega\lambda_c/c \approx 1$, and so we can conclude from (5.30) that the classical electron theory is inapplicable unless $\omega\tau \ll 1$, i.e., unless the radiation reaction force is small compared with the external force. Ginzburg (1979) writes that

> A lot of trouble has been taken to prove that it is possible to apply [equation (5.4)] and its relativistic generalization as an exact equation where one imposes some additional conditions, for instance, to remove the [runaway] solutions. Apart from anything else it is obscure why all this is necessary. We do not know any classical problems where the radiation force (in the frame of reference in which the electron is at rest) cannot be considered to be a perturbation. The idea of a classical point particle is inconsistent, and the construction of 'elementary' particles must be solved including quantum effects.

The assumption that the radiation reaction field is a small perturbation has already been used in the theory of spontaneous emission and the Lamb shift in Chapter 4.[6] The principal approximation there, either in the form of the Markovian approximation in the Heisenberg picture or the Weisskopf–Wigner approximation in the Schrödinger picture, is that the atom is only slightly perturbed by the radiation reaction and vacuum fields.

What if we could solve the quantum theory of radiation reaction *exactly*? Would there be runaways and divergences? Some interesting progress in this direction is discussed in Section 5.7.

5.6   Extended-Charge Theories

While the classical electron theory may be considered ultimately irrelevant from the standpoint of quantum theory, it is obviously of great interest nonetheless. It is important, if only for historical reasons, to know what modifications of the standard classical theory might make it more sensible when it comes to radiation reaction. We have already alluded to the Wheeler–Feynman absorber theory. Another modification is to reject the very notion of the classical electron as a *point* charge.

---

[6] Recall the discussion near equation (4.114).

Suppose the electron is regarded as a spherical shell of charge of radius a, so that

$$\rho(\mathbf{x}) = \frac{e}{4\pi a^2}\delta(x - a) \quad (x = |\mathbf{x}|). \tag{5.31}$$

Then all the coefficients A_n in (5.13) may be evaluated explicitly:

$$
\begin{aligned}
A_n &= \frac{(-1)^n}{e^2}\int d^3x\rho(x)\frac{e}{4\pi a^2}\int d^3x'\delta(x'-a)\left[\frac{|\mathbf{x}-\mathbf{x}'|}{a}\right]^{n-1} \\
&= \frac{(-1)^n}{e^2}\frac{e}{4\pi a^2}\int d^3x\rho(x)(2\pi a^2)2^{(n-1)/2} \\
&\quad \times \int_0^\pi d\theta' \sin\theta'(1-\cos\theta')^{(n-1)/2} \\
&= \frac{(-2)^n}{n+1}, \tag{5.32}
\end{aligned}
$$

and

$$
\begin{aligned}
\mathbf{F}_{RR} &= -\frac{2e^2}{3c^3}\sum_{n=}^{\infty}\frac{(-2)^n}{(n+1)!}\left(\frac{a}{c}\right)^{n-1}\frac{d^{n+1}\mathbf{v}}{dt^{n+1}} \\
&= -\frac{2e^2}{3c^3}\left[\frac{c}{a}\frac{d\mathbf{v}}{dt} - \frac{1}{2}\left(\frac{c}{a}\right)^2\left(\mathbf{v}(t-\frac{2a}{c}) - \mathbf{v}(t) + \frac{2a}{c}\frac{d\mathbf{v}}{dt}\right)\right]. \\
&\tag{5.33}
\end{aligned}
$$

Using this result in the equation $m_o\ddot{\mathbf{x}} = \mathbf{F}_{RR}$, with $\mathbf{x}$ the center of mass coordinate, we have the delay-differential equation

$$m\ddot{\mathbf{x}} = \frac{e^2}{3ca^2}\left[\mathbf{v}(t-\frac{2a}{c}) - \mathbf{v}(t) + \frac{2a}{c}\frac{d\mathbf{v}}{dt}\right], \tag{5.34}$$

where the observed mass

$$m = m_o + \delta m = m_o + \frac{2e^2}{3c^2a} = m_o + m\left(\frac{c\tau}{a}\right). \tag{5.35}$$

We can write (5.34) in the form

$$\ddot{\mathbf{x}}(t) = \xi\left[\dot{\mathbf{x}}(t-\frac{2a}{c}) - \dot{\mathbf{x}}(t)\right], \tag{5.36}$$

where

$$\xi \equiv \frac{(c/2a)(c\tau/a)}{1-c\tau/a}. \tag{5.37}$$

The electromagnetic mass δm appearing in (5.35) is the same as that implied by (5.17). Similarly (5.34) reduces to the Abraham–Lorentz force in the limit $a \to 0$. Equation (5.36), however, is applicable for all $a > 0$; i.e., it describes the motion of a *nonpoint* electron under the action of its own radiation reaction field.

Unlike the Abraham–Lorentz equation, (5.36) does not involve the third derivative of $\mathbf{x}$ with respect to time, unless of course we make a Taylor expansion of $\dot{\mathbf{x}}(t - 2a/c)$ in powers of the retardation time $2a/c$. Such an expansion brings in all the derivatives of $\dot{\mathbf{x}}(t)$, as in (5.13). The main question before us is whether (5.36) with $a > 0$ has substantially different properties than the Abraham–Lorentz equation.

Equation (5.36) for a uniformly charged spherical shell was derived by Bohm and Weinstein (1948), among others. Herglotz, and much later Wildermuth (1955), also considered the spherical shell model, and from their work it follows that *there are no runaway solutions unless the bare mass m_o is negative*. As emphasized by Erber (1961), this "Herglotz–Wildermuth theorem" also reveals the point-charge limit $a \to 0$ as the root of the runaway solutions, for the finite and positive value of the observed mass $m = m_o + \delta m$, with $\delta m \to \infty$ for $a \to 0$, implies $m_o < 0$.

The conclusion about the absence of runaways follows easily in the case of a uniformly charged spherical shell described by (5.36) (Levine, Moniz, and Sharp, 1977). Assuming a solution of the form $\mathbf{x}(t) = \mathbf{x}_o e^{\lambda t/\tau}$, we obtain for λ the solution $\lambda = 0$ as well as solutions determined by

$$\frac{\lambda}{\tau} = \xi[e^{-2\lambda a/c\tau} - 1]. \tag{5.38}$$

Runaways are absent if the real parts of all roots λ are negative. Let $\lambda = (c\tau/2a)(\mu + i\nu)$, with μ, ν real, so that (5.38) becomes

$$\mu = g[e^{-\mu}\cos\nu - 1], \tag{5.39}$$

$$\nu = -ge^{-\mu}\sin\nu, \tag{5.40}$$

with

$$g \equiv \left(\frac{a}{c\tau} - 1\right)^{-1}. \tag{5.41}$$

Now since $e^{-\mu}\cos\nu - 1 < 0$ for $\mu > 0$, equation (5.39) does not have a positive solution for μ when $g > 0$. That is, there are no runaway solutions when $a > c\tau$. For $a < c\tau$, however, $g < 0$ and (5.39) always has a positive solution for μ with $\nu > 0$.

We conclude, therefore, that equation (5.36) has no runaway solutions when the radius a of the uniformly charged shell is larger than $c\tau = 2r_o/3$.

Runaway solutions occur only for radii smaller than $2r_v/3$. In the latter case the electromagnetic mass $\delta m = m(c\tau/a) > m$, and (5.35) implies that the bare mass $m_o < 0$, in agreement with the Herglotz–Wildermuth theorem (Levine et al., 1977).

For most of the twentieth century the classical electron theory, based on the presumption of a *point* electron, has suffered from the runaway and preacceleration maladies, as well as the divergent electromagnetic mass. It is seldom acknowledged that the classical theory is free of runaways if the radius of an *extended* charged particle is larger than the radius for which its observed mass would be entirely electromagnetic.

It should be noted that the classical *relativistic* theory of a point charge also has runaway solutions that are suppressed by the assumption of preacceleration. The relativistic theory of an extended charged particle has evidently not been developed. Such a theory would be very complicated because the assumption of a *rigid* charge distribution as earlier is inconsistent with special relativity.[7]

Consider solutions of (5.36) of the form $\mathbf{x}(t) = \mathbf{x}_o e^{-i\omega t}$. Since $\delta m = 2e^2/3ac^2$ for a uniformly charged spherical shell, and $\xi = (c/2a)(\delta m/m_o)$, we obtain the formula

$$m_o = \frac{ie^2}{3\omega ca^2}[e^{2i\omega a/c} - 1] \tag{5.42}$$

relating the nonelectromagnetic mass m_o, the oscillation frequency ω, and the radius a of the spherical shell. The conditions for such an oscillatory solution are $m_o = 0$ and $\omega = n\pi c/a$, where n is an integer. Thus, in order to have a purely oscillatory solution, the nonelectromagnetic (bare) mass must vanish and the oscillation frequency must be an integral multiple of $\pi c/a$.

These results were obtained by Bohm and Weinstein (1948). Note that the Bohm–Weinstein "self-oscillations" are harmonic motions of the center of mass of the extended charged particle and are not necessarily associated with any internal dynamical properties of the charge distribution. The *quantized* energy levels for the harmonic motion of frequency ω are separated by

$$\Delta E = \hbar \omega = n\pi \frac{\hbar c}{a} \approx n\pi \left(\frac{\hbar c}{e^2}\right) mc^2 \approx (400n)mc^2 \tag{5.43}$$

if we take $a \approx e^2/mc^2$. The main point of the Bohm–Weinstein work may be appreciated from their remark that "the energy of the first excited state

---

[7]It is well-known that the concept of a rigid body, throughout which physical information can be transmitted instantaneously, is prohibited by the special theory of relativity.

with $n = 1$ is not far from the rest energy of a π meson ... The idea then suggests itself that perhaps some kinds of mesons are really excited states of the electron. The decay from one kind of meson to another, or from meson to electron would then correspond simply to the loss of this excitation energy." Needless to say, this suggestion can no longer be taken seriously. Contemporary theory holds that the π-mesons are composed of quarks, while the electron is treated in QED calculations as a pure point particle with no "self-oscillations."

5.7   The Moniz–Sharp Theory

The finite electromagnetic mass and the absence of runaways in the classical theory of an extended charge suggest that it might be very interesting to study the *quantum* theory of an extended charge and its point charge limit. This has been done by Moniz and Sharp (1974, 1977) for a nonrelativistic, rigid, spherical charge distribution.

Moniz and Sharp begin their analysis with the following Heisenberg equation of motion for the (mean) position operator $\mathbf{R}$ of a particle with charge density ρ:

$$m_o \ddot{\mathbf{R}}(t) = e\mathbf{E}_o(\mathbf{R}(t), t) - \frac{2}{3c^2} \sum_{n=0}^{\infty} \frac{(-1)^n}{n! c^n} \int d^3x \int d^3x'$$

$$\times \frac{1}{2} \left[\rho(\mathbf{x} - \mathbf{R}(t)) |\mathbf{x} - \mathbf{x}'|^{n-1}, \frac{\partial^{n+1}}{\partial t^{n+1}} \mathbf{J}(\mathbf{x}', t) \right]_{+} \quad (5.44)$$

where $\mathbf{E}_o$ is the source-free (vacuum) field and the anticommutator $[A, B]_{+} \equiv AB + BA$. Except for the symmetrization implied by the anticommutator, the radiation reaction field implied by (5.44) is basically of the same form as the classical expression (D.17) of Appendix D. In (5.44)

$$\mathbf{J}(\mathbf{x}, t) = \frac{1}{2}[\rho(\mathbf{x} - \mathbf{R}), \dot{\mathbf{R}}]_{+} = \rho(\mathbf{x} - \mathbf{R})\dot{\mathbf{R}} - \frac{1}{2}[\rho(\mathbf{x} - \mathbf{R}), \dot{\mathbf{R}}]$$

$$= \rho(\mathbf{x} - \mathbf{R})\dot{\mathbf{R}} - \frac{1}{2m_o}[\rho(\mathbf{x} - \mathbf{R}), \mathbf{P}]$$

$$= \rho(\mathbf{x} - \mathbf{R})\dot{\mathbf{R}} - \frac{i\hbar}{2m_o} \nabla_{\mathbf{R}} \rho(\mathbf{x} - \mathbf{R})$$

$$= \rho(\mathbf{x} - \mathbf{R})\dot{\mathbf{R}} + \frac{i\hbar}{2m_o} \nabla_{\mathbf{x}} \rho(\mathbf{x} - \mathbf{R}). \quad (5.45)$$

Consider the $n = 0$ term in (5.44). We have

$$\frac{\partial}{\partial t} \mathbf{J}(\mathbf{x}, t) = \frac{1}{i\hbar}[\mathbf{J}, H] = \rho(\mathbf{x} - \mathbf{R})\ddot{\mathbf{R}} + \frac{\partial \rho}{\partial t}\dot{\mathbf{R}} + \frac{i\hbar}{2m_o} \nabla_{\mathbf{x}} \frac{\partial \rho}{\partial t}. \quad (5.46)$$

If, as in the classical calculation of Appendix D, we drop terms nonlinear in $\mathbf{R}$ in this nonrelativistic theory and use the assumption of a spherically symmetric charge distribution, then only $\rho\ddot{\mathbf{R}}$ in (5.46) contributes to the $n = 0$ term in (5.44):

$$(m_o\ddot{\mathbf{R}})_{n=0} = -\left[\frac{2}{3c^2}\int d^3x \int d^3x' \frac{\rho(\mathbf{x})\rho(\mathbf{x}')}{|\mathbf{x}-\mathbf{x}'|}\right]\ddot{\mathbf{R}}, \qquad (5.47)$$

in agreement with the $n = 0$ term of the classical calculation of Appendix D. Similarly the $n = 1$ term may be evaluated to give

$$(m_o\ddot{\mathbf{R}})_{n=1} = \frac{2e^2}{3c^3}\dddot{\mathbf{R}}, \qquad (5.48)$$

in agreement with the $n = 2$ term of the classical theory.

Differences between the classical and quantum theories arise from non-vanishing commutators, and the lowest order in which a difference not non-linear in $\mathbf{R}$ emerges is at $n = 2$:

$$(m_o\ddot{\mathbf{R}})_{n=2} = -\left[\frac{1}{3c^4}\int d^3x\int d^3x'\rho(\mathbf{x})\rho(\mathbf{x}')|\mathbf{x}-\mathbf{x}'|\right]\ddddot{\mathbf{R}}$$
$$-\left[\frac{8\pi}{3c^2}\left(\frac{\hbar}{m_o c}\right)^2\int d^3x\rho^2(\mathbf{x})\right]\ddot{\mathbf{R}}. \qquad (5.49)$$

It is the second term on the right that represents the quantum correction to the classical $n = 2$ term. This term modifies the classical expression for the electromagnetic mass δm implied by (5.47). If a is a characteristic radius of the charge distribution, then the quantum correction is $\sim (\hbar/m_o c)^2/a^2$ times the classical δm and is small if $a \gg \hbar/m_o c$. In the point-charge limit $a \ll \hbar/m_o c$, however, the quantum correction is large, and in this case one must account as well for all terms with $n > 2$. Moniz and Sharp obtain the following expression for the electromagnetic mass including all quantum corrections:

$$\delta m = \frac{2}{3c^2}\left(1+\frac{\lambda}{6}\frac{\partial}{\partial\lambda}\right)\left(1+\lambda\frac{\partial}{\partial\lambda}\right)\Omega_o, \qquad (5.50)$$

$$\Omega_o \equiv \frac{2}{\pi}P\int_0^\infty \frac{dk\,\tilde{\rho}^2(k)}{1-\lambda^2 k^2/4}, \qquad (5.51)$$

where P denotes the Cauchy principal part, $\lambda = \hbar/m_o c$ is the Compton radius defined in terms of the bare mass m_o, and $\tilde{\rho}$ is the Fourier transform of $\mathbf{x}$ (Appendix D).

If $\lambda \to 0$,

$$\delta m = \frac{4}{3\pi c^2} \int_0^\infty dk \tilde{\rho}^2(k) \, , \qquad (5.52)$$

which is the classical result given by equation (D.21) of Appendix D. If, however, we first take the point-charge limit $\tilde{\rho}(k) \to 1$ *with* λ *fixed*, then

$$\Omega_o \to \frac{2}{\pi} \mathrm{P} \int_0^\infty \frac{dk}{1 - \lambda^2 k^2/4} = 0, \qquad (5.53)$$

so that the electromagnetic mass vanishes. That is, according to Moniz and Sharp, "the infinite electrostatic self-energy which occurs in the nonrelativistic *classical* calculation has no counterpart in the quantum theory." It is worth stressing, however, that for typical choices of $\rho(\mathbf{x})$ for an extended electron, such as a Yukawa form, the mass formula (5.50) gives results that are qualitatively and quantitatively reasonable ($\delta m \sim \alpha m$); $\delta m = 0$ holds for *point* particles.

Moniz and Sharp also find that, for the physical value of the fine structure constant, there are no runaway solutions of the Heisenberg equation of motion for $\mathbf{R}$, nor is there any preacceleration. This might perhaps have been anticipated from the classical theory of an extended charge, where runaways occur only if the bare mass is negative. However, the absence of runaways in the Moniz–Sharp theory holds for the *point* electron; the absence of runaways in their theory is therefore not so tightly coupled to the value of δm. (In the quantum theory even a point charge has an effective extent $\sim \lambda_c$, as noted earlier.) Note also that in the case $\delta m = 0$ found by Moniz and Sharp, the bare mass $m_0 = m > 0$.

As already described, the Moniz–Sharp theory gives different results depending on which limit, $\hbar \to 0$ or the limit of a point charge, is taken first. The classical limit $\hbar \to 0$ is singular in the sense of singular perturbation theory — the limit of the solution of an equation is not necessarily a solution of the limit of the equation.

The Moniz–Sharp results have been described as "most amazing" (Rohrlich, 1975) and "striking" (Pearle, 1982). However, it is not clear at this time what implications these results might have for specific observable effects, in particular the Lamb shift, for which mass renormalization is employed in standard, perturbative QED. The nonrelativistic electromagnetic mass obtained in perturbation theory is given by equation (3.59) as

$$\delta m = \frac{4e^2}{3\pi c^3} \int_0^\infty \frac{d\omega}{1 + \omega \lambda_c/2c} \, , \qquad (5.54)$$

where $\lambda_c = \hbar/mc$. (The term involving λ_c in the denominator appears only when retardation is included; without retardation we obtain the usual

linearly divergent δm of the nonrelativistic theory) Thus the Moniz–Sharp
result (5.53) does not reduce to the result (5.54) of standard perturbation
theory. This is not terribly surprising, as the Moniz–Sharp theory is not a
perturbative one. It does, however, illustrate the fact that the comparison
with standard QED calculations, even the simpler, nonrelativistic ones, is
nontrivial.

A serious concern with either (5.53) or (5.54) is that their interesting
features are produced by frequencies $\omega > c/\lambda_c = mc^2/\hbar$, where the nonrelativistic theory is inapplicable. In particular, (5.53) vanishes only because
such frequencies have been retained. At the present time, therefore, more
work is required before we can say exactly what the Moniz–Sharp results imply about conventional mass renormalization theory. However, as discussed
in the following section, it is possible at least to see where the theory differs
from an older quantum treatment of an extended electron by van Kampen.

With regard to runaways and preacceleration, it is worth noting that
Moniz and Sharp have shown that these do not occur when an external
force is applied, if the force changes by only a small amount in a time λ_c/c,
i.e., if the frequencies ω in the spectrum of the force are less than $\omega < c/\lambda_c$.
The model becomes inconsistent, in the sense of permitting runaways, if
$\omega > c/\lambda_c$. We refer the reader to Sharp (1980) for a more detailed overview
of the Moniz–Sharp theory.

5.8   Van Kampen's Thesis

Kramers, who perhaps more than anyone else emphasized the distinction
between observed and electromagnetic masses and the necessity of mass
renormalization, felt that point-electron theories "violate the spirit of the
original classical theory."[8] He "never believed that the quantum theory of
radiation, accepted since 1928, was anything but a first and crude approximation . . ."[9] and

> [his] whole philosophy was based on the belief that a necessary pre-
> requisite for the understanding of electromagnetic phenomena was a
> removal of the difficulties and ambiguities in the classical theory of
> electrons. It was for that reason that he had introduced the separa-
> tion of external and proper (self) fields. It was in this process that
> he stressed the important distinction between experimental mass and

---

[8] Dresden 1987, p. 377. Dresden (p. 391) points out that "the motivation for a good
share of the procedures employed [in the Schwinger–Feynman–Dyson QED] was due to
Kramers."

[9] Ibid., pp. 384–385.

observed mass. The analysis of the classical problem led Kramers to
the important notion of a structure-independent theory.

An approximately structure-independent theory, starting from a charge
distribution for an extended electron, was developed in the 1952 thesis
of Kramers's student, van Kampen (1951). Van Kampen showed that
Kramers's idea of a structure-independent quantum theory of electrons interacting with radiation could be consistently formulated, at least nonrelativistically. Here we will briefly describe this work and compare it with the
Moniz–Sharp theory.

The theory proceeds from the Hamiltonian

$$H = \frac{1}{2m_o} \left[\mathbf{P} - e\overline{\mathbf{A}}(\mathbf{R}) \right]^2 + V(\mathbf{R}) + \frac{1}{8\pi} \int d^3x \left[\mathbf{E}^2 + (\nabla \times \mathbf{A})^2 \right], \quad (5.55)$$

where $\mathbf{R}$ and $\mathbf{P}$ are the electron position and momentum, m_o is the bare
mass, and

$$e\overline{\mathbf{A}}(\mathbf{R}) \equiv \int d^3x \rho(\mathbf{x} - \mathbf{R})\mathbf{A}(\mathbf{x}) = \int d^3x \rho(\mathbf{x})\mathbf{A}(\mathbf{x} + \mathbf{R}), \quad (5.56)$$

with $\rho(\mathbf{x})$ being the charge density for the electron. (Following van Kampen,
we put $c = 1$.) An important approximation is made at the outset:

$$\overline{\mathbf{A}}(\mathbf{R}) \cong \int d^3x \rho(\mathbf{x})\mathbf{A}(\mathbf{x}) = \overline{\mathbf{A}}(0). \quad (5.57)$$

This is in effect the dipole approximation, the assumption that the variations in the electron coordinate $\mathbf{R}$ are small compared with the significant
wavelengths of the field (Section 4.3). The fields $\mathbf{E}$ and $\mathbf{A}$ in (5.55) are
transverse, with the potential V assumed to contain the effects of the longitudinal field $\mathbf{E}^{\parallel}$ in the Coulomb gauge (Section 4.2). In the approximation
(5.57),

$$\dot{\mathbf{P}} = -\nabla_{\mathbf{R}}V, \quad \dot{\mathbf{R}} = \frac{1}{m_o} \left[\mathbf{P} - e\overline{\mathbf{A}}(0) \right]. \quad (5.58)$$

The charge density $\rho(\mathbf{x})$ is taken to be spherically symmetric, so that
it is convenient to expand $\mathbf{A}(\mathbf{x})$ in modes appropriate to a large sphere of
radius L, with $\mathbf{A} = 0$ for $|\mathbf{x}| = L$:

$$\mathbf{A}(\mathbf{x}) = \text{Tr} \sum_{n=1}^{\infty} \sqrt{\frac{3}{L}} \mathbf{q}_n \frac{\sin k_n r}{r}, \quad (5.59)$$

where $k_n = n\pi/L$, $\mathbf{q}_n$ is the coordinate operator for the harmonic oscillator associated with the field mode n, and Tr means "transverse part of."

Similarly the transverse electric field is

$$\mathbf{E}(\mathbf{x}) = -\mathrm{Tr} \sum_{n=1}^{\infty} \sqrt{\frac{3}{L}} \mathbf{p}_n \frac{\sin k_n r}{r} , \tag{5.60}$$

where $\mathbf{p}_n$ is the momentum canonically conjugate to $\mathbf{q}_n$. From (5.57) it follows that the vector potential at the position of the electron is

$$\overline{\mathbf{A}}(0) = \frac{1}{e} \sum_{n=1}^{\infty} \epsilon_n \mathbf{q}_n , \tag{5.61}$$

$$\epsilon_n \equiv \sqrt{\frac{4}{3L}} 4\pi \int_0^{\infty} dr\, r \rho(r) \sin k_n r . \tag{5.62}$$

In terms of the canonical coordinates and momenta ($\mathbf{R}, \mathbf{P}$ and $\mathbf{q}_n, \mathbf{p}_n$), the Hamiltonian (5.55) is

$$\begin{aligned} H &= \frac{1}{2m_o} \mathbf{P}^2 + V(\mathbf{R}) - \frac{1}{m_o} \sum_{n=1}^{\infty} \epsilon_n \mathbf{q}_n \cdot \mathbf{P} + \frac{1}{2m_o} \left[\sum_{n=1}^{\infty} \epsilon_n \mathbf{q}_n \right]^2 \\ &\quad + \frac{1}{2} \sum_{n=1}^{\infty} \left(\mathbf{p}_n^2 + k_n^2 \mathbf{q}_n^2 \right) . \end{aligned} \tag{5.63}$$

The third and fourth terms correspond to the usual $\mathbf{A} \cdot \mathbf{p}$ and $\mathbf{A}^2$ interactions, respectively.

It is convenient to transform to new canonically conjugate variables $\mathbf{R}', \mathbf{P}'$ and $\mathbf{q}_n, \mathbf{p}_n$ defined by

$$\mathbf{R}' = \mathbf{R} - \sum_{n=1}^{\infty} \frac{\epsilon_n}{mk_n^2} \mathbf{P}_n' , \tag{5.64}$$

$$\mathbf{P}' = \mathbf{P}, \tag{5.65}$$

$$\mathbf{q}_n' = \mathbf{q}_n - \frac{\epsilon_n}{mk_n^2} \mathbf{P}' , \tag{5.66}$$

$$\mathbf{p}_n' = \mathbf{p}_n , \tag{5.67}$$

in terms of which (5.63) transforms to

$$\begin{aligned} H' &= \frac{1}{2m} \mathbf{P}'^2 + V\left(\mathbf{R}' + \sum_n \frac{\epsilon_n}{mk_n^2} \mathbf{P}_n' \right) + \frac{1}{2m_o} \left[\sum_n \epsilon_n \mathbf{q}_n' \right]^2 \\ &\quad + \frac{1}{2} \sum_n \left(\mathbf{p}_n'^2 + k_n^2 \mathbf{q}_n'^2 \right) , \end{aligned} \tag{5.68}$$

where

$$m \equiv m_o + \sum_{n=1}^{\infty} \frac{\epsilon_n}{k_n^2} = m_o + \frac{4e^2}{3L} \sum_{n=1}^{\infty} \delta_n^2 , \qquad (5.69)$$

$$\delta_n \equiv \frac{\epsilon_n}{k_n} \sqrt{\frac{3L}{4e^2}} = \frac{4\pi}{ek_n} \int_0^{\infty} dr r \rho(r) \sin k_n r . \qquad (5.70)$$

In the limit of a point electron, $\rho(\mathbf{x}) = e\delta^3(\mathbf{x})$, $\delta_n \to 1$ and the electromagnetic mass

$$\delta m = \sum_{n=1}^{\infty} \frac{\epsilon_n^2}{k_n^2} = \frac{4e^2}{3L} \sum_{n=1}^{\infty} \delta_n^2 \qquad (5.71)$$

diverges, as discussed later.

The third term in the Hamiltonian (5.68) derives from the $\mathbf{A}^2$ contribution to the original Hamiltonian (5.55). Together with the last term, it constitutes a contribution quadratic in the q_n'. H' can therefore be transformed to the form

$$H'' = \frac{1}{2m}\mathbf{P}'^2 + V\left(\mathbf{R}' + \frac{e}{m}\sum_n \sqrt{\frac{4}{3L_n}}\frac{\cos \eta_n}{K_n} \mathbf{P}_n''\right) + \frac{1}{2}\sum_n (\mathbf{p}_n''^2 + K_n^2 \mathbf{q}_n''^2) \qquad (5.72)$$

by a principal axis transformation with $q_n' = \sum_m c_{nm} q_m''$ and $\mathbf{p}_n' = \sum_m c_{nm} \times q_m''$. η_n and L_n are defined by

$$K_n L = \eta_n + n\pi, \quad 0 < \eta_n < \frac{\pi}{2}, \qquad (5.73)$$

$$L_n = L - c\tau \cos^2 \eta_n \cong L, \qquad (5.74)$$

and the K_n are the roots of a characteristic equation:

$$m = K^2 \sum_n \frac{\epsilon_n^2}{k_n^2(K^2 - k_n^2)} , \qquad (5.75)$$

which involves the assumed structure of the electron through the ϵ_n defined by (5.62). Once m is identified as the experimentally observed electron mass, however, the structure of the electron affects the Hamiltonian H'' only through the values of the K_n and can be expected to have a small effect on the physical predictions obtained with H''. That is, H'' is *approximately* the sort of structure-independent Hamiltonian long sought by Kramers (Dresden, 1987).

In the limit of a point electron, $\delta_n \to 1$ and we obtain from (5.71) the electromagnetic mass

$$\delta m = \frac{4e^2}{3L} \sum_{n=1}^{\infty} \to \frac{4e^2}{3L} \frac{L}{\pi} \int_0^{\infty} d\omega \to \frac{4e^2}{3\pi c^3} \int_0^{\infty} d\omega \qquad (5.76)$$

when we reinstate c in order to recover the electromagnetic mass (5.11) of a classical, nonrelativistic point electron. In this limit, $\epsilon_n \rightarrow k_n \sqrt{4e^2/3L}$ according to (5.70), and (5.75) becomes[10]

$$
\begin{aligned}
m &= K^2 \frac{4e^2}{3L} \sum_{n=1}^{\infty} \frac{n^2 \pi^2/L^2}{(n^2 \pi^2/L^2)(K^2 - n^2 \pi^2/L^2)} \\
&= K^2 \frac{4e^2}{3L} \frac{L^2}{\pi^2} \sum_{n=1}^{\infty} \frac{1}{(KL/\pi)^2 - n^2} \\
&= K^2 \frac{4e^2}{3L} \frac{L^2}{\pi^2} \frac{\pi^2}{2KL} \left(\cot KL - \frac{1}{KL} \right) \\
&= \frac{2e^2}{3} K \left(\cot KL - \frac{1}{KL} \right).
\end{aligned}
\tag{5.77}
$$

Since L can be taken to be large, this reduces approximately to

$$
\tan KL = \frac{2}{3} K r_o ,
\tag{5.78}
$$

where r_o is the classical electron radius (Section 5.2).

The solution of (5.78) for K includes two imaginary values, i.e., there are solutions leading to terms of the form $p_*''^2 - |K_*|^2 q_*''^2$ in the Hamiltonian (5.72). For a free point electron, this leads to runaway solutions, which van Kampen excludes by choosing the initial values of the p_*'' and q_*'' to be zero.

With regard to the electromagnetic mass, note that, for a spherically symmetric $\rho(\mathbf{x})$,

$$
\begin{aligned}
\delta_n &= \frac{1}{e} \int dr 4\pi r^2 \rho(r) \frac{\sin k_n r}{k_n r} = \frac{1}{e} \int d^3 x \rho(\mathbf{x}) e^{i \mathbf{k}_n \cdot \mathbf{r}} \\
&= \frac{1}{e} \tilde{\rho}(k_n).
\end{aligned}
\tag{5.79}
$$

Then

$$
\delta m = \frac{4}{3L} \sum_{n=1}^{\infty} \tilde{\rho}^2(k_n) \rightarrow \frac{4}{3L} \frac{L}{\pi} \int dk \tilde{\rho}^2(k)
\tag{5.80}
$$

in the limit $L \rightarrow \infty$, and this is equivalent to the classical result (5.52).

Van Kampen (1951) uses the Hamiltonian H'' to treat a variety of problems concerned primarily with the scattering of radiation.[11] In concluding

---

[10] We use the series expansion for the cotangent given in 1.421 of I. S. Gradshteyn and I. M. Rhyzhik, *Table of Integrals, Series, and Products* (Academic Press, New York, 1980).

[11] Van Kampen's transformations bear a familial resemblance to transformations employed earlier by Bloch and Nordsieck [*Phys. Rev.* **52**, 54 (1937)] and Pauli and Fierz

our summary of his theory we will confine ourselves to a few additional remarks about radiation reaction.

We note first that the radiation reaction field $\mathbf{A}_{RR}$ can be defined, using (5.59) and (5.66), by

$$\mathbf{A}_{RR} = \mathrm{Tr} \sum_n \sqrt{\frac{3}{L} \frac{\epsilon_n}{mk_n^2}} \mathbf{P}' \frac{\sin k_n r}{r} \ , \qquad (5.81)$$

implying that the *external* field is

$$\mathbf{A}_{ext} = \mathrm{Tr} \sum_n \sqrt{\frac{3}{L}} \mathbf{q}_n' \frac{\sin k_n r}{r} \ . \qquad (5.82)$$

From the definitions of ϵ_n and k_n it is readily shown that

$$\mathbf{A}_{RR} = \mathrm{Tr} \frac{\mathbf{P}'}{m} \int \frac{d^3 x' \rho(\mathbf{x}')}{|\mathbf{x} - \mathbf{x}'|} \ . \qquad (5.83)$$

Second, one can easily see that a Taylor series about $\mathbf{R}'$ of the second term in (5.72) will have a quadratic term involving $\cos^2 \eta_n$. The mean-square fluctuation in the electron coordinate resulting from the vacuum field fluctuations associated with the $\mathbf{p}_n''$ will therefore be similar to that calculated by Welton (Section 3.6), except that factors $\cos^2 \eta_n$ will appear. This leads to the replacement of Bethe's nonrelativistic Lamb shift formula (3.24) by

$$\Delta \tilde{E}_n^{obs} = \frac{2\alpha}{3\pi} \left(\frac{1}{mc} \right)^2 \sum_m |\mathbf{p}_{mn}|^2 (E_n - E_m) \int_0^\infty \frac{dE \cos^2 \eta(E)}{E_n - E_m - E} \ , \qquad (5.84)$$

which, owing to the factor $\cos^2 \eta$, is *convergent*. However, the effect of this factor is to cut the integral off at roughly $137mc^2$, as compared with Bethe's mc^2; not surprisingly, the convergent result for this nonrelativistic theory does not agree numerically with the experimentally observed Lamb shift (van Kampen, 1951).

Finally, it is easy to see where the van Kampen and Moniz–Sharp theories differ: van Kampen makes the dipole approximation (5.57), whereas Moniz and Sharp retain the $\mathbf{R}$ dependence in (5.56), leading to (5.44) under

---

[*Nuovo Cimento* **15**, 167 (1938)]. These transformations all involve the modification of particle coordinates by field variables, as in (5.64), and correspond physically to a "dressing" of the particle by a virtual "cloud" of photons. In the more recent literature such a transformation, the so-called Kramers–Henneberger transformation, has been employed in the treatment of atoms in very intense fields.

the appropriate symmetrization. This **R** dependence is obviously crucial to the Moniz–Sharp theory. Without it, no nontrivial commutators involving **R** and **P** ever have to be dealt with, and their results would have to be physically equivalent to van Kampen's.

5.9   Bibliography

Baylis, W. E. and J. Huschilt, "Nonuniqueness of Physical Solutions to the Lorentz-Dirac Equation," *Phys. Rev.* **D13**, 3237 (1976).

Bohm, D. and M. Weinstein, "The Self-Oscillations of a Charged Particle," *Phys. Rev.* **74**, 1789 (1948).

Candelas, P. and D. W. Sciama, "Is There a Quantum Equivalence Principle?," *Phys. Rev.* **D27**, 1715 (1983).

Coleman, S., "Classical Electron Theory from a Modern Standpoint," Rand Corporation Research Memorandum RM-2820-PR (September 1961); reprinted in *Electromagnetism: Paths to Research*, ed. D. Teplitz (Plenum Press, New York, 1982).

Dirac, P. A. M., "Classical Theory of Radiating Electrons," *Proc. Roy. Soc. Lond.* **167A**, 148 (1938).

Dresden, M., *H. A. Kramers: Between Tradition and Revolution* (Springer-Verlag, New York, 1987).

Erber, T., "The Classical Theories of Radiation Reaction," *Fortsch. d. Phys.* **9**, 343 (1961).

Feynman, R. P., R. B. Leighton, and M. Sands, *The Feynman Lectures on Physics* (Addison–Wesley, Reading, Mass., 1964), Volume 2, Chapter 28.

Fulton, T. and F. Rohrlich, "Classical Radiation from a Uniformly Accelerated Charge," *Ann. Phys.* (New York) **9**, 499 (1960).

Ginzburg, V. L., *Theoretical Physics and Astrophysics* (Pergamon Press, Oxford, 1979), Chapter 2.

Grandy, W. T., Jr. and A. Aghazadeh, "Radiative Corrections for Extended Charged Particles," *Ann. Phys.* (New York) **142**, 284 (1982).

Grandy, W. T., Jr., *Relativistic Quantum Mechanics of Leptons and Fields* (Kluwer, Dordrecht, 1991).

Grotch, H., E. Kazes, F. Rohrlich, and D. H. Sharp, "Internal Retardation," *Acta Phys. Austriaca* **54**, 31 (1982).

Jackson, J. D., *Classical Electrodynamics*, 2nd ed. (Wiley, New York, 1975), Chapter 17.

Klepikov, N. P., "Radiation damping forces and radiation from charged particles," *Sov. Phys. Usp.* **28**, 506 (1985).

Kwal, B., "Les Expressions de l'énergie et de l'impulsion du champ électromagétique propre de l'élektron en mouvement," *J. Phys. Radium* **10**, 103 (1949).

Levine, H., E. J. Moniz, and D. H. Sharp, "Motion of extended charges in classical electrodynamics," *Am. J. Phys.* **45**, 75 (1977).

Lorentz, H.A., *The Theory of Electrons* (Dover Books, New York, 1952).

Moniz, E. J. and D. II. Sharp, "Absence of Runaways and Divergent Self–Mass in Nonrelativistic Quantum Electrodynamics," *Phys. Rev.* D**10**, 1133 (1974).

Moniz, E. J. and D. H. Sharp, "Radiation Reaction in Nonrelativistic Quantum Electrodynamics," *Phys. Rev.* D**15**, 2850 (1977).

Pearle, P., "Classical Electron Models," in *Electromagnetism: Paths to Research*, ed. D. Teplitz (Plenum Press, New York, 1982).

Plass, G. N., "Classical Electrodynamic Equations of Motion with Radiative Reaction," *Rev. Mod. Phys.* **33**, 37 (1961).

Rohrlich, F., "Self-Energy and Stability of the Classical Electron," *Am. J. Phys.* **28**, 639 (1960).

Rohrlich, F., "A Lesson in the Construction of a Physical Theory," *Acta Physica Austriaca* **41**, 375 (1975).

Sciama, D. W., P. Candelas, and D. Deutsch, "Quantum Field Theory, Horizons and Thermodynamics," *Adv. Phys.* **30**, 327 (1981).

Sharp, D. H., "Radiation Reaction in Nonrelativistic Quantum Theory," in *Foundations of Radiation Theory and Quantum Electrodynamics*, ed. A.O. Barut (Plenum Press, New York, 1980).

Van Kampen, N. G., "Contribution to the Quantum Theory of Light Scattering," *Dan. Mat. Fys. Medd.* **26**, no. 15 (1951).

Wheeler, J. A. and R. P. Feynman, "Interaction with the Absorber as the Mechanism of Radiation," *Rev. Mod. Phys.* **17**, 157 (1945).

Wildermuth, K., "Zur physikalischen Interpretation der Elektronen-selbstbeschleunigung," *Z. Naturforsch.* **10A**, 450 (1955).

Chapter 6

The Vacuum in Quantum Optics

> ... it appears that Quantum Mechanics is not a bad preparation for optics. – Dennis Gabor (1956)

6.1 Introduction

Quantum optics is the study of the interaction of radiation with matter, especially under conditions where the field of interest is quasimonochromatic and resonant with some atomic or molecular transition, where the nonrelativistic approximation is adequate, and where one is concerned with questions of field coherence and photon statistics. In quantum optics one is usually faced with different calculational and conceptual problems than those of relativistic QED, and there are ample opportunities to compare theory and experiment.

In this chapter we consider some examples of the role of the vacuum field in quantum optics, chosen because of their fundamental importance and because they involve no specialized concepts. Sections 6.2 and 6.3 deal with the modification of the vacuum electromagnetic mode structure by reflecting surfaces, and in Section 6.4 we briefly discuss the interaction of an atom with a single mode. Both problems have been investigated experimentally in recent years. Sections 6.5 and 6.6 show how the vacuum field fluctuations contribute to the fundamental laser linewidth and to amplified spontaneous emission. In the remainder of the chapter we give a cursory overview of recent experimental studies of single trapped particles.

6.2   Spontaneous Emission near Mirrors

The quantized transverse vector potential associated with a single field mode is given by equation (2.34). The multimode generalization is

$$\mathbf{A}(\mathbf{r},t) = \sum_\alpha \left(\frac{2\pi\hbar c^2}{\omega_\alpha}\right)^{1/2} [a_\alpha(t)\mathbf{A}_\alpha(\mathbf{r}) + a_\alpha^\dagger(t)\mathbf{A}_\alpha^*(\mathbf{r})], \qquad (6.1)$$

where

$$\nabla^2 \mathbf{A}_\alpha(\mathbf{r}) + k_\alpha^2 \mathbf{A}_\alpha(\mathbf{r}) = 0, \quad k_\alpha^2 = \omega_\alpha^2/c^2, \qquad (6.2)$$
$$\nabla \cdot \mathbf{A}_\alpha(\mathbf{r}) = 0, \qquad (6.3)$$

and the mode functions $\mathbf{A}_\alpha(\mathbf{r})$ are chosen to form an orthonormal set:

$$\int d^3r \mathbf{A}_\alpha^*(\mathbf{r}) \cdot \mathbf{A}_\beta(\mathbf{r}) = \delta_{\alpha\beta}. \qquad (6.4)$$

As discussed in Section 2.4, the mode functions are ordinary *classical* vector functions determined by the Helmholtz equation (6.2), the transversality condition (6.3), and boundary conditions. The quantum properties of the field are determined by the annihilation and creation operators $a_\alpha(t)$ and $a_\alpha^\dagger(t)$ satisfying $[a_\alpha(t), a_\beta(t)] = 0, [a_\alpha(t), a_\beta^\dagger(t)] = \delta_{\alpha\beta}$.

Thus far we have restricted ourselves mainly to the field in free space (Section 2.5), in which case each mode α is characterized by a wave vector $\mathbf{k}$ and a polarization λ, and the mode functions are defined by (2.49) when periodic boundary conditions are imposed.

The theory of spontaneous emission presented in Chapter 4 is trivially extended from the case of free space to more general situations in which the vector potential is defined by (6.1). When (6.1) is used the spontaneous emission rate on a transition $2 \rightarrow 1$ of frequency ω_o is found to be

$$A_{21}(\mathbf{r}) = \frac{4\pi^2\omega_o^2}{\hbar} \sum_\alpha \frac{1}{\omega_\alpha} |\mathbf{A}_\alpha(\mathbf{r}) \cdot \mathbf{d}_{12}|^2 \delta(\omega_\alpha - \omega_o), \qquad (6.5)$$

where, as in Chapter 4, $\mathbf{d}_{12}$ is the transition dipole moment. For the free-space mode functions (2.49) we of course obtain $A_{21}(\mathbf{r}) = A_{21} = 4|\mathbf{d}_{12}|^2\omega_o^3/3\hbar c^3$, the Einstein A coefficient. In general, however, *the spontaneous emission rate varies with the position $\mathbf{r}$ of the atom.*

Consider as an example an atom near an infinite, perfectly conducting plane at $z = 0$. In this case the mode functions are those appropriate to the half-space $z > 0$ bounded at $z = 0$ by a plane on which the tangential components of the electric field, and therefore the mode functions $\mathbf{A}_\alpha(\mathbf{r})$, must vanish. To obtain such mode functions, we can combine a plane wave with wave vector $\mathbf{k} = k_1\hat{x} + k_2\hat{y} + k_3\hat{z} = \mathbf{k}_\| + k_3\hat{z}$ with its reflected wave with wave vector $\mathbf{k}_r = \mathbf{k}_\| - k_3\hat{z}$ to form

$$\mathbf{A}_{\mathbf{k}1}(\mathbf{r}) = \left(\frac{2}{V}\right)^{1/2} (\hat{k}_\| \times \hat{z}) \sin k_3 z e^{i\mathbf{k}_\| \cdot \mathbf{r}}, \tag{6.6}$$

which is normalized according to (6.4). Here V is once again a quantization volume, and a caret is used to denote a unit vector. Similarly we can form a class of mode functions

$$\mathbf{A}_{\mathbf{k}2}(\mathbf{r}) \propto (\hat{k}_\| \times \hat{z}) \times \hat{k} e^{i\mathbf{k} \cdot \mathbf{r}} + (\hat{k}_\| \times \hat{z}) \times \hat{k}_r e^{i\mathbf{k}_r \cdot \mathbf{r}}, \tag{6.7}$$

or

$$\mathbf{A}_{\mathbf{k}2}(\mathbf{r}) = \frac{1}{k}\left(\frac{2}{V}\right)^{1/2} [k_\|\hat{z}\cos k_3 z - ik_3\hat{k}_\|\sin k_3 z]e^{i\mathbf{k}_\| \cdot \mathbf{r}} \tag{6.8}$$

after normalization. The vector functions (6.6) and (6.8) form a complete set of mode functions for the half-space bounded by a conducting plane at $z = 0$. Equation (6.5) with these mode functions gives

$$\begin{aligned}
A_{21}(\mathbf{r}) &= \frac{4\pi^2\omega_o^2}{\hbar}\frac{2}{V}\sum_{\mathbf{k}}\frac{1}{\omega_k}\left[|(\hat{k}_\| \times \hat{z})\cdot\mathbf{d}_\||^2 \sin^2 k_3 z + \frac{k_\|^2}{k^2}|\mathbf{d}_\perp|^2\cos^2 k_3 z\right.\\
&\left. + \frac{k_3^2}{k^2}|\hat{k}_\| \cdot \mathbf{d}_\||^2 \sin^2 k_3 z\right]\delta(\omega_k - \omega_o),
\end{aligned} \tag{6.9}$$

where we have written $\mathbf{d}_{12} = d_\perp\hat{z} + d_x\hat{x} + d_y\hat{y} = d_\perp\hat{z} + \mathbf{d}_\| = \mathbf{d}_\perp + \mathbf{d}_\|$.

Consider the contribution to $A_{21}(\mathbf{r})$ from $\mathbf{d}_\perp$:

$$\begin{aligned}
A_{21}^\perp(\mathbf{r}) &= \frac{4\pi^2\omega_o^2}{\hbar}\frac{2}{V}\frac{c^2}{\omega_o^3}|\mathbf{d}_\perp|^2 \sum_{\mathbf{k}} k_\|^2\cos^2 k_3 z\delta(\omega_k - \omega_o)\\
&\rightarrow \frac{4\pi^2\omega_o^2}{\hbar}\frac{2}{V}\frac{c^2}{\omega_o^3}|\mathbf{d}_\perp|^2\frac{V}{8\pi^3}\int_{-\infty}^{\infty} dk_3\int d^2k_\| k_\|^2\cos^2 k_3 z\\
&\quad \times \frac{1}{c}\delta(\sqrt{k_\|^2 + k_3^2} - k_o),
\end{aligned} \tag{6.10}$$

where $k_o = \omega_o/c$ and we have gone to the usual mode continuum limit $V \rightarrow \infty$. Straightforward evaluation of (6.10) yields

$$A_{21}^\perp(z) = 3A_{21}\left[\frac{1}{3} - \frac{\cos 2k_o z}{(2k_o z)^2} + \frac{\sin 2k_o z}{(2k_o z)^3}\right] \tag{6.11}$$

for the spontaneous emission rate when an atom is a distance z from a conducting plane. Here again A_{21} is the ordinary free-space spontaneous emission rate. Similarly the contribution to $A_{21}(\mathbf{r})$ from $\mathbf{d}_{\parallel}$ is found to be

$$A_{21}^{\parallel}(z) = \frac{3}{2}A_{21}\left[\frac{2}{3} - \frac{\sin 2k_o z}{2k_o z} - \frac{\cos 2k_o z}{(2k_o z)^2} + \frac{\sin 2k_o z}{(2k_o z)^3}\right]. \qquad (6.12)$$

If the excited state is produced in such a way that all orientations of the transition dipole moment are equally likely, then the emission rate is the orientational average

$$\begin{aligned} A_{21}(z) &= \frac{1}{3}A_{21}^{\perp}(z) + \frac{2}{3}A_{21}^{\parallel}(z) \\ &= A_{21}\left[1 - \frac{\sin 2k_o z}{2k_o z} - \frac{2\cos 2k_o z}{(2k_o z)^2} + \frac{2\sin 2k_o z}{(2k_o z)^3}\right]. \quad (6.13) \end{aligned}$$

The modification of spontaneous emission rates by reflecting surfaces has been observed experimentally. Experiments by Drexhage (1970, 1974), for instance, involved the deposition of molecular monolayers on reflecting plates. The distance of a fluorescing molecule from a plate could be fixed according to the number of times the plate was dipped in a solution to coat it with a dielectric layer before the monolayer was added. The emission lifetime could then be monitored as a function of the distance z of the emitter from the plate. The oscillatory dependence of the lifetime on z, as predicted by (6.11)–(6.13), was confirmed. In fact rather good agreement between theory and experiment was obtained by modifying the previous idealized theory to account for effects such as the deviations from unity of both the mirror reflectivity and the refractive index of the dielectric layer.

Drexhage's results were at first regarded with suspicion in some quarters: it was argued that no modification of a spontaneous emission rate is possible, for how can the emission of a photon be affected by an atom's environment when the atom can only "see" its environment by emitting a photon in the first place?

Such an objection is invalid. As long as the emission lifetime is large compared with $2z/c$, the atom has ample opportunity to "see" its environment. But it need not emit a real photon to do so. One way to think about this is in terms of radiation reaction (Milonni, 1982). Using (6.1) and the mode functions (6.6) and (6.8), we obtain for the z-component of $\mathbf{E}_{\mathrm{RR}}$, for instance, the result

$$\begin{aligned} \hat{z}\cdot\mathbf{E}_{\mathrm{RR}}(z,t) &= \frac{2}{3c^3}\,\dddot{p}\,(t) - \frac{\delta m}{e}\ddot{p}(t) \\ &\quad - \frac{1}{2cz^2}\dot{p}(t - \frac{2z}{c}) - \frac{1}{4z^3}p(t - \frac{2z}{c}), \qquad (6.14) \end{aligned}$$

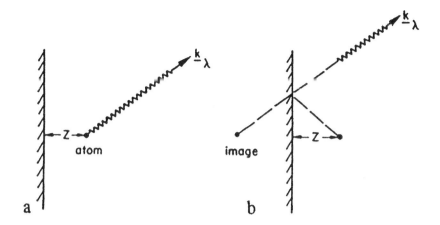

Figure 6.1: Two possible ways in which a photon can be emitted into the same mode by an atom near a mirror.

where $p(t)$ is the z-component of the atomic dipole operator. The radiation reaction electric field operator thus has the same form as the classical reaction field acting on a dipole near a mirror: it is just the free-space ($z \to \infty$) result plus the retarded dipole field from an image dipole at $-z$. Indeed the bracketed factors in (6.11)–(6.13) also describe the modification of the radiation rate of a *classical* dipole near a mirror, in which case there is no question whatsoever that the rate should be modified from its free-space value.

The classical image picture has a quantum analog (Milonni and Knight, 1973). Consider the two processes shown in Figure 6.1. In the first, the atom emits into the plane-wave mode $(\mathbf{k}, \lambda)$ without reflection of the field off the mirror. In the second a photon is emitted with the same wave vector and polarization after reflection and can be considered to have been emitted by a fictitious image atom. Thus we may describe the emission as coming from a *two-atom* system, with complete uncertainty as to which of the two atoms is initially excited and capable of emitting a photon. The initial states incorporating such uncertainty are

$$|\psi_\perp\rangle = \frac{1}{\sqrt{2}} \left[|2\rangle_\mathrm{A} |1\rangle_\mathrm{I} \perp |1\rangle_\mathrm{A} |2\rangle_\mathrm{I} \right], \qquad (6.15)$$

where A denotes the actual atom and I its image in the mirror, and 2 and 1 denote excited and unexcited states, respectively, of the emitting transition. The atom-image separation is $2z$. The state $|\psi_+\rangle$ is appropriate if the transition dipole moment of the atom is normal to the mirror, since in this case the dipole can be thought of as oscillating with the same phase as its image dipole. Similarly $|\psi_-\rangle$ is appropriate if the transition dipole moment is assumed to be parallel to the mirror surface, in which case the dipole and its image oscillate with a π phase difference.

As in the case of an atom in free space, we can also interpret (6.11)–(6.13), at least partly, in terms of the action of the vacuum field on the atom. The "vacuum field" in this case is the source-free, zero-point quantum field in the half-space bounded by the conducting plane, i.e., the field described by the vector potential

$$
\mathbf{A}(\mathbf{r}, t) = \sum_{\mathbf{k}\lambda} \left(\frac{2\pi\hbar c^2}{\omega_k} \right)^{1/2} [a_{\mathbf{k}\lambda}(0)e^{-i\omega_k t}\mathbf{A}_{\mathbf{k}\lambda}(\mathbf{r}) + a_{\mathbf{k}\lambda}^{\dagger}(0)e^{i\omega_k t}\mathbf{A}_{\mathbf{k}\lambda}^{*}(\mathbf{r})],
$$

(6.16)

with the mode functions $\mathbf{A}_{\mathbf{k}\lambda}(\mathbf{r})$ defined by (6.6) and (6.8). Note that the vacuum field in the half-space has a zero-point energy $\frac{1}{2}\hbar\omega_k$ per mode, as in the case of free space. The field operators such as (6.16), however, have spatial variations different from those of the field in unbounded free space. It might also be noted that the quantization of the field in the half-space leads automatically to retardation in the influence of the mirror on the atom (Milonni, 1983).

We can view the experiments on the modification of spontaneous emission by a mirror as confirming that the vacuum electromagnetic field has a mode structure determined by the solution of the *classical* electromagnetic problem defined by (6.2)–(6.4) plus boundary conditions. As in the case of unbounded free space, the quantum features of the field are associated with temporal, not spatial variations of the field.

Based on the theory of Chapter 4, it can be anticipated that the Lamb shift of a transition is also modified by the presence of a mirror. In fact the modification of the Lamb shift turns out to be just the Casimir–Polder interaction discussed in Section 3.12. This is discussed further in Chapter 8.

6.3   Cavity QED

Spontaneous emission near a mirror provides a simple example of *cavity quantum electrodynamics*, the study of the effects of mirrors and cavities on radiative corrections. Another example that immediately comes to mind is spontaneous emission *between* mirrors. Consider then an atom between two perfect, parallel plane mirrors at $z = \pm L/2$. The spontaneous emission rate in this case may be obtained from equation (6.5) using the mode functions (2.91)–(2.93) with $L_x, L_y \to \infty$ but $L_z = L$ finite. For an atom positioned at $z = z_o$ between the mirrors, and with transition dipole moment parallel to the mirror planes, the spontaneous emission rate is easily calculated to be (Barton 1970, 1987; Milonni and Knight, 1973; Philpott, 1973)

$$A_{21}^{\parallel}(z) = \left(\frac{3\pi}{2k_oL}\right) A_{21} \sum_{n=1}^{N} \left(1 + \frac{n^2\pi^2}{k_o^2L^2}\right) \sin^2\left(\frac{n\pi z_o}{L} - \frac{n\pi}{L}\right), \qquad (6.17)$$

where A_{21} is again the free-space emission rate and N is the greatest integer part of k_oL/π. For a transition dipole moment perpendicular to the mirror planes, similarly,

$$A_{21}^{\perp}(z) = \left(\frac{3\pi}{k_oL}\right) A_{21} \left[\frac{1}{2} + \sum_{n=1}^{N} \left(1 - \frac{n^2\pi^2}{k_o^2L^2}\right) \cos^2\left(\frac{n\pi z_o}{L} - \frac{n\pi}{L}\right)\right]. \quad (6.18)$$

Note that for $k_oL/\pi = 2L/\lambda < 1$, where λ is the transition wavelength, $A_{21}^{\parallel}(z) = 0$. In other words, when the spacing L of the mirrors is less than $\lambda/2$, the spontaneous emission rate at wavelength λ should be completely suppressed if the transition dipole moment is parallel to the mirror planes. For an atom near a single mirror such a complete suppression is not possible because the spectrum of allowed field frequencies is continuous.

Suppressed (or "inhibited") spontaneous emission has been observed using an atomic beam of highly excited cesium atoms passing between mirrors separated by about 0.2 mm (Hulet, Hilfer, and Kleppner, 1985). Before the atomic beam entered the cavity, an $n = 22$ state of cesium was prepared by excitation with two dye lasers, after which the atoms were put into a so-called circular state with $|m| = n - 1$, where n and m are the principal and magnetic quantum numbers.[1] (The quantization axis for the magnetic quantum number is defined by an applied electric field.) The circular state is such that the dipole moment for the $(n = 22, |m| = 21) \to (n = 21, |m| = 20)$ transition is parallel to the mirror planes, so that substantial suppression of spontaneous emission could be anticipated. The mirrors were separated by 230.1 μm$= 1.02(\lambda/2)$, where λ is the transition wavelength when there is no applied electric field. Application of an electric field was used to Stark–shift the wavelength over a tuning range $\Delta\lambda/\lambda - .04$. The state of the atoms emerging from the cavity was determined by field ionization, using the fact that the $n = 21$ and $n = 22$ levels have different ionization rates. A dramatic suppression of

---

[1] The circular state was obtained using an adiabatic passage technique. See R. G. Hulet and D. Kleppner, *Phys. Rev. Lett.* **51**, 1430 (1983).

spontaneous emission transitions - by a factor > 20   was observed when
the Stark-shifted wavelength was larger than $2L$.

Much smaller mirror spacings have been used in experiments with near-
infrared transitions (Jhe et al., 1987). In these experiments a beam of
cesium atoms prepared in the $5d$ level entered a cavity with spacing $L = 1.1$
μm, so that the $5d \rightarrow 6p$ transition wavelength of 3.5 μm was larger than
the cavity cutoff wavelength $2L = 2.2$ μm. It was observed that the atoms
passed through the cavity for > 10 (free-space) radiative lifetimes without
spontaneous emission. The orientation of the transition dipole moment
could be varied with an applied magnetic field, and it was confirmed that
spontaneous emission was no longer inhibited when the dipole moment had
a component perpendicular to the mirror surfaces.

Substantial inhibition of spontaneous emission at optical wavelengths
has been observed using a microscopic piezoelectrically tuned cavity and
a dye solution (DeMartini et al., 1987). Both inhibition and enhancement
of spontaneous emission in the optical regime have been observed using a
spherical Fabry–Perot resonator (Heinzen et al., 1987). In this case the rate
of spontaneous emission into the solid angle subtended by the Fabry-Perot
was varied by tuning the resonator through different resonances.

We refer the reader to the reviews by Haroche and Kleppner (1989) and
Hinds (1990) for more information and references on experiments in cavity
QED.

The remarks near the end of the preceding section apply to cavity QED
in general. In particular, experiments in cavity QED may be regarded
as confirmations of the modification of the electromagnetic vacuum from
its free-space structure. It might also be noted that the image method
mentioned earlier may be used also in the case of an atom between mirrors
(Milonni and Knight, 1973).

Generality of Cavity QED Effects

All radiative processes, not just spontaneous emission and the associated
frequency shifts, can in principle be modified by reflecting surfaces and
cavities. Perhaps the simplest and most general way to appreciate this is to
note that the propagator $D_{\mu\nu}(x', x)$ for the electromagnetic field (Chapters
10 and 12) is determined by the modal properties of the field. In traditional
QED the field modes are taken to be the plane waves of free, unbounded
space.

Of course the use of classically determined mode functions presupposes
that the particles constituting the reflecting media and cavity walls act
simply to enforce the boundary conditions for the Maxwell equations. Ob-

viously this cannot happen if the particle densities are too low. The Ewald–Oseen extinction theorem (Section 8.3) can be used to justify the macroscopic, "cavity QED" approach in cases where, for instance, the atoms constituting a dielectric medium are approximated by a continuous distribution. However, the "cavity QED" approximation of simply quantizing the field in terms of mode functions satisfying classical boundary conditions is not always valid (Sections 8.4 and 8.5).

There is a similar situation in the case of an atom embedded in a dielectric medium. The spontaneous emission rate of the embedded atom must be calculated in such a way as to include the effects of its neighboring particles, but in the continuum approximation to the dielectric we can simply characterize the dielectric by a (complex) dielectric constant.[2] Then, aside from local field corrections, the spontaneous emission rate for an electric dipole transition is found to be approximately $n(\omega_o)A_{21}$, where A_{21} is again the free-space emission rate and $n(\omega_o)$ is the real part of the refractive index at the unperturbed transition frequency ω_o (Ginzburg, 1979; Barnett, Huttner, and Loudon, 1992).[3]

An example of a fundamental radiative process modified by cavities is the change in the electromagnetic mass of an electron between parallel conducting plates. The effect in the nonrelativistic theory turns out to be rather simple and not of much interest (see Milonni, 1983, and references therein).

6.4 Single-Mode Interaction

In Chapter 4 we used the Hamiltonian (4.73) for a two-state atom coupled to the field to derive the Einstein A coefficient for spontaneous emission. If we take into account only the coupling of the atom to a single field mode, the Hamiltonian reduces to

$$ H = \frac{1}{2}\hbar\omega_o\sigma_z + \hbar\omega a^\dagger a + i\hbar C[a + a^\dagger][\sigma - \sigma^\dagger], \qquad (6.19) $$

where, as in Chapter 4, we drop the $\mathbf{A}^2$ term. Here a and $a^\dagger$ are the annihilation and creation operators for the single field mode with which the atom is assumed to interact, and C is the coupling constant for this interaction.

---

[2] The quantization of the field in the case of a lossy medium can be done in a variety of ways. One way is described in Section 7.3.

[3] A useful collection of formulas for emission and absorption, including the refractive index and local (Lorentz–Lorenz) field corrections, is given by B. DiBartolo, *Optical Interactions in Solids* (Wiley, New York, 1968), p. 405.

In the absence of atom–field coupling the Heisenberg equations of motion for $\sigma(t)$ and $a(t)$ have the solutions $\sigma(t) = \sigma(0)e^{-i\omega_o t}$ and $a(t) = a(0)e^{-i\omega t}$. If the coupling described by the last term in (6.19) is not too strong, the terms $a\sigma$ and $a^\dagger \sigma^\dagger$ for $\omega \approx \omega_o$ will be rapidly varying in time compared with the terms $a^\dagger \sigma$ and $a\sigma^\dagger$. If we choose to ignore the very rapidly oscillating terms, we can replace (6.19) by the RWA Hamiltonian (Section 4.9)

$$H_{\text{RWA}} = \frac{1}{2}\hbar\omega_o\sigma_z + \hbar\omega a^\dagger a + i\hbar C(a^\dagger \sigma - \sigma^\dagger a). \qquad (6.20)$$

This has the effect of removing rapidly oscillating terms in the Heisenberg equations of motion and is just another way of formulating the RWA introduced in Section 4.9. Note that the RWA ignores "energy nonconserving" terms; in the Schrödinger picture it leads to the essential-states approximation, as discussed in Section 4.14. Thus (6.20) describes the atom–field interaction in such a way that an upward atomic transition is always accompanied by the annihilation of a photon ($\sigma^\dagger a$), and a downward transition is always accompanied by the creation of a photon ($\sigma a^\dagger$).[4] The coupling "constant" C in general varies with position, as discussed in the preceding two sections. For simplicity we shall not bother to indicate this explicitly.

The single-mode atom–field model described by the Hamiltonian (6.20) is called the *Jaynes–Cummings model* (Jaynes and Cummings, 1963). There are various ways to calculate quantities of interest in this model, such as the dressed-state formalism originally used by Jaynes and Cummings (see also Knight and Milonni, 1980). The most direct approach is perhaps via the Heisenberg equations of motion for the atom and field operators. For $\sigma_z(t)$, for instance, one easily obtains

$$\ddot{\sigma}_z(t) + 2C^2\sigma_z(t) + 4C^2\sigma_z(t)N(t) = 0 \qquad (6.21)$$

in the case $\omega = \omega_o$ of exact resonance, where the operator

$$N(t) \equiv \frac{1}{2}\sigma_z(t) + a^\dagger(t)a(t) \qquad (6.22)$$

---

[4] The RWA is very often, but not always, an excellent approximation. For an atom in an externally applied field, the RWA ignores the so-called Bloch–Siegert shift. See L. Allen and J.H. Eberly, *Optical Resonance and Two-Level Atoms* (Dover Books, New York, 1987). The RWA does not take into account multiphoton processes, which become important at large field intensities and at field frequencies near multiphoton resonances, such as $\omega \cong \omega_o/3$ for a two-state atom. For a collection of two-state atoms interacting with the field, the RWA does not account for the fact that the semiclassical dynamics may be chaotic under certain circumstances. See P. W. Milonni, J. R. Ackerhalt, and H. W. Galbraith, *Phys. Rev. Lett.* **50**, 966 (1983); P. W. Milonni, M.-L. Shih, and J. R. Ackerhalt, *Chaos in Laser-Matter Interactions* (World Scientific, Singapore, 1987).

is a constant of motion (in the RWA).

Suppose, for instance, that at $t = 0$ the atom is in the upper state $|2\rangle$ and the field is in the photon number eigenstate $|n\rangle$, so that $a^\dagger(0)a(0)|n\rangle = n|n\rangle$. Then, for this assumed initial atom–field state $|\psi\rangle = |2\rangle|n\rangle$,

$$
\begin{aligned}
\langle \sigma_z(t)N(t)\rangle &= \langle\psi|\frac{1}{2}\sigma_z(t)\sigma_z(0)|\psi\rangle + \langle\sigma_z(t)a^\dagger(0)a(0)|\psi\rangle \\
&= (n + \frac{1}{2})\langle\sigma_z(t)\rangle,
\end{aligned}
\tag{6.23}
$$

and (6.21) gives

$$
\langle\ddot{\sigma}_z(t)\rangle + 4C^2(n+1)\langle\sigma_z(t)\rangle = 0,
\tag{6.24}
$$

with the solution

$$
\langle\sigma_z(t)\rangle = \cos(2C\sqrt{n+1})t \ .
\tag{6.25}
$$

Now recall from Chapter 4 that $\sigma_z = \sigma_{22} - \sigma_{11}$ is the operator corresponding to the population difference between the upper and lower states of the two–state atom. Writing $\langle\sigma_z(t)\rangle = P_2(t) - P_1(t)$, where P_2 and P_1 are the upper- and lower-state occupation probabilities, we obtain from (6.25) the Jaynes–Cummings results:

$$
P_2(t) = \cos^2(C\sqrt{n+1})t, \quad P_1(t) = \sin^2(C\sqrt{n+1})t,
\tag{6.26}
$$

and

$$
\langle a^\dagger(t)a(t)\rangle = \langle N(0)\rangle - \frac{1}{2}\langle\sigma_z(t)\rangle = n + \sin^2(C\sqrt{n+1})t.
\tag{6.27}
$$

When $n \gg 1$ we have $\langle a^\dagger(t)a(t)\rangle \cong n$ and $P_2(t) \cong \cos^2\Omega t$, $P_1(t) \cong \sin^2\Omega t$, with $\Omega \equiv C\sqrt{n}$. This is the solution predicted by semiclassical radiation theory, where the field is not quantized. The oscillation frequency Ω is called the *Rabi frequency*.[5] In this limit the atom is driven by an applied field of effectively constant amplitude. Of course this is a limit in which we expect semiclassical theory to apply, since if $n \gg 1$ the changes $n \to n \pm 1$ due to emission and absorption are negligible compared with n, and the field may as well be treated as a prescribed classical field with fixed intensity $\propto n$ and electric field amplitude $\propto \sqrt{n}$.

In the limit $n \to 0$, on the other hand, semiclassical theory fails because the standard classical treatment of the field does not account for the vacuum field. In other words, semiclassical radiation theory does not properly

---

[5] See, for instance, L. Allen and J. H. Eberly, ibid.

describe spontaneous emission (see Section 1.15 and Milonni, 1976). In this
limit, for the initial atom–field state $|\psi\rangle = |2\rangle|vac\rangle$,

$$P_2(t) = \cos^2 Ct, \quad P_1(t) = \sin^2 Ct, \tag{6.28}$$

$$\langle a^\dagger(t)a(t)\rangle = \sin^2 Ct = P_1(t). \tag{6.29}$$

The atom–field coupling constant C is therefore effectively a so-called vacuum Rabi frequency.

Of course equations (6.28) and (6.29) simply indicate that an initially
excited atom can spontaneously emit a photon into the initially unexcited
field mode, and that the atom and the field will sinusoidally exchange this
photon of energy. These results are simply the single-mode specialization of
the theory of spontaneous emission presented in Chapter 4, where the atom
was coupled to the infinity of modes of free space. As such, the single-mode
results do not say anything new about the vacuum.

What is noteworthy is that in recent years it has become possible to
experimentally test predictions of the idealized Jaynes–Cummings model.
In experiments of Rempe, Walther, and Klein (1987) a velocity-selected
beam of rubidium atoms was excited with laser radiation to the $63p_{3/2}$ level
and then passed through a superconducting microwave cavity operating on
a single mode at 21.6 GHz, near the $63p_{3/2}$–$61d_{5/2}$ transition frequency. The
damping rate for radiation in the superconducting cavity was sufficiently
small that the two-state "atom" consisting of the $63p_{3/2}$ and $61d_{5/2}$ levels
could interact with its spontaneously emitted photon. The atomic beam
flux could be made so small (500–3000 atoms/sec) that only a single atom
at a time was present in the cavity, and the cavity field could relax back to
the 2.5 K thermal equilibrium between successive atoms.

For $T = 2.5$ K the mean number of 21.6 GHz photons in a single mode
is $\bar{n} = (e^{\hbar\omega/kT} - 1)^{-1} \cong 2$, so that the field cannot be assumed to be well
described initially by the vacuum state. Rather, the field has a photon
probability distribution $P_n = \bar{n}^n(\bar{n} + 1)^{-(n+1)}$ [equation (2.133)]. The
upper-level probability given by (6.26) must be replaced by

$$P_2(t) = \sum_{n=0}^{\infty} P_n \cos^2(C\sqrt{n+1})t, \tag{6.30}$$

and similarly (6.27) is replaced by

$$\langle a^\dagger(t)a(t)\rangle = \bar{n} + \sum_{n=0}^{\infty} P_n \sin^2(C\sqrt{n+1})t. \tag{6.31}$$

In the experiments of Rempe et al. the atoms exiting the cavity were detected by field ionization with a field strength such that mainly the atoms in the $63p_{3/2}$ level were ionized. By selecting different atomic velocities, the interaction time of an atom with the field could be varied. The upper-level probability $P_2(t)$ as a function of the interaction time t could thus be determined, and the data for the limited range of accessible interaction times were found to be in quite good agreement with the prediction (6.30) of the Jaynes–Cummings model.

The results (6.28) and (6.29) show that there is no *irreversible* spontaneous emission when an atom is coupled to only a single mode of the vacuum field. When the atom is coupled to all the modes of free space, however, the upper-level probability is not sinusoidal but (approximately) exponentially damped (Section 4.14). Fermi (1932) illustrated this behavior with a coupled-oscillator model. Two coupled harmonic oscillators sinusoidally exchange the energy initially residing entirely in one of the oscillators. As more and more oscillators are coupled into the system, however, the time it takes for the one initially excited oscillator to recover a substantial portion of its initial energy increases; as the number of oscillators becomes very large, the energy of the initially excited oscillator is effectively lost irreversibly.

The discrete summations over n in single-mode results such as (6.30) and (6.31), of course, result from the quantization of the field, i.e., from the "granularity" associated with field quantization. A consequence of this discreteness is a "collapse and revival" behavior (Eberly, Narozhny, and Sanchez-Mondragon, 1980), evidence for which has been observed in the experiments of Rempe et al. Such behavior is related to the *quantum recurrence theorem* for quasiperiodic systems (see Milonni and Singh (1991) and references therein).

6.5   Laser Linewidth

It has been known since the earliest research on lasers that spontaneous emission prevents laser radiation from ever being perfectly monochromatic. In this section we consider this quantum limit to the laser linewidth and show how it may be associated, at least in part, with the vacuum field. The material presented here is a bit specialized, and in the interest of brevity we must refer the reader to the technical literature for details (see Milonni, 1991; Goldberg, Milonni, and Sundaram, 1991).

We assume that an amplifying medium fills the laser cavity defined by mirrors at $z = 0$ and $z = d$. The laser is assumed to operate on a single mode of the field, and for simplicity we describe the field in the plane-wave

approximation. We denote the annihilation part of the intracavity field
operator associated with propagation to the right (i.e., in the direction
from $z = 0$ to $z = d$) by $A_R(z,t)e^{-i\omega_o t}$, where A_R is slowly varying in time
compared with the oscillation at frequency ω_o, the frequency of the lasing
transition of the atoms making up the gain medium. The time dependence
of $A_R(z,t)$ allows for the nonmonochromaticity of the radiation.

Consider the field $A_R(d_<, t + 2d/c)$ at the right mirror:

$$
\begin{aligned}
A_R(d_<, t + \frac{2d}{c}) &= A_L(d_<, t)\sqrt{GR_1G} + A_{R,\mathrm{vac}}(-d, t)\sqrt{T_1G} \\
&\quad + A_{L,\mathrm{vac}}(d_>, t)\sqrt{T_2GR_1G} + A_{SP}(d_<, t). \quad (6.32)
\end{aligned}
$$

Here $d_<$ and $d_>$ denote points just inside and outside, respectively, the
mirror at d. R_j and T_j denote mirror power reflection and transmission
coefficients, and $\sqrt{G}$ is the amplitude amplification factor associated with
a single pass through the amplifying medium. Subscripts R and L label
right- and left-going fields, while vac labels source-free, vacuum fields.

The first term on the right side of (6.32) arises from the propagation of
the left-going field at $d_<$ through the gain medium ($\sqrt{G}$), reflection off the
mirror at $z = 0$ ($\sqrt{R_1}$), and a second pass through the gain medium ($\sqrt{G}$).
The second term arises from the transmission of the external vacuum field
through the mirror at $z = 0$ ($\sqrt{T_1}$), followed by amplification of this field as
it propagates to the mirror at $z = d$ ($\sqrt{G}$). The minus sign in $A_{R,\mathrm{vac}}(-d, t)$
merely indicates that the right-propagating field reaching $d_<$ at time $t + 2d/c$
is, except for the effects of transmission and gain, the right-propagating field
at $d_< - 2d = -d$ at the retarded time $(t + 2d/c) - 2d/c$. Similarly the third
term results from the transmission of the left-going vacuum field through
the mirror at $z = d$ ($\sqrt{T_2}$), amplification ($\sqrt{G}$) and reflection ($\sqrt{R_1}$), and
a second pass through the gain cell ($\sqrt{G}$). Finally A_{SP} stands for the
contribution from spontaneous emission, as opposed to the contributions
from stimulated emission involving the factor $\sqrt{G}$.

Equation (6.32) expresses a basic kinematical relationship. It has contributions only from vacuum fields that have passed at least once through the
medium. This expresses the assumption (which can be rigorously justified
by a mode expansion of the field) that the vacuum fields are transmitted
and reflected just as are fields of "real" photons.

It is convenient to convert (6.32) to a first-order differential equation by
making the approximation[6]

$$
A_R(d_<, t + \frac{2d}{c}) \cong A_R(d_<, t) + \frac{2d}{c}\dot{A}_R(d_<, t). \quad (6.33)
$$

---

[6] The neglect of higher derivatives in (6.33) may be justified a posteriori on the grounds
that $\gamma_c - c\bar{g}$ is small near steady-state oscillation.

In (6.32) we can also write $A_L(d_<,t) = \sqrt{R_2}A_R(d_<,t)$; note that the contribution from the transmitted vacuum field is already included in the third term on the right in equation (6.32). Thus we can replace (6.32) by

$$\dot{A}_R(d_<,t) \cong \frac{c}{2d}[G\sqrt{R_1R_2} - 1]A_R(d_<,t) + \frac{c}{2d}\left[A_{R,\mathrm{vac}}(-d,t)\sqrt{GT_1}\right.$$
$$\left. + A_{L,\mathrm{vac}}(d_>,t)G\sqrt{R_1T_2}\right] + \frac{c}{2d}A_{SP}(d_<,t). \quad (6.34)$$

In the limit of small output coupling (i.e., $T_1, T_2 << 1$) we have[7]

$$G\sqrt{R_1R_2} - 1 = \sqrt{R_1R_2}e^{\bar{g}d} - 1 \cong \sqrt{R_1R_2} + \bar{g}d\sqrt{R_1R_2} - 1$$
$$\cong \ln\sqrt{R_1R_2} + \bar{g}d$$
$$= -\frac{2d}{c}(\frac{1}{2}\gamma_c) + \bar{g}d = \frac{2d}{c}(\frac{1}{2}c\bar{g} - \frac{1}{2}\gamma_c). \quad (6.35)$$

Here $\bar{g} \equiv (1/d)\ln G$ is a "mean" power gain coefficient and $\gamma_c \equiv -(c/2d) \times \ln(R_1R_2)$ is the cavity power damping rate. Thus

$$\dot{A}_R(d_<,t) \cong \frac{1}{2}(c\bar{g} - \gamma_c)A_R(d_<,t) + \frac{c}{2d}[A_{R,\mathrm{vac}}(-d,t)\sqrt{GT_1}$$
$$+ A_{L,\mathrm{vac}}(d_>,t)G\sqrt{R_1T_2}] + \frac{c}{2d}A_{SP}(d_<,t). \quad (6.36)$$

Let us consider now the field correlation function $\langle A_R(t)A_R^\dagger(t+\tau)\rangle$ for the field at $d_<$, where the expectation value refers to an initial state $|\psi\rangle$ in which there are no photons in the field. This correlation function will determine the laser linewidth, as discussed later. Equation (6.36) implies

$$\langle A_R(t)A_R^\dagger(t+\tau)\rangle = \left(\frac{c}{2d}\right)^2\left[\int_0^t dt' \int_0^{t+\tau} dt'' \langle V(t')V^\dagger(t'')\rangle\right.$$
$$\times e^{\gamma(t'+t''-2t-\tau)}$$
$$+ \int_0^t dt' \int_0^{t+\tau} dt'' \langle A_{SP}(t')A_{SP}^\dagger(t'')\rangle e^{\gamma(t'+t''-2t-\tau)}\Bigg]. \quad (6.37)$$

Here

$$V(t) \equiv A_{R,\mathrm{vac}}(-d,t)\sqrt{GT_1} + A_{L,\mathrm{vac}}(d_>,t)G\sqrt{R_1T_2} \quad (6.38)$$

---

[7]When the output coupling and other losses are small, the steady-state gain required to overcome all the losses is small, i.e., $G \cong 1$. See, for instance, Goldberg et al., 1991.

and $\gamma = \frac{1}{2}(\gamma_c - c\bar{g})$. Since $\langle A_{R,\text{vac}}(-d,t)A_{R,\text{vac}}^\dagger(-d,t)\rangle$ and $\langle A_{L,\text{vac}}(d_>,t) \times A_{L,\text{vac}}^\dagger(d_>,t)\rangle$ are not zero, the vacuum field leaking into the cavity from the outside contributes *explicitly* to (6.37). We show later that

$$\langle A_{R,\text{vac}}(-d,t')A_{R,\text{vac}}^\dagger(-d,t'')\rangle = \langle A_{L,\text{vac}}(d_>,t')A_{L,\text{vac}}^\dagger(d_>,t'')\rangle = \frac{d}{c}\delta(t'-t''),$$

$$(6.39)$$

so that

$$\langle V(t')V^\dagger(t'')\rangle = [GT_1 + G^2R_1T_2]\frac{d}{c}\delta(t'-t''), \qquad (6.40)$$

while the corresponding normally ordered correlation function of course vanishes. Since they correspond to different, uncorrelated vacuum field modes, the left- and right-going vacuum fields do not make an "interference" contribution to (6.40). In the limit of small outcoupling [cf. (6.35)],

$$GT_1 = e^{\bar{g}d}(1 - R_1) \cong 1 - R_1 \cong -\ln R_1, \qquad (6.41)$$

$$G^2R_1T_2 = e^{2\bar{g}d}R_1(1 - R_2) \cong 1 - R_2 \cong -\ln R_2, \qquad (6.42)$$

and

$$\langle V(t')V^\dagger(t'')\rangle \cong -\frac{d}{c}(\ln R_1R_2)\delta(t'-t'') = \frac{1}{2}\left(\frac{2d}{c}\right)^2\gamma_c\delta(t'-t''). \quad (6.43)$$

It may also be shown that

$$\langle A_{SP}(t')A_{SP}^\dagger(t'')\rangle = \frac{1}{2}\left(\frac{2d}{c}\right)^2\frac{\gamma_cP_1}{P_2 - P_1}\delta(t'-t''), \qquad (6.44)$$

where P_2 and P_1 are respectively the steady-state upper- and lower-level occupation probabilities of the lasing transition. Then (6.37) gives

$$\langle A_R(t)A_R^\dagger(t+\tau)\rangle = \left[\frac{1}{2} + \frac{1}{2}\frac{P_1}{P_2 - P_1}\right]\left(\frac{\gamma_c}{2\gamma}\right)e^{-\gamma\tau} = \frac{1}{2}\frac{P_2}{P_2 - P_1}\left(\frac{\gamma_c}{2\gamma}\right)e^{-\gamma\tau},$$

$$(6.45)$$

where the two terms in brackets in the first equality arise from V and A_{SP}, respectively, i.e., from vacuum and source fields.

The laser spectrum is determined by the Fourier transform of (6.45). Thus the spectrum is predicted to be Lorentzian with linewidth (full width at half-maximum) $\Delta\omega = 2\gamma = \gamma_c - c\bar{g}$. Equation (6.45) also implies

$$\begin{aligned}\langle A_R(t)A_R^\dagger(t)\rangle &= \langle A_R^\dagger(t)A_R(t)\rangle + 1 = \frac{1}{2}n_{ss} + 1 \cong \frac{1}{2}n_{ss} \\ &= \frac{1}{2}\frac{P_2}{P_2 - P_1}\left(\frac{\gamma_c}{2\gamma}\right),\end{aligned} \qquad (6.46)$$

or

$$2\gamma \cong \frac{P_2}{P_2 - P_1} \frac{\gamma_c}{n_{ss}} \ , \tag{6.47}$$

where n_{ss} is the steady-state intracavity photon number.[8] Then

$$\Delta\omega = \frac{P_2}{P_2 - P_1} \frac{\gamma_c}{n_{ss}} \ , \tag{6.48}$$

or, since the output power $P_{\text{out}} = \gamma_c \hbar\omega n_{ss}$,

$$\Delta\omega = \frac{P_2}{P_2 - P_1} \left(\frac{\hbar\omega}{P_{\text{out}}}\right) \gamma_c^2 \ . \tag{6.49}$$

This lower limit to the laser linewidth is called the *Schawlow-Townes line-width*.[9]

It might be noted that the quantum limit to the laser linewidth is or-dinarily not of much practical concern, since other contributions to the linewidth (e.g., mirror jitter) are usually much larger. In semiconductor lasers, however, the linewidth is often dominated by quantum noise. This is due to the facts that $\Delta\omega$ is proportional to d^{-2} and that d is very small for semiconductor lasers.

As mentioned following equation (6.45), the laser linewidth in our calcu-lation has equal contributions [except for the factor $P_1/(P_2 - P_1)$] from the vacuum field entering the cavity from the outside world and the sponta-neously emitted radiation from atoms inside the cavity. If we had per-formed the calculation using the normally ordered correlation function $\langle A_R^\dagger(t)A_R(t + \tau)\rangle$, on the other hand, we would have found that the en-tire linewidth $\Delta\omega$ is attributable to spontaneous emission (Milonni, 1991; Goldberg et al., 1991). The choice of operator ordering in the field corre-lation function is dictated by the detection scheme employed to measure the linewidth (Appendix E), but has no influence on the actual value of the linewidth. The situation here is akin to the role of operator orderings in the interpretation of spontaneous emission and the Lamb shift in Chapter 4.

It is worth noting again that our calculation of the Schawlow–Townes linewidth assumes that the vacuum field is transmitted, reflected, and am-plified in the same way as fields of "real" photons, i.e., fields defined in terms of *excited* states of the field. The transmission and reflection properties of

---

[8] For small outcoupling $\langle A_R^\dagger(t)A_R(t)\rangle \cong \langle A_L^\dagger(t)A_L(t)\rangle$, so that $n_{ss} \equiv \langle A_R^\dagger(t)A_R(t)\rangle + \langle A_L^\dagger(t)A_L(t)\rangle \cong 2\langle A_R^\dagger(t)A_R(t)\rangle$.

[9] For numerical estimates of $\Delta\omega$ see, for instance, P. W. Milonni and J. H. Eberly, *Lasers* (Wiley, New York, 1988).

the vacuum field are consistent with the fact (see Sections 2.4 and 6.2) that the spatial variations of the quantized field are the same as those of the classical field subject to the same boundary conditions. The fact that the vacuum field may be amplified follows from the fact that its contribution to the laser linewidth is more or less interchangeable with the contribution from spontaneous emission: if spontaneously emitted radiation inside the cavity is amplified by the gain medium, then so too must the vacuum field entering the cavity. Another way to say this is that "quantum noise" may be amplified. Such noise amplification, which has been known for many years to electrical engineers engaged in maser research, for instance, appears also in other contexts in quantum optics.[10]

Evaluation of Vacuum Field Correlation Function

In our derivation of the laser linewidth we required the correlation functions (6.39) and (6.44). We shall now provide a derivation of (6.39). The derivation of (6.44) may be found in Goldberg et al. (1991).

Consider the operator

$$E_{R,\text{vac}}(t) = i \sum_k \left(\frac{2\pi\hbar\omega_k}{V} \right)^{1/2} a_k(0) e^{-i(\omega_k - \omega_o)t} , \qquad (6.50)$$

where ω_o is the central frequency of the lasing transition, V is a quantization volume, and $a_k(0)$ is the source-free photon annihilation operator for the mode k. The factor $e^{i\omega_o t}$ makes $E_{R,\text{vac}}(t)$ the slowly varying, positive–frequency part of the electric field operator that drives the slowly varying atomic dipole operator. The spatial dependence of the vacuum fields will be of no consequence in what follows and is therefore ignored here.

From (6.50) we have the vacuum expectation value

$$\langle E_{R,\text{vac}}(t') E_{R,\text{vac}}^\dagger(t'') \rangle = \sum_k \left(\frac{2\pi\hbar\omega_k}{V} \right) e^{-i(\omega_k - \omega_o)(t' - t'')} , \qquad (6.51)$$

since $\langle a_k(0) a_{k'}^\dagger(0) \rangle = \delta_{kk'}$. In the one-dimensional mode continuum limit appropriate for vacuum fields propagating along the direction defined by the optical axis of the laser we take $V = A_o L$, where A_o is some cross-sectional area and L is a length, and $\sum_k \to (L/2\pi) \int dk$. Then, in an approximation appropriate to our purposes,

$$\langle E_{R,\text{vac}}(t') E_{R,\text{vac}}^\dagger(t'') \rangle \quad \to \quad \frac{L}{2\pi c} \left(\frac{2\pi\hbar}{A_o L} \right) \int_0^\infty d\omega\, \omega\, e^{-i(\omega - \omega_o)(t' - t'')}$$

---

[10] See, for instance, P. W. Milonni, E. J. Bochove, and R. J. Cook, *Phys. Rev.* A**40**, 4100 (1989).

$$\cong \frac{\hbar\omega_o}{A_o c} \int_{-\infty}^{\infty} d\omega \, e^{-i(\omega-\omega_o)(t'-t'')}$$

$$= \frac{2\pi\hbar\omega_o}{A_o c} \delta(t'-t'') = \frac{d}{c}\left(\frac{2\pi\hbar\omega_o}{A_o d}\right)\delta(t'-t'').$$

(6.52)

In our approach to the laser linewidth we require the slowly varying field annihilation operator $A_{R,\text{vac}}(t)$, which may be defined by writing

$$E_{R,\text{vac}}(t) = i\left(\frac{2\pi\hbar\omega_o}{A_o d}\right)^{1/2} A_{R,\text{vac}}(t).$$

(6.53)

Comparison with (6.52) indicates that

$$\langle A_{R,\text{vac}}(t') A_{R,\text{vac}}^{\dagger}(t'') \rangle \cong \frac{d}{c}\delta(t'-t''),$$

(6.54)

and of course $\langle A_{R,\text{vac}}^{\dagger}(t') A_{R,\text{vac}}(t'') \rangle = 0$. Clearly $A_{L,\text{vac}}(t)$ has the same correlation properties.

Petermann and Purcell Effects for Lossy Cavities

For lossy cavities the Schawlow–Townes linewidth (6.49) is multiplied by a factor K, the "Petermann factor" (see, for instance, Milonni, 1991; Goldberg et al., 1991 and references therein). For mirror reflectivities $R_1 = R_2 = R$, the Petermann factor is $K = [(1-R^2)/2R\log R]^2$ if diffraction losses are ignored. This factor, which is near unity for $R \cong 1$, is attributable to the fact that the field vacuum fluctuations are amplified by the gain medium, *and* that this amplification in a lossy system cannot in general be derived under the approximation of uniformly distributed loss.

There is another effect of a lossy single-mode cavity that occurs even in the case of a *single* spontaneously emitting atom: if the cavity is "over-damped," in the sense that the photon loss rate γ_c is much larger than the "vacuum Rabi frequency" C (Section 6.4), then the spontaneous emission rate into the single mode is increased by the Q factor of the cavity[11] (see, for instance, Cook and Milonni, 1987, or Feng and Ujihara, 1990 and references therein). This effect was first predicted by Purcell in 1945, and has

---

[11] The "quality factor" Q is defined by writing the bandwidth, in this case $\delta\nu_c = (4\pi)^{-1}[(c/L)\log(R_1 R_2)^{-1}]$, as ν/Q. It was introduced in about 1920 by K. S. Johnson as a figure of merit for coils (inductors). According to E. I. Green, [*American Scientist* **43**, 584 (1955)] "His reason for choosing Q was quite simple. He says that it did not stand for 'quality factor' or anything else, but since the other letters of the alphabet had already been pre-empted for other purposes, Q was all he had left."

bccn obocrvcd in the experiments of several groups in more recent years (see, for instance, Goy et al., 1983).

6.6   Amplified Spontaneous Emission

To appreciate the interplay of spontaneous emission and amplified vacuum fields in gain media, we now consider in a heuristic and very simplified way the problem of amplified spontaneous emission (ASE). This problem arises when we have a gain cell without mirrors to provide feedback and multiple amplifications. Spontaneously emitted radiation can be amplified by stimulated emission in the gain medium and emerge as a directional beam of radiation resembling laser radiation. ASE is of practical importance because it can substantially deplete the gain available to an input signal to be amplified, and also because it can irradiate a target before the amplified signal pulse.

In the simplest model of propagation in a gain medium we write

$$\frac{dI}{dz} = gI, \tag{6.55}$$

where g is the gain coefficient, which for simplicity we take to be independent of I.[12] Under the assumption that most of the atoms of the gain medium are in the excited state of the lasing transition, we have $g = \sigma N$, where σ is the stimulated emission cross section and N is the number of atoms per unit volume in the excited state. Equation (6.55) assumes a steady-state situation in which the intensity I is independent of time.

Obviously (6.55) implies $I(z) = 0$ for all z if $I(z = 0) = 0$. In other words, if there is no input radiation at the entrance plane $z = 0$ of the gain cell, there will be no radiation anywhere in the cell. This is in conflict with the observation of ASE. To remedy this deficiency of our simple model we assume the existence of some effective input noise intensity I_{eff} at $z = 0$, in which case

$$I(z) = I_{\text{eff}} e^{gz} \ . \tag{6.56}$$

Since I_{eff} represents "quantum noise," it must be associated with the vacuum field. Recall from Chapter 1 that the vacuum field has a spectral energy density $\rho_o(\nu) = 4\pi h\nu^3/c^3$ such that $\rho_o(\nu)\Delta\nu$ is the zero-point field energy per unit volume in the narrow frequency interval $[\nu, \nu + \Delta\nu]$. Then

---

[12] This is the so-called small–signal or unsaturated limit in which the intensity is sufficiently small that it does not cause any substantial depletion of the gain.

the quantum noise intensity at the laser transition frequency ν is

$$I_{\text{eff}} = c\rho_o(\nu)\frac{\Omega}{4\pi} = \left(\frac{\pi hc^2\Omega}{\lambda^5}\right)\Delta\lambda , \qquad (6.57)$$

where we have inserted a factor $\Omega/4\pi$, with Ω the solid angle defined by the bore radius and length of the gain medium. This factor accounts for the fact that only the field modes within the solid angle Ω are effective noise sources. In the present context $\Delta\nu = c\Delta\lambda/\lambda^2$ should be on the order of the spectral width of the laser transition. For later purposes we have replaced $\Delta\nu$ by $\pi c\Delta\lambda/\lambda^2$ in (6.57). From (6.56), then,

$$I(z) = \left(\frac{\pi hc^2\Omega}{\lambda^5}\right)\Delta\lambda e^{gz} . \qquad (6.58)$$

This equation describes the buildup of intensity before the intensity becomes large enough to substantially deplete (saturate) the gain medium. It is used in simple estimates of ASE power.[13]

But now let us ignore the vacuum field and modify (6.55) in a different way, by including the contribution of spontaneous emission to the growth of intensity:

$$\frac{dI}{dz} = gI + \frac{h\nu A_{21}N\Omega}{4\pi} , \qquad (6.59)$$

where A_{21} is as usual the spontaneous emission rate and again we have included the solid angle factor $\Omega/4\pi$. In this case this factor is included because the spontaneous emission from a single atom is statistically isotropic, whereas we are interested only in the radiation into a solid angle Ω.

As earlier we assume that g and N are constants. Then (6.59) has the solution

$$I(z) = \left(\frac{h\nu A_{21}\Omega}{4\pi}\right)\frac{N}{g}(e^{gz} - 1), \qquad (6.60)$$

if we assume now that $I(0) = 0$. For a homogeneously broadened gain medium with Lorentzian lineshape of width (full width at half-maximum) $\Delta\nu$ we have[14]

$$\sigma = \frac{\lambda^2 A_{21}}{8\pi}\frac{2}{\pi\Delta\nu} = \frac{\lambda^4 A_{21}}{4\pi^2 c\Delta\lambda} \qquad (6.61)$$

and therefore (6.60) becomes

$$I(z) \cong \left(\frac{\pi hc^2\Omega}{\lambda^5}\right)\Delta\lambda e^{gz} \qquad (6.62)$$

---

[13] See, for instance, P. B. Corkum and R. S. Taylor, *IEEE J. Quantum Electron.* **QE-18**, 1962 (1982).

[14] See, for instance, P. W. Milonni and J. H. Eberly, *Lasers*.

in the approximation $c^{qz} \gg 1$. This is identical to the result (6.58) obtained from the perspective of quantum noise associated with the vacuum radiation field.

The equivalence of (6.58) and (6.62) in our highly simplified analysis shows that "quantum noise" due to the vacuum field is closely related to spontaneous emission, as in the theory of the laser linewidth presented in the preceding section. This circumstance may ultimately be traced to the fact that spontaneous emission itself may be viewed – except for a factor of two – as stimulated emission by the vacuum field (Sections 3.2 and 4.13).

6.7   Geonium

Single particles or small clouds of particles can now be trapped practically indefinitely in Penning traps. Dehmelt has called these trapped-particle systems *geonium atoms*, since earth–bound trapping apparatus plays a binding role analogous to that of an atomic nucleus (van Dyck, Schwinberg, and Dehmelt, 1978; Dehmelt, 1990; Brown and Gabrielse, 1986). With such traps, the electron $g - 2$ has been measured to an accuracy ≈ 900 times greater than was possible in previous measurements.

It is a well-known consequence of Gauss's law that charged particles cannot be trapped by electrostatic forces alone.[15] However, trapping is possible with a combination of a homogeneous magnetic field and an electrostatic quadrupole potential, as in a Penning trap. Consider first a particle in the homogeneous magnetic field $\mathbf{B} = B\hat{z}$. In this case trapping can be achieved radially, since in the xy plane the motion of a charge e of mass m is circular, with the cyclotron frequency $\omega_c = eB/mc$ and radius inversely proportional to B.

The charge can be trapped axially with an electrostatic potential $\phi(r, z)$ such that $F_z = -e\partial\phi/\partial z = -kz$, with $k > 0$. In this case $\partial^2\phi/\partial z^2 = k/e$, and so the Laplace equation, $\nabla^2\phi = \partial^2\phi/\partial z^2 + r^{-1}\partial/\partial r(r\partial\phi/\partial r) = 0$ in the case of azimuthal symmetry, can be satisfied by the quadrupole potential

$$\phi(r, z) = \frac{k}{2e}(z^2 - \frac{1}{2}r^2) = \phi_o \frac{z^2 - r^2/2}{2d^2} , \qquad (6.63)$$

where d is introduced as some characteristic trap dimension. This potential can be realized with charged conducting surfaces shaped so as to form equipotentials of ϕ, as in Figure 6.2. To be equipotentials of $\phi(r, z)$ these

---

[15] This is Earnshaw's theorem. See, for instance, R. P. Feynman, R. B. Leighton, and M. Sands, *The Feynman Lectures on Physics* (Addison–Wesley, Reading, Mass., 1964), Volume 2, p. 5-1.

surfaces must asymptotically approach $z = \pm r^2/2$. The two "endcaps" in Figure 6.2 are hyperboloids of revolution defined by

$$z^2 = z_o^2 + \frac{r^2}{2} \,, \tag{6.64}$$

where z_o is defined in the figure. On the two endcaps the potential is thus $\phi_o(z_o^2/2d^2)$. The ring electrode is a hyperboloid defined by

$$z^2 = \frac{1}{2}(r^2 - r_o^2), \tag{6.65}$$

where r_o is defined in the figure. On the ring electrode the potential is $-\phi_o(r_o^2/4d^2)$. The potential difference between the ring and endcap electrodes is therefore $\phi_o(z_o^2 + r_o^2/2)/2d^2 = \phi_o$ if we define $2d^2$ to be $z_o^2 + r_o^2/2$. The quadrupole potential in the Penning trap geometry of Figure 6.2 is thus

$$\phi(r, z) = \phi_o \frac{z^2 - r^2/2}{z_o^2 + r_o^2/2} \,, \tag{6.66}$$

with $e\phi_o > 0$.

The axial motion of a charged particle in an idealized Penning trap is a simple harmonic oscillation at frequency ω_z given by

$$\omega_z^2 = \frac{k}{m} = \frac{e\phi_o}{md^2} \,, \tag{6.67}$$

and typically $\omega_z << \omega_c$. Thus the trapping is primarily magnetic. For $B = 58.72$ kG, $\phi_o = 10.22$ V, and $d = z_o = r_o\sqrt{2} = 0.335$ cm (Gabrielse and Dehmelt, 1981), we have

$$\frac{\omega_c}{2\pi} \cong 164 \text{ GHz}, \quad \frac{\omega_z}{2\pi} \cong 64 \text{ MHz} \,. \tag{6.68}$$

The radial motion is described by the equation of motion

$$m\ddot{\mathbf{r}} = e(\mathbf{E} + \frac{1}{c}\dot{\mathbf{r}} \times \mathbf{B}) = e\left[\frac{\phi_o}{2d^2}\mathbf{r} + \frac{1}{c}\dot{\mathbf{r}} \times \mathbf{B}\right], \tag{6.69}$$

where $\mathbf{r}$ is the radial displacement. In terms of ω_z and $\omega_c \equiv -eB/mc = -eB\hat{z}/mc$,

$$\ddot{\mathbf{r}} = \frac{1}{2}\omega_z^2\mathbf{r} + \boldsymbol{\omega_c} \times \dot{\mathbf{r}} \,. \tag{6.70}$$

The radial dependence of the potential (6.66) gives rise to a *repulsive* electrostatic radial force associated with the first term on the right side of (6.70). We can write (6.70) as

$$\ddot{s} = \frac{1}{2}\omega_z^2 s - i\omega_c \dot{s}, \quad s \equiv x + iy. \tag{6.71}$$

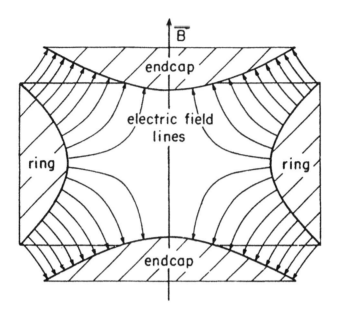

Figure 6.2: Electrode and field configurations in a Penning trap. From Brown and Gabrielse (1986), with permission.

The general solution of this equation has the form

$$s(t) \quad = \quad s_c e^{-i\omega_c' t} + s_m e^{-i\omega_m t} \,, \tag{6.72}$$

$$\omega_c' \quad \equiv \quad \frac{\omega_c}{2} + \sqrt{\left(\frac{\omega_c}{2}\right)^2 - \frac{\omega_z^2}{2}} = \omega_c - \omega_m \,, \tag{6.73}$$

$$\omega_m \quad \equiv \quad \frac{\omega_c}{2} - \sqrt{\left(\frac{\omega_c}{2}\right)^2 - \frac{\omega_z^2}{2}} \,. \tag{6.74}$$

The radial term in the potential (6.66) causes a decrease in the cyclotron frequency from ω_c to ω_c', and introduces a new frequency $\omega_m \ll \omega_c'$ called the *magnetron frequency*.[16] Thus the radial motion of a charged particle in a Penning trap consists of a cyclotron orbit superimposed on (much slower and larger) magnetron orbits, resulting in the epicyclical behavior depicted in Figure 6.3. This radial motion is superimposed on the harmonic axial oscillation.

---

[16] This terminology derives from the magnetron, the device used among other things to produce the radiation in a microwave oven. In a magnetron electrons from a heated cathode move under the influence of an axial magnetic field and a radial electric field.

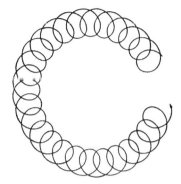

Figure 6.3: Motion of a particle in a Penning trap, projected onto the xy plane. The large circle represents a magnetron orbit, the small circles cyclotron orbits. Typical cyclotron orbit radii are actually about 100 times smaller than magnetron radii. The motion perpendicular to the plane of the paper is an axial oscillation with amplitude about 50 times larger than the magnetron radius. From Brown and Gabrielse (1986), with permission.

For the parameters used to obtain the estimates (6.68) we have

$$\frac{\omega_m}{2\pi} \cong 12.5 \text{ kHz}, \tag{6.75}$$

so that

$$\omega_m \ll \omega_z \ll \omega_c \tag{6.76}$$

for parameters typical of recent Penning trap $g - 2$ experiments. Typical dimensions for the cyclotron, axial, and (cooled) magnetron motions are $10^{-6}, 10^{-3}$, and 10^{-4} cm, respectively.

The magnetron motion arising from the radial term in (6.66) is actually unbounded: any decrease in the energy in the magnetron motion causes the magnetron radius to *increase* (Brown and Gabrielse, 1986). However, as a consequence of the relatively slow magnetron frequency, the damping time of the magnetron energy due to radiation is very large – on the order of years – and so as a practical matter the magnetron motion is quite stable enough.

In the geonium experiments of Dehmelt and his collaborators the Penning "electron cage," about 4 cm in diameter, is immersed in liquid helium and contained within the core of a superconducting magnet. The trap is pumped to obtain a good vacuum, and electrons are injected into the trap by emission from a hot filament. An applied RF voltage causes the electrons to oscillate axially, inducing oscillating image charges in the endcaps and therefore an oscillating current through a resistor in an external circuit. The IR drop through the resistor, together with noise, is the voltage between an endcap electrode and the ring electrode, and serves as the signal for monitoring the axial motion of the electrons.[17] Steplike reductions in the signal, separated in time by several minutes, indicate the one-by-one escape of electrons from the trap. Reduction of the amplitude of the applied RF voltage when one electron is left in the trap allows the single electron to be trapped for weeks.

Single-electron trapping times on the order of 10 months have been realized by "motional sideband cooling" (Wineland and Dehmelt, 1975; van Dyck et al., 1978). In essence this effect involves the absorption of RF photons with an energy defect made up for at the expense of energy in the magnetron motion (see also Brown and Gabrielse, 1986). Cooling of the magnetron motion ensures that the trapped electrons move in small orbits, for which broadening and shifting of spectral lines due to electrostatic and magnetic field inhomogeneities are minimized. (The axial and cyclotron motions are cooled "automatically" by resistive and radiative damping.)

Geonium experiments have required enormous effort and ingenuity, and the reader is urged to consult the cited literature and the references therein for clear and authoritative accounts. We now turn briefly to the greatest accomplishment (and the original purpose) of these experiments, namely the remarkably precise measurement of the electron magnetic moment anomaly.

Measurement of the Spin Anomaly

The interaction Hamiltonian for a magnetic dipole moment $\boldsymbol{\mu}$ in a magnetic field $\mathbf{B}$ is $-\boldsymbol{\mu} \cdot \mathbf{B}$. Associated with the electron spin angular momentum s is a magnetic dipole moment $\boldsymbol{\mu} = g\mu_o \mathbf{s}$, where $\mu_o = e\hbar/2mc$ is the Bohr magneton and $g \cong 2$ (Section 3.13). Thus, for $\mathbf{B} = B\hat{z}$,

$$H_s = -g\mu_o s_z B = -g\mu_o \frac{1}{2}\sigma_z B, \qquad (6.77)$$

and the allowed energy levels are $\pm g\mu_o B/2$. The frequency for transitions between these two levels is therefore

$$\omega_s = g\frac{\mu_o B}{\hbar} = \frac{g}{2}\omega_c , \qquad (6.78)$$

---

[17]It is worth noting that $I^2 R$ is only about 10^{-17} W!

where again $\omega_c = eB/mc$ is the cyclotron frequency. The spin "anomaly" (Section 3.13) is then

$$\frac{g-2}{2} = \frac{\omega_s - \omega_c}{\omega_c} \equiv \frac{\omega_a}{\omega_c} , \qquad (6.79)$$

where ω_a is the so-called anomaly frequency.

Equation (6.79) shows that the spin anomaly can be determined from measurements of the spin precession frequency ω_s and the cyclotron frequency ω_c. But $\omega_s - \omega_c$ is the difference between two large numbers, and so a determination of $g - 2$ from measurements of ω_s and ω_c would entail large errors. It would obviously be much better to measure the "anomaly frequency" $\omega_a \cong 10^{-3}\omega_c$ directly.

In fact a key to the great precision of the geonium $g-2$ experiments is the fact that a *ratio* of two measured frequencies, as in (6.79), is determined.[18] This eliminates the need to know $\mu_o B/\hbar$, as would be the case, say, if g itself were determined using (6.78) and a measurement of ω_s. A similar approach was used in the earlier $g - 2$ experiments of Crane et al. (see Rich and Wesley, 1972).

It should be recalled that, due to the electrostatic quadrupole potential, the cyclotron frequency ω_c is shifted to ω_c' according to equation (6.73), and so ω_a is likewise shifted to $\omega_a' = \omega_s - \omega_c'$. Then

$$\frac{g-2}{2} = \frac{\omega_s - (\omega_c' - \omega_m)}{\omega_c' + \omega_m} = \frac{\omega_a' - \omega_m}{\omega_c' + \omega_m} , \qquad (6.80)$$

or, since $\omega_m = \omega_z^2/2\omega_c'$ from (6.73),

$$\frac{g-2}{2} = \frac{\omega_a' - \omega_z^2/2\omega_c'}{\omega_c' + \omega_z^2/2\omega_c'} , \qquad (6.81)$$

which is the formula used in the paper of van Dyck, Schwinberg, and Dehmelt (1987), for instance.

In the experiments some very important modifications of the basic Penning trap design are introduced. One of the most crucial ones is the use of a ferromagnetic (nickel) ring that distorts the magnetic field, causing field lines to bow outward slightly at the midplane of the trap. This reinforces the effect of the electrostatic field in "binding" the motion to the midplane, resulting in a positive shift of the axial frequency. This addition to the Penning trap is called a *magnetic bottle*.

---

[18] The determination of $g - 2$ involves an average of measurements made with several electrons, but results for different electrons agree out to the last decimal place of accuracy, confirming that all "copies" of the electron appear to be identical (Ekstrom and Wineland, 1980).

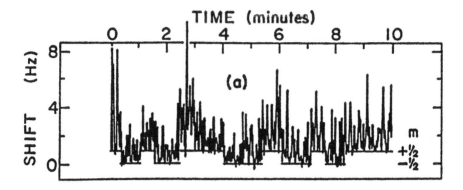

Figure 6.4: A typical record of the axial frequency shift. Spin flips are signalled by the change in the "root level" of the "cyclotron grass." From Dehmelt (1990), with permission.

The small axial frequency shift due to the magnetic bottle depends on the energy of the cyclotron orbit and on the electron spin. With n the quantum number describing the quantized (harmonic oscillator) cyclotron motion, and m $(= \pm1/2)$ the spin quantum number, the axial frequency shift is

$$\Delta\nu_z \cong (m + n + \frac{1}{2})\delta, \qquad (6.82)$$

where δ is independent of the charge but inversely proportional to the mass of the trapped particle. For typical experimental parameters, $\delta \approx 1$ Hz (Dehmelt, 1990) compared with the unperturbed axial frequency $\nu_z \approx 60$ MHz.

The axial frequency shift is used directly in the determination of the cyclotron and anomaly frequencies and therefore of $g - 2$. A signal near ω_c is introduced via a microwave inlet to the trap in order to induce transitions among cyclotron states. A typical record of the measured axial frequency shift is shown in Figure 6.4. The "grass" is due to the thermal fluctuations at the ambient temperature of about 4 K; the "root level" of the grass corresponds to the lowest rung of the ladder associated with (6.82). The cyclotron frequency is determined by scanning the applied signal frequency to find that frequency giving the tallest grass (Ekstrom and Wineland, 1980).

A second signal is used to induce transitions in which both the cyclotron orbit and the spin orientation change (Figure 6.5). The axial frequency shift (6.82) is smaller when the spin is down than when it is up, and so when

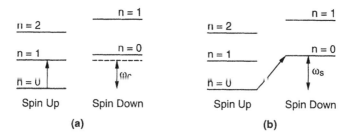

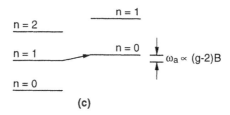

Figure 6.5: Transitions between (a) two cyclotron orbits, with no spin flip; (b) two spin states, with no change in the cyclotron orbit; and (c) two cyclotron states, accompanied by a spin flip. The last transition occurs at the anomaly frequency $\propto g - 2$.

the electron is in the down state the root level is different from what it is when the spin is up, as can be seen in Figure 6.4. The anomaly frequency is determined as that frequency producing the the most frequent changes in the grass root level.[19]

As discussed in Chapter 3, these experiments have determined g with remarkable precision. The agreement with *eighth*-order QED computations, which are so complicated that they required about three months of super-computer CPU time, is superb [compare equations (3.96) and (3.97)]. At the present time g is the most precisely measured property of any elementary particle, and this is not likely to change soon. Moreover, it has been confirmed to a high degree of accuracy that the electron and positron g factors are identical (van Dyck et al., 1987).

The main obstacle at this time to even greater precision in the $g - 2$ experiments is the "cavity shift." We remarked at the end of Section 6.2

---

[19] The driving frequencies for measuring both the cyclotron and anomaly frequencies can be determined very accurately with an atomic clock.

that the Lamb shift of a transition, like the spontaneous emission rate, is generally modified by the presence of a conducting surface. In general any radiative transition frequency and emission rate will be modified, including, for instance, those for the cyclotron transitions in geonium. In this case there is an effective microwave cavity defined by the electrodes of the Penning trap. Measurements of the cyclotron radiative lifetime by Gabrielse and Dehmelt (1985) gave $\gamma_c^{-1} \cong 0.3$ sec, whereas a value $\cong 1$ sec was found for a trap whose electrodes had fewer slits (van Dyck, Schwinberg, and Dehmelt, 1984). The calculated lifetime is .08 sec (Brown and Gabrielse, 1986). There is evidently an inhibition of spontaneous emission due to the absence of any cavity mode frequency at the natural emission frequency, just as in the cavity QED experiments discussed earlier.

A cavity shift of the cyclotron frequency is obviously cause for concern in the geonium $g - 2$ experiments. Unfortunately the shape of the Penning trap electrodes, as well as the presence of holes and slits, makes quantitative assessments of the role of cavity shifts extremely difficult. Some progress, however, has been made for a much simpler (cylindrical) geometry by Brown and Gabrielse (1986). Their estimates indicate that "an experimental search for this systematic effect should be made to confirm the value of the g factor of the electron."

Thus the modification of the electromagnetic vacuum field by conducting surfaces appears to be an important consideration in attempts to push the $g - 2$ measurements to still higher accuracy. Of course $g - 2$ itself is, at least in part, a *consequence* of the coupling of the electron to the vacuum field (Section 3.13).

6.8   Quantum Jumps

What are now often referred to as *quantum jumps* were first discussed by Dehmelt in 1975 in connection with a single-atom spectroscopic scheme for detecting weak transitions. Dehmelt's proposal has been realized experimentally with single trapped *ions* (Nagourney, Sandberg, and Dehmelt, 1986; Sauter et al., 1986; Bergquist et al., 1986; see also Cook and Kimble, 1985).

The basic idea is simple and is illustrated in Figure 6.6, which shows a ground atomic state 0 that has a large transition dipole moment connecting it to level 1, but a very small dipole moment connecting it to level 2. In other words, the transitions $0 \leftrightarrow 1$ and $0 \leftrightarrow 2$ are strong and weak transitions, respectively. If radiation at the $0 \leftrightarrow 1$ transition frequency is applied to the atom, the excitation of level 1 leads to fluorescence of, say, $\approx 10^8$ photons/sec. Level 2 is far off resonance and is not excited

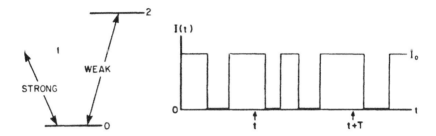

Figure 6.6: (a) An energy-level scheme for the observation of quantum jumps, and (b) the expected strong-transition fluorescence intensity versus time for a single atom, with "dark periods" due to excitation of the weak transition. From Cook and Kimble (1985), with permission.

by the field. If a field at the $0 \leftrightarrow 2$ transition frequency is also applied, however, there is a small but finite $0 \leftrightarrow 2$ transition rate. As long as level 2 is not populated, there continue to be $\approx 10^8$ fluorescence photons/sec at the $1 \to 0$ emission frequency. But when level 2 is populated, the atom cannot undergo $0 \to 1$ transitions, and so the fluorescence at the $1 \to 0$ frequency is temporarily shut off. If the $0 \leftrightarrow 2$ transition is very weak, the atomic electron is "shelved" in level 2 for a relatively long time. During such times there are "dark" periods in the fluorescence from the $1 \to 0$ transition, as indicated in Figure 6.6. The strong-transition fluorescence is "off" whenever the weak transition is excited, and "on" whenever it is not.

Thus the *strong*-transition fluorescence intensity $I(t)$ provides a measure of the excitation probability of the weak transition. This is the essence of Dehmelt's proposal. Note that, aside from such an application, the strong transition fluorescence can be regarded as a monitor of the "quantum jumps" associated with spontaneous emission on the weak transition. Since these quantum jumps are random (God is playing dice!), $I(t)$ is presumably among Nature's truest random processes (Erber et al., 1989; Cook, 1990).

6.9 Remarks

The phenomena described in this chapter are interesting not because they reveal anything particularly new about the electromagnetic vacuum, but because technological developments have made it possible to study them experimentally. These studies have supported the theoretical prediction that the vacuum electromagnetic field, and therefore emission rates and frequencies, for instance, are modified by conducting surfaces. They indicate that vacuum field effects are relevant to a variety of situations, ranging from the fundamental linewidth of a laser to the experiments measuring the electron magnetic moment with unprecedented accuracy.

Nonrelativistic vacuum electromagnetic effects are all associated with zero-point "motion" of the field variables, i.e., with the zero-point energy of the harmonic oscillators describing the field modes. In Chapter 1 we discussed some of the early history of ideas about zero-point energy, but there is little in that history to "prove" the existence of zero-point energy associated with harmonic oscillators. In closing this chapter we note that recent experiments with trapped ions have provided strong evidence for the reality of this zero-point energy (Diedrich et al., 1989). In the experiments with trapped particles the zero-point energy sets a lower limit to freezing of the particle motion.

6.10   Bibliography

Barnett, S. M., B. Huttner, and R. Loudon, "Spontaneous Emission in Absorbing Dielectric Media," *Phys. Rev. Lett.* **68**, 3698 (1992).

Barton, G., "Quantum Electrodynamics of Spinless Particles Between Conducting Plates," *Proc. Roy. Soc. Lond.* **A320**, 251 (1970).

Barton, G., "Quantum-Electrodynamic Level Shifts Between Parallel Mirrors: Analysis," *Proc. Roy. Soc. Lond.* **A410**, 141 (1987); "Quantum- Electrodynamic Level Shifts Between Parallel Mirrors: Applications, Mainly to Rydberg States," *Proc. Roy. Soc. Lond.* **A410**, 175 (1987).

Bergquist, J. C., R. G. Hulet, W. M. Itano, and D. J. Wineland, "Observation of Quantum Jumps in a Single Atom," *Phys. Rev. Lett.* **57**, 1699 (1986).

Brown, L. S. and G. Gabrielse, "Geonium Theory: Physics of a Single Electron or Ion in a Penning Trap," *Rev. Mod. Phys.* **58**, 233 (1986).

Cook, R. J. and H. J. Kimble, "Possibility of Direct Observation of Quantum Jumps," *Phys. Rev. Lett.* **54**, 1023 (1985).

Cook, R. J. and P. W. Milonni, "Quantum Theory of an Atom Near Partially Reflecting Walls," *Phys. Rev.* **A35**, 5081 (1987).

Cook, R. J., in *Progress in Optics*, ed. E. Wolf (North-Holland, Amsterdam, 1990), Volume 28.

Dehmelt, H., "Experiments with an Isolated Subatomic Particle at Rest," *Rev. Mod. Phys.* **62**, 525 (1990).

DeMartini, F., G. Innocenti, G. R. Jacobovitz, and P. Mataloni, "Anomalous Spontaneous Emission Time in a Microscopic Optical Cavity," *Phys. Rev. Lett.* **59**, 2955 (1987).

Diedrich, F., J. C. Bergquist, W. M. Itano, and D. J. Wineland, "Laser Cooling to the Zero-Point Energy of Motion," *Phys. Rev. Lett.* **62**, 403 (1989)

Drexhage, K. H., "Monomolecular Layers and Light," *Sci. Am.* **222**, 108 (March 1970).

Drexhage, K. H., in *Progress in Optics*, ed. E. Wolf (North-Holland, Amsterdam, 1974), Volume 12.

Eberly, J. H., N. B. Narozhny, and J. J. Sanchez–Mondragon, "Periodic Spontaneous Collapse and Revival in a Simple Quantum Model," *Phys. Rev. Lett.* **44**, 1323 (1980).

Ekstrom, P. and D. Wineland, "The Isolated Electron," *Sci. Am.* **243**, 105 (February 1980).

Erber, T., P. Hammerling, G. Hockney, M. Porrati, and S. Putterman, "Resonance Fluorescence and Quantum Jumps in Single Atoms: Testing the Randomness of Quantum Mechanics," *Ann. Phys.* (New York) **190**, 254 (1989).

Feng, X.-P. and K. Ujihara, "Quantum Theory of Spontaneous Emission in a One-Dimensional Optical Cavity with Two-Sided Output Coupling," *Phys. Rev.* **41**, 2668 (1990).

Fermi, E., "Quantum Theory of Radiation," *Rev. Mod. Phys.* **4**, 87 (1932).

Gabor, D., "Light and Information," *Proceedings of a Symposium on Astronomical Optics and Related Subjects*, ed. Z. Kopal (North-Holland, Amsterdam, 1956), p. 17.

Gabrielse, G. and H. G. Dehmelt, in *Precision Measurements and Fundamental Constants II*, ed. B. N. Taylor and W. D. Phillips (U.S. National Bureau of Standards, Spec. Pub. 617, 1981).

Gabrielse, G. and H. G. Dehmelt, "Observation of Inhibited Spontaneous Emission," *Phys. Rev. Lett.* **55**, 67 (1985).

Gabrielse, G., H. Dehmelt, and W. Kells, "Observation of a Relativistic Bistable Hysteresis in the Cyclotron Motion of a Single Electron," *Phys. Rev. Lett.* **54**, 537 (1985).

Ginzburg, V. L., *Theoretical Physics and Astrophysics* (Pergamon Press, Oxford, 1979).

Goldberg, P., P. W. Milonni, and B. Sundaram, "Theory of the Fundamental Laser Linewidth," *Phys. Rev.* **A44**, 1969 (1991).

Goy, P., J. M. Raimond, M. Gross, and S. Haroche, "Observation of Cavity-Enhanced Single-Atom Spontaneous Emission," *Phys. Rev. Lett.* **50**, 1903 (1983).

Haroche, S. and D. Kleppner, "Cavity Quantum Electrodynamics," *Physics*

Today (January 1989), 25.

Heinzen, D. J., J. J. Childs, J. F. Thomas, and M. S. Feld, "Enhanced and Inhibited Visible Spontaneous Emission by Atoms in a Confocal Resonator," *Phys. Rev. Lett.* **58**, 1320 (1987).

Hinds, E. A., "Cavity Quantum Electrodynamics," in *Advances in Atomic, Molecular, and Optical Physics*, ed. D.R. Bates and B. Bederson (Academic Press, Boston, 1990), Volume 28.

Hulet, R. G., E. S. Hilfer, and D. Kleppner, "Inhibited Spontaneous Emission by a Rydberg Atom," *Phys. Rev. Lett.* **55**, 2137 (1985).

Jaynes, E. T. and F. W. Cummings, "Comparison of Quantum and Semiclassical Radiation Theories with Application to the Beam Maser," *Proc. IEEE* **51**, 89 (1963).

Jhe, W., A. Anderson, E. A. Hinds, D. Meschede, L. Moi, and S. Haroche, "Suppression of Spontaneous Decay at Optical Frequencies: Test of Vacuum-Field Anisotropy in Confined Space," *Phys. Rev. Lett.* **58**, 666 (1987).

Knight, P. L. and P. W. Milonni, "The Rabi Frequency in Optical Spectra," *Phys. Rep.* **66**, 21 (1980).

Milonni, P. W. and P. L. Knight, "Spontaneous Emission Between Mirrors," *Opt. Commun.* **9**, 119 (1973).

Milonni, P. W., "Semiclassical and Quantum-Electrodynamical Approaches in Nonrelativistic Radiation Theory," *Phys. Rep.* **25**, 1 (1976).

Milonni, P. W., "Casimir Forces Without the Vacuum Radiation field," *Phys. Rev.* **A25**, 1315 (1982).

Milonni, P. W., "Does the Electromagnetic Mass of an Electron Depend on Where It Is?," *Int. J. Theor. Phys.* **22**, 323 (1983).

Milonni, P. W., "QED Theory of Excess Spontaneous Emission Noise," in *Laser Noise*, ed. R. Roy (SPIE, Washington, D.C., 1991).

Milonni, P. W. and S. Singh, "Some Recent Developments in the Fundamental Theory of Light," in *Advances in Atomic, Molecular, and Optical Physics*, Volume 28, ed. D.R. Bates and B. Bederson (Academic Press, Cambrideg, Mass., 1991).

Nagourney, W., J. Sandberg, and H. Dehmelt, "Shelved Optical Electron Amplifier: Observation of Quantum Jumps," *Phys. Rev. Lett.* **56**, 2797 (1986).

Philpott, M. R., "Fluorescence of Molecules Between Mirrors," *Chem. Phys. Lett.* **19**, 435 (1973).

Rempe, G., H. Walther, and N. Klein, "Observation of Quantum Collapse and Revival in a One-Atom Maser," *Phys. Rev. Lett.* **58**, 353 (1987).

Rich, A. and J. C. Wesley, "Current Status of the Lepton *g* Factors,"

Rev. Mod. Phys. **44**, 250 (1972).

Sauter, Th., W. Neuhauser, R. Blatt, and P. E. Toschek, "Observation of Quantum Jumps," *Phys. Rev. Lett.* **57**, 1696 (1986).

Van Dyck, R. S., Jr., P. B. Schwinberg, and H. G. Dehmelt, in *New Frontiers in High Energy Physics*, ed. B. Kursunoglu, A. Perlmutter, and L. Scott (Plenum Press, New York, 1978).

Van Dyck, R. S., Jr., P. B. Schwinberg, and H. G. Dehmelt, in *Atomic Physics 9*, ed. R. S. van Dyck, Jr. and E. N. Forston (World Scientific, Singapore, 1984).

Van Dyck, R. S., Jr., P. B. Schwinberg, and H. G. Dehmelt, "New High–Precision Comparison of Electron and Positron g Factors," *Phys. Rev. Lett.* **59**, 26, (1987).

Wineland, D. J. and H. G. Dehmelt, "Line Shifts and Widths of Axial, Cyclotron and $G - 2$ Resonances in Tailored, Stored Electron (Ion) Clouds," *Int. J. Mass Spectrom. Ion Phys.* **16**, 338 (1975); **19**, 251(E) (1975).

Wineland, D. J., W. M. Itano, and R. S. van Dyck, Jr., "High-Resolution Spectroscopy of Stored Ions," in *Advances in Atomic and Molecular Physics*, ed. D.R. Bates and B. Bederson (Academic Press, New York, 1983), Volume 19.

Chapter 7

Casimir and van der Waals Forces: Prelude

> I mentioned my results to Niels Bohr, during a walk. That is nice, he said, that is something new. I told him that I was puzzled by the extremely simple form of the expressions for the interaction at very large distances and he mumbled something about zero-point energy. That was all, but it put me on a new track.
> — H. B. G. Casimir (private communication, March 1992)

7.1 Introduction

The Casimir force between conducting plates is often cited as proof for the reality of the vacuum electromagnetic field (Itzykson and Zuber, 1980). It should be clear to the reader by now that there are *many* observable consequences of the vacuum field, including spontaneous emission, the Lamb shift, the anomalous magnetic moment, van der Waals forces, and the fundamental laser linewidth, all of which may be attributed at least in part to the vacuum field. In this chapter and the next we describe various "Casimir effects," including macroscopic manifestations of van der Waals forces.

Casimir's work actually had its origin in a problem of colloidal chemistry, namely, the stability of hydrophobic suspensions of particles in dilute aqueous electrolytes (Sparnaay, 1989). Such suspensions are "stable" if the particles ($\approx$ 0.1-1 μm in size) do not coagulate. The particles in a stable suspension are charged; coagulation occurs when the electrolyte concentration is increased beyond a critical value.

This behavior is explained as a consequence of the interplay between

217

repulsive and attractive forces. The repulsive force arises because each charged particle is surrounded by ions of opposite charge; the "layer" of ions is characterized by a Debye length L_D. When two particles come within about L_D of each other there is a repulsion caused by these ionic layers. The attraction is due to the integrated effect of van der Waals interatomic forces, each particle consisting of typically billions of atoms.

A quantitative model for the stability of such colloidal suspensions assumes two parallel plates separated by a distance d (Verwey and Overbeek, 1948). The repulsive force per unit area between two such "particles" is found to be $F_{\text{rep}} \approx 100 n_s kT e^{-d/L_D}$, where n_s is the ion number density and $L_D^2 \approx kT/8\pi n_s e^2$. The attractive force per unit area is obtained by integrating the pairwise forces between atoms, assuming an interatomic force given by the London–van der Waals interaction (3.66). It is found that $F_{\text{attr}} = -A/d^3$, where A is a positive constant. Coagulation occurs when $|F_{\text{attr}}| > F_{\text{rep}}$ for interparticle separations d, e.g., if n_s is increased beyond a certain critical value.

In 1946 Overbeek inferred from experiments that for relatively large colloidal particles the attractive force decreases more rapidly than d^{-3} (Verwey and Overbeek, 1948). He conjectured that the r^{-6} interatomic energy calculated by London might actually fall off more rapidly with r if the finite velocity of light were accounted for. Casimir and Polder (1948) then calculated that, due to retardation associated with the finite value of c, the van der Waals interaction actually varies as r^{-7} at large interatomic separations.[1] This "retarded" van der Waals interaction was derived in Section 3.11.

In attempting to better understand this result, Casimir found that the van der Waals interaction could be attributed to the change in the zero-point energy of the field due to the presence of the two atoms, much as in Feynman's later argument for the Lamb shift (Section 3.7). He then considered the simpler example of two parallel conducting plates. The same sort of argument involving the change in the zero-point energy of the field due to the presence of the plates led to his prediction of the "Casimir force" discussed in Chapters 2 and 3.

---

[1] The modification of the van der Waals interaction at large separations was perhaps first considered by J. A. Wheeler, *Phys. Rev.* **59**, 928 (1941). In Wheeler's work the modification at large interatomic separations is identified simply with the fact that the field from an oscillating dipole has contributions that fall off more slowly than r^{-3}. See, for instance, equation (3.78).

7.2   Force Between Dielectrics

In Section 2.7 we derived the Casimir force per unit area between two infinite, parallel, perfectly conducting plates separated by a distance d:

$$F = -\frac{\pi^2 hc}{240 d^4} . \tag{7.1}$$

We followed Casimir's original approach of calculating the zero-point energy of the electromagnetic field when the plates are separated by d, then subtracting the energy for $d \rightarrow \infty$. In Section 3.10 we showed that the Casimir force can be attributed to the radiation pressure of the zero-point field.

With d the plate separation in microns, the Casimir force (7.1) is

$$F = \frac{.013}{d^4} \text{ dynes/cm}^2 \ (d \text{ in } \mu\text{m}). \tag{7.2}$$

Let us compare this with the Coulomb force between two oppositely charged conducting plates. In that case the attractive force per unit area is $F_{\text{Coul}} = 2\pi\sigma^2$, where σ is the surface charge density. Writing $\sigma = CV/A = (A/4\pi d) \times V/A$, where C is the capacitance, V is the potential difference between the plates, and A is the cross-sectional area, we have

$$F_{\text{Coul}} = \frac{1}{8\pi}\frac{V^2}{d^2} , \tag{7.3}$$

and for $d = 1\mu\text{m}$, F_{Coul} is equal to the Casimir force (7.2) when $V = 17\text{ mV}$. This shows that small differences in surface potentials between the plates must be avoided in experimental measurements of the tiny Casimir force.

For the interpretation of experiments on the forces between two plates, the simplifying assumption of perfect conductivity (reflectivity) at all field frequencies is unrealistic, and the Casimir expression (7.1) must be replaced by one that includes the dielectric properties of the media. An obvious way to accomplish this is to add the van der Waals forces of attraction between the molecules of the two plates. Consider first a single molecule at a distance d from a half-space $z \geq d$ filled with N_1 identical molecules per unit volume. If the intermolecular potential is $V(r) = -B/r^\gamma$, then the interaction energy between the single molecule and the half-space is

$$
\begin{aligned}
V(d) &= -N_1 B \int_{-\infty}^{\infty} dx \int_{-\infty}^{\infty} dy \int_{d}^{\infty} dz [z^2 + x^2 + y^2]^{-\gamma/2} \\
&= -\frac{2\pi N_1 B}{(\gamma - 2)(\gamma - 3)} d^{3-\gamma} .
\end{aligned} \tag{7.4}
$$

If the single molecule is now replaced by a half-space $z \leq 0$ of N_2 molecules per unit volume, then the interaction energy becomes

$$u(d) = -\frac{2\pi N_1 N_2 B}{(\gamma - 2)(\gamma - 3)} \int_0^\infty dR(R + d)^{3-\gamma} = \frac{-2\pi N_1 N_2 B}{(\gamma - 2)(\gamma - 3)(\gamma - 4)} \frac{1}{d^{\gamma - 4}}$$

(7.5)

per unit area, and so the force per unit area between the two dielectrics is

$$F(d) = -\frac{2\pi N_1 N_2 B}{(\gamma - 2)(\gamma - 3)} \frac{1}{d^{\gamma - 3}} .$$

(7.6)

Assuming the intermolecular London–van der Waals potential (3.88), we have $\gamma = 6, B = 3\hbar\omega_o\alpha^2/4$, and, for $N_1 = N_2 \equiv N$,

$$F(d) = -\frac{\pi N^2 \hbar\omega_o\alpha^2}{8d^3} \equiv -\frac{A}{6\pi d^3} ,$$

(7.7)

where $A \equiv 3\pi^2 \hbar\omega_o\alpha^2/4$ is called the *de Boer–Hamaker constant* of the material.[2] If instead we assume the retarded van der Waals interaction (3.91), then $\gamma = 7, B = 23\hbar c\alpha^2/4\pi$, and

$$F(d) = -\frac{23N^2 \hbar c\alpha^2}{40d^4}$$

(7.8)

for identical, semi-infinite dielectric slabs. Note the obvious but important fact: the force between macroscopic bodies can be very much more long-ranged than the forces between individual molecules.

The earliest experiments on the forces between dielectric plates indicated that the forces acting were roughly an order of magnitude larger than expected from estimated de Boer–Hamaker constants. It was recognized, of course, that the de Boer–Hamaker approach of integrating over pairwise interactions applied strictly to rarefied media. Specifically, the derivation of (7.6) assumes that the intermolecular forces are additive in the sense that the force between two molecules is independent of the presence of a third molecule. However, the van der Waals forces are not in fact simply additive (Section 8.2).

The discrepancy between microscopic theories assuming additive intermolecular forces, and experimental results reported in the early 1950s, evidently motivated Lifshitz (1956) to develop a *macroscopic* theory of the forces between dielectrics.[3] His results reduce to those of the microscopic

---

[2] J. H. de Boer, *Trans. Faraday Soc.* **32**, 10 (1936); H. C. Hamaker, *Physica* **4**, 1058 (1937).

[3] See also L. D. Landau and E. M. Lifshitz, *Electrodynamics of Continuous Media* (Pergamon Press, Oxford, 1966), Section 90.

theory with additive intermolecular forces when the dielectric constants are close to unity, but are substantially different otherwise. In the limiting case of perfect conductors, the predicted force between two plates reduces to the Casimir force (7.1). Except for a few details discussed in the following chapter, the Lifshitz theory is now generally accepted and, as we shall see, has been supported by some careful experiments.

The Lifshitz theory is rather complicated, and doubts have occasionally been raised concerning its validity. Lifshitz assumed in effect that the dielectrics are characterized by randomly fluctuating sources (see Section 7.3). From the assumed delta-function correlation of these sources, correlation functions for the field were calculated, and from these in turn the Maxwell stress tensor was determined. The force per unit area acting on the two dielectrics was then calculated as the zz-component of the stress tensor.

Rather than presenting the details of Lifshitz's derivation here, we shall obtain his results following the Casimir approach based on the zero-point energy of the quantized field. This approach in the case of dielectric slabs was evidently first used by van Kampen, Nijboer, and Schram (1968) for the case of small separations, where retardation can be ignored. In this case the magnetic field $\mathbf{B}$ can be assumed in effect to vanish, and this simplifies the determination of the mode frequencies.

We consider the case of a medium with dielectric constant $\epsilon_3(\omega)$ sandwiched between two semi-infinite media with dielectric constants $\epsilon_1(\omega)$ and $\epsilon_2(\omega)$. These media are assumed to occupy the regions $0 \leq z \leq d, z < 0$, and $z > d$, respectively, as shown in Figure 7.1. We will calculate the force per unit area between the two semi-infinite slabs from the total zero-point field energy $\sum_n \frac{1}{2}\hbar\omega_n$, where the ω_n are the mode frequencies for the situation depicted in Figure 7.1.

Our first task is to determine the mode frequencies. For this purpose we consider solutions of the form

$$\mathbf{E}(\mathbf{r},t) = \mathbf{E}_o(\mathbf{r})e^{-i\omega t} , \tag{7.9}$$

$$\mathbf{B}(\mathbf{r},t) = \mathbf{B}_o(\mathbf{r})e^{-i\omega t} , \tag{7.10}$$

of the Maxwell equations

$$\nabla \cdot \mathbf{D} = 0, \tag{7.11}$$

$$\nabla \times \mathbf{E} = -\frac{1}{c}\frac{\partial \mathbf{B}}{\partial t} , \tag{7.12}$$

$$\nabla \cdot \mathbf{B} = 0, \tag{7.13}$$

$$\nabla \times \mathbf{B} = \frac{1}{c}\frac{\partial \mathbf{D}}{\partial t} , \tag{7.14}$$

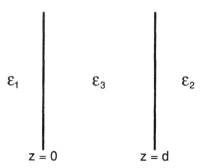

$$\varepsilon_1 \qquad\qquad \varepsilon_3 \qquad\qquad \varepsilon_2$$

$$z = 0 \qquad\qquad z = d$$

Figure 7.1: Geometry for the derivation of the Lifshitz expression for the force between two semi-infinite dielectric slabs separated by a layer with dielectric constant ϵ_3.

for nonmagnetic media with zero net charge density ρ. We assume isotropic media such that the electric displacement vector $\mathbf{D}(\mathbf{r},t) = \epsilon(\omega)\mathbf{E}_o(\mathbf{r})e^{-i\omega t}$. We can satisfy the Maxwell equations if, in each medium in Figure 7.1, $\nabla \cdot \mathbf{E}_o = \nabla \cdot \mathbf{B}_o = 0$,

$$\nabla^2 \mathbf{E}_o + \frac{\omega^2}{c^2}\epsilon(\omega)\mathbf{E}_o = 0, \qquad (7.15)$$

$$\nabla^2 \mathbf{B}_o + \frac{\omega^2}{c^2}\epsilon(\omega)\mathbf{B}_o = 0, \qquad (7.16)$$

and the appropriate boundary conditions are satisfied. Let us assume solutions of the form

$$\mathbf{E}_o(\mathbf{r}) = [e_x(z)\hat{x} + e_y(z)\hat{y} + e_z(z)\hat{z}]e^{i(k_x x + k_y y)} , \qquad (7.17)$$

$$\mathbf{B}_o(\mathbf{r}) = [b_x(z)\hat{x} + b_y(z)\hat{y} + b_z(z)\hat{z}]e^{i(k_x x + k_y y)} , \qquad (7.18)$$

so that

$$\frac{d^2 e_x}{dz^2} - K^2 e_x = 0, \qquad (7.19)$$

$$\frac{d^2 b_x}{dz^2} - K^2 b_x = 0, \qquad (7.20)$$

and likewise for the y, z components of e and b, where we define

$$K^2 = k_x^2 + k_y^2 - \epsilon(\omega)\frac{\omega^2}{c^2} . \qquad (7.21)$$

In the case of free space, $\epsilon(\omega) = 1$ everywhere for all ω, and the usual plane-wave solutions are obtained for $K^2 < 0$: $k_x^2 + k_y^2 + k_z^2 = \omega^2/c^2, K^2 = -k_z^2$.

For the case illustrated in Figure 7.1, let us first consider modes with $K^2 > 0$ in all three media. By a particular choice of the coordinate system we can make $k_y = 0$ and

$$K^2 - k^2 \quad \epsilon(\omega)\frac{\omega^2}{c^2} \quad , \quad k = k_x \; . \tag{7.22}$$

The condition that the normal component of $\mathbf{D}$ be continuous implies that $\epsilon(\omega)e_z(z)$ is continuous for each ω, while $\nabla \cdot \mathbf{E}_o = 0$ implies

$$ike_x + \frac{de_z}{dz} = 0. \tag{7.23}$$

From $\nabla \times \mathbf{E}_o = i(\omega/c)\mathbf{B}_o$, furthermore, we have

$$\mathbf{B}_o(\mathbf{r}) = \left[i\frac{c}{\omega}\frac{de_y}{dz}\hat{x} - \frac{c}{\omega}(ke_z + i\frac{de_x}{dz})\hat{y} + \frac{c}{\omega}ke_y\hat{z} \right] e^{ikx} \; . \tag{7.24}$$

Hence, $\nabla \cdot \mathbf{B}_o = 0$ is satisfied identically, and the continuity of the normal component of $\mathbf{B}$ implies that e_y must be continuous. Continuity of the tangential component of $\mathbf{E}$ is then guaranteed if e_x is continuous, and this condition is satisfied if, from (7.23), de_z/dz is continuous. Finally the tangential component of $\mathbf{H} \, (= \mathbf{B})$ is continuous if de_y/dz and $ke_z + ide_x/dz$ are continuous. But, from (7.23), (7.19), and (7.21),

$$
\begin{aligned}
ke_z + i\frac{de_x}{dz} &= ke_z - \frac{1}{k}\frac{d^2e_z}{dz^2} = -\frac{1}{k}\left[\frac{d^2e_z}{dz^2} - k^2e_z \right] \\
&= -\frac{1}{k}[K^2 - k^2]e_z = \frac{1}{k}\left[\epsilon(\omega)\frac{\omega^2}{c^2} \right]e_z \; ,
\end{aligned} \tag{7.25}
$$

and we have noted that the continuity of this quantity is already required by the continuity of the normal component of $\mathbf{D}$. Thus all the boundary conditions are satisfied if (1) ϵe_z and de_z/dz are continuous and (2) e_y and de_y/dz are continuous.

Now since $d^2e_z/dz^2 - K^2e_z = 0$ we have, ignoring unphysical, exponentially growing solutions,

$$
\begin{aligned}
e_z(z) &= Ae^{K_1 z} \; , \quad z < 0 \\
&= Be^{K_3 z} + Ce^{-K_3 z} \; , \quad 0 \leq z \leq d \\
&= De^{-K_2 z} \; , \quad z > d
\end{aligned} \tag{7.26}
$$

where $K_j \equiv \sqrt{k^2 - \epsilon_j(\omega)\omega^2/c^2}$. The preceding boundary conditions (1), that $\epsilon e_z(z)$ and $de_z(z)/dz$ are continuous at $z = 0$ and $z = d$, then yield four linear algebraic equations for $A, B, C,$ and D. The condition that nontrivial solutions of these equations exist, i.e., that the determinant of the matrix of coefficients vanishes, yields after straightforward algebra the expression

$$\frac{(\epsilon_3 K_1 + \epsilon_1 K_3)(\epsilon_3 K_2 + \epsilon_2 K_3)}{(\epsilon_3 K_1 - \epsilon_1 K_3)(\epsilon_3 K_2 - \epsilon_2 K_3)} e^{2K_3 d} - 1 = 0. \qquad (7.27)$$

It is easily shown in the same fashion that the boundary conditions (2) are satisfied across the boundaries at $z = 0$ and $z = d$ if

$$\frac{(K_1 + K_3)(K_2 + K_3)}{(K_1 - K_3)(K_2 - K_3)} e^{2K_3 d} - 1 = 0. \qquad (7.28)$$

Equations (7.27) and (7.28) are conditions on the allowed frequencies ω. They cannot in general be satisfied simultaneously. However, both boundary conditions (1) and (2) can be satisfied if (a) (7.27) is satisfied and $e_y \equiv 0$ or (b) (7.28) is satisfied and $e_z \equiv 0$. Thus we have two types of modes. Those of type (a) have $e_y = 0$ and mode frequencies ω determined by the solutions of (7.27). Those of type (b) have $e_z = 0$ and mode frequencies determined by the solutions of (7.28).

Zero-Point Energy of the Surface Modes

The modes we have obtained under the assumption that the K_j are real are exponentially decaying functions of z for $z < 0$ and $z > d$, and for this reason are called *surface modes* (Barton, 1979). We will consider now the zero-point energy associated with these surface modes:

$$E(d) = \sum_n \frac{1}{2} \hbar \omega_{na} + \sum_n \frac{1}{2} \hbar \omega_{nb} , \qquad (7.29)$$

where the ω_{na} and ω_{nb} are the frequencies associated with the modes of types (a) and (b), respectively.

The sums over modes in (7.29) include the continuum of values of k_x and k_y:

$$\sum_n \to \left(\frac{L}{2\pi}\right)^2 \int dk_x \int dk_y \sum_N = \left(\frac{L}{2\pi}\right)^2 \int 2\pi k \, dk \sum_N , \qquad (7.30)$$

where L is a length for the x, y sides of our "quantization box" and $\sum_N$ denotes the sum over all solutions of (7.27) or (7.28) for $\omega(k)$. Thus

$$E(d) = \frac{\hbar L^2}{4\pi} \int_0^\infty dk \, k \left[\sum_N \omega_{Na}(k) + \sum_N \omega_{Nb}(k) \right]. \qquad (7.31)$$

For the purpose of performing the summations we first recall the "argument theorem" from the theory of functions of a complex variable: for a function $f(z)$ that is analytic except for poles (i.e., is meromorphic) on and inside a simple closed curve C,

$$\frac{1}{2\pi i} \oint_C \frac{f'(z)}{f(z)} dz = N - P, \qquad (7.32)$$

where N is the number of zeros and P the number of poles of $f(z)$ inside C.[4] A straightforward generalization of the argument theorem states that

$$\frac{1}{2\pi i} \oint_C z \frac{f'(z)}{f(z)} dz = \left[\sum_i z_i\right]_{f(z_i)=0} - \left[\sum_i z_i\right]_{f(z_i)=\infty} = \text{(sum of zeros}$$

$$\text{of } f(z) \text{ inside } C) - \text{(sum of poles of } f(z) \text{ inside } C). \qquad (7.33)$$

Let $F_a(\omega)$ and $F_b(\omega)$ denote the left-hand sides of (7.27) and (7.28), respectively, so that

$$\sum_N \omega_{N\alpha}(k) = \text{sum of zeros of } F_\alpha(\omega), \quad \alpha = a, b. \qquad (7.34)$$

Obviously the poles of $F_\alpha(\omega)$, considered as a function of a complex variable ω, are independent of d. Therefore we can write

$$\frac{1}{2\pi i} \oint_C \omega \frac{F_\alpha'(\omega)}{F_\alpha(\omega)} d\omega = \sum_N \omega_{N\alpha}(k) - (\ldots), \qquad (7.35)$$

where $(\ldots)$ is independent of d. Here C is the closed curve defined by the imaginary axis of the complex ω plane and a semicircle in the right half of this plane, with the radius of the semicircle extending to infinity.[5] Since d-independent contributions to (7.31) do not contribute to any force, we can for our purposes write

$$E(d) = \frac{\hbar L^2}{4\pi} \left(\frac{1}{2\pi i}\right) \int_0^\infty dk\, k \left[\oint_C \omega \frac{F_a'(\omega)}{F_a(\omega)} d\omega + \oint_C \omega \frac{F_b'(\omega)}{F_b(\omega)} d\omega\right]. \qquad (7.36)$$

---

[4] A zero of order n is counted as n zeros, and likewise for a pole of order n. The reader who does not recall the argument theorem or who is unable to quickly reconstruct its proof based on the residue theorem can find it in elementary textbooks such as M. L. Boas, *Mathematical Methods in the Physical Sciences* (Wiley, New York, 1983), p. 609.

[5] The $\omega_{N\alpha}(k)$ of physical interest are of course those lying along the positive real axis. It may be shown that all the zeros of $F_\alpha(\omega)$ do in fact lie on the positive real axis when the $\epsilon_j(\omega)$ are real and vary with ω in the way predicted by elementary dispersion theory. Thus the contour integral (7.35) does not have contributions from any frequencies other than those of physical interest. However, it is not obvious how to extend this particular approach to the case of absorbing media, where the $\epsilon_j(\omega)$ are complex, see Section 7.3.

Recalling the definition of the curve C, we can write each contour integral in the preceding as the part along the imaginary axis plus the part along the semicircle. In the required limit of infinite semicircle radius, the latter integral is d-independent and makes no contribution to a force. The integral along the imaginary axis is

$$
\begin{aligned}
\int_{\infty}^{-\infty} (i\xi) \frac{1}{F_\alpha(i\xi)} \frac{\partial F_\alpha(i\xi)}{\partial(i\xi)} (id\xi) &= -i \int_{-\infty}^{\infty} d\xi\xi \frac{G'_\alpha(\xi)}{G_\alpha(\xi)} \\
&= -i \int_{-\infty}^{\infty} d\xi\xi \frac{d}{d\xi} \log G_\alpha(\xi) \\
&= i \int_{-\infty}^{\infty} d\xi \log G_\alpha(\xi), \quad (7.37)
\end{aligned}
$$

for $\alpha = 1, 2$, where in the last step we have performed a partial integration. We have defined $G_\alpha(\xi) = F_\alpha(i\xi)$; explicitly,

$$
G_a(\xi) = \frac{(\epsilon_3 K_1 + \epsilon_1 K_3)(\epsilon_3 K_2 + \epsilon_2 K_3)}{(\epsilon_3 K_1 - \epsilon_1 K_3)(\epsilon_3 K_2 - \epsilon_2 K_3)} e^{2K_3 d} - 1, \quad (7.38)
$$

$$
G_b(\xi) = \frac{(K_1 + K_3)(K_2 + K_3)}{(K_1 - K_3)(K_2 - K_3)} e^{2K_3 d} - 1, \quad (7.39)
$$

where now $\epsilon_j = \epsilon_j(i\xi)$ and

$$
K_j^2 = k^2 + \epsilon_j(i\xi)\xi^2/c^2 . \quad (7.40)
$$

Equation (7.36) is then

$$
E(d) = \frac{\hbar L^2}{8\pi^2} \int_0^{\infty} dk\, k \left[\int_{-\infty}^{\infty} d\xi \log G_a(\xi) + \int_{-\infty}^{\infty} d\xi \log G_b(\xi) \right]. \quad (7.41)
$$

Force Between the Dielectrics

We will now calculate the force, associated with the zero-point energy of the surface modes, between the dielectric slabs 1 and 2 of Figure 7.1. From (7.38) and (7.39) we see that $\partial G_\alpha/\partial d = 2K_3(G_\alpha + 1)$, and therefore that

$$
\begin{aligned}
\frac{\partial}{\partial d} \int_{-\infty}^{\infty} d\xi \log G_\alpha(\xi) &= \int_{-\infty}^{\infty} d\xi \frac{1}{G_\alpha} 2K_3(G_\alpha + 1) \\
&= \int_{-\infty}^{\infty} d\xi K_3 + 2 \int_{-\infty}^{\infty} d\xi \frac{K_3}{G_\alpha} . \quad (7.42)
\end{aligned}
$$

The first integral is independent of the presence of the dielectrics 1 and 2, and therefore is not related to any force between them. This force per unit area is therefore

$$F(d) = -\frac{\partial}{\partial d}E(d) = -\frac{\hbar}{2\pi^2}\int_0^\infty dk\,k\int_0^\infty d\xi\,K_3\left[\frac{1}{G_a(k,\xi)}+\frac{1}{G_b(k,\xi)}\right],$$
(7.43)

where we have used the fact that the K_j and ϵ_j, and therefore G_a and G_b, are even functions of ξ. More explicitly, the force between the dielectric slabs 1 and 2 of Figure 7.1 is

$$\begin{aligned}
F(d) = {}& -\frac{\hbar}{2\pi^2}\int_0^\infty dk\,k\int_0^\infty d\xi\,K_3\\
& \times\left(\left[\frac{\epsilon_3 K_1+\epsilon_1 K_3}{\epsilon_3 K_1-\epsilon_1 K_3}\frac{\epsilon_3 K_2+\epsilon_2 K_3}{\epsilon_3 K_2-\epsilon_2 K_3}e^{2K_3 d}-1\right]^{-1}\right.\\
& \left.+\left[\frac{K_1+K_3}{K_1-K_3}\frac{K_2+K_3}{K_2-K_3}e^{2K_3 d}-1\right]^{-1}\right),
\end{aligned}$$
(7.44)

where $\epsilon_j = \epsilon_j(i\xi)$ and the K_j are defined by (7.40).

Comparison with the Lifshitz Theory

We now use in equation (7.44) the variable p defined by writing

$$k^2 = \epsilon_3\frac{\xi^2}{c^2}(p^2-1).$$
(7.45)

Then $K_3^2 = k^2+\epsilon_3\xi^2/c^2 = \epsilon_3[\xi^2(p^2-1)/c^2+\xi^2/c^2]$, or

$$K_3 = \sqrt{\epsilon_3}\frac{\xi}{c}p,$$
(7.46)

and similarly

$$\begin{aligned}
K_{1,2}^2 &= k^2+\epsilon_{1,2}\frac{\xi^2}{c^2} = \epsilon_3\frac{\xi^2}{c^2}[p^2-1+\frac{\epsilon_{1,2}}{\epsilon_3}]\\
&\equiv \epsilon_3\frac{\xi^2}{c^2}s_{1,2}^2 .
\end{aligned}$$
(7.47)

Using the variables p and $s_{1,2}$ in (7.44), and $dk\,k = (\xi^2/c^2)\epsilon_3 dp\,p$, we have

$$\begin{aligned}
F(d) = {}& -\frac{\hbar}{2\pi^2 c^3}\int_1^\infty dp\,p^2\int_0^\infty d\xi\,\xi^3\epsilon_3^{3/2}\left(\left[\frac{\epsilon_3 s_1+\epsilon_1 p}{\epsilon_3 s_1-\epsilon_1 p}\frac{\epsilon_3 s_2+\epsilon_2 p}{\epsilon_3 s_2-\epsilon_2 p}\right.\right.\\
& \left.\times e^{2\xi p\sqrt{\epsilon_3}d/c}-1\right]^{-1}+\left.\left[\frac{s_1+p}{s_1-p}\frac{s_2+p}{s_2-p}e^{2\xi p\sqrt{\epsilon_3}d/c}-1\right]^{-1}\right).
\end{aligned}$$
(7.48)

This result for the force per unit area between the dielectric slabs 1 and 2 (Figure 7.1) agrees exactly with that of Lifshitz (1956) when we consider with Lifshitz the case of vacuum between the slabs ($\epsilon_3 = 1$).[6] In the more general case of a dielectric medium 3, our result is the same as that of Schwinger, DeRaad, and Milton (1978).

Force Between Perfectly Conducting Plates

In the special case of two perfectly conducting plates separated by vacuum we take $\epsilon_{1,2} \to \infty$ and $\epsilon_3 \to 1$. Then (7.48) reduces to

$$
\begin{aligned}
F(d) &= -\frac{\hbar}{2\pi^2 c^3} \int_1^\infty dp\, p^2 \int_0^\infty d\xi\, \xi^3 \frac{2}{e^{2\xi p d/c} - 1} \\
&= -\frac{\hbar c}{16\pi^2 d^4} \int_1^\infty dp\, p^{-2} \int_0^\infty \frac{dx\, x^3}{e^x - 1} \\
&= -\frac{\pi^2 \hbar c}{240 d^4} \,,
\end{aligned}
\tag{7.49}
$$

which is the familiar Casimir force (Sections 2.7 and 3.10).

Imperfectly Conducting Plates

From the form of expressions like (7.48) and (7.49) it is clear that the dominant contribution to the Casimir force between conductors comes from frequencies ξ in the range $\xi \sim c/d$. For $d \approx 1$ μm, therefore, the dominant frequencies are in the infrared and visible regions of the electromagnetic spectrum. For such frequencies the dielectric constant[7]

$$
\epsilon(\omega) \cong 1 - \frac{\omega_p^2}{\omega^2} \,,
\tag{7.50}
$$

where ω_p is the plasma frequency: $\omega_p^2 = 4\pi N e^2/m$, with N the number density of free electrons. We will assume identical imperfect conductors separated by vacuum, in which case (7.48) becomes

$$
\begin{aligned}
F(d) &= -\frac{\hbar}{2\pi^2 c^3} \int_1^\infty dp\, p^2 \int_0^\infty d\xi\, \xi^3 \left(\left[X e^{2\xi p d/c} - 1 \right]^{-1} \right. \\
&\quad \left. + \left[Y e^{2\xi p d/c} - 1 \right]^{-1} \right),
\end{aligned}
\tag{7.51}
$$

---

[6] See equation (2.9) of Lifshitz's 1956 paper.
[7] For lower frequencies the dependence of $\epsilon(\omega)$ on the static conductivity σ must be included.

where

$$X \equiv \left(\frac{s + \epsilon p}{s - \epsilon p}\right)^2, \quad Y \equiv \left(\frac{s + p}{s - p}\right)^2. \tag{7.52}$$

For perfect conductors, $X = Y = 1$. Writing $X = 1 + \Delta X$ and $Y = 1 + \Delta Y$, with $\Delta X, \Delta Y$ small, we have

$$F(d) = F_C(d) + \frac{\hbar}{2\pi^2 c^3} \int_1^\infty dp\, p^2 \int_0^\infty d\xi\, \xi^3 [\Delta X + \Delta Y] \frac{e^{2\xi pd/c}}{[e^{2\xi pd/c} - 1]^2}, \tag{7.53}$$

where $F_C(d)$ is the Casimir force (7.49) for the case of perfect conductors. To calculate ΔX and ΔY, we first note that

$$s = [p^2 - 1 + \epsilon(i\xi)]^{1/2} = \left[p^2 + \frac{\omega_p^2}{\xi^2}\right]^{1/2}$$

$$\cong \frac{\omega_p}{\xi} + \frac{p^2 \xi}{2\omega_p} \tag{7.54}$$

to first order in ξ/ω_p. Then

$$Y \cong \left(\frac{\omega_p/\xi + p^2\xi/2\omega_p + p}{\omega_p/\xi + p^2\xi/2\omega_p - p}\right)^2 \cong 1 + \frac{4p\xi}{\omega_p}, \tag{7.55}$$

or $\Delta Y \cong 4p\xi/\omega_p$. Similarly we obtain $\Delta X \cong 4\xi/p\omega_p$ and therefore

$$\begin{aligned}
F(d) &\cong F_C(d) + \frac{\hbar}{2\pi^2 c^3} \int_1^\infty dp\, p^2 \int_0^\infty d\xi\, \xi^3 \frac{4\xi}{\omega_p} [p + p^{-1}] \frac{e^{2\xi pd/c}}{[e^{2\xi pd/c} - 1]^2} \\
&= F_C(d) + \frac{\pi^2 \hbar c^2}{45\omega_p d^5} \\
&= F_C(d) \left[1 - \frac{16}{3} \frac{c}{\omega_p d}\right], \tag{7.56}
\end{aligned}$$

in agreement with Hargreaves (1965) and Schwinger et al. (1978).[8]

The effect of imperfect conductivity is therefore to diminish the Casimir force of attraction. For a metal with $N \approx 10^{23}$ cm^{-3}, the plasma frequency $\omega_p \approx 10^{16}$s^{-1}, and $16c/3\omega_p d \approx 0.2/d(\mu m)$. For very small separations, therefore, the Casimir expression for the force is a poor approximation.

---

[8] Lifshitz (1956) used the same approximations for the case of imperfect conductors, but obtained, apparently incorrectly, the numerical factor $15\sqrt{\pi}$ instead of the 16/3 of (7.56).

Dielectric Plates with Small Separations

The general expression (7.48) for the force involves the dielectric constants at all frequencies. However, a simplification can be achieved for dielectrics at large and small separations by recognizing that the dominant contribution to the force comes from values of p and ξ such that $2p\xi d/c \sim 1$. Suppose d is small compared with c/ω_o, where ω_o is a resonance (absorption) frequency of the dielectrics 1 and 2, which we will assume to be identical and separated by vacuum ($\epsilon_1 = \epsilon_2 = \epsilon, \epsilon_3 = 1$). Then $2p\xi d/c << 2p\xi/\omega_o$, and $2p\xi d/c \sim 1$ implies $p\xi >> \omega_o$. Since $\epsilon(i\xi) \to 1$ for $\xi >> \omega_o$, we can assume that the dominant contribution to (7.48) at small separations comes from $p >> 1$. In this case $s_{1,2} \cong p$ and

$$
\begin{aligned}
F(d) &\cong -\frac{\hbar}{2\pi^2 c^3} \int_0^\infty dp\, p^2 \int_0^\infty d\xi\, \xi^3 \left[\left(\frac{1+\epsilon}{1-\epsilon}\right) \left(\frac{1+\epsilon}{1-\epsilon}\right) e^{2\xi p d/c} - 1 \right]^{-1} \\
&\cong -\frac{\hbar}{16\pi^2 d^3} \int_0^\infty dx \int_0^\infty d\xi\, \frac{x^2}{\left(\frac{\epsilon+1}{\epsilon-1}\right)^2 e^x - 1}\,,
\end{aligned}
\tag{7.57}
$$

in agreement with Lifshitz (1956). Note that the force between dielectrics at small separations varies as d^{-3}.

Dielectric Plates with Large Separations

Large separations are defined by $d >> c/\omega_o$, where ω_o is again a frequency at which significant dielectric absorption occurs. Assuming again $\epsilon_1 = \epsilon_2 = \epsilon$ and $\epsilon_3 = 1$, and defining the new variable x by writing $\xi = (c/2pd)x$, we write (7.48) as

$$
\begin{aligned}
F(d) = &-\frac{\hbar}{2\pi^2 c^3} \left(\frac{c}{2d}\right)^4 \int_1^\infty dp\, p^2 \int_0^\infty dx\, \frac{x^3}{p^4} \\
&\times \left(\left[\left(\frac{s+\epsilon p}{s-\epsilon p}\right)^2 e^x - 1 \right]^{-1} + \left[\left(\frac{s+p}{s-p}\right)^2 e^x - 1 \right]^{-1} \right),
\end{aligned}
\tag{7.58}
$$

$$
\epsilon = \epsilon(icx/2pd), \quad s = \sqrt{p^2 - 1 + \epsilon}\,.
\tag{7.59}
$$

Since the main contribution to the force arises from values of $x \sim 1$, and since $p \geq 1$, we can effectively replace $\epsilon(icx/2pd)$ by the electrostatic dielectric constant $\epsilon(0) \equiv \epsilon_o$.[9] Then

$$
\begin{aligned}
F(d) \cong\ &-\frac{\hbar c}{32\pi^2 d^4} \int_1^\infty dp\, p^{-2} \int_0^\infty dx\, x^3 \\
&\times \left(\left[\left(\frac{s_o + \epsilon_o p}{s_o - \epsilon_o p}\right)^2 e^x - 1 \right]^{-1} + \left[\left(\frac{s_o + p}{s_o - p}\right)^? e^x - 1 \right]^{-1} \right),
\end{aligned}
$$

$$(7.60)$$

where $s_o \equiv \sqrt{p^2 - 1 + \epsilon_o}$. In this case the force varies as d^{-4} and depends only on the *electrostatic* dielectric constant ϵ_o.

Rarefied Media and Intermolecular Potentials

The dielectric constant is given in terms of the atomic (or molecular) polarizability α and number density N by

$$
\epsilon(\omega) = 1 + 4\pi N \alpha(\omega) \tag{7.61}
$$

for media in which Lorentz–Lorenz corrections are unimportant. For rarefied media, such that $\epsilon(\omega) \cong 1$, we can approximate the short-separation force (7.57) by

$$
\begin{aligned}
F(d) &\cong -\frac{\hbar}{64\pi^2 d^3} \int_0^\infty dx\, x^2 e^{-x} (4\pi N)^2 \int_0^\infty d\xi\, \alpha^2(i\xi) \\
&= -\frac{\hbar N^2}{2\pi^2 d^3} \int_0^\infty d\xi\, \alpha^2(i\xi) .
\end{aligned} \tag{7.62}
$$

In this limit $F(d)$ can be obtained by pairwise integration over the intermolecular potentials, as in equation (7.6). Comparison of (7.62) to (7.6) indicates that $\gamma = 6$ and that the intermolecular potential

$$
U(r) = -\frac{B}{r^6} = -\frac{3\hbar}{\pi r^6} \int_0^\infty d\xi\, \alpha^2(i\xi) , \tag{7.63}
$$

which is the London–van der Waals interaction.[10]

For rarefied media the large-separation force (7.60) becomes

$$
F(d) \cong -\frac{\hbar c}{32\pi^2 d^4} \int_1^\infty dp\, p^{-2} \int_0^\infty dx\, x^3 e^{-x} \left[\left(\frac{s_o - \epsilon_o p}{s_o + \epsilon_o p}\right)^2 + \left(\frac{s_o - p}{s_o + p}\right)^2 \right]
$$

---

[9] For perfect conductors, $\epsilon_o \to \infty$ and (7.60) reduces to the Casimir force.

[10] See the second line of equation (3.87) and recall that we can replace $e^{-2ur/c}$ by 1 in that equation for small separations r.

$$\cong -\frac{12\hbar c(\epsilon_o - 1)^2}{(16)(32\pi^2 d^4)} \int_1^\infty dp \frac{1 - 2p^2 + 2p^4}{p^6}$$

$$= -\frac{23\hbar c(\epsilon_o - 1)^2}{640\pi^2 d^4} , \tag{7.64}$$

where we have used the approximations $\epsilon_o \cong 1$ and

$$s_o = \sqrt{p^2 + \epsilon_o - 1} \cong p + \frac{\epsilon_o - 1}{2p} . \tag{7.65}$$

In this case the comparison of (7.64) to (7.6) gives $\gamma = 7$ and

$$B = \frac{(20)(23)\hbar c(\epsilon_o - 1)^2}{(2\pi N^2)(640\pi^2)} = \frac{23\hbar c\alpha^2}{4\pi} , \tag{7.66}$$

where we have used (7.61) and defined α as the static molecular polarizability. Then we infer for large separations the intermolecular potential

$$U(r) = -\frac{B}{r^7} = -\frac{23\hbar c\alpha^2}{4\pi r^7} , \tag{7.67}$$

which is the Casimir–Polder result (3.91) for the retarded (long-range) van der Waals interaction.

Thus we can derive both the retarded and unretarded van der Waals interactions between two molecules from the *macroscopic* theory of the force between two dielectric plates. This theory requires basically only the Maxwell equations and the assumption that each mode of the field has a zero-point energy $\frac{1}{2}\hbar\omega$.

Remarks

We have derived the Lifshitz expression (7.48) for the force between two dielectric plates in a conceptually simple way based on the zero-point energy of the electromagnetic field. The first step in the derivation was to obtain the "dispersion" equations (7.27) and (7.28), which determine the allowed field frequencies in the presence of the dielectrics. Using these results, we then obtained equation (7.41) for the part of the total zero-point field energy depending on the distance d between the dielectrics. Differentiation of that expression with respect to d produced exactly the Lifshitz result for the force per unit area.

As noted earlier, the validity of the Lifshitz theory has sometimes been questioned, although the results now seem to be generally accepted. These results have been obtained in a variety of ways. Schwinger et al. (1978), for instance, obtained equation (7.48) "by adopting far superior and more physically transparent methods for computing the force," namely using Schwinger's source theory, "where the vacuum is regarded as truly a state with all physical properties equal to zero."

Before commenting on these and other methods, we should acknowledge that the approach we have followed, which is in the spirit of Casimir's origi nal work for perfect conductors and extends the approach of van Kampen et al. (1968) to include retardation, is not entirely rigorous (Langbein, 1973; Schram, 1973). The weak point in the analysis is that it ignores the fact that the functions $F_\alpha(\omega)$ have branch points. However, this shortcoming is easily circumvented by assuming that the dielectric slabs are contained within a finite cavity, summing over the zero-point energy of *all* the modes of this cavity, and passing to the infinite-cavity limit only at the end of the calculation. Then one incurs no branch points. In addition such an approach, though somewhat more complicated mathematically than the one we have presented, shows in the end that only the surface modes contribute, as assumed without formal justification in our calculation (Langbein, 1973). We also note that Barton (1979) has shown that "bulk modes" make no contribution to the small-separation force (7.57) when there is no spatial dispersion, i.e., when ϵ is independent of $\mathbf{k}$. (Our entire discussion assumes no spatial dispersion, as in nearly all the vast literature on the subject.)

But what are more interesting to us are approaches based on different *physical* ideas. We next turn our attention to some of these.

7.3   Lifshitz and Barash–Ginzburg Theories

Since we have reproduced Lifshitz's result for the force between two parallel dielectric slabs, it is obviously of interest to compare our calculations with his. Unfortunately, as noted earlier, the details of Lifshitz's calculation are fairly complicated.[11] Nevertheless, the physical basis of Lifshitz's calculation is not so difficult to understand.

The principal equations at the start of Lifshitz's analysis are the Maxwell equations

$$\nabla \times \mathbf{E}_o \;=\; i\frac{\omega}{c}\mathbf{B}_o \,, \tag{7.68}$$

$$\nabla \times \mathbf{B}_o \;=\; -i\frac{\omega}{c}\epsilon(\omega)\mathbf{E}_o - i\frac{\omega}{c}\mathbf{K} \tag{7.69}$$

for nonmagnetic dielectric media. Here $\epsilon(\omega) = \epsilon'(\omega) + i\epsilon''(\omega)$ is the complex dielectric constant, allowing for both dispersion and absorption, and $\mathbf{K}$ is

---

[11] Ginzburg (1979) writes that the calculations are "so cumbersome that they were not even reproduced in the relevant Landau and Lifshitz volume where, as a rule, all important calculations are given."

a random field corresponding to some randomly fluctuating current. We already see a difference between Lifshitz's original approach and the one we have followed: Lifshitz takes proper account of the fact that $\epsilon(\omega)$ is in general complex, whereas in our approach based on zero-point electromagnetic energy, $\epsilon(\omega)$ was purely real.[12] Furthermore there was no random field $\mathbf{K}$ in our approach.

Actually the use of a complex $\epsilon(\omega)$ in Lifshitz's theory *requires* the random field $\mathbf{K}$. Recall our discussion in Section 2.6 where the dissipative effect of radiation reaction had to be balanced by the fluctuating vacuum (zero-point) field in order to preserve commutation relations; formal consistency of the theory demanded such a fluctuation–dissipation relation. The situation is similar in the case of an absorbing medium: the dissipative influence of the medium must be balanced by a fluctuating source term whose correlations are related to the form of the dissipation, and in particular to the imaginary (absorptive) part of the dielectric constant. At zero temperature this fluctuation–dissipation relation as employed by Lifshitz is

$$\langle K_i(\mathbf{r})K_j(\mathbf{r'})\rangle = 2\hbar\epsilon''(\omega)\delta_{ij}\delta^3(\mathbf{r}-\mathbf{r'}). \tag{7.70}$$

Lifshitz solves equations (7.68) and (7.69) subject to the appropriate boundary conditions, as in our approach, and obtains spatial Fourier components of $\mathbf{E}_o$ and $\mathbf{B}_o$ in terms of the Fourier components $\mathbf{g}(\mathbf{k})$ of $\mathbf{K}$. He then calculates the force between the dielectrics 1 and 2 in terms of the Maxwell stress tensor, using the correlation function $\langle g_i(\mathbf{k})g_j(\mathbf{k'})\rangle$ that follows from (7.70). After some consideration of contour integrals he is led to the formula (7.48) for the special case $\epsilon_3 = 1$ he assumes. It is perhaps worth noting that Lifshitz acknowledges at the outset that his approach is connected with the notion of the vacuum field:

> ... the interaction of the objects is regarded as occurring through the medium of the fluctuating electromagnetic field which is always present in the interior of any absorbing medium, and also extends beyond its boundaries, — partially in the form of travelling waves radiated by the body, partially in the form of standing waves which are damped exponentially as we move away from the surface of the body. It must be emphasized that this field does not vanish even at absolute zero, at which point it is associated with the zero point vibrations of the radiation field.

It is remarkable that our approach based on zero-point field energy reproduces Lifshitz's result for the force: we did not allow for any possibility

---

[12] The concept of a dispersive but nonabsorbing medium violates the Kramers–Kronig dispersion relations and is an approximation that cannot be realized in practice.

of absorption, and yet the Lifshitz force (7.48) we derived applies when there is absorption, and indeed Lifshitz explicitly *required* absorption ($\epsilon''(\omega) \neq 0$) in his derivation. How could we obtain Lifshitz's result by restricting ourselves to completely transparent media? And how can the notion of zero-point field energy be used in the case of absorbing media, where it is not obvious how we might define eigenfrequencies such as ω_{na}, ω_{nb} in (7.29)?

Such questions were answered by Barash (1973) and Barash and Ginzburg (1975). To appreciate the essentials of their approach, it is useful to consider first the simple example of the driven, damped harmonic oscillator described by the equation

$$\ddot{x} + \alpha\dot{x} + \omega_o^2 x = \frac{1}{m}F(t). \tag{7.71}$$

Defining the Fourier components F_ω, x_ω by writing

$$F(t) = \int_0^\infty d\omega[F_\omega e^{-i\omega t} + F_\omega^* e^{i\omega t}] \,, \tag{7.72}$$

$$x(t) = \int_0^\infty d\omega[x_\omega e^{-i\omega t} + x_\omega^* e^{i\omega t}] \,, \tag{7.73}$$

we obtain the steady-state relation

$$x_\omega = \frac{1}{m}\frac{F_\omega}{\omega_o^2 - \omega^2 - i\alpha\omega} \equiv \chi(\omega)F_\omega \,, \tag{7.74}$$

from which it follows that the kinetic and potential energies of the oscillator are respectively

$$K = \frac{1}{2}m\dot{x}^2 = \frac{1}{2}m\int_0^\infty d\omega\omega \int_0^\infty d\omega'\omega'[\chi(\omega)\chi^*(\omega')F_\omega F_{\omega'}^* e^{-i(\omega-\omega')t}$$
$$- \chi(\omega)\chi(\omega')F_\omega F_{\omega'} e^{-i(\omega+\omega')t}] + \text{c.c.} \,, \tag{7.75}$$

$$V = \frac{1}{2}m\omega_o^2 x^2 = \frac{1}{2}m\omega_o^2 \int_0^\infty d\omega \int_0^\infty d\omega'[\chi(\omega)\chi^*(\omega')F_\omega F_{\omega'}^* e^{-i(\omega-\omega')t}$$
$$+ \chi(\omega)\chi(\omega')F_\omega F_{\omega'} e^{-i(\omega+\omega')t}] + \text{c.c.} \,. \tag{7.76}$$

Now suppose that $F(t)$ is a randomly varying force, such that the expectation values

$$\langle F_\omega F_{\omega'}\rangle = 0, \quad \langle F_\omega F_{\omega'}^*\rangle = G(\omega)\delta(\omega - \omega'). \tag{7.77}$$

Then the expectation value of the oscillator energy is

$$\langle E \rangle = \langle K + V \rangle = m \int_0^\infty d\omega (\omega^2 + \omega_o^2) |\chi(\omega)|^2 G(\omega)$$

$$= \frac{1}{m} \int_0^\infty \frac{d\omega (\omega^2 + \omega_o^2) G(\omega)}{(\omega^2 - \omega_o^2)^2 + \alpha^2 \omega^2} . \qquad (7.78)$$

The simplest way to arrive at the desired result is to assume that $G(\omega)$ is a relatively flat function of ω compared with the remainder of the integrand, and then to use the fact that the integrand is sharply peaked at $\omega = \omega_o$ for $\alpha \ll \omega_o$:

$$\langle E \rangle \cong \frac{1}{m} (2\omega_o^2) G(\omega_o) \frac{1}{4\omega_o^2} \int_0^\infty \frac{d\omega}{(\omega - \omega_o)^2 + \alpha^2/4} = \frac{\pi}{m\alpha} G(\omega_o). \qquad (7.79)$$

Therefore, if the oscillator is to reach thermal equilibrium in the presence of the fluctuation force $F(t)$ and the dissipation force $m\alpha\dot{x}$, we must have the fluctuation–dissipation relation

$$\frac{\pi}{m\alpha} G(\omega_o) = \frac{1}{2}\hbar\omega_o + \frac{\hbar\omega_o}{e^{\hbar\omega_o/kT} - 1} \equiv f(\omega_o, T), \qquad (7.80)$$

where we have included the zero-point energy $\frac{1}{2}\hbar\omega_o$. In other words, $G(\omega)$ must be proportional to the dissipation coefficient α:

$$G(\omega) = \frac{m}{\pi} \alpha f(\omega, T). \qquad (7.81)$$

This is an example of the fluctuation–dissipation theorem relating the spectrum of the fluctuating force to the dissipation coefficient. Our derivation of the relation (7.81) has been couched in the language of classical physics, with $\hbar$ arising from the assumed form (7.80) of the average oscillator energy. However, the derivation is easily extended to the quantum domain, and in fact the $T = 0$ fluctuation–dissipation relation may be derived from the requirement that the fluctuation and dissipation forces should be related in just such a way as to preserve the commutation relation between x and its conjugate momentum (see Section 2.6 and Milonni, 1981). Using (7.81) in (7.78), we have

$$\langle E \rangle = \frac{\alpha}{2\pi} \int_{-\infty}^\infty \frac{d\omega (\omega^2 + \omega_o^2) f(\omega, T)}{(\omega^2 - \omega_o^2)^2 + \alpha^2 \omega^2} , \qquad (7.82)$$

where we have used the fact that $f(\omega, T)$ is an even function of ω for $T > 0$.

Now the most important feature of the Barash–Ginzburg theory, the notion of an "auxiliary system," may be introduced in this simple example as follows. Define the "auxiliary" oscillator equation

$$\ddot{x} + \alpha\frac{\omega}{\omega_1}\dot{x} + \omega_o^2 x = 0, \tag{7.83}$$

where ω_1 is a function of ω, which is now regarded as a parameter appearing in the definition of the auxiliary oscillator. Equation (7.83) has solutions of the form $\exp[-i\omega_1(\omega)t]$, where

$$\omega_1^2(\omega) = \omega_o^2 - i\alpha\omega . \tag{7.84}$$

In terms of the frequency $\omega_1(\omega)$ we may write (7.74) and (7.82) as

$$x_\omega = \frac{1}{m}\frac{F_\omega}{\omega_1^2 - \omega^2} , \tag{7.85}$$

$$
\begin{aligned}
\langle E \rangle &= -\frac{i}{\pi}\int_{-\infty}^{\infty}\frac{d\omega\,\omega f(\omega, T)}{\omega_1^2(\omega) - \omega^2} + \frac{i}{2\pi}\int_{-\infty}^{\infty}\frac{d\omega(-i\alpha)f(\omega, T)}{\omega_1^2(\omega) - \omega^2} \\
&= -\frac{i}{\pi}\int_{-\infty}^{\infty}\frac{d\omega\,\omega f(\omega, T)}{\omega_1^2(\omega) - \omega^2} + \frac{i}{2\pi}\int_{-\infty}^{\infty}\frac{d\omega\,f(\omega, T)d\omega_1^2(\omega)/d\omega}{\omega_1^2(\omega) - \omega^2} .
\end{aligned}
\tag{7.86}
$$

The equivalence of this expression to (7.82) is easily demonstrated by simple algebra, together with the fact that $f(\omega, T)$ is an even function of ω for $T > 0$. We can obviously rewrite $\langle E \rangle$ in the form

$$\langle E \rangle = \frac{i}{2\pi}\int_{-\infty}^{\infty}d\omega\,f(\omega, T)\frac{\partial}{\partial\omega}\log[\omega_1^2(\omega) - \omega^2]. \tag{7.87}$$

Note that the integrand has no singularity along the path of integration (i e , the real ω axis), since $\omega_1(\omega)$ is complex.

Of course nothing is really gained in the present example by the replacement of (7.82) by the equivalent expression (7.87). In the case of electromagnetic waves, the auxiliary system corresponding to (7.68) and (7.69) is defined by

$$\nabla \times \mathbf{E}_a = i\frac{\omega_a}{c}\mathbf{B}_a , \tag{7.88}$$

$$\nabla \times \mathbf{B}_a = -i\frac{\omega_a}{c}\epsilon(\omega)\mathbf{E}_a , \tag{7.89}$$

where, as in (7.83), ω is regarded as a parameter in the definition of the auxiliary modes labelled by a. These modes may be shown to be orthogonal

(Barash, 1973), so that the field energy has the form of a sum of contributions of the type (7.87) — each auxiliary mode is analogous to the auxiliary oscillator (7.83):

$$
\begin{aligned}
\langle E \rangle &= \frac{i}{2\pi} \sum_a \int_{-\infty}^{\infty} d\omega\, f(\omega, T) \frac{\partial}{\partial \omega} \log[\omega_a^2(\omega) - \omega^2] \\
&= \frac{i}{2\pi} \int_{-\infty}^{\infty} d\omega\, f(\omega, T) \frac{\partial}{\partial \omega} \log \Pi_a[\omega_a^2(\omega) - \omega^2] \\
&\to \frac{i}{2\pi} \int_{-\infty}^{\infty} d\omega\, f(\omega, T) \int d\beta \rho(\beta) \frac{\partial}{\partial \omega} \log D(\beta, \omega), \qquad (7.90)
\end{aligned}
$$

where $\rho(\beta)$ is a density of states and $D(\beta, \omega)$ determines the mode frequencies through the equation $D(\beta, \omega) = 0$. Thus $D(\beta, \omega) \to D(\omega) = \omega_1^2(\omega) - \omega^2$ for the single oscillator, whereas $D(\beta, \omega) \to D(k_x, k_y, \omega) = F_a(\omega)$ or $F_b(\omega)$ for the Lifshitz problem, where F_a, F_b are again defined by the left sides of (7.27) and (7.28).

It is an easy matter to show that (7.90) leads directly to the Lifshitz force (7.48). The Barash–Ginzburg approach has several advantages, not the least of which is that it shows how the zero-point field energy may be employed in the calculation of the force even when absorption is allowed for. It thus establishes a useful bridge between the Lifshitz theory and the approach of the preceding section, where zero-point field energy was used but absorption was not accounted for. In this connection the orthogonality of the auxiliary modes is crucial. The Barash–Ginzburg approach also avoids the technical difficulty of branch points mentioned earlier, and the final result is applicable for $T > 0$, as discussed in the following chapter. For further details of the theory the interested reader is referred to the review by Barash and Ginzburg (1975).

As already noted, the result obtained from changes in zero-point field energy, without taking absorption into account, happens to be correct also when $\epsilon''(\omega) \neq 0$, i.e., when the possibility of absorption is allowed in the final expression for the force. This circumstance is perhaps less surprising when we recognize that the force (7.48) in fact involves $\epsilon_j(i\xi)$, i.e., the dielectric constant on the imaginary axis, and that this quantity is always real.[13] In particular, it follows from the Kramers–Kronig dispersion relations that

$$
\epsilon(i\xi) = 1 + \frac{2}{\pi} \int_0^{\infty} \frac{x\epsilon''(x)}{x^2 + \xi^2} dx . \qquad (7.91)
$$

---

[13] See L. D. Landau and E. M. Lifshitz, *Electrodynamics of Continuous Media*, Section 62.

7.4   Source Theory

The most unconventional approach to the force between dielectrics, and the Casimir force between perfect conductors, is that of Schwinger et al. (1978) in which, as mentioned earlier, "the vacuum is regarded as truly a state with all physical properties equal to zero." Rather than attempting to describe their source theory on which this approach is based, we shall show how the relevant expression of Schwinger et al. may be understood from the standpoint of ordinary QED. The basic idea here will be that the Casimir force may be derived from source fields alone even in completely conventional QED, and that the derivation becomes identical after a certain point to that of Schwinger et al. The connection with the vacuum field interpretation of the Casimir force will be made in the following section.

We begin by recalling that an induced dipole $\mathbf{d}$ in a field $\mathbf{E}$ has an energy $-\frac{1}{2}\mathbf{d} \cdot \mathbf{E}$ classically, and that for N dipoles per unit volume defining a polarization $\mathbf{P} = N\mathbf{d}$, the expectation value of the energy in quantum theory is

$$\langle E \rangle = -\frac{1}{2} \int d^3r \langle \mathbf{P} \cdot \mathbf{E} \rangle. \tag{7.92}$$

We have already employed such an expression in Section 3.8 in connection with the interpretation of the Lamb shift as a Stark shift due to the zero-point field. Since in the present discussion we wish to emphasize the role of the source fields, we will use a normal ordering of field operators in (7.92):

$$\langle E \rangle = -\frac{1}{2} \int d^3r \langle \mathbf{E}^{(-)} \cdot \mathbf{P} + \mathbf{P} \cdot \mathbf{E}^{(+)} \rangle. \tag{7.93}$$

As in Chapter 4, for instance, we can write

$$\mathbf{E}^{(+)}(\mathbf{r}, t) = \mathbf{E}_o^{(+)}(\mathbf{r}, t) + \mathbf{E}_s^{(+)}(\mathbf{r}, t), \tag{7.94}$$

where $\mathbf{E}_o^{(+)}(\mathbf{r}, t)$ is the source-free (vacuum) part of $\mathbf{E}^{(+)}(\mathbf{r}, t)$ and $\mathbf{E}_s^{(+)}(\mathbf{r}, t)$ is the part due to any sources. Then, since $\mathbf{E}_o^{(+)}(\mathbf{r}, t)|\text{vac}\rangle = \langle \text{vac}|\mathbf{E}_o^{(-)}(\mathbf{r}, t) = 0$,

$$\langle E \rangle = -\frac{1}{2} \int d^3r \langle \mathbf{P}(\mathbf{r}, t) \cdot \mathbf{E}_s^{(+)}(\mathbf{r}, t) \rangle + \text{c.c.}, \tag{7.95}$$

where vacuum expectation values are implied.

To obtain an expression for $\mathbf{E}_s^{(+)}(\mathbf{r}, t)$, we first recall from Chapter 2 that we can write the electric field operator as an expansion in mode functions $\mathbf{A}_\alpha(\mathbf{r})$:

$$\mathbf{E}(\mathbf{r}, t) = i \sum_\alpha (2\pi\hbar\omega_\alpha)^{1/2} [a_\alpha(t)\mathbf{A}_\alpha(\mathbf{r}) - a_\alpha^\dagger(t)\mathbf{A}_\alpha^*(\mathbf{r})], \tag{7.96}$$

where the field modes, which are assumed to form a complete set, are labelled by the subscript α. Then from the interaction term

$$H_{\text{INT}} = - \int d^3r \mathbf{P}(\mathbf{r}, t) \cdot \mathbf{E}(\mathbf{r}, t) \tag{7.97}$$

in the Hamiltonian we obtain the Heisenberg equation of motion

$$\dot{a}_\alpha(t) = -i\omega_\alpha a_\alpha(t) + \left(\frac{2\pi\omega_\alpha}{\hbar} \right)^{1/2} \int d^3r \mathbf{A}_\alpha^*(\mathbf{r}) \cdot \mathbf{P}(\mathbf{r}, t), \tag{7.98}$$

and therefore

$$a_{\alpha s}(t) = \left(\frac{2\pi\omega_\alpha}{\hbar} \right)^{1/2} \int_0^t dt' e^{i\omega_\alpha(t'-t)} \int d^3r \mathbf{A}_\alpha^*(\mathbf{r}) \cdot \mathbf{P}(\mathbf{r}, t') \tag{7.99}$$

for the source part of $a_\alpha(t)$. Thus the "positive-frequency" (photon annihilation) part of $\mathbf{E}_s(\mathbf{r}, t)$ is

$$
\begin{aligned}
\mathbf{E}_s^{(+)}(\mathbf{r}, t) &= 2\pi i \sum_\alpha \omega_\alpha \mathbf{A}_\alpha(\mathbf{r}) \int_0^t dt' e^{i\omega_\alpha(t'-t)} \int d^3r' \mathbf{A}_\alpha^*(\mathbf{r}') \cdot \mathbf{P}(\mathbf{r}', t') \\
&= 2\pi i \int d^3r' \int_0^t dt' \sum_\alpha \omega_\alpha \mathbf{A}_\alpha(\mathbf{r}) \mathbf{A}_\alpha^*(\mathbf{r}') e^{i\omega_\alpha(t'-t)} \cdot \mathbf{P}(\mathbf{r}', t') \\
&\equiv 8\pi \int d^3r' \int_0^t dt' \overset{\leftrightarrow(+)}{G}(\mathbf{r}, \mathbf{r}'; t, t') \cdot \mathbf{P}(\mathbf{r}', t'), \tag{7.100}
\end{aligned}
$$

where $\overset{\leftrightarrow(+)}{G}$ is a dyadic Green function.

Equations (7.95) and (7.100) give

$$\langle E \rangle = -8\pi \text{Re} \int d^3r \int d^3r' \int_0^t dt' G_{ij}^{(+)}(\mathbf{r}, \mathbf{r}'; t, t') \langle P_j(\mathbf{r}, t) P_i(\mathbf{r}', t') \rangle, \tag{7.101}$$

where a summation over repeated indices is understood. This is the energy of the induced dipoles in a medium *due to the source fields produced by the dipoles*.

To calculate the force between dielectrics for the configuration shown in Figure 7.1, we consider a slightly different situation. We imagine making a small change in the dielectric constant by adding more polarizable particles (atoms) to the dielectric media. The polarization density will be changed from $\mathbf{P}$ to $\mathbf{P} + \mathbf{P}'$, and the change in the energy due to the interaction of the added dipoles with each other will have the form[14]

$$\langle \delta E \rangle = -8\pi \text{Re} \int d^3r \int d^3r' \int_0^t dt' G_{ij}^{(+)}(\mathbf{r}, \mathbf{r}'; t, t') \langle P_i'(\mathbf{r}, t) P_j'(\mathbf{r}', t') \rangle .$$

(7.102)

From Maxwell's equations for the Heisenberg-picture field operators we have

$$-\nabla \times \nabla \times \mathbf{E} - \frac{1}{c^2} \frac{\partial^2 \mathbf{E}}{\partial t^2} = \frac{4\pi}{c^2} \frac{\partial^2 \mathbf{P}}{\partial t^2} + \frac{4\pi}{c^2} \frac{\partial^2 \mathbf{P}'}{\partial t^2} ,$$

(7.103)

or

$$-\nabla \times \nabla \times \mathbf{E} - \frac{1}{c^2} \frac{\partial^2 \mathbf{D}}{\partial t^2} = \frac{4\pi}{c^2} \frac{\partial^2 \mathbf{P}'}{\partial t^2} .$$

(7.104)

Of course this equation has exactly the same form as the corresponding classical equation (recall the discussion in Section 4.6). As in classical theory we can write the dyadic Green function $\overset{\leftrightarrow}{G}$ for (7.104) as

$$\overset{\leftrightarrow}{G}(\mathbf{r}, \mathbf{r}'; t, t') = \frac{1}{2\pi} \int_{-\infty}^{\infty} d\omega \, \overset{\leftrightarrow}{\Gamma}(\mathbf{r}, \mathbf{r}', \omega) e^{-i\omega(t-t')} ,$$

(7.105)

and use the constitutive relation $\mathbf{D}(\mathbf{r}, \omega) = \epsilon(\mathbf{r}, \omega)\mathbf{E}(\mathbf{r}, \omega)$ between the Fourier components of $\mathbf{D}$ and $\mathbf{E}$ to obtain for $\overset{\leftrightarrow}{\Gamma}(\mathbf{r}, \mathbf{r}', \omega)$ the equation

$$-\nabla \times \nabla \times \overset{\leftrightarrow}{\Gamma} + \frac{\omega^2}{c^2} \epsilon(\omega) \overset{\leftrightarrow}{\Gamma} = -\frac{\omega^2}{c^2} \overset{\leftrightarrow}{1} \delta^3(\mathbf{r} - \mathbf{r}').$$

(7.106)

The positive-frequency Green function $\overset{\leftrightarrow}{G}^{(+)}$ is

$$\overset{\leftrightarrow}{G}^{(+)}(\mathbf{r}, \mathbf{r}'; t, t') = \frac{1}{2\pi} \int_0^{\infty} d\omega \, \overset{\leftrightarrow}{\Gamma}(\mathbf{r}, \mathbf{r}', \omega) e^{-i\omega(t-t')} ,$$

(7.107)

so that

$$\langle \delta E \rangle = -4\text{Re} \int d^3r \int d^3r' \int_0^t dt' \int_0^{\infty} d\omega \Gamma_{ij}(\mathbf{r}, \mathbf{r}', \omega) \langle P_j'(\mathbf{r}, t) P_i'(\mathbf{r}', t') \rangle$$
$$\times e^{-i\omega(t-t')} .$$

(7.108)

Before proceeding we wish to emphasize an important point in connection with equation (7.108). This equation is similar to (7.101), and in fact

---

[14] Terms of the type $\langle P_i(\mathbf{r}, t) P_j'(\mathbf{r}', t') \rangle$ do not contribute under the assumption that the new atoms we introduce are independent and uncorrelated from the original atoms composing the dielectric media.

it is the *change* in the energy (7.101) due to the addition of the atoms associated with the polarization density $\mathbf{P}'$. The fields associated with these added atoms propagate with a velocity determined by the dielectric constant produced by the *unperturbed* polarization density $\mathbf{P}$; this is why the dielectric constant $\epsilon(\omega)$ appears on the left side of (7.106).

We wish now to express the expectation value in the integrand of (7.108) in terms of the dielectric constant associated with $\mathbf{P}'$. We first write

$$P_i'(\mathbf{r},t) = \sum_\beta d_{\beta i}(t)\delta^3(\mathbf{r}-\mathbf{r}_\beta), \qquad (7.109)$$

where $d_{\beta i}(t)$ is the i-component of the dipole moment operator $\mathbf{d}_\beta(t)$ for an atom at position $\mathbf{r}_\beta$. Thus

$$\langle P_i'(\mathbf{r},t)P_j'(\mathbf{r}',t')\rangle = \sum_\beta \sum_\gamma \langle d_{\beta i}(t)d_{\gamma j}(t')\rangle \delta^3(\mathbf{r}-\mathbf{r}_\beta)\delta^3(\mathbf{r}'-\mathbf{r}_\gamma). \quad (7.110)$$

We assume all the atoms are, to a good approximation, in their ground states $|g\rangle$ at all times, and that the multiatomic wave function has the uncorrelated form $|g\rangle = |g\rangle_\beta|g\rangle_\gamma \cdots$, so that

$$\langle d_{\beta i}(t)d_{\gamma j}(t')\rangle = \delta_{\beta\gamma}\langle d_{\beta i}(t)d_{\beta j}(t')\rangle \qquad (7.111)$$

and

$$\begin{aligned}
\langle P_i'(\mathbf{r},t)P_j'(\mathbf{r}',t')\rangle &= \sum_\beta \langle d_{\beta i}(t)d_{\beta j}(t')\rangle \delta^3(\mathbf{r}-\mathbf{r}_\beta)\delta^3(\mathbf{r}'-\mathbf{r}_\beta) \\
&\to \langle d_i(t)d_j(t')\rangle \int d^3r'' N(\mathbf{r}'')\delta^3(\mathbf{r}-\mathbf{r}'')\delta^3(\mathbf{r}'-\mathbf{r}'') \\
&= \langle d_i(t)d_j(t')\rangle N(\mathbf{r})\delta^3(\mathbf{r}-\mathbf{r}'), \qquad (7.112)
\end{aligned}$$

where we have gone to the continuum approximation in which we assume $N(\mathbf{r})$ atoms per unit volume at the position $\mathbf{r}$. We have also assumed that all the atoms are identical, so that $\langle d_{\beta i}(t)d_{\beta j}(t')\rangle \to \langle d_i(t)d_j(t')\rangle$ for all β. Then (7.108) becomes

$$\langle \delta E\rangle = -4\mathrm{Re}\int d^3r N(\mathbf{r})\int_0^\infty d\omega \Gamma_{ij}(\mathbf{r},\mathbf{r}',\omega)\int_0^t dt'\langle d_i(t)d_j(t')\rangle e^{-i\omega(t-t')}. \qquad (7.113)$$

Since the Heisenberg-picture operator $d_i(t)$ evolves in time according to the equation $d_i(t) = U^\dagger(t)d_iU(t)$, where $d_i = d_i(0)$ is the Schrödinger-picture operator and $U(t)$ is the time evolution operator satisfying $i\hbar U =$

HU, $U(0) = 1$, we have

$$
\begin{aligned}
\langle d_i(t)d_j(t')\rangle &= \langle g|U^\dagger(t)d_iU(t)U^\dagger(t')d_jU(t')|g\rangle \\
&= \sum_k \langle g|U^\dagger(t)d_iU(t)|k\rangle\langle k|U^\dagger(t')d_jU(t')|g\rangle \ . \ (7.114)
\end{aligned}
$$

We have inserted the unit operator in the guise of $\sum_k |k\rangle\langle k|$, where $\{|k\rangle\}$ is the complete set of eigenstates of the Hamiltonian H_A for a single atom. If we approximate $U(t)$ by $e^{-iH_At/\hbar}$, then

$$
\begin{aligned}
\langle d_i(t)d_j(t')\rangle &\cong \sum_k e^{i(E_g-E_k)t/\hbar}e^{i(E_k-E_g)t'/\hbar}\langle g|d_i|k\rangle\langle k|d_j|g\rangle \\
&= \sum_k e^{i\omega_{kg}(t'-t)}(d_i)_{gk}(d_j)_{kg} \qquad (7.115)
\end{aligned}
$$

to order e^2, and

$$
\begin{aligned}
\langle \delta E\rangle &= -4\mathrm{Re}\int d^3r N(\mathbf{r})\int_0^\infty d\omega \Gamma_{ij}(\mathbf{r},\mathbf{r},\omega)\sum_k (d_i)_{gk}(d_j)_{kg} \\
&\quad \times \int_0^t dt'e^{i(\omega+\omega_{kg})(t'-t)} \\
&\to 4\mathrm{Re}\left[i\int d^3r N(\mathbf{r})\int_0^\infty d\omega \Gamma_{ij}(\mathbf{r},\mathbf{r},\omega)\sum_k \frac{(d_i)_{gk}(d_j)_{kg}}{\omega+\omega_{kg}}\right],
\end{aligned}
$$
$$(7.116)$$

where we ignore the rapidly oscillating term $e^{-i(\omega+\omega_{kg})t}$ associated with an artificial turn-on of the atom–field interaction at $t = 0$.

Let us rewrite the frequency integration in (7.116) by changing the integration contour in such a way that $\omega \to i\xi$, ξ real:

$$
\langle \delta E\rangle = -4\int d^3r N(\mathbf{r})\int_0^\infty d\xi \Gamma_{ij}(\mathbf{r},\mathbf{r},i\xi)\sum_k \frac{\omega_{kg}(d_i)_{gk}(d_j)_{kg}}{\xi^2+\omega_{kg}^2} \ . \ (7.117)
$$

Now recall the definition of the polarizability tensor for the ground state g:[15]

$$
\alpha_{ij}(\omega) = \frac{1}{\hbar}\sum_k \frac{2\omega_{kg}(d_i)_{gk}(d_j)_{kg}}{\omega_{kg}^2-\omega^2} \ . \qquad (7.118)
$$

---

[15]See, for instance, A. S. Davydov, *Quantum Mechanics* (Pergamon Press, Oxford, 1965), pp. 316–321.

For a spherically symmetric system (e g , an atom), $\alpha_{ij}(\omega) = \alpha(\omega)\delta_{ij}$, where $\alpha(\omega)$ is the usual polarizability.[16] Then

$$\alpha_{ij}(i\xi) = \frac{1}{\hbar}\sum_k \frac{2\omega_{kg}(d_i)_{gk}(d_j)_{kg}}{\omega_{kg}^2 + \xi^2} = \delta_{ij}\frac{1}{3\hbar}\sum_k \frac{2\omega_{kg}|d_{kg}|^2}{\omega_{kg}^2 + \xi^2} = \delta_{ij}\alpha(i\xi),$$

(7.119)

which allows us to rewrite (7.117) as

$$\begin{aligned}
\langle\delta E\rangle &= -2\hbar\int d^3r\,N(\mathbf{r})\int_0^\infty d\xi\,\alpha(i\xi)\Gamma_{jj}(\mathbf{r},\mathbf{r},i\xi)\\
&= -\frac{\hbar}{2\pi}\int d^3r\int_0^\infty d\xi[\epsilon(\mathbf{r},i\xi)-1]\Gamma_{jj}(\mathbf{r},\mathbf{r},i\xi), \quad (7.120)
\end{aligned}$$

when we employ the relation $\epsilon - 1 = 4\pi N\alpha$ between the dielectric constant and the polarizability.[17]

To calculate the force between the dielectric plates 1 and 2 of Figure 7.1, we consider the change in $\langle\delta E\rangle$ as a result of an infinitesimal change δd in d. For this purpose we imagine that the atoms we have added to change the energy according to (7.120) constitute a layer of width δd at $z = d$. This results in a change $\epsilon \rightarrow \epsilon + \delta\epsilon$ in the dielectric constant, and a force

$$f = \frac{\hbar}{2\pi}\int d^3r\int_0^\infty d\xi\frac{\delta\epsilon}{\delta d}\Gamma_{jj}(\mathbf{r},\mathbf{r},i\xi),$$

(7.121)

where, since

$$\epsilon(\mathbf{r},\omega) = \epsilon_1(\mathbf{r}_\perp,z,\omega)\theta(-z) + \epsilon_3(\mathbf{r}_\perp,z,\omega)\theta(z)\theta(d-z) + \epsilon_2(\mathbf{r}_\perp,z,\omega)\theta(z-d)$$

(7.122)

and $d\theta(x)/dx = \delta(x)$,

$$\frac{\delta\epsilon}{\delta d} = (\epsilon_3 - \epsilon_2)\delta(z-d).$$

(7.123)

Here $\mathbf{r}_\perp$ is the component of $\mathbf{r}$ in the xy plane.

Since the dielectric constant varies only in the z direction in the present example, it is natural to write

$$\overset{\leftrightarrow}{\Gamma}(\mathbf{r},\mathbf{r}',\omega) = \left(\frac{1}{2\pi}\right)^2\int d^2k_\perp\,\overset{\leftrightarrow}{\Gamma}(z,z',\mathbf{k}_\perp,\omega)e^{i\mathbf{k}_\perp\cdot(\mathbf{r}-\mathbf{r}')},$$

(7.124)

---

[16] See equation (3.51).

[17] Our final results are unchanged when we use the more general Clausius–Mossotti relation $4\pi N\alpha = 3(\epsilon-1)/(\epsilon+2)$.

so that

$$f(d) = -i\hbar \left(\frac{1}{2\pi}\right)^3 \int d^3r \int_0^\infty d\omega (\epsilon_3 - \epsilon_2)\delta(z-d) \int d^2k_\perp \Gamma_{jj}(z,z,\mathbf{k}_\perp,\omega),$$
$$(7.125)$$

where we have reverted back to $\omega = i\xi$ as the frequency integration variable. Equation (7.125) implies the force per unit area

$$F(d) = i\hbar \left(\frac{1}{2\pi}\right)^3 \int_0^\infty d\omega \int d^2k_\perp (\epsilon_2 - \epsilon_3)\Gamma_{jj}(d,d,\mathbf{k}_\perp,\omega). \qquad (7.126)$$

The calculation of $\Gamma_{jj} = \Gamma_{xx} + \Gamma_{yy} + \Gamma_{zz}$ is a straightforward exercise in *classical* electromagnetic theory: $\overleftrightarrow{\Gamma}$ satisfies (7.106) plus the electric field boundary conditions at the dielectric interfaces at $z = 0$ and $z = d$. The result is[18]

$$
\begin{aligned}
(\epsilon_2 - \epsilon_3)\Gamma_{jj}(d,d,\mathbf{k}_\perp,\omega) &= 2(K_3 - K_2) \\
&+ 2K_3 \left[\left(\frac{K_1 + K_3}{K_1 - K_3}\frac{K_2 + K_3}{K_2 - K_3}e^{2K_3 d} - 1 \right)^{-1} \right. \\
&+ \left. \left(\frac{\epsilon_3 K_1 + \epsilon_1 K_3}{\epsilon_3 K_1 - \epsilon_1 K_3}\frac{\epsilon_3 K_2 + \epsilon_2 K_3}{\epsilon_3 K_2 - \epsilon_2 K_3}e^{2K_3 d} - 1 \right)^{-1} \right],
\end{aligned}
$$
$$(7.127)$$

where $K^2 = k_\perp^2 - \omega^2 \epsilon(\omega)/c^2$. The first term on the right corresponds to a change in the volume energy of the system and does not depend on d. The remaining terms give the force between the dielectrics:

$$
\begin{aligned}
F(d) &= 2i\hbar \left(\frac{1}{2\pi}\right)^3 \int_0^\infty d\omega \int d^2k_\perp K_3[...] \\
&= \frac{i\hbar}{2\pi^2}\int_0^\infty d\omega \int_0^\infty dk k K_3[...],
\end{aligned}
$$
$$(7.128)$$

where $[...]$ denotes the bracketed factor in (7.127).

This result is identical to (7.44), and so we have succeeded in obtaining the Lifshitz force (and therefore the Casimir force when the limits $\epsilon_1, \epsilon_2 \rightarrow \infty, \epsilon_3 \rightarrow 1$ are taken) *in terms of source fields.* Our derivation, which obviously follows ordinary QED, bears some similarity to the derivation based on Schwinger's source theory (Schwinger et al., 1978). By considering

---

[18] For details of the calculation see Schwinger et al. (1978).

a variation of an appropriate action expression, Schwinger et al. infer an effective product of polarization sources

$$i\mathbf{P}(x)\mathbf{P}(x')|_{\text{eff}} = \overset{\leftrightarrow}{1}\,\delta(x - x'), \qquad (7.129)$$

where x denotes a four-dimensional coordinate $(\mathbf{r}, t)$ and $\delta\epsilon$ is an infinitesimal change in the dielectric constant. They then infer a change in energy[19]

$$\delta E = \frac{i}{2}\left(\frac{1}{2\pi}\right)\int d^3r \int_{-\infty}^{\infty} d\omega\,\delta\epsilon(\mathbf{r},\omega)\Gamma_{kk}(\mathbf{r},\mathbf{r},\omega), \qquad (7.130)$$

and a force per unit area

$$F = \frac{i}{2}\left(\frac{1}{2\pi}\right)^3 \int_{-\infty}^{\infty} d\omega \int d^2k_\perp(\epsilon_2 - \epsilon_3)\Gamma_{kk}(d, d, \mathbf{k}_\perp, \omega), \qquad (7.131)$$

which is seen to be the same as (7.126) in their units in which $\hbar = c = 1$. The Green dyadic $\overset{\leftrightarrow}{\Gamma}$ is determined by the classical Maxwell equations plus the boundary conditions at the dielectric interfaces, in much the same way as the determination of the fields in Section 7.2. It is therefore the same in Schwinger et al.'s source theory as in conventional classical or quantum electrodynamics.

7.5   Vacuum and Source Theories

In Section 4.10 we said that the possibility of attributing the radiative level shift to the *source* field of an atom was remarkable in view of the various ways in which the shift could be attributed so naturally to the *vacuum* field. The fact that the Casimir effect can be attributed to source fields is equally remarkable in view of its standard derivation as a vacuum field effect (Section 2.7). The reconciliation of these two viewpoints for the Lamb shift involved the commutativity of equal-time atom–field operators. This same commutativity explains also the possibility of deriving the Casimir force in terms of either vacuum or source fields, as we will now show.

Consider the symmetrically ordered form (Section 4.12) of the energy expectation value (7.92):

$$\begin{aligned}
\langle E \rangle &= -\frac{1}{2}\int d^3r\langle\frac{1}{2}\mathbf{P}\cdot[\mathbf{E}^{(+)} + \mathbf{E}^{(-)}] + \frac{1}{2}[\mathbf{E}^{(+)} + \mathbf{E}^{(-)}]\cdot\mathbf{P}\rangle \\
&= \langle E \rangle_{\text{VF}} + \langle E \rangle_{\text{S}} ,
\end{aligned} \qquad (7.132)$$

---

[19]Equations (7.129)–(7.131) correspond respectively to equations (2.7), (2.10), and (2.29) of Schwinger et al.

where

$$\langle E \rangle_{\rm VF} \equiv -\frac{1}{2}\int d^3r \langle \frac{1}{2}\mathbf{P}\cdot[\mathbf{E}_o^{(+)}+\mathbf{E}_o^{(-)}]+\frac{1}{2}[\mathbf{E}_o^{(+)}+\mathbf{E}_o^{(-)}]\cdot\mathbf{P}\rangle \quad (7.133)$$

and

$$\begin{aligned}\langle E \rangle_{\rm S} &\equiv -\frac{1}{2}\int d^3r\langle\frac{1}{2}\mathbf{P}\cdot[\mathbf{E}_s^{(+)}+\mathbf{E}_s^{(-)}]+\frac{1}{2}[\mathbf{E}_s^{(+)}+\mathbf{E}_s^{(-)}]\cdot\mathbf{P}\rangle\\ &= -\frac{1}{4}\int d^3r\langle\mathbf{P}\cdot\mathbf{E}_s+\mathbf{E}_s\cdot\mathbf{P}\rangle \end{aligned} \quad (7.134)$$

are vacuum-field and source-field contributions, respectively.

The source field operator $\mathbf{E}_s = \mathbf{E}_s^{(+)} + \mathbf{E}_s^{(-)}$ can be written as [see equations (7.100) and (7.105)]

$$\begin{aligned}\mathbf{E}_s(\mathbf{r},t) &= 8\pi\int d^3r'\int_0^t dt'\,\overset{\leftrightarrow}{G}(\mathbf{r},\mathbf{r}';t,t')\cdot\mathbf{P}(\mathbf{r}',t')\\ &= 4\int d^3r'\int_{-\infty}^{\infty}d\omega\,\overset{\leftrightarrow}{\Gamma}(\mathbf{r},\mathbf{r}',\omega)\cdot\int_0^t dt'\mathbf{P}(\mathbf{r}',t')e^{-i\omega(t-t')}\,, \end{aligned} \quad (7.135)$$

and so

$$\begin{aligned}\langle E \rangle_{\rm S} &= -\int d^3r\int d^3r'\int_{-\infty}^{\infty}d\omega\Gamma_{ij}(\mathbf{r},\mathbf{r}',\omega)\int_0^t dt'[\langle P_i(\mathbf{r},t)P_j(\mathbf{r}',t')\rangle\\ &\quad + \langle P_j(\mathbf{r}',t')P_i(\mathbf{r},t)\rangle]e^{-i\omega(t-t')}\,. \end{aligned} \quad (7.136)$$

Equations (7.112) and (7.115) then give

$$\begin{aligned}\langle E \rangle_{\rm S} &= -\int d^3r N(\mathbf{r})\int_{-\infty}^{\infty}d\omega\Gamma_{jj}(\mathbf{r},\mathbf{r},\omega)\frac{1}{3}\sum_k|\mathbf{d}_{kg}|^2\\ &\quad\times\int_0^t dt'[e^{i(\omega+\omega_{kg})(t'-t)}+e^{i(\omega-\omega_{kg})(t'-t)}]\\ &\to \int d^3r N(\mathbf{r})\int_{-\infty}^{\infty}d\omega\Gamma_{jj}(\mathbf{r},\mathbf{r},\omega)\frac{i}{3}\sum_k|\mathbf{d}_{kg}|^2\\ &\quad\times\left[\frac{1}{\omega+\omega_{kg}}+\frac{1}{\omega-\omega_{kg}}\right] \end{aligned} \quad (7.137)$$

as in (7.116). Then, since $\Gamma_{jj}(\mathbf{r},\mathbf{r},\omega)$ is an even function of ω,

$$\langle E \rangle_{\rm S} = 0 \quad (7.138)$$

when the field operators are symmetrically ordered.

To calculate $\langle E \rangle_{VF}$ we note that $\mathbf{E}_o$ is independent of the electron charge e, and so to obtain an energy to order e^2 we use

$$\mathbf{P}(\mathbf{r}, t) = N(\mathbf{r}) \int_0^\infty d\omega \alpha(\omega) [\mathbf{F}_o^{(+)}(\mathbf{r}, \omega) e^{-i\omega t} + \mathbf{F}_o^{(-)}(\mathbf{r}, \omega) e^{i\omega t}], \quad (7.139)$$

where $\mathbf{F}_o^{(\pm)}(\mathbf{r}, \omega)$ are defined by writing

$$\mathbf{E}_o^{(\pm)}(\mathbf{r}, t) = \int_0^\infty d\omega \mathbf{F}_o^{(\pm)}(\mathbf{r}, \omega) e^{\mp i\omega t} \quad (7.140)$$

and α is again the polarizability. In other words, we will use in (7.133) the polarization induced by the vacuum field:

$$\langle E \rangle_{VF} = -\frac{1}{2} \int d^3 r N(\mathbf{r}) \int_0^\infty d\omega \alpha(\omega) \langle \mathbf{F}_o^{(+)}(\mathbf{r}, \omega) \cdot \mathbf{F}_o^{(-)}(\mathbf{r}, \omega) \rangle . \quad (7.141)$$

Now equations (7.100), (7.96), (7.140), and (7.107) imply

$$\overleftrightarrow{\Gamma}(\mathbf{r}, \mathbf{r}', \omega) = \frac{i}{4\hbar} \langle \mathbf{F}_o^{(+)}(\mathbf{r}, \omega) \mathbf{F}_o^{(-)}(\mathbf{r}', \omega) \rangle \quad (7.142)$$

and

$$\Gamma_{jj}(\mathbf{r}, \mathbf{r}', \omega) = \frac{i}{4\hbar} \langle \mathbf{F}_o^{(+)}(\mathbf{r}, \omega) \cdot \mathbf{F}_o^{(-)}(\mathbf{r}', \omega) \rangle, \quad (7.143)$$

which allows us to write (7.141) as

$$\begin{aligned} \langle E \rangle_{VF} &= 2i\hbar \int d^3 r N(\mathbf{r}) \int_0^\infty d\omega \alpha(\omega) \Gamma_{jj}(\mathbf{r}, \mathbf{r}, \omega) \\ &= -\frac{\hbar}{2\pi} \int d^3 r \int_0^\infty d\xi [\epsilon(\mathbf{r}, i\xi) - 1] \Gamma_{jj}(\mathbf{r}, \mathbf{r}, i\xi). \quad (7.144) \end{aligned}$$

This is identical to (7.120) and therefore leads to the same force when we consider a small change in the dielectric constant.

We can summarize the situation here as follows. In the preceding section we derived the Casimir force by considering the change in the dipole energy (7.92) due to the field produced by the same dipoles. We were led to the expression (7.120) for the variation in this energy associated with an infinitesimal change in the distance d between the plates, and from this variation we obtained the Lifshitz expression for the force. This calculation was based on a normal ordering of field operators, in which case there is no explicit contribution from the vacuum field, and the force could be attributed solely to source fields. In the calculation leading to (7.144), however, we employed a symmetrical ordering in which there is no explicit contribution from *source* fields. The result for the change in the dipole energy due to the interaction of the dipoles of the medium with the source-free, vacuum field agrees exactly with the energy variation (7.120), and therefore implies exactly the same force. The agreement between the two approaches is a consequence of the commutativity of equal-time atom and field operators, as discussed in Chapter 4 in the context of spontaneous emission and the Lamb shift. There, as in the case of the Casimir force, a normal ordering led to the interpretation of the Lamb shift in terms of the source field, whereas a symmetric ordering of field operators led just as naturally to the interpretation in terms of the vacuum field.

7.6   Discussion

The late 1940s were seminal years for the development of our present concept of the quantum vacuum. The Lamb–Retherford experiments and the interpretation of the Lamb shift as an effect of zero-point electromagnetic field fluctuations indicated that the vacuum field is a real physical entity with *observable* consequences. It was precisely Planck's feeling that zero-point energy in particular would have no observable consequences that led to his abandonment of the concept four decades earlier, and later Pauli, among others, had the "gravest hesitations" against zero-point energy[20] (Enz, 1974).

At the same time and, remarkably enough, completely independent of attempts to physically interpret the Lamb shift,[21] Casimir interpreted the retarded van der Waals interaction in terms of zero-point field energy, and this then led to his prediction of the Casimir force between conducting plates. This effect has been called "startling" (DeWitt, 1989), and Schwinger et al. (1978), for instance, have referred to it as "one of the least intuitive consequences of quantum electrodynamics." As remarkable as the effect itself, perhaps, is the extent to which interest in it has endured over nearly half a century; indeed, this interest seems if anything to be increasing as the importance of the quantum vacuum is becoming more and more a part of the mindset of physicists working in various fields. In the next chapter we will describe aspects of the Casimir force that were investigated only many years after its discovery, as well as some of the experimental

---

[20] In his autobiography Casimir (1983) recalls that, " ... I explained to [Pauli] my results on van der Waals forces and their relation to field fluctuations in empty space. He began by bluntly telling me it was all nonsense, but was obviously amused when I did not give in. Finally, after I had countered all his arguments, he agreed ... "

[21] H. B. G. Casimir, private communication, 12 March 1992.

work on Casimir effects.

Whether the Casimir force should be regarded as startling or nonintuitive is, of course, arguable. If we regard it is a macroscopic manifestation of van der Waals forces between molecules, there is hardly any reason for surprise. Consider again the potential energy (7.4) for a neutral polarizable particle at a distance d from a dielectric wall. Using the Casimir–Polder results $\gamma = 7$ and $B = 23\hbar c\alpha^2/4\pi$ for the intermolecular potential, we obtain

$$
\begin{aligned}
V(d) &= -\frac{2\pi}{(5)(4)}\frac{1}{d^4}N_1\frac{23\hbar c\alpha^2}{4\pi} = -\frac{23}{40}\frac{\alpha\hbar c}{d^4}N_1\alpha \\
&= -\frac{69}{160\pi}\frac{\alpha\hbar c}{d^4}\frac{\epsilon-1}{\epsilon+1} ,
\end{aligned}
\tag{7.145}
$$

where we have employed the Clausius–Mosotti relation, $(\epsilon - 1)/(\epsilon + 1) = 4\pi N_1\alpha/3$, between the dielectric constant ϵ and the polarizability α of the individual molecules. In the limit $\epsilon \to \infty$ of a perfect conductor, therefore,

$$
V(d) \to -\frac{69}{160\pi}\frac{\alpha\hbar c}{d^4} ,
\tag{7.146}
$$

in reasonably good agreement with the Casimir–Polder result (3.95). In the same fashion we have, from (7.8),

$$
F(d) = -\frac{23\hbar c}{40d^4}\left(\frac{3}{4\pi}\right)^2\left(\frac{\epsilon-1}{\epsilon+1}\right)^2 \to -\frac{23\hbar c}{40d^4}\left(\frac{3}{4\pi}\right)^2
\tag{7.147}
$$

in the limit of perfect conductors. This is about 80% of the Casimir result (7.1). In other words, we can obtain fairly reasonable estimates of these Casimir effects by simply adding up the *pairwise* intermolecular forces.

On the other hand, the way in which Casimir derived the force between perfect conductors would certainly command the attention of anyone who believed that zero-point energy is merely an additive constant to a Hamiltonian and can never be of any physical consequence [recall the remarks following equation (2.87)].

The calculation of the Casimir force in terms of changes in zero-point electromagnetic energy seems so natural that, as noted earlier, the Casimir effect has, with few exceptions (Schwinger et al., 1978; Milonni, 1982), been regarded as "proof" for the reality of vacuum, zero-point field energy. We have shown, however, that, as in the case of the Lamb shift, the interpretation of the Casimir force in terms of the vacuum field is largely a matter of taste: underlying this interpretation is a particular and arbitrary choice of ordering of field operators. Different orderings reveal that the vacuum-field picture is only one of many ways to describe the effect, and in particular a normal ordering allows us to attribute the Casimir force entirely to *source* fields. Indeed, Schwinger et al. (1978) have been able to derive the Casimir force from the standpoint of a theory in which, contrary to prevailing ideas, there are no nontrivial vacuum fields.

Why has it taken so long to recognize that the Casimir effect and other vacuum field effects have equivalent derivations in terms of source fields? We have no ready explanation for this circumstance. According to Jaynes (1978), "For many years, starting with Einstein's relation between diffusion coefficient and mobility, theoreticians have been discovering a steady stream of close mathematical connections between stochastic problems and dynamical problems. It has taken us a long time to recognize that QED was just another example of this."

7.7 Bibliography

Agarwal, G. S., "Quantum Electrodynamics in the Presence of Dielectrics and Conductors. II. Theory of Dispersion Forces," *Phys. Rev.* A11, 243 (1975).

Barash, Yu. S., "Energy of an Equilibrium Fluctuational Electromagnetic Field in a Medium," *Radiofizika* 16, 1086 (1973).

Barash, Yu. S. and V. F. Ginzburg, "Electromagnetic Fluctuations in Matter and Molecular (van der Waals) Forces Between Them," *Sov. Phys. Usp.* 18, 305 (1975).

Barton, G., "Some Surface Effects in the Hydrodynamical Model of Metals," *Rep. Prog. Phys.* 42, 963 (1979).

Casimir, H. B. G. and D. Polder, "The Influence of Retardation on the London-van der Waals Forces," *Phys. Rev.* 73, 360 (1948).

Casimir, H. B. G., "On the Attraction Between Two Perfectly Conducting Plates," *Proc. Kon. Ned. Akad. Wetenschap* 51, 793 (1948).

Casimir, Hendrik, *Haphazard Reality* (Harper and Row, New York, 1983), p. 247.

DeWitt, B., "The Casimir Effect in Field Theory," in *Physics in the Making*, ed. A. Sarlemijn and M. J. Sparnaay (Elsevier, Amsterdam, 1989).

Enz, C. P., "Is the Zero-Point Energy Real?," in *Physical Reality and Mathematical Description*, ed. C. P. Enz and J. Mehra (Reidel, Dordrecht, 1974).

Gerlach, E., "Equivalence of van der Waals Forces Between Solids and the Surface-Plasmon Interaction," *Phys. Rev.* B4, 393 (1971).

Ginzburg, V. F., *Theoretical Physics and Astrophysics* (Pergamon Press,

Oxford, 1070), Chapter 13.

Hargreaves, C. M., "Corrections to the Retarded Dispersion Force Between Metal Bodies," *Proc. Kon. Ned. Akad. Wetensch.* **68**B, 231 (1965).

Hinds, E. A., C. I. Sukenik, M. G. Boshier, and D. Cho, "Deflection of an Atomic Beam by the Casimir Force," in *Atomic Physics 12*, ed. J. Zorn and R. Lewis (American Institute of Physics, New York, 1991).

Itzykson, C. and J.-B. Zuber, *Quantum Field Theory* (McGraw-Hill, New York, 1980).

Jaynes, E. T., "Electrodynamics Today," in *Coherence and Quantum Optics IV*, ed. L. Mandel and E. Wolf (Plenum Press, New York, 1978).

Langbein, D., "The Macroscopic Theory of van der Waals Attraction," *Solid State Commun.* **12**, 853 (1973).

Lifshitz, E. M., "The Theory of Molecular Attractive Forces Between Solids," *Sov. Phys. JETP* **2**, 73 (1956).

Milonni, P. W., "Radiation Reaction and the Nonrelativistic Theory of the Electron," *Phys. Lett.* **82**A, 225 (1981).

Milonni, P. W., "Casimir Forces Without the Vacuum Radiation Field," *Phys. Rev.* A**25**, 1315 (1982).

Milonni, P. W. and M.-L. Shih, "Source Theory of Casimir Force," *Phys. Rev.* A**45**, 4241 (1992).

Schram, K., "On the Macroscopic Theory of Retarded van der Waals Forces," *Phys. Lett.* **43**A, 282 (1973).

Schwinger, J., L. L. DeRaad, Jr., and K. A. Milton, "Casimir Effect in Dielectrics," Ann. Phys. (New York) **115**, 1 (1978).

Sparnaay, M. J., "The Historical Background of the Casimir Effect," in *Physics in the Making*, ed. A. Sarlemijn and M. J. Sparnaay (Elsevier, Amsterdam, 1989).

Van Kampen, N. G., B. R. A. Nijboer, and K. Schram, "On the Macroscopic Theory of van der Waals Forces," *Phys. Lett.* **26**A, 307 (1968).

Verwey, E. J. W. and J. T. G. Overbeek, *Theory of the Stability of Lyophobic Colloids* (Elsevier, Amsterdam, 1948).

Chapter 8

Casimir and van der Waals Forces: Elaborations

> ... of special interest and difficulty is the process which takes place in a physical body when many molecules interact simultaneously, the oscillations of the latter being interdependent owing to their proximity. If the solution of this problem ever becomes possible we shall be able to calculate in advance the values of the intermolecular forces due to intermolecular interradiation, deduce the laws of their temperature dependence, solve the fundamental problem of molecular physics whether all so-called "molecular forces" are confined to the already known mechanical action of light radiation, to electromagnetic forces, or whether some forces of hitherto unknown origin are involved ...
>
> – P. N. Lebedev [*Wied. Ann.* **52**, 621 (1894)][1]

8.1 Introduction

In the preceding chapter we discussed several theoretical approaches to the van der Waals Casimir forces between dielectrics, and the connections among them. In this chapter we will address a few further aspects of these forces, as well as some of the experiments that have been carried out to test the theoretical predictions. We will also consider the possibility of deriving

---

[1] Quotation from Derjaguin and Abrikosova, 1957.

vacuum electromagnetic effects such as the van der Waals–Casimir forces from the standpoint of a purely classical theory of the vacuum field called *stochastic electrodynamics*.

8.2   Nonadditivity of Dispersion Forces

As noted in Chapter 7, the van der Waals dispersion forces[2] are not additive: the force between two molecules depends in general on the presence of other molecules. To understand the origin of this nonadditivity, we begin with the formula [cf. equation (7.92)]

$$\langle E \rangle = -\frac{1}{2}\langle \mathbf{p} \cdot \mathbf{E}(\mathbf{r}, t) \rangle \tag{8.1}$$

for the expectation value of the potential energy of an induced dipole $\mathbf{p}$ at point $\mathbf{r}$ in an electric field $\mathbf{E}$. The quantized electric field has the form

$$\mathbf{E}(\mathbf{r}, t) = i \sum_\beta (2\pi\hbar\omega_\beta)^{1/2} a_\beta(0) \mathbf{A}_\beta(\mathbf{r}) e^{-i\omega_\beta t} + \text{h.c.}, \tag{8.2}$$

where $a_\beta(0)e^{-i\omega_\beta t}$ is the source-free, Heisenberg-picture photon annihilation operator for mode β of the field, with associated (c-number) mode function $\mathbf{A}_\beta(\mathbf{r})$. The dipole moment (operator) induced by $\mathbf{E}$ is

$$\mathbf{p}(t) = i \sum_\beta (2\pi\hbar\omega_\beta)^{1/2} \alpha(\omega_\beta) a_\beta(0) \mathbf{A}_\beta(\mathbf{r}) e^{-i\omega_\beta t} + \text{h.c.}, \tag{8.3}$$

where, for frequencies ω_β away from any absorption resonances, the polarizability $\alpha(\omega_\beta)$ may be assumed to be real. Equations (8.1)–(8.3), together with the vacuum field expectation values $\langle a_\beta(0)a_{\beta'}(0)\rangle = \langle a_\beta^\dagger(0)a_{\beta'}(0)\rangle = 0, \langle a_\beta(0)a_{\beta'}^\dagger(0)\rangle = \delta_{\beta\beta'}$, imply

$$\langle E \rangle = -\frac{1}{2} \sum_\beta (2\pi\hbar\omega_\beta)\alpha(\omega_\beta)|\mathbf{A}_\beta(\mathbf{r})|^2. \tag{8.4}$$

Note that we are assuming for simplicity here an isotropic polarizability, i.e., $\alpha_{ij} = \alpha\delta_{ij}$, as is the case for atoms but not necessarily for molecules.

In Chapter 7 it was shown that van der Waals dispersion forces, and their macroscopic (Casimir) manifestations, can be obtained from either source fields or source-free fields, depending on how field annihilation and

---

[2] Recall the discussion of the dispersion interaction in Section 3.11.

creation operators are ordered. Here we are taking the source-free (vacuum-field) approach, corresponding to a symmetric ordering of annihilation and creation operators. This ordering is already implicit in equation (8.1), since **E** is the (symmetric) sum of annihilation and creation parts. However, whichever approach we adopt, we must use the appropriate mode functions $\mathbf{A}_\beta(\mathbf{r})$ for the field. For a collection of atoms, this means that the $|\mathbf{A}_\beta(\mathbf{r})|^2$ appearing in equation (8.4) for the energy of an atom at **r** *must account for the presence of all the other atoms*. That is, the quantization of the field as in equation (8.2) must be performed subject to the presence of all the atoms.

The mode functions $\mathbf{A}_\beta(\mathbf{r})$ in the presence of polarizable matter are determined by purely *classical* electromagnetic considerations. Let $\mathbf{A}_\beta^{(o)}(\mathbf{r})$ be a mode function corresponding to frequency ω_β in the absence of any particles:

$$\mathbf{A}_\beta^{(o)}(\mathbf{r}) = \frac{1}{\sqrt{V}} e^{i\mathbf{k}_\beta \cdot \mathbf{r}} \mathbf{e}_\beta , \tag{8.5}$$

or

$$\mathbf{A}_\beta^{(o)}(\mathbf{r}) \rightarrow \mathbf{A}_{\mathbf{k}\lambda}^{(o)}(\mathbf{r}) = \frac{1}{\sqrt{V}} \mathbf{e}_{\mathbf{k}\lambda} e^{i\mathbf{k}\cdot\mathbf{r}} , \tag{8.6}$$

where V is a quantization volume and $\mathbf{e}_{\mathbf{k}\lambda}$, $\lambda = 1, 2$, is a polarization unit vector. The form $e^{i\mathbf{k}\cdot\mathbf{r}}$ in this case is dictated by the requirement that the energy (8.4) for an atom in an otherwise perfect vacuum must be independent of the position **r** of the atom. The equation (8.4) then yields the nonrelativistic expression for the Lamb shift (Section 3.8).

To determine the modification of the mode functions (8.6) due to the presence of identical atoms at the positions $\mathbf{r}_j$, we use the superposition principle for electromagnetic fields, which in this case states that the total field $\mathbf{A}_{\mathbf{k}\lambda}(\mathbf{r})$ at **r** corresponding to the mode $(\mathbf{k}, \lambda)$ is the unperturbed field $\mathbf{A}_{\mathbf{k}\lambda}^{(o)}(\mathbf{r})$ plus the fields from all the atoms. The field from each atom is a dipole field associated with the dipole moment induced by the total field $\mathbf{A}_{\mathbf{k}\lambda}(\mathbf{r}_j)e^{-i\omega_k t}$ at that atom. This dipole moment is $\alpha(\omega_k)\mathbf{A}_{\mathbf{k}\lambda}(\mathbf{r}_j)e^{-i\omega_k t}$. Thus

$$\mathbf{A}_{\mathbf{k}\lambda}(\mathbf{r}) = \mathbf{A}_{\mathbf{k}\lambda}^{(o)}(\mathbf{r}) + \sum_j \alpha(\omega_k)\nabla \times \nabla \times \frac{\mathbf{A}_{\mathbf{k}\lambda}(\mathbf{r}_j)e^{ik|\mathbf{r}-\mathbf{r}_j|}}{|\mathbf{r} - \mathbf{r}_j|} , \tag{8.7}$$

where we have used the fact that the electric field produced by a dipole moment $\mathbf{p}(t)$ at $\mathbf{r}_j$ is (Born and Wolf, 1970)

$$\mathbf{E}(\mathbf{r}, t) = \nabla \times \nabla \times \frac{\mathbf{p}(t - |\mathbf{r} - \mathbf{r}_j|/c)}{|\mathbf{r} - \mathbf{r}_j|} . \tag{8.8}$$

(Here and in what follows, $\nabla \times \nabla \times \mathbf{A}$ means $\nabla \times [\nabla \times \mathbf{A}]$ for any vector field $\mathbf{A}$.) The field at the atom $\mathbf{r}_i$ is given by

$$\mathbf{A}_{\mathbf{k}\lambda}(\mathbf{r}_i) = \mathbf{A}_{\mathbf{k}\lambda}^{(o)}(\mathbf{r}_i) + \sum_{j \neq i} \alpha(\omega_k) \nabla_i \times \nabla_i \times \frac{\mathbf{A}_{\mathbf{k}\lambda}(\mathbf{r}_j) e^{ikr_{ij}}}{r_{ij}} , \qquad (8.9)$$

where $r_{ij} \equiv |\mathbf{r}_i - \mathbf{r}_j|$, and the solution of this (multiple scattering) equation then yields the potential energy

$$\langle E_i \rangle = -\frac{1}{2} \sum_{\beta} (2\pi \hbar \omega_\beta) \alpha(\omega_\beta) |\mathbf{A}_\beta(\mathbf{r}_i)|^2 \qquad (8.10)$$

for an atom at $\mathbf{r}_i$.

Now in the lowest order of approximation $\mathbf{A}_{\mathbf{k}\lambda}(\mathbf{r}_i) \cong \mathbf{A}_{\mathbf{k}\lambda}^{(o)}(\mathbf{r}_i)$ in (8.9) and then, as already noted, equation (8.10) gives the nonrelativistic Lamb shift. In the next order of approximation

$$\mathbf{A}_{\mathbf{k}\lambda}^{(1)}(\mathbf{r}_i) \cong \mathbf{A}_{\mathbf{k}\lambda}^{(o)}(\mathbf{r}_i) + \sum_{j \neq i} \alpha(\omega_k) \nabla_i \times \nabla_i \times \frac{\mathbf{A}_{\mathbf{k}\lambda}^{(o)}(\mathbf{r}_j) e^{ikr_{ij}}}{r_{ij}} \qquad (8.11)$$

and the $\mathbf{r}_i$-dependent part of (8.10) is

$$
\begin{aligned}
\langle E_i^{(2)} \rangle &= -\frac{1}{2} \sum_{\mathbf{k}\lambda} (2\pi \hbar \omega_k) \alpha(\omega_k) |\mathbf{A}_{\mathbf{k}\lambda}^{(1)}(\mathbf{r}_i)|^2 \\
&\cong -\mathrm{Re} \sum_{\mathbf{k}\lambda} (2\pi \hbar \omega_k) \alpha^2(\omega_k) \sum_{j \neq i} \mathbf{A}_{\mathbf{k}\lambda}^{(o)}(\mathbf{r}_i)^* \cdot \nabla_i \times \nabla_i \times \\
&\qquad \frac{\mathbf{A}_{\mathbf{k}\lambda}^{(o)}(\mathbf{r}_j) e^{ikr_{ij}}}{r_{ij}} \\
&= -\frac{2\pi \hbar}{V} \sum_{j \neq i} \mathrm{Re} \sum_{\mathbf{k}\lambda} k^3 \omega_k \alpha^2(\omega_k) e^{-i\mathbf{k}\cdot\mathbf{r}_{ij}} e^{ikr_{ij}} \\
&\qquad \times \left([1 - (\mathbf{e}_{\mathbf{k}\lambda} \cdot \mathbf{s}_{ij})^2] \frac{1}{kr_{ij}} + [3(\mathbf{e}_{\mathbf{k}\lambda} \cdot \mathbf{s}_{ij})^2 - 1] \right. \\
&\qquad \times \left. \left[\frac{1}{(kr_{ij})^3} - \frac{i}{(kr_{ij})^2} \right] \right) \qquad (8.12)
\end{aligned}
$$

to second order in $\alpha(\omega_k)$. Here $\mathbf{s}_{ij}$ is the unit vector pointing from atom i to atom j, $\mathbf{s}_{ij} = \mathbf{r}_{ij}/r_{ij}$. We obtain[3]

$$\langle E_i^{(2)} \rangle = -\frac{\hbar}{\pi c^6} \sum_{j \neq i} \int_0^\infty d\omega \omega^6 \alpha^2(\omega) G(\omega r_{ij}/c), \qquad (8.13)$$

---

[3] See equation (3.82) and the discussion following it.

$$G(x) \equiv \frac{\sin 2x}{x^2} + \frac{2\cos 2x}{x^3} - \frac{5\sin 2x}{x^4} - \frac{6\cos 2x}{x^5} + \frac{3\sin 2x}{x^6} . \tag{8.14}$$

For small r_{ij} (see note 2),

$$\langle E_i^{(2)} \rangle \cong -\frac{\hbar}{\pi c^6} \sum_{j \neq i} \frac{3c^6}{r_{ij}^6} \int_0^\infty d\omega\, \alpha^2(\omega) \sin\frac{2\omega r_{ij}}{c}$$

$$= -\sum_{j \neq i} \frac{3\hbar}{\pi r_{ij}^6} \int_0^\infty du\, \alpha^2(iu) e^{-2ur_{ij}/c}$$

$$= -\sum_{j \neq i} \frac{3\hbar}{\pi r_{ij}^6} \left(\frac{2}{3\hbar}\right)^2 \sum_m \sum_p \omega_{mg}\omega_{pg} |d_{mg}|^2 |d_{pg}|^2$$

$$\times \int_0^\infty \frac{du\, e^{-2ur_{ij}/c}}{(u^2 + \omega_{mg}^2)(u^2 + \omega_{pg}^2)} , \tag{8.15}$$

where we have used the expression

$$\alpha(\omega) = \frac{2}{3\hbar} \sum_m \frac{\omega_{mg} |d_{mg}|^2}{\omega_{mg}^2 - \omega^2} \tag{8.16}$$

for the polarizability of an atom in the ground state $|g\rangle$, where ω_{mg} and d_{mg} are the $m \to g$ (angular) transition frequency and the electric dipole moment, respectively. If $r_{ij} \ll c/\omega_{mg}$ for all transitions $m \leftrightarrow g$, we may replace $e^{-2ur_{ij}/c}$ by 1 in (8.15), and this gives the r_{ij}^{-6} form of the van der Waals interaction derived by London. If we assume furthermore that one particular transition is dominant, then

$$\langle E_i^{(2)} \rangle \cong -\sum_{j \neq i} \frac{3\hbar}{\pi r_{ij}^6} \left(\frac{2}{3\hbar}\right)^2 \omega_o^2 |d|^4 \int_0^\infty \frac{du}{(u^2 + \omega_o^2)^2}$$

$$= -\sum_{j \neq i} \frac{3\hbar\omega_o \alpha^2}{4r_{ij}^6} , \tag{8.17}$$

where ω_o and d correspond to the dominant transition and $\alpha = \frac{2}{3}|d|^2/\hbar\omega_o$ is the static ($\omega = 0$) polarizability in the two-level approximation. This is the London result (see note 2).

For large separations ($r_{ij} \gg 137a_o$, where a_o is the Bohr radius) we can approximate (8.13) by

$$\langle E_i^{(2)} \rangle \cong -\frac{\hbar}{\pi c^6} \sum_{j \neq i} \alpha^2 \int_0^\infty d\omega\, \omega^6 G(\frac{\omega r_{ij}}{c}) = -\sum_{j \neq i} \frac{23\hbar c\alpha^2}{4\pi r_{ij}^7} , \tag{8.18}$$

which is the well known, long range form of the dispersion interaction obtained by Casimir and Polder (see note 2).

These results, based on the approximation (8.11), give $\langle E_i^{(2)} \rangle$ as a sum of *pairwise* ("additive") interactions.

In the next order of approximation we replace (8.9) by

$$
\begin{aligned}
\mathbf{A}_{\mathbf{k}\lambda}^{(2)}(\mathbf{r}_i) \;\cong\;& \mathbf{A}_{\mathbf{k}\lambda}^{(o)}(\mathbf{r}_i) + \sum_{j \neq i} \alpha(\omega_k) \nabla_i \times \nabla_i \times \frac{\mathbf{A}_{\mathbf{k}\lambda}^{(1)}(\mathbf{r}_j) e^{ik r_{ij}}}{r_{ij}} \\[2mm]
=\;& \mathbf{A}_{\mathbf{k}\lambda}^{(o)}(\mathbf{r}_i) + \sum_{j \neq i} \alpha(\omega_k) \nabla_i \times \nabla_i \times \frac{\mathbf{A}_{\mathbf{k}\lambda}^{(o)}(\mathbf{r}_j) e^{ik r_{ij}}}{r_{ij}} \\[2mm]
+\;& \sum_{j \neq i} \sum_{p \neq j} \alpha^2(\omega_k) \nabla_i \times \nabla_i \times \left[\frac{e^{ik r_{ij}}}{r_{ij}} \nabla_j \times \nabla_j \times \frac{\mathbf{A}_{\mathbf{k}\lambda}^{(o)}(\mathbf{r}_p) e^{ik r_{jp}}}{r_{jp}} \right]
\end{aligned}
$$

$$(8.19)$$

and obtain a correction to $\langle E_i \rangle$ that is of third order in the polarizability. This correction is associated with nonadditive, three-body contributions to $\langle E_i \rangle$. Although the calculation of the detailed form of the nonadditive contributions is somewhat complicated, we can understand their general form from (8.19). $\nabla_i \times \nabla_i$ and $\nabla_j \times \nabla_j$ give rise to (near-field) terms varying as r_{ij}^{-3} and r_{jp}^{-3}, respectively. Cross products of the form $\mathbf{A}_{\mathbf{k}\lambda}^{(o)}(\mathbf{r}_i)^* \cdot \mathbf{A}_{\mathbf{k}\lambda}^{(o)}(\mathbf{r}_p) \propto e^{-i\mathbf{k}\cdot\mathbf{r}_{ip}}$ give rise similarly to terms varying as r_{ip}^{-3} after the integration over all solid angles about $\mathbf{k}$ is carried out as required by (8.10). Then we obtain a three-body (nonretarded) contribution to $\langle E_i \rangle$ of the form

$$
\langle E_{ijp}^{(3)} \rangle \propto \frac{\alpha^3}{r_{ij}^3 r_{jp}^3 r_{ip}^3} \tag{8.20}
$$

associated with the three-atom triplet i, j, p. Detailed forms of such nonadditive, nonpairwise interaction energies have been derived from standard perturbation theory by Axilrod and Teller (1943), and, more recently, by Power and Thirunamachandran (1985) using the Heisenberg picture. These authors note that $\langle E_{ijp}^{(3)} \rangle$ may be attractive or repulsive, depending on the geometrical arrangement of the atoms.

The three-body interaction is roughly a factor α/r^3 smaller than the usual (two-body) van der Waals interaction, where r is a characteristic interatomic spacing. This point will play an important role in Section 8.5.

8.3   Extinction Theorem

Formally, assuming the polarizability $\alpha(\omega)$ is known exactly, the main problem in the calculation of dispersion forces is the solution of the self-consistent scattering equation (8.9). One approach to the solution of this equation is to assume a continuous, uniform distribution of N atoms per unit volume. In this macroscopic approach (8.9) is replaced by the integro-differential equation

$$\mathbf{A}'_{\mathbf{k}\lambda}(\mathbf{r}) = \mathbf{A}^{(o)}_{\mathbf{k}\lambda}(\mathbf{r}) + N\alpha(\omega_k) \int d^3r' \nabla \times \nabla \times \frac{\mathbf{A}'_{\mathbf{k}\lambda}(\mathbf{r}')e^{ik|\mathbf{r}-\mathbf{r}'|}}{|\mathbf{r}-\mathbf{r}'|} . \quad (8.21)$$

Here it is to be understood that a small volume about $\mathbf{r}$ must be excluded from the integration, owing to the restriction $j \neq i$ in (8.9). As it stands equation (8.21) is also satisfied by the mode functions at points $\mathbf{r}$ *outside* the region occupied by the atoms, this equation being a general statement of the superposition principle. For such points $|\mathbf{r}-\mathbf{r}'|$ cannot vanish and consequently there is no restriction on the integration in (8.21).

The Ewald–Oseen extinction theorem (Ewald, 1912, 1916; Oseen, 1915; Born and Wolf, 1970) for isotropic or crystalline media states that the integral in (8.21) has two parts, one of which satisfies the wave equation in vacuum and exactly cancels ("extinguishes") the incident field $A^{(o)}_{\mathbf{k}\lambda}$.[4] The other part satisfies the wave equation with propagation velocity $c/n(\omega_k)$, where $n(\omega_k)$ is the refractive index and is related to the polarizability through the relation

$$\frac{4\pi}{3}N\alpha(\omega) = \frac{n^2(\omega)-1}{n^2(\omega)+2} = \frac{\epsilon(\omega)-1}{\epsilon(\omega)+2} . \quad (8.22)$$

In the proof of the theorem (Born and Wolf, 1970) it is shown that $\mathbf{A}'_{\mathbf{k}\lambda}$ is related to the average field $\mathbf{A}_{\mathbf{k}\lambda}$ by $\mathbf{A}'_{\mathbf{k}\lambda} - \mathbf{A}_{\mathbf{k}\lambda} + 4\pi\mathbf{P}/3 = [n^2(\omega_k) + 2]\mathbf{A}_{\mathbf{k}\lambda}/3$, so that (8.21) may be written as

$$\frac{1}{3}[\epsilon(\omega_k)+2]\mathbf{A}_{\mathbf{k}\lambda}(\mathbf{r}) = \mathbf{A}^{(o)}_{\mathbf{k}\lambda}(\mathbf{r}) + \frac{1}{4\pi}[\epsilon(\omega_k)-1] \int d^3r' \nabla \times \nabla \times \frac{\mathbf{A}_{\mathbf{k}\lambda}(\mathbf{r}')e^{ik|\mathbf{r}-\mathbf{r}'|}}{|\mathbf{r}-\mathbf{r}'|} ,$$
$$(8.23)$$

or equivalently

$$\nabla \times \nabla \times \mathbf{A}_{\mathbf{k}\lambda} + \frac{\omega^2}{c^2}\epsilon(\omega)\mathbf{A}_{\mathbf{k}\lambda} = 0. \quad (8.24)$$

---

[4] See also Wolf (1976) and references therein; Hynne and Bullough (1990) and references therein.

The fact that the continuum limit of equation (8.9) leads via the extinction theorem to the wave equation (8.24) plus appropriate (macroscopic) boundary conditions is of course well known (Born and Wolf, 1970), but does not appear to have been previously stated in the context of van der Waals forces. This is somewhat unfortunate, for the extinction theorem provides a foundation for the macroscopic theory of van der Waals forces due originally to Lifshitz, which, as we have noted in Chapter 7, is rather complicated and has sometimes been subject to doubts as to its validity. In the present version of the theory, the macroscopic approach to van der Waals forces reduces to (a) the solution of the *classical* Maxwell boundary-value problem (8.24), and (b) the evaluation of the energy (8.10). An example of this procedure, which underlies the theory of "cavity QED," for instance (Section 6.3), is given in the following section.

There is a great deal more that can be said about the extinction theorem and the microscopic theory of dispersion. We restrict ourselves to two remarks. First, we note that $A'_{k\lambda}$ is the Lorentz–Lorenz *local* field, as opposed to the *average* field $A_{k\lambda}$. The difference arises essentially from the fact that the dipoles (atoms) of the medium are not in fact continuously distributed, but have spaces between them. For nonisotropic media the relation between the local and average fields is generally slightly different from that given previously. Textbooks usually do not indicate that there is not always a practical difference between the local and average fields; loosely speaking, the local field correction arises only when the particles are in fact spatially localized. Thus, whereas there is a local field correction in a dielectric gas or in impurity atom absorption in a host crystal, there is essentially no correction necessary for a plasma or for the conduction electrons of a metal, and usually none necessary for exciton transitions in pure crystals.[5]

We note also that the extinction theorem may be regarded as a nonlocal boundary condition that the field must satisfy (Wolf, 1976). The cancellation of the incident field is often regarded as "caused by the dipoles on the *boundary* of the medium,"[6] because in the classical macroscopic approach the term that cancels the incident field can be cast in the form of a surface integral over dipole sources. Of course, from a microscopic viewpoint, *all* the dipoles act to cancel the incident field and produce the reflected and transmitted fields. A simple, fully quantum-mechanical model shows that the cancellation is effectively due to dipoles within a depth at the surface

---

[5] See, for instance, P. Nozières and D. Pines, *Phys. Rev.* **109**, 762 (1958), and references therein especially to the work of C. G. Darwin.

[6] J. D. Jackson, *Classical Electrodynamics*, 2nd ed. (Wiley, New York, 1975), p. 513.

approximately equal to the field wavelength λ.[7] In particular, "The [reflected] radiation comes from everywhere in the interior, but it turns out that the total effect is equivalent to a reflection from the surface."[8]

8.4   Latent Heat: Macroscopic Theory

Latent heat, the energy required to vaporize a unit mass of material without a change of temperature, is a consequence of intermolecular attractive forces. For many substances, including liquid helium, inert gas solids, and many organic crystals, latent heat results primarily from dispersion forces. In this section we will consider a macroscopic theory of latent heat following ideas due originally to Schwinger, DeRaad, and Milton (1978).

We will consider first the case of an atom at a distance $z > 0$ from a half-space filled with identical atoms. The half-space ($z \leq 0$) will be treated macroscopically according to the wave equation (8.24) plus the boundary conditions at the interface $z = 0$ between vacuum and the half-space with dielectric constant $\epsilon(\omega)$.

For an atom at $z > 0$ we require, according to (8.10), the mode functions in the vacuum to the right of the dielectric half-space. Consider a plane-wave electric field

$$A_{\mathbf{k}\lambda 1}^{(I)}(\mathbf{r}) \equiv \frac{1}{\sqrt{2V}} \mathbf{e}_{\mathbf{k}\lambda} e^{i\mathbf{k}\cdot\mathbf{r}} \tag{8.25}$$

incident on the interface from the right. Such an incident field leads to a reflected field

$$A_{\mathbf{k}\lambda 1}^{(R)}(\mathbf{r}) \equiv \frac{1}{\sqrt{2V}} \mathbf{e}_{\mathbf{k}\lambda} \frac{k_3 - k_3'}{k_3 + k_3'} e^{i\mathbf{k}^{(R)}\cdot\mathbf{r}} , \tag{8.26}$$

where

$$\mathbf{k} = (k_1, k_2, k_3), \tag{8.27}$$

$$\mathbf{k}^{(R)} = (k_1, k_2, -k_3), \tag{8.28}$$

$$k_3 = [\frac{\omega^2}{c^2} - k_1^2 - k_2^2]^{1/2}, \tag{8.29}$$

$$k_3' = [\epsilon(\omega)\frac{\omega^2}{c^2} - k_1^2 - k_2^2]^{1/2}. \tag{8.30}$$

These fields correspond to transverse electric (TE) modes. We also have transverse magnetic (TM) modes with incident and reflected electric fields

$$A_{\mathbf{k}\lambda 2}^{(I)}(\mathbf{r}) = \frac{1}{\sqrt{2V}} (\mathbf{e}_{\mathbf{k}\lambda} \times \mathbf{k}/k) e^{i\mathbf{k}\cdot\mathbf{r}} , \tag{8.31}$$

---

[7]R. J. Cook and P. W. Milonni, *Phys. Rev.* **A35**, 5081 (1987).

[8]R. P. Feynman, R. B. Leighton, and M. Sands, *The Feynman Lectures on Physics* (Addison–Wesley, Reading, Mass., 1964), Volume 1, p. 31-2.

$$A_{k\lambda 2}^{(R)}(\mathbf{r}) = \frac{1}{\sqrt{2}V}(e_{k\lambda} \times \mathbf{k}^{(R)}/k)\left(\frac{\epsilon(\omega)k_3 - k_3'}{\epsilon(\omega)k_3 - k_3'}\right)e^{i\mathbf{k}^{(n)}\cdot\mathbf{r}} . \qquad (8.32)$$

The factors of $1/\sqrt{2}$ are introduced for mode normalization (Carniglia and Mandel, 1971). The addition of (8.25) and (8.26) gives a TE mode, and addition of (8.31) and (8.32) gives a TM mode:

$$\mathbf{A}_{k\lambda 1}(\mathbf{r}) \equiv \frac{1}{\sqrt{2}V}e_{k\lambda}e^{i\mathbf{k}_\perp\cdot\mathbf{r}}\left[e^{ik_3 z} + \left(\frac{k_3 - k_3'}{k_3 + k_3'}\right)e^{-ik_3 z}\right], \qquad (8.33)$$

$$\mathbf{A}_{k\lambda 2}(\mathbf{r}) \equiv \frac{1}{\sqrt{2}V}e^{i\mathbf{k}_\perp\cdot\mathbf{r}}[(e_{k\lambda} \times \mathbf{k}/k)e^{ik_3 z} + (e_{k\lambda} \times \mathbf{k}^{(R)}/k)$$
$$\times \left(\frac{\epsilon(\omega)k_3 - k_3'}{\epsilon(\omega)k_3 + k_3'}\right)e^{-ik_3 z}], \qquad (8.34)$$

where $\mathbf{k}_\perp = (k_1, k_2)$. Thus

$$|\mathbf{A}_{k\lambda 1}(\mathbf{r}_i)|^2 = \frac{1}{2V}\left[1 + \left(\frac{k_3 - k_3'}{k_3 + k_3'}\right)^2 + 2\left(\frac{k_3 - k_3'}{k_3 + k_3'}\right)\cos 2k_3 z\right] \qquad (8.35)$$

and

$$|\mathbf{A}_{k\lambda 2}(\mathbf{r}_i)|^2 = \frac{1}{V}\left[\frac{c^2}{\omega^2}(2k_\perp^2 - \frac{\omega^2}{c^2})\left(\frac{\epsilon(\omega)k_3 - k_3'}{\epsilon(\omega)k_3 + k_3'}\right)\right]\cos 2k_3 z$$
$$+ \frac{1}{2V}\left[1 + \left(\frac{\epsilon(\omega)k_3 - k_3'}{\epsilon(\omega)k_3 + k_3'}\right)^2\right], \qquad (8.36)$$

where z is the distance of the atom from the dielectric. The z-dependent part of (8.10) is therefore[9]

$$\langle E(z) \rangle = -\frac{1}{V}\sum_{k\lambda}(2\pi\hbar\omega_k)\alpha(\omega_k)\left[\frac{k_3 - k_3'}{k_3 + k_3'} + \frac{c^2}{\omega_k^2}(2k_\perp^2 - \frac{\omega_k^2}{c^2})\right.$$
$$\times \left.\left(\frac{\epsilon(\omega)k_3 - k_3'}{\epsilon(\omega)k_3 + k_3'}\right)\right]\cos 2k_3 z$$
$$\rightarrow -\frac{1}{V}\frac{V}{8\pi^3}\sum_\lambda \int dk_3 \int d^2k_\perp(2\pi\hbar\omega)\alpha(\omega)\left[\frac{k_3 - k_3'}{k_3 + k_3'}\right.$$
$$+ \frac{c^2}{\omega^2}(2k_\perp^2 - \frac{\omega^2}{c^2})\left.\left(\frac{\epsilon(\omega)k_3 - k_3'}{\epsilon(\omega)k_3 + k_3'}\right)\right]\cos 2k_3 z \qquad (8.37)$$

---

[9]It is necessary to account for modes resulting from reflections of only *right-going* waves. The modes resulting from left-going incident waves in the right half-space involve Fresnel transmission coefficients times plane waves, and therefore their squared moduli are spatially independent and they do not contribute to a position-dependent energy.

in the mode continuum limit. The change of integration path as in (8.15) (see note 2) allows us to write $\langle E(z)\rangle$ as

$$
\langle E(z)\rangle = -\frac{\hbar c}{2\pi}\int_0^\infty d\xi \int_0^\infty dk\, k\frac{1}{\kappa}\alpha(i\xi c)\left[-\xi^2\frac{\kappa-\kappa_1}{\kappa+\kappa_1}+(2k^2+\xi^2)\right]
$$
$$
\times\frac{\epsilon(i\xi c)\kappa-\kappa_1}{\epsilon(i\xi c)\kappa+\kappa_1}\right]e^{-2\kappa z}\;, \tag{8.38}
$$

which is identical to the result of Schwinger et al. (1978) Here[10]

$$
\kappa \equiv \sqrt{\xi^2+k^2}, \tag{8.39}
$$

and

$$
\kappa_1 \equiv \sqrt{\epsilon(i\xi c)\xi^2+k^2}. \tag{8.40}
$$

Note that the limiting case of a perfect conductor may be obtained by taking $\epsilon\to\infty$:

$$
\langle E(z)\rangle \rightarrow -\frac{\hbar c}{2\pi}\int_0^\infty d\xi\int_0^\infty dk\, k\frac{\alpha(i\xi c)}{\sqrt{\xi^2+k^2}}[\xi^2+2k^2+\xi^2]e^{-2\sqrt{\xi^2+k^2}\,z}
$$
$$
= -\frac{\hbar c}{8\pi z^4}\int_0^\infty du\,\alpha(\frac{iuc}{2z})(1+u+\frac{1}{2}u^2)e^{-u}\;. \tag{8.41}
$$

For long distances of the atom from the conductor we can replace $\alpha(iuc/2z)$ by $\alpha(0)=\alpha$ and obtain the well-known Casimir–Polder interaction

$$
\langle E(z)\rangle = -\frac{3\alpha\hbar c}{8\pi z^4}\;. \tag{8.42}
$$

For $\epsilon(i\xi c)=1+4\pi N\alpha(i\xi c)\cong 1$, (8.41) reduces to the result of Schwinger et al.:

$$
\langle E(z)\rangle \cong -\frac{\hbar c}{8\pi^2 N}\int_0^\infty d\xi[\epsilon(i\xi c)-1]^2\int_0^\infty dk\, k\frac{1}{\kappa^3}[k^4+k^2\xi^2+\frac{1}{2}\xi^4]e^{-2\kappa z}\;. \tag{8.43}
$$

For small z the leading contribution to (8.43) is

$$
\langle E(z)\rangle \cong -\frac{\hbar N}{2z^3}\int_0^\infty du\,\alpha^2(iu)\equiv -\frac{N}{z^3}\left(\frac{\pi}{6}\right)C, \tag{8.44}
$$

---

[10] Recall that $\epsilon(i\xi c)$ is real. See L. D. Landau and E. M. Lifshitz, *Electrodynamics of Continuous Media* (Pergamon Press, Oxford, 1966), Section 62.

where, according to (8.17), the short range interaction between two atoms a distance r apart is $-C/r^6$. The result (8.44) can also be derived by integrating over pairwise interactions between the atom outside the dielectric and all the atoms making up the dielectric:

$$\langle E(z) \rangle = -NC \int_z^\infty dz' \int_0^\infty dr(2\pi r) \frac{1}{[(r^2 + z'^2)^{1/2}]^6} = -\frac{\pi}{6} NC \frac{1}{z^3} \ . \quad (8.45)$$

Schwinger et al. have used the result (8.43) in a calculation of the latent heat of liquid helium. If the latent heat q is due primarily to the dispersion force between atoms, then

$$q = -E(0)N/\rho = \frac{\hbar c}{8\pi^2 \rho} \int_0^\infty d\xi [\epsilon(i\xi c) - 1]^2 \int_0^\infty dk k \frac{1}{\kappa^3} [k^4 + k^2 \xi^2 + \frac{1}{2}\xi^4] \ . \quad (8.46)$$

where $E(0)$ is the $z \to 0$ limit of (8.43), i.e., the limit in which the atom is at the surface of the dielectric, and N and ρ are respectively the number and mass densities. For liquid He Schwinger et al. use the approximation (Sabisky and Anderson, 1973)

$$\epsilon(\omega) \cong 1 + \frac{.05}{1 - \omega^2/\omega_o^2} \quad (8.47)$$

with $\omega_o = 3.5 \times 10^{16} \ \sec^{-1}$. The divergence of (8.46) is avoided by cutting off the upper limit of integration over the transverse photon momentum k at $k_c = \omega_c/c$ to obtain (Schwinger et al., 1978)

$$q \cong \frac{k_c^3}{96\pi\rho} (.05)^2 \hbar\omega_o \ . \quad (8.48)$$

Taking $\omega_c = 10^{18} \ \sec^{-1}$, corresponding to $k_c = 3.3 \times 10^7 \ cm^{-1}$, and using $\rho = 0.15 \ g/cm^3$ for liquid He, they obtain

$$q = 7 \ J/g, \quad (8.49)$$

which they refer to as the *Casimir contribution to the latent heat*. The experimental value they compare this to is about twice as large:

$$q_{exp} = 15 \ J/g \ . \quad (8.50)$$

The same cutoff k_c gives a predicted surface tension of liquid helium that is about three times larger than the experimental value. Schwinger et al. state that "we can fairly conclude that the Casimir effect, a manifestation of van der Waals forces, is responsible for a significant part of these phenomena."

The divergence of the calculated latent heat and surface tension in the theory of Schwinger et al. is due to the macroscopic nature of the theory, which does not account for the finite distances between atoms. (Note that the transverse momentum cutoff k_c is roughly on the order of an inverse atomic spacing.) That is, the theory treats the dielectric medium as a perfect continuum. Obviously the divergence of the macroscopic theory for $z \to 0$ can be anticipated from the fact that, for $\epsilon \cong 1$, the interaction of the atom with the dielectric is simply an integral over all the pairwise interactions with the atoms forming the dielectric [equations (8.44) and (8.45)]. Since these interactions diverge as the distance between the atoms approaches 0, a continuum theory cannot give a finite latent heat or surface tension.

8.5   Simple Microscopic Theory

Slater and Kirkwood (1931) obtained the following expression for the energy of interaction between two He atoms separated by r(Å):

$$E^{(2)}(r) = [7.7e^{-4.58r} - \frac{.015}{r^6}] \times 10^{-10} \text{ erg.} \qquad (8.51)$$

The first (repulsive) term is actually a fit to a more complicated repulsion term resulting from wavefunction overlap at short distances. The second, dispersion interaction term is about 30% larger than the London approximation $-3\hbar\omega_o\alpha^2/4r^6$, with $\alpha \cong 2.0 \times 10^{-24}$ cm^3 the static polarizability of He.[11] For $r \cong 3.16$ Å, however, corresponding to the peak of the pair distribution function for liquid He^4 at $T = 0$ K,[12] the London approximation differs from (8.51) by less than 10%.

X-ray scattering data indicate that each atom in liquid He^4 at $T = 0$ K has six nearest neighbors at separations $r = 3.16$ Å.[12] In the approximation of retaining only nearest-neighbor interactions, each atom therefore participates in six pairs of interactions. Since there are $\frac{1}{2}(6.023 \times 10^{23})$ pairs of atoms in a mole, the total cohesive energy per mole is expected to be

$$6\frac{1}{2}(6.023 \times 10^{23})E^{(2)}(3.16 \text{ Å}) = 200 \text{ J/mole} = 50 \text{ J/g} \qquad (8.52)$$

on the basis of this simple model. This is more than three times the experimental value of 15 J/g. However, London (1930) noted that each atom in liquid He can be regarded as vibrating in the "cage" formed by its nearest

---

[11] C. W. Allen, *Astrophysical Quantities* (Athlone Press, University of London, 1955).

[12] W. E. Keller, *Helium-3 and Helium-4* (Plenum Press, New York, 1969).

neighbors and made the semiempirical estimate of 30 J/g for the zero-point energy of this vibration. Following London, we subtract this effectively repulsive energy from (8.52) to obtain

$$q = 20 \text{ J/g} , \qquad (8.53)$$

which is in fair agreement with the observed value.

For heavier atoms the zero-point energy contribution to the latent heat is negligible and the simple estimate of the dispersion energy as in (8.52) can by itself provide a fairly accurate estimate of latent heat. Consider, for instance, the example of solid Ne, where $\alpha = 3.96 \times 10^{-25}$ cm^3, $r = 3.1$ Å, and the ionization potential $\hbar\omega_o = 21.56$ eV. Assuming a close-packed fcc structure with 12 nearest neighbors,[13] we estimate

$$q = \frac{1}{2}(12)(6.023 \times 10^{23})\frac{3\hbar\omega_o\alpha^2}{4r^6} = 1.7 \text{ kJ/mole}, \qquad (8.54)$$

in good agreeement with the experimental value of 2.1 kJ/mole for the latent heat of sublimation of solid Ne. Similarly good agreement is obtained for the other inert gas solids. Surface tensions may be estimated in the same fashion (Milonni and Lerner, 1992). For large molecules such estimates fail, mainly because the molecular radii can be comparable to or larger than intermolecular separations and consequently the London approximation $(-3\hbar\omega_o\alpha^2/4r^6)$ fails.

Simple estimates of this type, due originally to London (1930), shed considerable light on the conclusion of Schwinger et al. (1978) that "the Casimir effect, a manifestation of van der Waals forces, is responsible for a significant part of [latent heat and surface tension]." First, the conclusion about the importance of van der Waals forces is undoubtedly correct and has in fact been well accepted for many years since the original work of London. Second, the example of liquid He is unfortunately not a good one on which to base the conclusion that specifically macroscopic (Casimir) manifestations of these forces are important, since the repulsive zero-point contribution is not included in the macroscopic theory of Schwinger et al. or that in the preceding section. Indeed it appears from the previous estimates that this contribution by itself is *larger* than the attractive energy calculated by Schwinger et al.[14]

We emphasize again that Casimir effects, referring specifically to *macroscopic* manifestations of van der Waals forces, cannot without qualification

---

[13] C. Kittel, *Introduction to Solid State Physics* (Wiley, New York, 1966).

[14] Detailed microscopic theories are available. Variational calculations of the ground state of liquid He^4 by W. L. McMillan, *Phys. Rev.* **138**, A442 (1965) and D. Schiff and L. Verlet, *Phys. Rev.* **160**, 208 (1967), yield $q \cong 12$ J/g.

be said to be responsible for cohesive properties such as latent heat and surface tension. For when simple, pairwise London-type microscopic models are accurate, we can reasonably presume that nonadditive contributions to the van der Waals interactions are small, in which case the macroscopic theory, with its requirement of a transverse-momentum cutoff, is unnecessary.

For liquid He^4 and solid Ne we have, respectively, $\alpha/r^3 \sim .006$ and $.013$. As noted in Section 8.2, this implies that the nonadditive contributions requiring a macroscopic approach (or many-body theory) are small. We conclude that "Casimir effects" then reduce to ordinary, pairwise van der Waals interactions. For substances in which nonadditive effects are large, the Casimir effect may indeed provide an accurate estimate of cohesive energies, but we are unaware of any calculations along these lines.

Another point concerning the Ewald–Oseen extinction theorem in this context is worth emphasizing. In the original Lifshitz theory, and in all the subsequent work on "Casimir effects" I am aware of, the force between dielectric media is obtained from electromagnetic modes determined by Maxwell equations together with macroscopic boundary conditions. This procedure, as opposed to a completely atomistic treatment of the dielectrics, is justified if the most significant virtual photon wavelengths determining the interaction are large compared with the spacing of the atoms in the dielectric. In this case the continuum approximation is appropriate and the extinction theorem, as originally obtained by Oseen (1915) for continuous media, is applicable. The effect of all the multiple dipole scatterings by the atoms in the dielectrics is then simply to enforce the laws of reflection and refraction, giving the modes of the macroscopic theory. In the case of two dielectric plates, for instance, the significant wavelengths are those on the order of the spacing between the plates (Section 7.2), and if this is large compared with interatomic distances, the macroscopic theory can be used with impunity. Obviously the same kind of assumption underlies "cavity QED" (Section 0.3).

Finally it should be noted that one of the first successes of the Lifshitz theory was in explaining the wetting properties of liquid helium (Dzyaloshinskii, Lifshitz, and Pitaevskii, 1961). It turns out that, because the dielectric constant of liquid helium is so small ($\epsilon \cong 1.057$), the Casimir force across an adsorbed liquid helium film is generally repulsive, and therefore tends to *thicken* the film. This explains the remarkable property of liquid helium of climbing the walls of a beaker. The repulsive nature of the force, in this case as a consequence of retardation, is also believed to account for the spreading of pentane on water, whereas many other hydrocarbons experience attractive forces and consequently form lenslike globules on water.

8.6   Casimir Forces at Finite Temperature

In our derivation of the Lifshitz expression for the force between two dielectric slabs in Chapter 7 we assumed the vacuum state for the field. For
finite temperatures the appropriate state of the field is the thermal state
with mean number of photons (Section 2.11)

$$n(\omega) = \frac{1}{e^{\hbar\omega/k_B T} - 1} \tag{8.55}$$

for each field mode of frequency ω.[15] To generalize the Lifshitz expression
to this case, let us employ the symmetric ordering of the field operators in
which the source field makes no contribution (Section 7.5). Then the force
is due solely to the "external" field, which in Section 7.5 was actually the
vacuum field, for which $\langle \mathbf{F}_o^{(-)}(\mathbf{r},\omega) \cdot \mathbf{F}_o^{(+)}(\mathbf{r},\omega) \rangle = 0$, where $\mathbf{F}_o^{(\pm)}$ are defined
by (7.140). The generalization of (7.141) with this expectation value not
equal to zero, i.e., with the field not in the vacuum state, is easily seen to
be

$$\begin{aligned} \langle E \rangle \;=\; & -\frac{1}{2} \int d^3 r\, N(\mathbf{r}) \int_0^\infty d\omega\, \alpha(\omega) \langle \mathbf{F}_o^{(+)}(\mathbf{r},\omega) \cdot \mathbf{F}_o^{(-)}(\mathbf{r},\omega) \\ & + \mathbf{F}_o^{(-)}(\mathbf{r},\omega) \cdot \mathbf{F}_o^{(+)}(\mathbf{r},\omega) \rangle. \end{aligned} \tag{8.56}$$

Here, $\langle \mathbf{F}_o^{(+)}(\mathbf{r},\omega) \cdot \mathbf{F}_o^{(-)}(\mathbf{r},\omega) \rangle$ is proportional to $\langle a_\omega(0) a_\omega^\dagger(0) \rangle = n(\omega) + 1$
and likewise $\langle \mathbf{F}_o^{(-)}(\mathbf{r},\omega) \cdot \mathbf{F}_o^{(+)}(\mathbf{r},\omega) \rangle$ is proportional to $\langle a_\omega^\dagger(0) a_\omega(0) \rangle = n(\omega)$
(see, for instance, Appendix E). The spatially dependent mode functions
determining the dyadic Green function are the same regardless of the state
of the field. The generalization of (7.144) to the case $n(\omega) \neq 0$ is thus

$$\langle E \rangle = 2i\hbar \int d^3 r\, N(\mathbf{r}) \int_0^\infty d\omega\, \alpha(\omega) [2n(\omega) + 1] \Gamma_{jj}(\mathbf{r},\mathbf{r},\omega). \tag{8.57}$$

As in Section 7.5 it is implicitly assumed that $\langle \mathbf{F}_o^{(+)}(\mathbf{r},\omega) \cdot \mathbf{F}_o^{(-)}(\mathbf{r},\omega') \rangle$ and
$\langle \mathbf{F}_O^{(-)}(\mathbf{r},\omega) \cdot \mathbf{F}_O^{(+)}(\mathbf{r},\omega') \rangle$ vanish unless $\omega = \omega'$, i.e., that different frequency
components of the field are uncorrelated. This is the case for both the
vacuum state of the field and the thermal state. For the latter state $n(\omega)$
is given by (8.55) and accordingly

$$\langle E \rangle \;=\; 2i\hbar \int d^3 r\, N(\mathbf{r}) \int_0^\infty d\omega\, \alpha(\omega) \coth\left(\frac{\hbar\omega}{2k_B T}\right) \Gamma_{jj}(\mathbf{r},\mathbf{r},\omega)$$

---

[15] To avoid confusion with $k = \sqrt{k_x^2 + k_y^2}$ we denote the Boltzmann constant by k_B.

$$= \frac{i\hbar}{2\pi} \int d^3r \int_0^\infty d\omega [\epsilon(\mathbf{r}, \omega) - 1] \coth\left(\frac{\hbar\omega}{2k_BT}\right) \Gamma_{jj}(\mathbf{r}, \mathbf{r}, \omega).$$

(8.58)

As noted in the preceding chapter, the Casimir force between dielectric slabs comes predominantly from field frequencies $\omega \sim c/d$, where d is the separation between the dielectrics. The finite-temperature correction to Casimir forces will therefore be negligible when

$$\frac{1}{e^{\hbar\omega/k_BT} - 1} = \frac{1}{e^{\hbar c/k_BTd} - 1} \ll 1,$$

(8.59)

i.e., when $T \ll \hbar c/k_B d$. For $d = 1$ μm, this condition becomes $T \ll 2300$ K. We conclude, therefore, that for $d \lesssim 1$ μm we can in effect assume $T = 0$.

For larger separations, or for large temperatures, the temperature corrections may be significant. The quickest way to arrive at a more explicit expression for the force in this case is to use (7.44) with $\xi = -i\omega$ and with the factor $\coth(\hbar\omega/2k_BT)$ included in the integrand. For simplicity we will assume $\epsilon_3 = 1$, i.e., that the dielectrics are separated by vacuum. Then $K_3 = -i(\omega/c)p$ [see equation (7.46)] and we obtain the force per unit area

$$
\begin{aligned}
F(d) &= -\frac{\hbar}{2\pi^2 c^3} \int_1^\infty dp\, p^2 \int_0^\infty d\omega\, \omega^3 \\
&\quad \times \left(\left[\frac{s_1 + \epsilon_1 p}{s_1 - \epsilon_1 p} \frac{s_2 + \epsilon_2 p}{s_2 - \epsilon_2 p} e^{-2i\omega pd/c} - 1 \right]^{-1} \right. \\
&\quad + \left. \left[\frac{s_1 + p}{s_1 - p} \frac{s_2 + p}{s_2 - p} e^{-2i\omega pd/c} - 1 \right]^{-1} \right) \coth\left(\frac{\hbar\omega}{2k_BT}\right).
\end{aligned}
$$

(8.60)

The change of integration involving $\omega \to i\xi$ in Chapter 7 is useful here as well.[16] Now, however, we must account for the fact that $\coth(\hbar\omega/2k_BT)$ has poles on the imaginary axis at

$$\omega_n = 2\pi i n(k_BT/\hbar) \equiv i\xi_n$$

(8.61)

for all integers n. The deformation of the integration path so that it runs along the imaginary ω axis, therefore, involves semicircles about the poles (8.61), as shown in Figure 8.1. The integration along the semicircle about ω_n contributes $i\pi$ times the residue of the integrand at the pole ω_n for

---

[16] The advantage of this transformation, of course, is that it eliminates the rapid oscillations associated with $e^{-2i\omega pd/c}$.

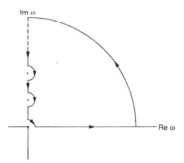

Figure 8.1: The integration of (8.60) along the real ω axis may be transformed to an integration along a quarter-circle, whose radius extends to infinity and makes no contribution, plus an integration along the imaginary axis. In the latter the poles at ω_n are avoided by integrating along semicircles centered at the $\omega_n \neq 0$ and along a quarter-circle at $\omega = 0$.

$n \neq 0$, whereas for $n = 0$ the contribution from the quarter-circle is $i\pi/2$. Thus

$$
\begin{aligned}
F(d) &= -\frac{\hbar}{2\pi^2 c^3}\left(\frac{2\pi i k_B T}{\hbar}\right) i^3 \sum_{n=0}^{\infty}{}' \xi_n^3 \int_1^{\infty} dp\, p^2 \\
&\times \left(\left[\frac{s_{1n}+\epsilon_{1n}p}{s_{1n}-\epsilon_{1n}p}\frac{s_{2n}+\epsilon_{2n}}{s_{2n}-\epsilon_{2n}p}e^{2p\xi_n d/c}-1\right]^{-1}\right. \\
&\left. + \left[\frac{s_{1n}+p}{s_{1n}-p}\frac{s_{2n}+p}{s_{2n}-p}e^{2p\xi_n d/c}-1\right]^{-1}\right),
\end{aligned}
\qquad (8.62)
$$

where

$$
s_{jn} = \sqrt{p^2 - 1 + \epsilon_{jn}}\,, \quad \epsilon_{jn} = \epsilon_j(i\xi_n), \quad j = 1, 2, \qquad (8.63)
$$

and the prime on the summation sign indicates that a factor $1/2$ must be included in the $n = 0$ term. This result for finite temperature was derived by Lifshitz (1956).[17]

The case of two perfectly conducting plates follows by taking $\epsilon_1, \epsilon_2 \to \infty$, as in Section 7.2:

$$
F(d) = -\frac{2k_B T}{\pi c^3}\sum_{n=0}^{\infty}{}' \xi_n^3 \int_0^{\infty}\frac{dp\, p^2}{e^{2p\xi_n d/c}-1}
$$

[17]See also L. D. Landau and E. M. Lifshitz, *Electrodynamics of Continuous Media*, Section 90.

$$= -\frac{k_B T}{4\pi d^3} \sum_{n=0}^{\infty}{}' \int_{nx}^{\infty} \frac{dy\, y^2}{e^y - 1} \,, \qquad (8.64)$$

where $x \equiv 4\pi k_B T d / \hbar c$. At high temperatures, $x \gg 1$, the dominant contribution comes from the $n = 0$ term:

$$F(d) = -\frac{k_B T}{4\pi d^3} \int_0^{\infty} \frac{dy\, y^2}{e^y - 1} = -\frac{2.4 k_B T}{4\pi d^3} \,. \qquad (8.65)$$

Note the $1/d^3$ dependence in this case. For low temperatures, $x \ll 1$, it may be shown that (Schwinger et al., 1978)

$$F(d) \cong -\frac{\pi^2 \hbar c}{240 d^4} \left[1 + \frac{16}{3} \left(\frac{k_B T d}{\hbar c} \right)^4 - \frac{240}{\pi} \left(\frac{k_B T d}{\hbar c} \right) e^{-\pi \hbar c / k_B T d} \right]. \quad (8.66)$$

In either case the magnitude of the force per unit area is increased from its $T = 0$ value.

8.7 Experiments

As discussed in Section 7.2, experiments on macroscopic van der Waals forces between dielectrics provided the stimulus for the Lifshitz theory. These experiments unveiled the inadequacy of theories based on pairwise additive intermolecular van der Waals forces.

The literature on experimental studies of van der Waals forces is enormous, and here we cannot begin to even summarize it. The Lifshitz theory bears on many areas of colloid science, for instance, and "The impetus given to colloid science by the appearance of the Lifshitz theory, and the subsequent developments that it led to, can hardly be overestimated" (Israelachvili and Ninham, 1977). We have already discussed in very simple terms how the latent heat of some substances provides a direct measure of the intermolecular van der Waals forces (Sections 8.4 and 8.5). Here we will focus primarily on perhaps the most directly measurable manifestation of van der Waals forces, namely the force between dielectric bodies. The Lifshitz theory was of course developed to deal precisely with such forces.

Discussion of the many other manifestations of van der Waals forces would take us much too far afield from our study of the electromagnetic vacuum. The force between two dielectric plates is close enough to the example "every physicist" knows is relevant to electromagnetic zero-point energy — the Casimir force in the case of perfectly conducting plates — that it is not difficult to keep the relevance to the vacuum field in mind.

The derivation of the Lifshitz force given in Section 7.2 brings out in similar fashion the relevance of zero-point field energy to the case of dielectrics. Of course all the manifestations of van der Waals forces can be traced to the electromagnetic zero-point field, since the van der Waals force between any two molecules can be regarded as a direct consequence of the zero-point, vacuum field.[18]

The two essential quantities that must be measured to determine macroscopic van der Waals interactions between dielectrics are the force and the distance between the dielectrics. These are hardly easy to measure, since the forces are tiny and perceptible only at very small separations. The first successful experiments were reported by Derjaguin and Abrikosova in 1951 (see Derjaguin and Abrikosova, 1957; Derjaguin, 1960). In these experiments the measured forces were between a glass plate (4×7 mm) and spherical lenses with radii of curvature $R = 10$ cm and $R = 25$ cm. Such a configuration is more easily adjustable than that of two flat plates and also allows the closest distance of separation to be measured optically from the diameter of Newton rings. From the measurement of the force between the lens and the plate, one can infer the interaction energy between two flat plates using the "Derjaguin approximation," which we now briefly derive.

Consider two spheres with radii R_1 and R_2 and separated by a distance $H << R_1, R_2$ (Figure 8.2). The force between them is

$$F(H) \cong \int_{z=H}^{z=\infty} 2\pi r dr f(z), \qquad (8.67)$$

where $f(z)$ is the force per unit area between two *flat* surfaces. We are assuming that the force on the circular area $2\pi r dr$ on one sphere is due to locally flat surfaces on the other sphere at distances $z = H + z_1 + z_2$ away (Figure 8.2); this is permissible so long as $R_1, R_2 >> H$. Now $r^2 \cong 2R_1 z_1 \cong 2R_2 z_2$, so that $z \cong H + \frac{1}{2}r^2(R_1^{-1} + R_2^{-1})$ and $rdr \cong (R_1^{-1} + R_2^{-1})^{-1}dz = [R_1 R_2/(R_1 + R_2)]dz$. Thus we have the Derjaguin approximation

$$F(H) \cong 2\pi \left(\frac{R_1 R_2}{R_1 + R_2} \right) \int_H^\infty dz f(z) = 2\pi \left(\frac{R_1 R_2}{R_1 + R_2} \right) u(H), \qquad (8.68)$$

where $u(H)$ is the interaction energy per unit area between two flat surfaces separated by H. For $R_1 = R$ and $R_2 = \infty$, corresponding to a sphere of radius R and a flat plate, we have (Derjaguin and Abrikosova, 1957)

$$F(H) = 2\pi R u(H). \qquad (8.69)$$

---

[18] Recall the discussions in Chapters 3 and 7, and especially the remarks following equation (3.92).

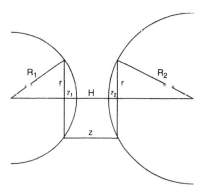

Figure 8.2: Geometry for deriving the Derjaguin approximation to the force between two spheres of radii R_1 and R_2 separated by a distance $H \ll R_1, R_2$.

This result may also be derived under the assumption of pairwise additive intermolecular forces, but it has a more general validity so long as $R_1, R_2 \gg H$. The importance of the Derjaguin formula (8.69) is that it relates the force between a sphere and a flat plate to the theoretically simpler force between two flat plates. The force between the latter is given approximately by the Lifshitz theory.

The measurement of the force is complicated by its short range and the tendency of the surfaces to adhere when H is very small. The balance used to measure the force must therefore have both high sensitivity and a large restoring torque. In the early experiments of Derjaguin and others the separations H ranged between .07 and 0.5 μm. The force was measured by an electrical negative feedback technique in which a displacement of a knife-edge balance beam produces a current acting to restore the beam to its equilibrium position, and a measurement of the current then determines the force. The negative feedback mechanism also reduces the vibrational period of the balance, so that equilibrium is established nearly instantaneously (Derjaguin and Abrikosova, 1957).

The data of Derjaguin, Rabinovich, and Churaev conflicted with estimates based on pairwise additive intermolecular forces and provided the first "good quantitative agreement" with the predictions of the Lifshitz theory in the retarded regime (Derjaguin et al., 1978). Precise quantitative comparison with the Lifshitz theory was complicated by the appearance in the theory of the dielectric constant over the entire range of frequencies, and in particular by a lack of data on high-frequency absorption spectra.

Early experiments between flat glass plates (area 1 cm^2, separations between 0.6 and 1.5 μm) were carried out by Overbeek and Sparnaay (1954). In these experiments one of the plates was attached to a spring and the displacement Δx was determined by a capacitive method. The force was then calculated from the force constant K of the spring ($F = K\Delta x$). Due to the effects of residual charges,[19] however, the forces were overestimated (Sparnaay, 1989).

Sparnaay (1958) performed the first experiments to test the Casimir theory of the force between *conducting* plates. The force between cleansed metal plates (chromium, chromium steel, and aluminum) inside an evacuated chamber filled with nitrogen gas was again determined from the deflection of a spring; the spring in these experiments was connected to an aluminum beam attached to one of the plates. The deflection was determined from the change in the capacitance of a capacitor connected to the beam (see Sparnaay (1958), Figure 2). Great care was taken to eliminate the effects of dust particles, which appeared to give rise in some cases to repulsive forces. For the experimental plate separations between 0.5 and 2 μm, forces between about 0.2 dyne/cm^2 and about .003 dyne/cm^2 were measured, "in qualitative agreement with Casimir's prediction" [equation (7.1)]. Although Sparnaay's experiments are often cited as experimental confirmation of the Casimir force between conducting plates, Sparnaay himself was more cautious, writing in the abstract of his paper that "The observed attractions do not contradict Casimir's theoretical prediction."

Tabor and Winterton (1969) reported the first measurements of forces between dielectrics for separations so small that retardation effects are negligible. For $H < 500$ A the polished surfaces used in previous experiments were still sufficiently rough that their variations were comparable to H. Tabor and Winterton used extremely smooth thin sheets of mica on glass cylinders of radii $\cong 1$ cm. They were able to probe the transition between retarded and nonretarded forms of the Lifshitz force, which they determined to occur between $H = 120$ and 500 A.

Hunklinger, Geisselmann, and Arnold (1972) developed a "dynamic method" for measuring macroscopic van der Waals forces, or actually the derivative of the force with respect to the distance of separation. We quote from their clear presentation:

> A small plate (0.1 mm thick, 2 mm^2 area) and a plano-convex lens (radius of the spherical surface $R = 250$ mm) are used as specimens. The plate is cemented to the membrane of a modified condenser microphone (see Fig. 1) and the lens is mounted on a low frequency mechanical vibrator of a construction similar to that of a loudspeaker. When the lens is oscillating a periodically modulated van der Waals

---

[19] Recall the remarks near equation (7.3).

force acts on the plate. This dynamic force causes oscillations of the membrane of the microphone, producing an alternating output voltage which is amplified and recorded. The distance between the plate and the lens can be determined interferometrically by illuminating the specimens and observing the Newton fringe pattern which is generated.

The amplified signal was related to the derivative of the force with respect to distance. Results "in good agreement with theory" for borosilicate glass were reported for distances between .08 and 1.2 μm.

Van Blokland and Overbeek (1978) measured forces between a chromium plate and a chromium sphere at distances between 0.13 and 0.67 μm. The separations were determined from the capacitance *produced by the objects*. Of particular interest here is the comparison that van Blokland and Overbeek made of their data to the predictions of the Lifshitz theory. For this purpose they used the following expression for the dielectric constant of a metal:

$$\epsilon(i\xi) = 1 + \frac{\omega_p^2}{\xi^2 + \xi\omega_2} + \sum_i \frac{\omega_{ip}^2}{\omega_i^2 + \xi^2 + \xi g_i} , \qquad (8.70)$$

where ω_p is the plasma frequency, $\omega_2 = \omega_p^2/4\pi\sigma_o$, σ_o is the specific conductivity, and the sum is over the absorption bands of central frequency ω_i, halfwidth g_i, and strength ω_{ip}^2. For chromium one absorption band (actually two overlapping bands) at $\omega_i = 3.0 \times 10^{15}$ sec^{-1} provided a sufficient approximation to the sum. Van Blokland and Overbeek concluded that "at distances between 132 and 670 nm the measured force and the calculated force are in excellent agreement when the absorption band of chromium is taken into account in the calculation of the force. Surface roughness exists and it prevents measurements at small separations but has hardly any influence on the measured forces."

The reader is referred to the surveys by Derjaguin et al. (1978) and Sparnaay (1080), and the references therein, for details about the many beautiful experiments on van der Waals forces between macroscopic bodies.

8.8   The Casimir–Polder Force: Experiments

An example considered by Casimir and Polder (1948) to illustrate the effect of retardation is the force on an atom at a distance r from a conducting plane. At short distances the potential $V(r)$ varies as r^{-3} and is easily understood from the interaction of the atomic dipole with its image. For large r, however, the interaction falls off as r^{-4}. As shown in Section 3.12, the retarded Casimir–Polder interaction may be interpreted as the change, due to the presence of the conducting plane, of the Stark shift of the atom in the vacuum field.

The r^{-3} interaction at short distances is often referred to in the atom–surface context as the *van der Waals interaction*.[20] The interaction between the dipole $\mathbf{p}$ and its image dipole $\mathbf{p}_i$ is

$$V(r) = \frac{1}{2} \frac{\mathbf{p} \cdot \mathbf{p}_i - 3(\hat{z} \cdot \mathbf{p})(\hat{z} \cdot \mathbf{p}_i)}{(2r)^3} . \tag{8.71}$$

The factor $\frac{1}{2}$ appears because the forces involved in bringing the dipole from ∞ to r act only on the dipole, not on its image. For dipole orientations normal to the mirror surface, i.e., for $\mathbf{p} = p_z \hat{z} \equiv \mathbf{p}_\perp$, we have $\mathbf{p}_i = \mathbf{p} = \mathbf{p}_\perp$, whereas for orientations parallel to the surface, $\mathbf{p}_i = -\mathbf{p} = -\mathbf{p}_\parallel$. Thus

$$V(r) = \frac{1}{16r^3}(-p_\parallel^2 - 2p_\perp^2), \tag{8.72}$$

and for an atom in its ground state $|g\rangle$ this becomes

$$V(r) = -\frac{1}{16r^3} \left[\langle g|p_\parallel^2|g\rangle + 2\langle g|p_\perp^2|g\rangle \right], \tag{8.73}$$

where $\mathbf{p} = \mathbf{p}_\parallel + \mathbf{p}_\perp$ is the dipole moment operator.

The potential (8.73) implies a force $\propto r^{-4}$ attracting the atom to the surface. The van der Waals interaction holds, provided the atom is not so close to the surface that complications associated with the microscopic surface structure come into play. This typically requires r to be a few tens of Ångstroms. Application of (8.73) also assumes that r is small compared with characteristic transition wavelengths of the atom, so that retardation is negligible.

Shih and Parsegian (1975) studied experimentally the deflection of Cs, Rb, and K atomic beams by gold surfaces, and obtained results consistent with the r^{-3} van der Waals interaction.

Anderson et al. (1988) achieved very much larger van der Waals forces using highly excited Cs and Na atoms: the electric dipole moment of such a (Rydberg) atom scales as the square of the effective principal quantum number n, and therefore the van der Waals force scales as n^4. Furthermore the Rydberg atoms have large transition wavelengths, and so the range of their van der Waals interaction with the surface is substantially greater than that for ground-state atoms.

The experiment of Anderson et al. involved passage of a beam of Rydberg atoms through an 8 mm channel formed by two gold mirrors separated

---

[20]It is also called the *Lennard-Jones interaction*, since it was considered by J. E. Lennard–Jones, *Trans. Faraday Soc.* **28**, 333 (1932).

by a variable distance between 2.1 and 8.5 μm. Atoms with larger n values are attracted to the mirrors more strongly than atoms with smaller n values, and so fewer of them escape the channel without sticking to one of the mirror surfaces.[21] One can define a maximal n for which an atom can pass through the tunnel. To deduce how this n_m scales with the mirror separation w, consider an atom of mass M and effective principal quantum number n initially at a distance z_o from a single mirror surface. From energy conservation we have

$$\frac{1}{2}M\left(\frac{dz}{dt}\right)^2 + V(z) = V(z_o) \tag{8.74}$$

if $dz/dt = 0$ at $t = 0$. Since $V(z) \sim -e^2 a_o^2 n^4/16 z^3$, (8.74) implies that the time it takes for the atom to reach the mirror surface is

$$t \sim \frac{\sqrt{M}z_o^{5/2}}{n^2 e a_o} \ . \tag{8.75}$$

Now if the atom has an initial velocity v parallel to the mirror surface, it will move a longitudinal distance

$$\ell = vt \sim \frac{\sqrt{Mv^2}z_o^{5/2}}{n^2 e a_o} \tag{8.76}$$

before striking the surface. This implies the relation

$$\frac{n^4\ell^2}{z_o^5} \lesssim \frac{Mv^2}{e^2 a_o^2} \tag{8.77}$$

for atoms passing a mirror of length ℓ without colliding with the mirror. For two mirrors separated by a distance w, forming a channel of length ℓ, we replace z_o by $w/2$ and Mv^2 by kT in (8.77) to obtain (Anderson et al., 1988)

$$\frac{n_m^4\ell^2}{w^5} \leq \eta\frac{kT}{(ea_o)^2} \ , \tag{8.78}$$

where η is a dimensionless factor of order unity and T (~ 350 K) is the temperature of the oven source for the atomic beam. Thus $n_m \propto w^{5/4}$, a scaling law confirmed by numerical simulations assuming classical trajectories for the atoms.

---

[21] It is important for the analysis of the experimental results that, independent of their state of excitation, atoms striking the mirrors adhere, without "bouncing." This assumption is supported by other experimental evidence.

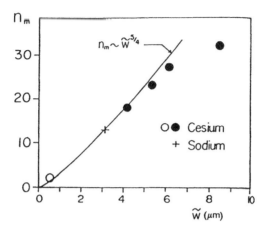

Figure 8.3: Maximal effective principal quantum number n_{m} vs. the scaled channel width $\tilde{w}= w[8 \text{ mm}/\ell_{\mathrm{m}}]^{2/5}[T/(350 \text{ K})]^{1/5}$, where ℓ_{m} is the largest channel length for which there is transmission. The solid curve is theoretical, and the circles are experimental values for Cs atoms, corrected for radiative decay in the channel. The open circle is estimated from ground-state experimental results of D. Raskin and P. Kusch, *Phys. Rev.* A9, 652 (1974). From A. Anderson et al. (1988), with permission.

In the experiments, Rydberg Cs or Na atomic beams were excited to prescribed n levels using two-step excitation with two lasers. The maximal principal quantum number n_{m} was defined as the value of n for which the fraction of atoms escaping from the channel was reduced to 1%. (The n values of escaping atoms were determined by static electric field ionization, whose probability depends upon n, and detection with an electron multiplier.) Figure 8.3 shows experimental results for n_{m} for different channel widths w. The results are compatible with the $n_{\mathrm{m}} \propto w^{5/4}$ scaling law, but are not accurate enough to rule out, say, an $n_{\mathrm{m}} \propto w$ scaling. The deviation from the van der Waals theory at the larger gap separations in Figure 8.3 is believed to be due to stray electric fields (Anderson et al., 1988).

Deflections of ground-state atomic beams in similar experiments have been observed by Hinds et al. (1991). The atoms escaping the channel in these experiments were detected by excitation to the $n = 14$ level with two lasers, followed by field ionization and ion counting. Figure 8.4 compares experimentally measured transmission factors vs. the channel width w with theoretical calculations based on the van der Waals r^{-3} and Casimir–Polder ("QED") r^{-4} interactions with the walls. Also shown is the geometrical

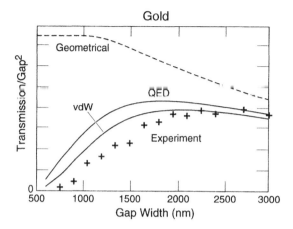

Figure 8.4: Relative transmission factors divided by the square of the channel width obtained by Hinds *et al.* for ground-state Na atoms. The dashed curve is based on purely geometrical considerations, ignoring atom-wall interactions. The curves labelled vdW and QED are theoretical predictions based on the van der Waals and Casimir–Polder atom-wall interactions, respectively. The crosses are the experimental results, normalized to agree exactly with the QED curve for a channel width of 2700 nm. From Hinds *et al.* (1991), with permission.

transmission factor obtained assuming that the atoms do not interact with the walls. It is clear from the experimental data that there *is* an atom-wall interaction, but the data are not sufficiently accurate to distinguish between the van der Waals and Casimir–Polder interactions.

Recent experiments of the type just described have measured the atom–surface interaction spectroscopically as a position-dependent level shift (Sandoghdar et al., 1992), and have succeeded in providing the first detailed quantitative confirmation of both the van der Waals and Casimir–Polder interactions (Sukenik et al., 1993). In the latter experiments the escaping ground-state sodium atoms are excited to the $12s$ level with two lasers and field-ionized and counted as in the earlier experiments (Hinds et al., 1991). The lasers are focused sufficiently tightly that all the excited atoms have traversed the cavity within a width of about 2–3 nm. For cavity widths below about 1.2 μm, the data fit the theoretical predictions based on the Casimir–Polder interaction very well, as shown in Figure 8.5, and confirm that the interaction indeed varies as r^{-4} rather than r^{-3} in the retarded regime. It is important to note that both the van der Waals (or Lennard-Jones) and Casimir–Polder interactions are confirmed with essentially no

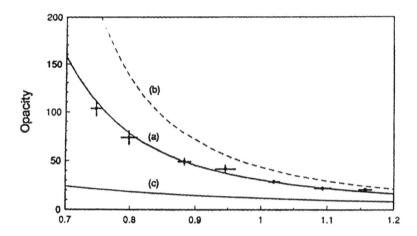

Figure 8.5: Opacity (reciprocal of transmittance) measured by Sukenik et al. The three curves are theoretical opacities assuming (a) the Casimir–Polder interaction, (b) the nonretarded van der Waals interaction, and (c) no atom–wall interaction. The results are plotted relative to the values at a 0.6 μm cavity width. From Sukenik et al. (1993), with permission.

adjustable parameters in these experiments and their theoretical simulations as outlined previously.[22]

Experiments of a different type, that might also enable a study of the van der Waals and Casimir–Polder forces, involve the interaction with a glass surface of slowly moving ($T \sim 1$ μK) atoms from an "atomic fountain" (Kasevich et al., 1991). In these experiments the atom–surface interaction includes a repulsive potential due to an evanescent field produced by total internal reflection at the glass surface, leading to a bouncing probability that depends on the frequency of the evanescent wave. Atoms in the atomic fountain are velocity-selected so that they collide with the glass surface as their trajectories begin to turn over due to gravity, and the final velocity of the bounced atoms can be determined using an ionizing laser beam.

---

[22] The experiments involve a number of complications, perhaps the most difficult of which are stray electric fields. The reader is referred to the cited literature for discussions of how the Yale group overcame these difficulties.

8.9 Casimir Effects in Atomic Spectroscopy

The Lamb shifts in the energy levels of an atom can be ascribed in large measure to vacuum field fluctuations (Chapter 3). Spruch and Kelsey have described another effect of vacuum field fluctuations on energy levels of a single atom. As in the case of the Lamb shift and van der Waals interactions, the effect was first obtained by conventional perturbation-theoretic methods (Kelsey and Spruch, 1978) and was later rederived more simply from the perspective of vacuum field fluctuations (Spruch and Kelsey, 1978).

Imagine a stationary electron at a distance r from an atom with polarizability $\alpha(\omega)$. From the energy $-\frac{1}{2}\alpha \mathbf{E}^2$ (Section 3.8) with $\mathbf{E} = er/r^3$ the Coulomb field of the electron, we obtain the polarization potential

$$V(r) = -\frac{1}{2}\frac{\alpha e^2}{r^4} , \qquad (8.79)$$

where $\alpha = \alpha(0)$ is the static polarizability. Bernabéu and Tarrach (1976) found with a dispersion-theoretic approach that there is an additional contribution to the polarization potential due to retardation; for large r they obtained the leading correction

$$V_{\mathrm{ret}}(r) \cong \frac{11\hbar e^2 \alpha}{4\pi mcr^5} . \qquad (8.80)$$

Using perturbation theory, Kelsey and Spruch (1978) obtained the same retardation correction to the polarization potential between an electron and a *charged* polarizable system. It is important to note that, in contrast to the effect of retardation on the van der Waals interaction, $V_{\mathrm{ret}}(r)$ is an *addition* to the nonretarded potential (8.79) rather than a replacement at large r.

Of special interest here is Spruch and Kelsey's derivation of V_{ret} from the perspective of zero-point electromagnetic energy (Spruch and Kelsey, 1978). In fact Spruch and Kelsey have described a simple approach applicable to other retarded interactions associated with vacuum field fluctuations. To describe this approach we first recall the expression (3.74) for the interaction between two polarizable systems. The contribution to this interaction from a single field mode of frequency ω is

$$U(\omega, \mathbf{r}) \sim \alpha_2(\omega)\mathbf{E}_b(\omega, \mathbf{r}_2) \cdot \mathbf{E}_{1\to 2}(\omega, \mathbf{r}_2) \qquad (8.81)$$

in the notation of Spruch and Kelsey, where $\alpha_2(\omega)$ is the polarizability of system 2, $\mathbf{E}_b(\omega, \mathbf{r}_2)$ is a background field, $\mathbf{E}_{1\to 2}(\omega, \mathbf{r}_2)$ is the ω-component of the field from system 1 at the point $\mathbf{r}_2$ of system 2, and $\mathbf{r} = \mathbf{r}_1 - \mathbf{r}_2$. The interaction due to a continuum of field modes, with $N(\omega)d\omega$ modes in the interval $[\omega, \omega + d\omega]$, is

$$
\begin{aligned}
U(\mathbf{r}) &\sim \int d\omega\, N(\omega) U(\omega, \mathbf{r}) \\
&= \int d\omega\, N(\omega) \alpha_2(\omega) \mathbf{E}_b(\omega, \mathbf{r}_2) \cdot \mathbf{E}_{1\to2}(\omega, \mathbf{r}_2).
\end{aligned}
\tag{8.82}
$$

Following closely the approach of Spruch and Kelsey (1978), we ignore signs and factors of order unity, and assume for $\mathbf{E}_{1\to2}(\omega, \mathbf{r}_2)$ the general form

$$
\mathbf{E}_{1\to2}(\omega, \mathbf{r}_2) \sim [\alpha_1(\omega)\mathbf{E}_b(\omega, \mathbf{r}_1)/r^3]g(\omega r/c).
\tag{8.83}
$$

This expresses $\mathbf{E}_{1\to2}(\omega, \mathbf{r}_2)$ as the field from the dipole moment $\alpha_1(\omega) \times \mathbf{E}_b(\omega, \mathbf{r}_1)$ induced by the background field, with $g(\omega r/c)$ a function of order unity over the significant range of $\omega r/c$, and such that $g(x)/x$ remains finite as $x \to 0$. Thus

$$
U(\mathbf{r}) \sim \frac{1}{r^3} \int d\omega\, N(\omega) \alpha_1(\omega) \alpha_2(\omega) \mathbf{E}_b(\omega, \mathbf{r}_2) \cdot \mathbf{E}_b(\omega, \mathbf{r}_1) g(\omega r/c).
\tag{8.84}
$$

For $r > c/\omega$ we can expect there to be substantial cancellations associated with the oscillations with $\omega r/c$ of the dipole fields, and so we replace $\mathbf{E}_b(\omega, \mathbf{r}_1) \cdot \mathbf{E}_b(\omega, \mathbf{r}_2)$ by 0 for $\omega > c/r$ and by $\mathbf{E}_b^2(\omega, \mathbf{r}_2)$ for $\omega < c/r$ (Spruch and Kelsey, 1978). Then

$$
U(r) \sim \frac{1}{r^3} \int_0^{c/r} d\omega\, N(\omega) \alpha_1(\omega) \alpha_2(\omega) \mathbf{E}_b^2(\omega, \mathbf{r}_2) g(\omega r/c).
\tag{8.85}
$$

Now the case of interest is that in which the "background" field is the zero-point, vacuum field with energy density $\mathbf{E}_b^2(\omega, \mathbf{r}) = \hbar\omega/V$ for each mode and, for free space, $N(\omega) = V\omega^2 d\omega/c^3$. In this case we obtain the Spruch–Kelsey approximation

$$
U(r) \sim \frac{\hbar}{c^3 r^3} \int_0^{c/r} d\omega\, \omega^3 \alpha_1(\omega) \alpha_2(\omega) g(\omega r/c).
\tag{8.86}
$$

This simple result, based on the interaction of two polarizable systems in the polarizing vacuum field, allows us to obtain various retarded interactions in a neatly unified fashion. For two neutral atoms, for instance, we have the retarded dipole–dipole interaction

$$
\begin{aligned}
U_{dd}(r) &\sim \frac{\hbar}{c^3 r^3} \alpha_1 \alpha_2 \int_0^{c/r} d\omega\, \omega^3 g(\omega r/c) \\
&= \frac{\hbar c}{r^7} \alpha_1 \alpha_2 \int_0^1 dx\, x^3 g(x),
\end{aligned}
\tag{8.87}
$$

where we have replaced the polarizabilities $\alpha_j(\omega)$ by the static polarizabilities $\alpha_j = \alpha_j(0)$ under the assumption that c/r is much smaller than any transition frequency contributing significantly to the polarizability. This assumption, of course, defines a long-range, "retarded" interaction. If we assume also that

$$\int_0^1 dx\, x^3 g(x) \sim 1, \qquad (8.88)$$

then

$$U_{dd}(r) \sim \frac{\hbar c}{r^7} \alpha_1 \alpha_2 , \qquad (8.89)$$

which is the retarded van der Waals interaction (3.91) except for the factor $23/4\pi$. We can compare (8.87) to the expression (3.89) from which (3.91) follows:

$$V(r) = -\frac{\hbar c}{\pi r^7} \alpha_1 \alpha_2 \int_0^\infty dx\, x^6 G(x), \qquad (8.90)$$

where $G(x)$ is defined by (3.86). We can see, then, that the Spruch–Kelsey approximation (8.86) is fully justified except, as stated, for factors of order unity. Indeed the argument leading to equation (3.91) can easily be seen to be a more detailed version of that of Spruch and Kelsey, accounting for the numerical factors arising from angular integrations.

Consider next the interaction between a polarizable system and a free point charge q of mass m. For the latter the polarizability $\alpha_2(\omega) = -q^2/m\omega^2$, and (8.86) becomes

$$\begin{aligned} U_{df}(r) &\sim \frac{\hbar q^2}{mc^3 r^3} \int_0^{c/r} d\omega\, \omega \alpha_1(\omega) g(\omega r/c) \\ &\cong \frac{\hbar q^2 \alpha_1}{mcr^5} \int_0^1 dx\, x g(x) \end{aligned} \qquad (8.91)$$

if we again replace $\alpha_1(\omega)$ by its low-frequency approximation α_1. Assuming

$$\int_0^1 dx\, x g(x) \sim 1, \qquad (8.92)$$

we have (Spruch and Kelsey, 1978)

$$U_{df}(r) \sim \frac{\hbar q^2 \alpha_1}{mcr^5} \qquad (8.93)$$

for the retarded interaction between a polarizable system and a free charge. Except for the factor $11/4\pi$, this is just the interaction $V_{\text{ret}}(r)$ given by (8.80).

Spruch and Kelsey have given similar derivations, based on (8.86), of other retarded interactions involving free charges and neutral or charged polarizable systems. Here we will focus attention on the interaction (8.80), since it provides a novel example of a Casimir-type effect in atomic spectroscopy. Kelsey and Spruch (1978) suggested that this interaction might be observed in the energy levels of highly excited Rydberg states of atoms. In this case the (charged) polarizable system is the ionic core of the atom and the "free" charge is an electron in a highly excited state. Kelsey and Spruch considered the case of a helium–like ionic core with the outer electron in a state with $n > \ell \gg 1$, where n and ℓ are the principal and orbital angular momentum quantum numbers. From a detailed analysis of a perturbation expansion they concluded that the effective potential between an electron and a polarizable system of charge $(Z-1)e$ is

$$V(r) = -\frac{(Z-1)e^2}{r} - \frac{1}{2}\frac{\alpha e^2}{r^4}\left[1 - \frac{11}{2\pi}\frac{\hbar/mc}{r}\right] + \ \cdots \ . \qquad (8.94)$$

The first term is the Coulomb interaction, and ... denotes terms of order r^{-6} and higher. Of interest here is the second term, which is the polarization potential modified by the retarded potential $V_{\rm ret}(r)$. We note again that $V_{\rm ret}(r)$ is an additive correction to the polarization potential (8.79). The shift due to $V_{\rm ret}(r)$ of the energy of an electron in state $|n\ell m\rangle$ is

$$\Delta E = \frac{11}{4\pi}\frac{\hbar e^2 \alpha}{mc}\langle n\ell m|\frac{1}{r^5}|n\ell m\rangle, \qquad (8.95)$$

and for a Rydberg electron with large n and ℓ (say, $n = 15$ and $\ell = 14$) this shift is of order $(e^2/\hbar c)^3 mc^2/n^{10}$ (Kelsey and Spruch, 1978). The retarded contribution to the polarization potential has been confirmed by dispersion-theoretic methods (Feinberg and Sucher, 1983; Feinberg, Sucher, and Au, 1989).

As noted in Chapter 3, the condition for retardation to be important is roughly that r exceed $137a_o$. At a separation of $137a_o$ the correction to the polarization potential due to the second term in brackets in (8.94) is

$$\frac{11}{2\pi}\left(\frac{\hbar}{mc}\right)\left(\frac{e^2}{\hbar c}\right)\frac{1}{a_o} \sim 10^{-4} \ , \qquad (8.96)$$

and so the effect of $V_{\rm ret}(r)$ is obviously very small. Nevertheless Lundeen et al. have been able to measure certain level shifts to such high precision that a contribution from $V_{\rm ret}(r)$ can be inferred.[23]

---

[23] For a description of the experimental method we refer the reader to Lundeen (1991) and Hessels et al. (1992). The latter article summarizes the comparison of theory and experiment.

These experiments measure fine-structure intervals of high angular momentum states of Rydberg He atoms with $n = 10$, for which the separations of states of different orbital angular momentum have been calculated very accurately without the long-range, retarded Kelsey–Spruch interaction (Drachman 1982, 1985, 1988). The differences between the measured and calculated values were compared with the energy shifts expected from the Kelsey–Spruch interaction. It was found that these differences are about an order of magnitude smaller than simple estimates of the Kelsey–Spruch energy shifts (Palfrey and Lundeen, 1984). In other words, the experimental results suggested that any effect of a retarded polarization potential is substantially smaller than indicated by the Kelsey–Spruch theory.

However, further theoretical work revealed that the complete interaction $V_{ret}(r)$ that approaches the Kelsey–Spruch form asymptotically is in fact about an order of magnitude smaller than the asymptotic interaction for the electron-core separation $r \sim 100a_o$ appropriate to $n = 10$ states of He (Au, Feinberg, and Sucher, 1984, 1987; Babb and Spruch, 1988, 1989). It was then noted that the values of $V_{ret}(r)$ for small r were large enough to ruin the agreement between theory and experiment for low-lying states (Lundeen, 1991). This led to the realization that the first two terms of the short-range expansion of $V_{ret}(r)$ are in fact already included in the "standard atomic theory" of helium (Au, 1989; Au and Mesa, 1990; see also Lundeen, 1991).

The most accurate theory for comparison with experiment appears to be that of Drake, which includes both a short-range approximation to V_{ret} and radiative corrections (Drake, 1990; Goldman and Drake, 1992). In addition to high-precision calculations obtained by variational methods, Goldman and Drake (1992) show that the Lamb shift is about an order of magnitude larger than the Casimir effect, and must be well understood for the comparison of theory and experiment. There are statistically significant differences between theory and experiment that are not yet understood, but nevertheless the data confirm, to an accuracy better than 10%, the effect of retardation in Rydberg atoms (Hessels et al., 1992).

Regarding the interpretation of the experiments, it is worth mentioning that Goldman and Drake (1992) derive the asymptotic expression for the Lamb shift by assuming the Rydberg electron to produce a field that acts to modify the Lamb shift of the $1s$ electron. From this point of view, the experiments of Lundeen et al. probe the effects of external electric fields on the one-electron Lamb shift.

It is also interesting that the level shifts have been interpreted and calculated from two points of view — the "standard atomic theory" that treats the Rydberg helium atoms as just another two-electron system, and the "long-range interaction" picture just outlined. Although the asymptotic expansions used in the latter are not sufficiently accurate for $\ell \sim 7$, the long-range interaction picture does give predictions that are fairly close to the calculations of Drake et al.

The Kelsey–Spruch suggestion that the long-range "vacuum fluctuation" contribution to the polarization potential might be observed in the spectroscopy of Rydberg atoms has thus stimulated not only precise experimental studies of the fine structure of Rydberg helium, but also a clearer understanding of the more traditional aspects of the theory. At present there are no experiments that correspond to electron–core separations r large enough to measure the original asymptotic contribution obtained by Kelsey and Spruch. However, the extension of the theory to smaller separations has led to related predictions (Au et al., 1984, 1987; Babb and Spruch, 1988, 1989) that have in fact been corroborated experimentally to $\sim 10\%$ precision (Lundeen, 1991).

8.10   Casimir's Electron Model

Consider a spherical conducting shell of radius a. The zero-point field energy $E(a) = \frac{1}{2} \sum_\beta \hbar \omega_\beta$ can be expected to come predominantly from frequencies $\omega \sim c/a$, so that

$$E(a) = -C \frac{\hbar c}{2a} \ . \tag{8.97}$$

Here C is some dimensionless factor which, if positive, means that there is an inward, attractive force on the shell due to the vacuum field, analogous to the Casimir force between parallel conducting plates.

Casimir (1953) suggested that an electron might be envisioned as a spherical shell of charge e and that the attractive force associated with (8.97) might exactly balance the outward, repulsive force associated with the electrostatic self-energy [equation (5.18)]

$$U = \frac{e^2}{2a} \ . \tag{8.98}$$

The force per unit area associated with (8.97) is $F(a) = -C(\hbar c/8\pi a^4)$, and the force per unit area associated with (8.98) is $e^2/8\pi a^4$. The condition that these two forces balance each other is then

$$C = \frac{e^2}{\hbar c} \ . \tag{8.99}$$

In other words, Casimir suggested that an attractive force due to electromagnetic zero-point energy might provide the so-called Poincare' stress necessary, from a classical perspective, to hold the electron together. The particular value e of the electron charge according to this model is just the value that allows the repulsive and attractive forces to balance each other; the vacuum field and the electrostatic self-interaction together account for charge quantization. In particular, the value of the fine structure constant according to this "admittedly very crazy" model (Casimir, 1953) is just the number C appearing in (8.97), and is independent of any presumed electron radius.[24]

Unfortunately this fascinating model appears to fail because the factor C in (8.97) turns out to be *negative*, i.e., the force associated with the zero-point field energy in this case is repulsive rather than attractive. This result was first obtained by Boyer (1968). Using mode functions appropriate to the spherical symmetry of the problem, Boyer obtained [see also Davies (1972)]

$$C \cong -0.09 \ . \tag{8.100}$$

Thus C not only has the "wrong" sign, but is also about 12 times larger in magnitude than the fine structure constant. Similarly, Balian and Duplantier (1978) and Milton, DeRaad, and Schwinger (1978) have obtained $C = -0.09235$. Candelas (1982) has argued that the more recent calculations are in error to the extent that they omit a cutoff-dependent contribution to the zero-point energy. He obtains

$$E(a) = 0.09235 \frac{\hbar c}{2a} - \frac{\hbar}{4\pi} \int_0^\infty d\nu \ , \tag{8.101}$$

where it is implicit in the second term that a high-frequency cutoff is determined by the internal structure of the shell, which is not dealt with in the "macroscopic" electromagnetic theory (recall the remarks in Sections 8.3 and 8.4). Of course the second term in (8.101) is independent of the radius a and so does not affect the calculation of the force.

These calculations are rather complicated, and consequently will not be reproduced here. There does not appear to be any simple explanation for the repulsive character of the zero-point energy of a spherical shell.

Casimir has remained "reluctant entirely to give up the idea that the value of $\hbar c/e^2$ has something to do with a compensation of zero-point energy and electrostatic energy ... ," and has speculated on another possibility (Casimir, 1978):

---

[24] See the remarks and questions by Pais, Rosenfeld, Peierls, Dirac, Fierz, Heisenberg, and Belinfante appearing after the brief discussion by Casimir (1953).

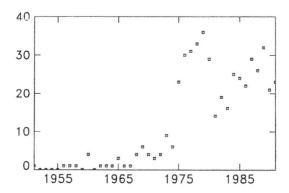

Figure 8.6: Number of papers per year, according to *Science Citation Index*, citing Casimir's original paper "On the Attraction between Two Perfectly Conducting Plates," *Proc. Kon. Ned. Akad. Wetenschap* **51**, 793 (1948).

Suppose geometry involves somehow a shortest length R and also a largest wavenumber k. The product kR would have to be of order unity, but its exact value would depend on the details of the theory. Then an attempt to create a point-singularity where the electromagnetic field disappears will create disappearance in a region with volume R^3 and the reduction of self-energy will be $\cong \hbar c R^3 k^4$. On the other hand, a charge will have to occupy a volume $\cong R^3$ too, and this yields a potential energy e^2/R. So there will be a compensation if

$$\hbar c/e^2 \cong 1(Rk)^4. \qquad (8.102)$$

Is this an indication that a theory involving a shortest length (or perhaps a theory involving some "fuzziness" ...) might automatically lead to a definite value for the fine structure constant? Or is what I have written no more than a roundabout way of saying that $\hbar c$ and e^2 have the same dimension?

8.11   Remarks

There is a large literature on Casimir effects, and the subject in all its aspects could easily fill a large book. As noted at the end of the preceding chapter, interest in the Casimir force between perfectly conducting plates has endured over many years (Figure 8.6). In this chapter and the previous one we have dealt with perhaps the most important aspects of Casimir effects. But there are important aspects of the subject that we have not broached, and in this section we will briefly mention a few of them.

The simple interpretation of the Casimir force between conducting plates as a consequence of vacuum radiation pressure (Section 3.10) suggests an approach based on the vacuum electromagnetic stress tensor. Such an approach to the Casimir force between conducting plates was taken by Brown and Maclay (1969), who calculated the stress tensor using an image method for both zero and finite temperatures. Their finite-temperature results agree with those obtained in Section 8.6.

Barton (1991) has pointed out that the stress tensor appropriate to the Casimir force between conducting plates does not commute with the Hamiltonian of the quantized field and therefore that it is subject to fluctuations. He calculated mean-square fluctuations, including averages over finite times and areas appropriate to some measurement apparatus and found them to be too small to be measured directly.

Shakeshaft and Spruch (1980) have considered radiative level shifts of an electron bound by its image to a conducting wall, a system they refer to as *murium* [see also Barton (1970)]. This system, sometimes called for obvious reasons a *one-dimensional hydrogen atom,* can of course be solved exactly when the vacuum field is ignored, and for an electron bound to a surface of liquid helium there have been experiments in which Rydberg level structure has been observed. Shakeshaft and Spruch interpret the rather large shift (0.2 % for the ground state in the case of an ideal wall[25]) in terms of the electron acquiring, "through vacuum fluctuations, a zero-point kinetic energy whose magnitude depends on the distance of the electron from the wall." The effect may be viewed as a Casimir–Polder interaction in the limiting case of a free electron, and its calculation is certainly "one of the simplest calculations in all of quantum electrodynamics" (Spruch, 1986).

Barut and his collaborators have shown how effects attributable to vacuum field fluctuations can be derived with a theory in which there are source fields but no nontrivial vacuum field; in this theory the electromagnetic field is not quantized[26] (Barut and Van Huele, 1985). The theory

---

[25] For a realistic model of the wall the order of the effect changes from $e^2/\hbar c$ to $(e^2/\hbar c)^3$ (Spruch, 1986).

[26] This theory has been criticized on the grounds that an assumed completeness relation leads in effect to an approximate version of field quantization, and that this version will not agree with QED to all orders of a loop expansion. See the comment by Bialynicki-Birula (1986) and the reply by Barut (1986). Grandy (1991) has also criticized the presumed completeness relation. The fate of Barut's formalism will most likely be decided ultimately by a comparison of higher order iterations within that formalism to QED calculations beyond one-loop perturbations.

has been applied to problems of cavity QED (Barut and Dowling, 1987a), Casimir–Polder and van der Waals forces (Barut and Dowling, 1987b), non-relativistic and relativistic calculations of $g - 2$ (Barut and Dowling, 1988, 1989a,b), and to a derivation of the Unruh–Davies effect via source fields (Barut and Dowling, 1990).

We indicated in Section 8.7 that the Lifshitz theory has played a major role in the study of colloidal and other dielectric systems where van der Waals forces are dominant. The Lifshitz theory has also been corroborated in some elegant experiments of Sabisky and Anderson (1973) involving acoustical interferometric measurements of the thickness of liquid helium films. The comparison with the Lifshitz theory is excellent, although it involves a somewhat phenomenological expression for the dielectric constant as a function of frequency.

There is a considerable literature on the application of the Lifshitz theory to the theory of biological membranes (see, for instance, Mitchell, Ninham, and Richmond, 1973). Here again one must resort to phenomenological models for $\epsilon(\omega)$.

8.12 Stochastic Electrodynamics

In QED we cannot arbitrarily set to zero the homogeneous, source-free solution of the Maxwell operator equations in the Heisenberg picture. This "vacuum" field is necessary for the formal consistency of QED: without it, it would not be possible to preserve canonical commutation relations and all they entail (recall, for instance, the discussion in Section 2.6).

These equations are formally the same as the classical Maxwell equations. Classically, however, we generally assume implicitly that the homogeneous solution of the Maxwell equations is that in which the electric and magnetic fields vanish identically. That is, we assume that there are no fields in the absence of any sources. In the absence of sources the vacuum field is simply zero, with no energy or fluctuations whatsoever.

This difference between classical and quantum electrodynamics, together with the evident importance of the fluctuating vacuum field in QED, suggests the adoption of a different boundary condition in classical electrodynamics: instead of assuming that the classical field vanishes in the absence of sources, we can assume that there is a fluctuating *classical* field with zero-point energy $\frac{1}{2}\hbar\omega$ per mode. As long as this field satisfies the Maxwell equations there is no a priori inconsistency in this assumption. Whether it is a better working assumption than the standard, "obvious" one is a matter to be decided ultimately by comparison with experiment. The appearance of $\hbar$ in this modification of classical electrodynamics implies no deviation from conventional classical ideas, for h is regarded as nothing more than a number chosen to obtain consistency of the predictions of the theory with experiment. In this theory $\hbar$ has nothing directly to do with such quantum-mechanical notions as a fundamental limitation on the precision to which position and momentum variables can be measured

This modification of classical electrodynamics to include a zero-point field of energy $\frac{1}{2}\hbar\omega$ per mode is called *stochastic electrodynamics*. Since its original proposal by Marshall (1963, 1965) it has been of considerable interest among a small but active group of researchers.

In stochastic electrodynamics both the field and the particles with which it interacts are treated classically. Marshall derived the form of the spectral energy density of the vacuum field in stochastic electrodynamics by requiring the mean-square displacement of a charged harmonic oscillator to be identical to that given by quantum theory. In this approach the fluctuations in the displacement are *caused* by the interaction with the fluctuating vacuum field. Marshall showed that the resulting spectral energy density,

$$\rho_o(\omega) = \frac{\hbar\omega^3}{2\pi^2 c^3} \, , \qquad (8.103)$$

is Lorentz–invariant. This is identical to the result (2.73) of quantum theory. Boyer (1969) also demonstrated the Lorentz invariance of the spectral density (8.103). The proportionality of $\rho_o(\omega)$ to ω^3 is in fact *required* by the condition of Lorentz invariance, and Boyer takes this as a foundation for stochastic electrodynamics. Thus one can arrive at the form $\rho_o(\omega) \propto \omega^3$ by imposing the general requirement of Lorentz invariance, and $\hbar$ then enters the theory as a multiplicative constant chosen by comparison of theoretical predictions with experiment. This entry of $\hbar$ into the theory is no less "fundamental" than the way it is introduced in quantum theory.

In classical Coulomb-gauge electrodynamics the transverse fields may be obtained from the vector potential satisfying

$$\nabla^2 \mathbf{A} - \frac{1}{c^2}\ddot{\mathbf{A}} = -\frac{4\pi}{c}\mathbf{J}^\perp, \qquad (8.104)$$

where $\mathbf{J}$ is the current density. The solution of this equation has the form

$$\mathbf{A}(\mathbf{r}, t) = \mathbf{A}_o(\mathbf{r}, t) + \mathbf{A}_s(\mathbf{r}, t), \qquad (8.105)$$

where $\mathbf{A}_s(\mathbf{r}, t)$ is the solution of the inhomogeneous wave equation and $\mathbf{A}_o(\mathbf{r}, t)$ is a solution of the homogeneous equation

$$\nabla^2 \mathbf{A}(\mathbf{r}, t) - \frac{1}{c^2}\ddot{\mathbf{A}} = 0. \qquad (8.106)$$

The solution $\mathbf{A}_s(\mathbf{r}, t)$ may be written formally in terms of a Green function. The traditional approach is to use the retarded Green function and to take $\mathbf{A}_o$ to be identically zero. The use of the retarded Green function satisfies our intuitive notion of causality, and the *choice* $\mathbf{A}_o \equiv 0$ is reasonable in that we expect no field when there is no source. But we can choose any (transverse) solution of the homogeneous wave equation for $\mathbf{A}_o$ and still satisfy the wave equation for the total vector potential (8.105). The choice $\mathbf{A}_o \equiv 0$ defines "traditional electrodynamics," whereas the choice of a random field of zero mean defines stochastic electrodynamics.

In free space we can expand the electric and magnetic fields in transverse plane waves as

$$\mathbf{E}_o(\mathbf{r}, t) = i \sum_{\mathbf{k}\lambda} [C_{\mathbf{k}\lambda} \mathbf{e}_{\mathbf{k}\lambda} e^{i(\mathbf{k}\cdot\mathbf{r}-\omega_k t+\theta_{\mathbf{k}\lambda})} - C_{\mathbf{k}\lambda}^* \mathbf{e}_{\mathbf{k}\lambda}^* e^{-i(\mathbf{k}\cdot\mathbf{r}-\omega_k t+\theta_{\mathbf{k}\lambda})}],$$

(8.107)

with $\mathbf{B}_o$ determined from the Maxwell equation $-c^{-1}\dot{\mathbf{B}}_o(\mathbf{r}, t) = \nabla \times \mathbf{E}_o(\mathbf{r}, t)$. As in Chapter 2 the $\mathbf{e}_{\mathbf{k}\lambda}$ are unit polarization vectors, with $\mathbf{k} \cdot \mathbf{e}_{\mathbf{k}\lambda} = 0$. The "randomness" of the field is contained in the phases $\theta_{\mathbf{k}\lambda}$, which are assumed to be uniformly and independently distributed over the interval $[0, 2\pi]$. Similar random classical fields were used by Planck, Einstein, and Hopf to represent *thermal* fields [cf. equation (1.49)]. The (classical) expectation value of the energy density of the random field is $(8\pi)^{-1}\langle \mathbf{E}_o^2(\mathbf{r}, t)+\mathbf{B}_o^2(\mathbf{r}, t)\rangle_\theta$, where $\langle ... \rangle_\theta$ denotes an average with respect to the random variables $\{\theta_{\mathbf{k}\lambda}\}$. It follows easily from the form of $\mathbf{E}_o(\mathbf{r}, t)$ and $\mathbf{B}_o(\mathbf{r}, t)$ that

$$\frac{1}{8\pi}\langle \mathbf{E}_o^2 + \mathbf{B}_o^2\rangle_\theta = \frac{1}{2\pi}\sum_{\mathbf{k}\lambda}|C_{\mathbf{k}\lambda}|^2 \equiv \sum_{\mathbf{k}\lambda}\frac{1}{2}\hbar\omega_k,$$

(8.108)

so that

$$|C_{\mathbf{k}\lambda}|^2 = \frac{\pi\hbar\omega_k}{V}.$$

(8.109)

This result can also be obtained from the requirement that the spectral energy density of the zero-point field be Lorentz invariant. Then $\hbar$ appears as a suitably chosen constant in the theory (Boyer, 1975). Thus we can write

$$\mathbf{E}_o(\mathbf{r}, t) = i \sum_{\mathbf{k}\lambda} \left(\frac{\pi\hbar\omega_k}{V}\right)^{1/2} [e^{i(\mathbf{k}\cdot\mathbf{r}-\omega_k t+\theta_{\mathbf{k}\lambda})} - e^{-i(\mathbf{k}\cdot\mathbf{r}-\omega_k t+\theta_{\mathbf{k}\lambda})}]\mathbf{e}_{\mathbf{k}\lambda}$$

(8.110)

if, without any loss of generality, we take the polarization unit vectors $\mathbf{e}_{\mathbf{k}\lambda}$ to be real.

The factors $\exp(i\theta_{\mathbf{k}\lambda})$ and $\exp(-i\theta_{\mathbf{k}\lambda})$ in (8.110) replace the photon annihilation and creation operators $a_{\mathbf{k}\lambda}(0)$ and $a^{\dagger}_{\mathbf{k}\lambda}(0)$ in the corresponding QED expression for the vacuum electric field in free space. The expectation values $\langle\text{vac}|a_{\mathbf{k}\lambda}(0)|\text{vac}\rangle = \langle\text{vac}|a^{\dagger}_{\mathbf{k}\lambda}(0)|\text{vac}\rangle = 0$ are replaced in stochastic electrodynamics (SED) by

$$\langle e^{i\theta}\mathbf{k}\lambda\rangle_{\theta} = \langle e^{-i\theta}\mathbf{k}\lambda\rangle_{\theta} = 0. \tag{8.111}$$

Similarly the QED expectation value $\langle\text{vac}|a_{\mathbf{k}\lambda}(0)a^{\dagger}_{\mathbf{k}\lambda}(0)|\text{vac}\rangle = 1$ corresponds in SED to

$$\langle e^{i\theta}\mathbf{k}\lambda e^{-i\theta}\mathbf{k}\lambda\rangle_{\theta} = 1. \tag{8.112}$$

However, the expectation value $\langle\text{vac}|a^{\dagger}_{\mathbf{k}\lambda}(0)a_{\mathbf{k}\lambda}(0)|\text{vac}\rangle = 0$ has no analog in SED. This explains the factor $(\pi\hbar\omega_k/V)^{1/2}$ in (8.110) as opposed to the factor $(2\pi\hbar\omega_k/V)^{1/2}$ appearing in the QED expression for the electric field. The $1/\sqrt{2}$ in SED is necessary in order to give the same zero-point energy $\frac{1}{2}\hbar\omega_k$ per mode. Another way to say this is that the positive- and negative-frequency parts of the field in SED contribute symmetrically to the vacuum field energy density. For this reason calculations in SED are analogous to QED calculations with a symmetric ordering of photon annihilation and creation operators (Milonni and Smith, 1975; Boyer, 1975).

If one looks back over the derivation in Section 2.7 of the Casimir force between conducting plates, it becomes clear that the argument can be couched in the language of SED rather than QED. That is, all that is really required in that derivation is the zero-point energy $\frac{1}{2}\hbar\omega_k$ per mode of the electromagnetic field. Whether this zero-point energy is of quantum or classical origin is irrelevant for the purpose of deriving the Casimir force — SED accounts perfectly well for the Casimir force. It is similarly able to account for the Casimir–Polder force between an atom and a conducting wall, the retarded and unretarded van der Waals interactions between two atoms, and, subject to certain assumptions, the Planck spectrum (Boyer, 1969; Milonni, 1981). Being a purely classical theory of radiation *and matter*, SED is unable to derive quantitative expressions for the polarizabilities appearing in these forces, or for that matter to account for the discrete energy levels of the interacting atoms.

The Unruh–Davies effect for accelerated detectors is also accounted for by SED (Boyer, 1984). Indeed the SED derivations of this and other vacuum phenomena parallel so closely the Heisenberg-picture derivations we have given in the preceding chapters that there is little point in going through them here. We refer the reader to the surveys by Milonni (1976) and Boyer

(1980) and the references therein, as well as to the many more recent papers in this field.

In spite of the successes of SED, it cannot at this time be considered a serious alternative to QED. For one thing, no *classical* theory of the electromagnetic field can account for such experimentally observed phenomena as the photon polarization correlations in a cascade radiative decay of atomic states, the correlations that have been studied in the context of Bell's theorem (Clauser, 1972; Clauser and Shimony, 1978; Aspect, 1984; Milonni and Singh, 1991). That is, the strictly classical nature of the electromagnetic field in SED, with or without external fields, rules it out from consideration as a fundamental theory of the electromagnetic field. In addition, SED runs into difficulties when one considers the thermal equilibrium properties of *nonlinear* dipole oscillators (Boyer, 1980). We refer the reader to the reviews by Boyer (1980) and de la Peña and Cetto (1991, de la Peña, 1983), and the references therein, for a detailed account of SED.

8.13   Concluding Remarks

Maxwell wrote that[27]

> The facts of electromagnetism are so complicated and various, that the explanation of any number of them by several different hypotheses must be interesting, not only to physicists, but to all who desire to understand how much evidence the explanation of phenomena leads to the credibility of a theory, or how far we ought to regard a coincidence in the mathematical expression of two sets of phenomena as an indication that these phenomena are of the same kind.

The "facts of electromagnetism," after all these years, are as "complicated and various" as ever. In the preceding pages we have described phenomena that are explainable from the hypothesis that the quantum theory of electromagnetism is correct in its prediction of a fluctuating vacuum field. The variety of phenomena that can be explained with this hypothesis is certainly impressive. Spontaneous radiation, the phenomenon responsible for nearly all the light we see and for bringing energy from the sun to our planet, is explainable at least in part in terms of fluctuations of the electromagnetic vacuum field, as are the universal van der Waals forces that are crucial to our understanding of a wide variety of systems of physical, chemical, and biological interest. Macroscopic, Casimir-type manifestations of these forces are interpretable in terms of the effects of boundary conditions on the vacuum field, as are other cavity QED effects such as

---

[27]From Aitchison (1991).

the environmental modification of spontaneous radiation rates. The "best theory we have," quantum electrodynamics, finds its most precise tests in phenomena such as the Lamb shift and the anomalous magnetic moment of the electron, where the vacuum field fluctuations are the basis of beautiful heuristic interpretations. These fluctuations appear to be "promotable" to the level of thermal fluctuations when our detectors are accelerated, an effect too small to be measured in the laboratory but important to our understanding of quantum field theory and general relativity. And then there are "applications" of vacuum field fluctuations to the fundamental linewidths and coherence lengths of semiconductor lasers of interest for optical communications. It is up to the reader to decide to what extent the explanation of these diverse phenomena lends "credibility" to the concept of the vacuum field.

These effects attributable to vacuum electromagnetic fluctuations, however, can be explained by "several different hypotheses." We have noted repeatedly, particularly in Chapters 4 and 7, that the vacuum fluctuation effects we have described can be explained, equally well in most cases, in terms of sources fields, or in terms of some combination of source and vacuum fields. In other words, for many purposes we can choose to think in terms of either the fluctuation or the dissipation aspect of the interaction of charged particles with the electromagnetic field. And both the vacuum and source fields are absolutely essential for the formal consistency of quantum electrodynamics.

At the opposite extreme one can take a stand "against interpretation," and argue that none of these effects *require* us to think in terms of vacuum fields, or source fields, and that for the purpose of calculations all we need to know about is the Schrödinger equation and the other tenets of quantum theory. Such an approach, though perfectly rational, appears to me to be not only uninteresting, but also contrary to the way physics has for the most part developed – intuitively and with physical images rather than deductively from the formalisms, when they exist, that happen at any given time to be fashionable. Moreover, most physicists would agree on the value of a single concept that provides intuitive explanations for the "complicated and various facts of electromagnetism." We shall soon see that the concept of the quantum vacuum is just as valuable when we broaden our perspective to include relativistic effects.

In Chapter 10 we discuss the Casimir effect for the quantized Dirac field, which is of interest in connection with the "MIT bag model." However, there are many examples and implications of Casimir effects that we cannot address in any sort of detail in this book.[28] An important implication

---

[28] Other aspects of Casimir effects are reviewed by Mostepanenko and Trunov (1988).

arises in cosmology, in connection with the cosmological constant problem. The reality of zero-point energies suggested by the existence of Casimir forces evidently means that zero-point energies should be taken seriously in general relativity. When this is done the total zero-point energy density of the vacuum acts in effect as a cosmological constant of the type introduced by Einstein in order to have static solutions of his field equations. However, astronomical data indicate that any such cosmological constant must be many orders of magnitude smaller than predicted by quantum field theory (Weinberg, 1989). This difficulty remains unresolved.

8.14   Bibliography

Aitchison, I. J. R., "The Vacuum and Unification," in *The Philosophy of Vacuum*, ed. S. Saunders and H. R. Brown (Clarendon Press, Oxford, 1991).

Anderson, A., S. Haroche, E. A. Hinds, W. Jhe, and D. Meschede, "Measuring the van der Waals Forces Between a Rydberg Atom and a Metallic Surface," *Phys. Rev.* A37, 3594 (1988).

Aspect, A., in *The Wave-Particle Dualism*, ed. S. Diner, D. Fargue, G. Lochak, and F. Selleri (Reidel, Dordrecht, 1984).

Au, C. K., G. Feinberg, and J. Sucher, "Retarded Long-Range Interaction in He Rydberg States," *Phys. Rev. Lett.* 53, 1145 (1984).

Au, C. K., G. Feinberg, and J. Sucher, "A Quantum-Field Theory Approach to the Calculation of Energy Levels in Helium-like Rydberg Atoms," *Ann. Phys.* (New York) 173, 355 (1987).

Au, C. K., "Effects of Transverse Photon Exchange in Helium Rydberg States: Corrections Beyond the Coulomb-Breit Interaction," *Phys. Rev.* A39, 2789 (1989).

Au, C. K. and M. A. Mesa, "Addendum to 'Effects of Transverse Photon Exchange in He Rydberg States,' " *Phys. Rev.* A41, 2848 (1990).

Axilrod, B. M. and E. Teller, "Interaction of the van der Waals Type Between Three Atoms," *J. Chem. Phys.* 11, 299 (1943).

Babb, J. F. and L. Spruch, "Evaluation of Retardation Energy Shifts in a Rydberg Helium Atom," *Phys. Rev.* A38, 13 (1988).

Babb, J. F. and L. Spruch, "Retardation (Casimir) Effect for a Multielectron Core System and a Rydberg Electron," *Phys. Rev.* A40, 2917 (1989).

Balian, R. and B. Duplantier, "Electromagnetic Waves Near Perfect Conductors. II. Casimir Effect," *Ann. Phys.* (New York) 112, 165 (1978).

Barton, G., "Quantum Electrodynamics of Spinless Particles Between

Conducting Plates," *Proc. Roy. Soc. Lond.* A**320**, 251 (1970).

Barton, G., "On the Fluctuations of the Casimir Force," *J. Phys.* A**24**, 991 (1991); "On the Fluctuations of the Casimir Force: II. The Stress-Correlation Function," *J. Phys.* A**24**, 5533 (1991).

Barut, A. O., "Quantum Electrodynamics Based on Self-Energy versus Quantization of Fields: Illustration by a Simple Model," *Phys. Rev.* A**34**, 3502 (1986).

Barut, A. O. and J. P. Dowling, "Quantum Electrodynamics Based on Self-Energy: Spontaneous Emission in Cavities," *Phys. Rev.* A**36**, 649 (1987a).

Barut, A. O. and J. P. Dowling, "Quantum Electrodynamics Based on Self-Energy, Without Second Quantization: The Lamb shift and Long-Range Casimir–Polder van der Waals Forces Near Boundaries," *Phys. Rev.* A**36**, 2550 (1987b).

Barut, A. O. and J. P. Dowling, "Quantum Electrodynamics Based on Self-Fields, Without Second Quantization: A Nonrelativistic Calculation of $g - 2$," *Phys. Rev.* A**38**, 4405 (1988).

Barut, A. O. and J. P. Dowling, "Quantum Electrodynamics Based on Self-Fields, Without Second Quantization: Apparatus Dependent Contributions to $g - 2$," *Phys. Rev.* A**39**, 2796 (1989a).

Barut, A. O. and J. P. Dowling, "QED Based on Self-Fields: A Relativistic Calculation of $g - 2$," *Z. Naturforsch.* **44a**, 1051 (1989b).

Barut, A. O. and J. P. Dowling, "Quantum Electrodynamics Based on Self- Fields: On the Origin of Thermal Radiation Detected by an Accelerated Observer," *Phys. Rev.* A**41**, 2277 (1990).

Barut, A. O. and J. F. Van Huele, "Quantum Electrodynamics Based on Self- Energy: Lamb Shift and Spontaneous Emission Without Field Quantiz- ation," *Phys. Rev.* A**32**, 3187 (1985).

Bernabéu, J. and R. Tarrach, "Long-Range Potentials and the Electromagnetic Polarizabilities," *Ann. Phys.* (New York) **102**, 323 (1976).

Bialynicki-Birula, I., "Comment on 'Quantum Electrodynamics Based on Self-Energy: Lamb shift and Spontaneous Emission Without Field Quantization,'" *Phys. Rev.* A**34**, 3500 (1986).

Born, M. and E. Wolf, *Principles of Optics* (Pergamon Press, Oxford, 1970).

Boyer, T. H., "Quantum Electromagnetic Zero-Point Energy of a Conducting Spherical Shell and the Casimir Model for a Charged Particle," *Phys. Rev.* **174**, 1764 (1968).

Boyer, T. H., "Derivation of the Blackbody Radiation Spectrum without Quantum Assumptions," *Phys. Rev.* **182**, 1374 (1969).

Boyer, T. H., "General Connection Between Random Electrodynamics and Quantum Electrodynamics for Free Electromagnetic Fields and for Dipole

Oscillator Systems," *Phys. Rev.* D11, 809 (1975).

Boyer, T. H., "A Brief Survey of Stochastic Electrodynamics," in *Foundations of Radiation Theory and Quantum Electrodynamics*, ed. A.O. Barut (Plenum Press, New York, 1980).

Boyer, T. H., "Thermal Effects of Acceleration for a Classical Dipole Oscillator in Classical Electromagnetic Zero-Point Radiation," *Phys. Rev.* D29, 1089 (1984).

Brown, L. S. and G. J. Maclay, "Vacuum Stress Between Conducting Plates: An Image Solution," *Phys. Rev.* 184, 1272 (1969).

Candelas, P., "Vacuum Energy in the Presence of Dielectric and Conducting Surfaces," *Ann. Phys.* (New York) 143, 241 (1982).

Carniglia, C. K. and L. Mandel, "Quantization of Evanescent Electromagnetic Waves," *Phys. Rev.* D3, 280 (1971).

Casimir, H. B. G. and D. Polder, "The Influence of Retardation on the London– van der Waals Forces," *Phys. Rev.* 73, 360 (1948).

Casimir, H. B. G., "Introductory Remarks on Quantum Electrodynamics," *Physica* 19, 846 (1953).

Casimir, H. B. G., "Remarks on the Fine-Structure Constant," *Rev. Roumaine de Physique* 23, 723 (1978).

Clauser, J. F., "Experimental Limitations to the Validity of Semiclassical Radiation Theories," *Phys. Rev.* A6, 49 (1972).

Clauser, J. F. and A. Shimony, "Bell's Theorem: Experimental Tests and Implications," *Rep. Prog. Phys.* 41, 1881 (1978).

Davies, B., "Quantum Electromagnetic Zero-Point Energy of a Conducting Spherical Shell," *J. Math. Phys.* 13, 1324 (1972).

Derjaguin, B. V., "The Force Between Molecules," *Sci. Am.* 203, 47 (July 1960).

De la Peña, L., "Stochastic Electrodynamics: Its Development, Present Situation and Perspectives," in *Stochastic Processes Applied to Physics and Other Related Fields* (World Scientific, Singapore, 1983), 428– 581.

De la Peña, L. and A. M. Cetto, "Teorías estocásticas de la mecánica cuántica," *Rev. Mex. Fis.* 37, 17 (1991).

Derjaguin, B. V. and I. I. Abrikosova, "Direct Measurement of the Molecular Attraction of Solid Bodies. I. Statement of the Problem and Method of Measuring Forces by Using Negative Feedback," *Sov. Phys. JETP* 3, 819 (1957).

Derjaguin, B. V., Y. I. Rabinovich, and N. V. Churaev, "Direct Measurement of Molecular Forces," *Nature* 272, 313 (1978).

Drachman, R., "Rydberg States of Helium: An Optical-Potential Analysis," *Phys. Rev.* A26, 1228 (1982).

Drachman, R., "Rydberg States of Helium: Relativistic and Second-Order

Corrections," *Phys. Rev.* A**31**, 1253 (1985).

Drachman, R., "Rydberg States of Helium: A New Recoil Term," *Phys. Rev.* A**37**, 1979 (1988).

Drake, G. W. F., "Eigenvalues and Retardation Effects in the $n = 10$ States of Helium," *J. Phys.* B**22**, L651 (1989); *J. Phys.* B**23**, 1943 (1990) (corrigendum).

Drake, G. W. F., "Variational Eigenvalues for the Rydberg States of Helium: Comparison with Experiment and with Asymptotic Expansions," *Phys. Rev. Lett.* **65**, 2769 (1990).

Dzyaloshinskii, I. E., E. M. Lifshitz, and L. P. Pitaevskii, "The General Theory of Van der Waals Forces," *Adv. Phys.* **10**, 165 (1961).

Ewald, P. P., Dissertation, University of Munich, 1912. An English translation by L. M. Hollingsworth is available as U.S. Air Force Report AFCRL-70- 0580.

Ewald, P. P., "Zur Begründung der Kristalloptik," *Ann. d. Physik* **49**, 1 (1916).

Feinberg, G. and J. Sucher, "Long-Range Forces Between a Charged and Neutral System," *Phys. Rev.* A**27**, 1958 (1983).

Feinberg, G., J. Sucher, and C. K. Au, "The Dispersion Theory of Dispersion Forces," *Phys. Rep.* **180**, 83 (1989).

Goldman, S. P. and G. W. F. Drake, "Asymptotic Lamb Shifts for Helium Rydberg States," *Phys. Rev. Lett.* **68**, 1683 (1992).

Grandy, W. T., Jr., "The Explicit Nonlinearity of Quantum Electrodynamics," in *The Electron — 1990*, ed. D. Hestenes and A. Weingartshofer (Kluwer, Dordrecht, 1991).

Hessels, E. A., P. W. Arcuni, F. J. Deck, and S. R. Lundeen, "Microwave Spectroscopy of High-L, N=10 Rydberg States of Helium," *Phys. Rev.* A**46**, 2622 (1992).

Hinds, E. A., C. I. Sukenik, M. G. Boshier, and D. Cho, "Deflection of an Atomic Beam by the Casimir Force," in *Atomic Physics 12*, ed. J. Zorn and R. Lewis (American Institute of Physics, New York, 1991).

Hunklinger, S., H. Geisselmann, and W. Arnold, "A Dynamic Method for Measuring the van der Waals Forces Between Macroscopic Bodies," *Rev. Sci. Instr.* **43**, 584 (1972).

Hynne, F. and R. K. Bullough, "The Scattering of Light. 3. External Scattering from a Finite Molecular Fluid," *Phil. Trans. Roy. Soc.* A**330**, 253 (1990).

Israelachvili, J. N. and B. W. Ninham, "Intermolecular Forces — the Long and Short of It," *J. Colloid. Interface Sci.* **58**, 14 (1977).

Israelachvili, J. N. and D. Tabor, "The Measurement of van der Waals Dispersion Forces in the Range 1.5 to 130 nm," *Proc. Roy. Soc.* A**331**,

19 (1972).

Kasevich, M., K. Moler, E. Riis, E. Sunderman, D. Weiss, and S. Chu, "Applications of Laser Cooling and Trapping," in *Atomic Physics 12*, ed. J. Zorn and R. Lewis (*American Institute of Physics*, New York, 1991).

Kelsey, E. J. and L. Spruch, "Retardation Effects on High Rydberg States: A Retarded R^{-5} Polarization Potential," *Phys. Rev.* **A18**, 15 (1978).

Lifshitz, E. M., "The Theory of Molecular Attractive Forces Between Solids," *Sov. Phys. JETP* **2**, 73 (1956).

London, F., "Zur Theorie und Systematik der Molekularkräfte," *Z. Phys.* **63**, 245 (1930).

Lundeen, S. R., "Precision Spectroscopy of High-L Rydberg States of Helium," in *Atomic Physics 12*, ed. J. Zorn and R. Lewis (*American Institute of Physics*, New York, 1991).

Marshall, T. W., "Random Electrodynamics," *Proc. Roy. Soc. Lond.* **A276**, 475 (1963).

Marshall, T. W., "Statistical Electrodynamics," *Proc. Camb. Philos. Soc.* **61**, 537 (1965).

Maxwell, J. C., "On Physical Lines of Force. Part II: The Theory of Molecular Forces Applied to Electric Currents," *Phil. Mag.* **21**, 338 (1861).

Milonni, P. W., "Semiclassical and Quantum-Electrodynamical Approaches in Nonrelativistic Radiation Theory," *Phys. Rep.* **25**, 1 (1976).

Milonni, P. W., "Quantum Mechanics of the Einstein–Hopf Model," *Am. J. Phys.* **49**, 177 (1981).

Milonni, P. W. and P. B. Lerner, "Extinction Theorem, Dispersion Forces, and Latent Heat," *Phys. Rev.* **A46**, 1185 (1992).

Milonni, P. W. and M.-L. Shih, "Source Theory of Casimir Force," *Phys. Rev.* **A45**, 4241 (1992).

Milonni, P. W. and S. Singh, "Some Recent Developments in the Fundamental Theory of Light," in *Advances in Atomic, Molecular, and Optical Physics*, Volume 28, ed. D.R. Bates and B. Bederson (Academic Press, Cambridge, Mass., 1991).

Milonni, P. W. and W. A. Smith, "Radiation Reaction and Vacuum Fluctuations in Spontaneous Emission," *Phys. Rev.* **A11**, 814 (1975).

Milton, K. A., L. L. DeRaad, Jr., and J. Schwinger, "Casimir Self-Stress on a Perfectly Conducting Spherical Shell," *Ann. Phys.* (New York) **115**, 388 (1978).

Mitchell, D., B. W. Ninham, and P. Richmond, "Van der Waals Forces Between Cylinders. 1. Nonretarded Forces between Thin Isotropic Rods and Finite Size Corrections," *Biophys. J.* **13**, 359 (1973).

Mostepanenko, V. M. and N. N. Trunov, "The Casimir Effect and Its

Applications," *Sov. Phys. Usp.* **31**, 965 (1988).

Oseen, C. W., "Über die Wechselwirkung zwischen zwei elektischen Dipolen der Polarisationsebene in Kristallen und Flüssigkeiten," *Ann. d. Physik* **48**, 1 (1915).

Overbeek, J. T. G., and M. J. Sparnaay, "London–van der Waals Attraction Between Macroscopic Objects," *Disc. Faraday Soc.* **18**, 12 (1954).

Palfrey, S. L. and S. R. Lundeen, "Measurement of High-Angular-Momentum Fine Structure in Helium: An Experimental Test of Long-Range Electromagnetic Forces," *Phys. Rev.* **53**, 1141 (1984).

Power, E. A. and T. Thirunamachandran, "The Nonadditive Dispersion Energies for N Molecules: a Quantum-Electrodynamical Theory," *Proc. Roy. Soc. Lond.* A**401**, 267 (1985).

Sabisky, E. S. and C. H. Anderson, "Verification of the Lifshitz Theory of the van der Waals Potential Using Liquid-Helium Films," *Phys. Rev.* A**7**, 790 (1973).

Sandoghdar, V., C. I. Sukenik, E. A. Hinds, and S. Haroche, "Direct Measurement of the van der Waals Interaction Between an Atom and Its Images in a Micron-Sized Cavity," *Phys. Rev. Lett.* **68**, 3432 (1992).

Schwinger, J., L. L. DeRaad, Jr., and K. A. Milton, "Casimir Effect in Dielectrics," *Ann. Phys.* (New York) **115**, 1 (1978).

Shakeshaft, R. and L. Spruch, "Radiative Corrections to the Energy Levels of 'Murium,' an Electron Bound by Its Image Charge to a Wall," *Phys. Rev.* A**22**, 811 (1980).

Shih, A. and V. A. Parsegian, "Van der Waals Forces Between Heavy Alkali Atoms and Gold Surfaces: Comparison of Measured and Predicted Values," *Phys. Rev.* A**12**, 835 (1975).

Slater, J. C. and J.G. Kirkwood, "The van der Waals Forces in Gases," *Phys. Rev.* **37**, 682 (1931).

Sparnaay, M. J., "Measurements of Attractive Forces Between Flat Plates," *Physica* **24**, 751 (1958).

Sparnaay, M. J., "The Historical Background of the Casimir Effect," in *Physics in the Making*, ed. A. Sarlemijn and M. J. Sparnaay (Elsevier, Amsterdam, 1989).

Spruch, L. and E. J. Kelsey, "Vacuum Fluctuation and Retardation Effects on Long-Range Potentials," *Phys. Rev.* A**18**, 845 (1978).

Spruch, L., "Retarded, or Casimir, Long-Range Rotentials," *Physics Today* (November 1986), 37.

Sukenik, C. I., M. G. Boshier, D. Cho, V. Sandoghdar, and E. A. Hinds, "Measurement of the Casimir–Polder Force," *Phys. Rev. Lett.* **70**,

560 (1993).

Tabor, D. and R. H. S. Winterton, "Direct Measurement of Normal and Retarded Van der Waals Forces," *Proc. Roy. Soc. Lond.* **A312**, 435 (1969).

Van Blokland, P. H. G. M. and J. T. G. Overbeek, "Van der Waals Forces Between Objects Covered with a Chromium Layer," *J. Chem. Soc. Faraday Trans.* **74**, 2637 (1978).

Weinberg, S., "The Cosmological Constant Problem," *Rev. Mod. Phys.* **61**, 1 (1989).

Wolf, E., "Electromagnetic Scattering as a Non-Local Boundary Value Problem," in *Symposia Mathematica*, Volume 18 (Academic Press, London, 1976), p. 333.

Chapter 9

The Dirac Equation

> Previously, people have thought of the vacuum as a region of space that is completely empty, a region of space that does not contain anything at all. Now we must adopt a new picture. We may say that the vacuum is a region of space where we have the lowest possible energy. Now, to get the lowest energy we must fill up all the states of negative energy ... Thus we must set up a new picture of the vacuum in which all the negative energy states are occupied and all the positive energy states are unoccupied.
> — P. A. M. Dirac (1978)

9.1 Introduction

The Schrödinger equation and the Heisenberg equations of motion as used in the preceding chapters are nonrelativistic. When quantum theory is formulated relativistically, there appear phenomena such as pair creation and vacuum polarization that could not be anticipated in purely nonrelativistic theory. Such phenomena force a revision of the concept of the vacuum considered thus far, where vacuum fluctuations have been associated primarily with the electromagnetic field. In this chapter we describe some of these relativistic phenomena.

The principle of relativity was stated by Isaac Newton as follows: "The motions of bodies included in a given space are the same among themselves whether that space is at rest or moves uniformly forward in a straight line" (Feynman, Leighton, and Sands, 1964). Thus no experiment performed inside a closed spaceship in uniform motion can be used to infer its velocity. All physical phenomena appear exactly the same as when the vehicle

is at rest; there is nothing special about zero velocity.[1] Newton's laws
of motion satisfy the principle of relativity under the "obvious," Galilean
transformation

$$x' = x - vt, \quad y' = y, \quad z' = z, \quad t' = t. \tag{9.1}$$

However, Maxwell's equations change form under this transformation of
coordinates, such that the velocity of light would be different in different
reference frames. Einstein postulated that Maxwell's equations should have
the same form in all coordinate systems, and in particular that the velocity
of light c is the same in all systems. The coordinate transformation that
leaves the Maxwell equations invariant in form, and under which *all* the
laws of physics are invariant according to the theory of special relativity, is
the Lorentz transformation:

$$x' = \frac{x - vt}{\sqrt{1 - v^2/c^2}}, \quad y' = y, \quad z' = z, \quad t' = \frac{t - vx/c^2}{\sqrt{1 - v^2/c^2}}. \tag{9.2}$$

We will for the most part follow the convention of relativistic quantum
theory and use "natural" units in which $\hbar = c = 1$. In these units the
Compton wavelength of a particle of mass m is $1/m$, i.e., mass becomes a
unit of inverse length. A mass corresponding to 1 fm (1 fm $= 10^{-15}$m $= 1$
femtometer $= 1$ fermi) corresponds to an energy of about 200 MeV, i.e., 1
fm $\cong 5$ (GeV)$^{-1}$.

Not very much about the theory of relativity will be required for our
purposes, but it may be useful to briefly review some notation. Space–time
coordinates $(t, x, y, z) = (t, \mathbf{x})$ are denoted (x^0, x^1, x^2, x^3), or simply by
the contravariant component symbol x^μ. This applies to all four-vectors,
such as the energy–momentum $p^\mu \equiv (E, \mathbf{p})$. The corresponding covariant
components are $x_\mu = g_{\mu\nu}x^\nu = (t, -\mathbf{x})$ and $p_\mu = g_{\mu\nu}p^\nu = (E, -\mathbf{p})$, where
the metric tensor[2]

$$g_{\mu\nu} = \begin{pmatrix} 1 & 0 & 0 & 0 \\ 0 & -1 & 0 & 0 \\ 0 & 0 & -1 & 0 \\ 0 & 0 & 0 & -1 \end{pmatrix}. \tag{9.3}$$

---

[1] Ideas bearing on the principle of relativity go back at least as far as Galileo, who
posed the question of whether someone sleeping on an anchored boat, and waking to see
only a clear sky, could discern that the boat had begun drifting with a uniform velocity.
I am indebted to Professor Nandor Balazs for this remark.

[2] This choice of metric is widespread but not universal. The reader who does not
recall the general distinction between covariant and contravariant components of a vector
should understand that for our purposes here the distinction is trivial and involves only
different signs. The origin of the sign differences is simply the minus sign in the quantity
$x^2 - c^2t^2$ left invariant by the Lorentz transformation of coordinates.

Scalar products of four-vectors A^μ and B^μ are denoted by $A \cdot B = A^\mu B_\mu = A_\mu B^\mu = g_{\mu\nu} A^\mu B^\nu = A^0 B^0 - \mathbf{A} \cdot \mathbf{B}$. The gradient $\nabla^\mu = \delta^\mu = \partial/\partial x_\mu = g^{\mu\nu} \partial/\partial x^\nu = (\partial/\partial t, -\nabla)$, so that the four-vector momentum operator in the coordinate representation is $p^\mu = i\partial/\partial x_\mu = i(\partial/\partial t, -\nabla)$. The divergence of a four-vector A^μ is thus

$$\partial^\mu A_\mu = g^{\mu\nu} \frac{\partial A_\mu}{\partial x^\nu} = g^{\mu\nu} g_{\mu\eta} \frac{\partial A^\eta}{\partial x^\nu} = \delta^\nu_\eta \frac{\partial A^\eta}{\partial x^\nu} = \frac{\partial A^\nu}{\partial x^\nu} = \frac{\partial A^0}{\partial t} + \nabla \cdot \mathbf{A}. \quad (9.4)$$

9.2 The Dirac Equation

The theory of "relativity" is really a theory of *invariance*: the fundamental equations of physics must be Lorentz covariant, i.e., they must have the same form when Lorentz transformations are applied to the space–time variables. The first attempts to derive a relativistic quantum-mechanical wave equation were made by Gordon, Klein, and Schrödinger in 1926-27. For a free particle of rest mass m, the replacement $\mathbf{p} \to -i\nabla$ in the relativistic energy–momentum relation $E^2 = \mathbf{p}^2 + m^2$, together with the "squared" Schrödinger equation $H^2\psi = -\partial^2\psi/\partial t^2$, produces the Klein–Gordon equation

$$\left(\frac{\partial^2}{\partial t^2} - \nabla^2 \right) \psi(\mathbf{x}, t) + m^2\psi(\mathbf{x}, t) = 0. \quad (9.5)$$

The Klein–Gordon equation is in fact a valid Lorentz-covariant wave equation for spinless particles. However, until 1934 it was considered unacceptable for two reasons. One reason is that it allows negative-energy solutions. For instance, the plane wave $\psi(\mathbf{x}, t) = Ae^{-i(Et - \mathbf{k} \cdot \mathbf{x})}$ is a solution of (9.5) with $E^2 = \mathbf{k}^2 + m^2$ or $E = \pm\sqrt{\mathbf{k}^2 + m^2}$; the negative-energy solutions cannot be simply discarded — as they can be in classical (relativistic) mechanics — because they are required for the completeness of the eigensolutions of the Klein–Gordon equation. With negative energies, the energy spectrum is not bounded from below, and therefore it is possible in principle to extract energy indefinitely by applying some perturbation that induces downward transitions.

The other reason for the original dissatisfaction with the Klein–Gordon equation has to do with the definition of a probability current. These difficulties were actually a consequence of the attempt to formulate a *single-particle* theory. The field $\psi(\mathbf{x}, t)$ should itself be quantized, and when this is done the negative-energy solutions are associated in a natural way with antiparticles. That is, quantum field theory and Lorentz covariance *require* negative energies and the existence of antiparticles. We shall return to these points in Chapter 10.

Historically, of course, things evolved first with attempts to formulate "better" single-particle relativistic wave equations. Since the negative-energy solutions of the Klein–Gordon equation are a consequence of having a second derivative with respect to time, Dirac (1928) sought an equation having only a first time derivative. The requirement of Lorentz covariance suggests that such an equation should also have only first derivatives with respect to the spatial coordinates, so that the desired wave equation would have the form

$$i \left(\frac{\partial}{\partial t} + \boldsymbol{\alpha} \cdot \nabla \right) \psi = \beta m \psi. \tag{9.6}$$

We also want to satisfy the relativistic energy–momentum relation $E^2 = \mathbf{p}^2 + m^2$ with $E \to i\partial/\partial t$ and $\mathbf{p} \to -i\nabla$. Equation (9.6) implies

$$-\frac{\partial^2 \psi}{\partial t^2} = (-i\boldsymbol{\alpha} \cdot \nabla + \beta m)^2 \psi. \tag{9.7}$$

Then $E^2 = \mathbf{p}^2 + m^2$ will hold, i.e., ψ will satisfy the Klein–Gordon equation, if

$$\beta^2 = 1, \quad \alpha_i \beta + \beta \alpha_i \equiv \{\alpha_i, \beta\} = 0, \quad \{\alpha_i, \alpha_j\} = 2\delta_{ij} . \tag{9.8}$$

Thus (9.6) will not be consistent with the relativistic energy–momentum relation unless β and the α_i are matrices and ψ is therefore a column vector.

Equations (9.8) imply that $\alpha_i^2 = 1$ and therefore that α_i and β have eigenvalues ± 1. Furthermore, from the cyclic property of the trace of a matrix, and the anticommutation relations (9.8), we have

$$\text{tr}\beta = \text{tr}\beta\alpha_i^2 = \text{tr}\alpha_i\beta\alpha_i = -\text{tr}\alpha_i^2\beta = -\text{tr}\beta = 0. \tag{9.9}$$

Likewise $\text{tr}\alpha_i = 0$. Since the α_i and β are traceless and have eigenvalues ± 1, they must be even-dimensional matrices. Dimension 2 is too small because it is only possible to construct three anticommuting 2×2 matrices, which may be taken to be the Pauli spin matrices σ_1, σ_2, and σ_3 defined conventionally as

$$\sigma_1 = \begin{pmatrix} 0 & 1 \\ 1 & 0 \end{pmatrix}, \quad \sigma_2 = \begin{pmatrix} 0 & -i \\ i & 0 \end{pmatrix}, \quad \sigma_3 = \begin{pmatrix} 1 & 0 \\ 0 & -1 \end{pmatrix}. \tag{9.10}$$

The algebra (9.8) can be satisfied with 4×4 matrices. For instance, this can be done with the Dirac representation:

$$\alpha_i = \begin{pmatrix} 0 & \sigma_i \\ \sigma_i & 0 \end{pmatrix}, \quad \beta = \begin{pmatrix} 1 & 0 \\ 0 & -1 \end{pmatrix}, \tag{9.11}$$

where all the entries in these matrices are of course themselves 2×2 matrices. It is conventional to work instead with the matrices[3]

$$\gamma^0 \equiv \beta, \quad \gamma^i \equiv \beta\alpha_i , \tag{9.12}$$

which satisfy

$$\{\gamma^\mu, \gamma^\nu\} = 2g^{\mu\nu} \tag{9.13}$$

and $(\gamma^i)^\dagger = -\gamma^i, (\gamma^i)^2 = -1, (\gamma^0)^\dagger = \gamma^0, (\gamma^0)^2 = 1$. In terms of the γ matrices we can write the Dirac equation (9.6) for a free particle as

$$i \left(\beta\frac{\partial}{\partial t} + \beta\boldsymbol{\alpha} \cdot \nabla \right) \psi = i \left(\gamma^0 \frac{\partial}{\partial x^0} + \gamma^i \frac{\partial}{\partial x^i} \right) \psi = \beta^2 m\psi \tag{9.14}$$

or

$$\left(i\gamma^\mu \frac{\partial}{\partial x^\mu} - m \right) \psi = 0. \tag{9.15}$$

In Feynman's slash notation, where $\displaystyle{\not{A} \equiv g_{\mu\nu}\gamma^\mu A^\nu = \gamma^\mu A_\mu}$ and in particular $\not{\partial} \equiv \gamma^\mu \partial/\partial x^\mu$, we can write the Dirac equation for a free particle as

$$(i\not{\partial} - m)\psi = 0, \tag{9.16}$$

or

$$(\not{p} - m)\psi = 0. \tag{9.17}$$

Taking the Hermitian conjugate of both sides of (9.6), using the fact that $\boldsymbol{\alpha}$ and β are Hermitian and defining

$$\overline{\psi} \equiv \psi^\dagger \beta, \tag{9.18}$$

we obtain the conjugate form of the Dirac equation:

$$\overline{\psi}(i \overleftarrow{\not{\partial}} + m) = 0, \tag{9.19}$$

where the arrow over $\not{\partial}$ reminds us that the derivative acts to the left. Now we can define a conserved current in the same fashion as for the nonrelativistic Schrödinger equation. Multiply (9.16) on the left by $\overline{\psi}$, (9.19) on the right by ψ, and add:

$$\overline{\psi}\gamma^\mu \frac{\partial\psi}{\partial x^\mu} + \gamma^\mu \frac{\partial\overline{\psi}}{\partial x^\mu}\psi = \frac{\partial}{\partial x^\mu}(\overline{\psi}\gamma^\mu \psi) = 0. \tag{9.20}$$

---

[3]This representation for the γ matrices follows Bjorken and Drell (1964) and other textbooks, but differs from that used in Dirac's original paper and in Sakurai (1976), where the γ matrices are all Hermitian. This difference is inconsequential, as the physical predictions of the theory are independent of the choice of representation for the γ matrices and depend only on their algebra.

Thus the conserved current density, in a representation such as the "standard" one (9.11) in which β and α are Hermitian, is

$$j^\mu \equiv \overline{\psi}\gamma^\mu\psi = (\psi^\dagger\psi, \psi^\dagger\boldsymbol{\alpha}\psi), \tag{9.21}$$

i.e., the density ρ is $\psi^\dagger\psi$, the current density $\mathbf{j} = \psi^\dagger\boldsymbol{\alpha}\psi$, and $\partial\rho/\partial t + \nabla\cdot\mathbf{j} = 0$.

It is worth noting that any set of 4×4 matrices satisfying $\{\gamma^\mu, \gamma^\nu\} = 2g^{\mu\nu}$, with $\gamma^{0\dagger} = \gamma^0$ and $\gamma^{i\dagger} = -\gamma^i$, $i = 1, 2, 3$, may be used in writing the Dirac equation. In particular, if $\{\gamma^{\mu\prime}, \gamma^{\nu\prime}\} = \{\gamma^\mu, \gamma^\nu\} = 2g^{\mu\nu}$, then the γ^μ and $\gamma^{\mu\prime}$ matrices are equivalent up to a unitary transformation: $\gamma^\mu = S^{-1}\gamma^{\mu\prime}S$.[4] It then follows that $\psi' = S\psi$ satisfies the Dirac equation with the γ^μ replaced by $\gamma^{\mu\prime}$. Since S is unitary, this leads easily to the conclusion that the physical predictions of the theory are independent of the specific representation chosen for the gamma matrices.

This representation independence may be used to prove the Lorentz covariance of the Dirac equation. For this purpose, suppose for the moment that γ^μ transforms as a four-vector under Lorentz transformations. Then since $\partial/\partial x^\mu$ is a four-vector, $\gamma^\mu\partial/\partial x^\mu$ is a scalar under Lorentz transformations. But the supposition that γ^μ transforms as a four-vector amounts only to different choices of representation in different frames, and so the Dirac equation in the transformed frame will be satisfied by a wave function $\psi' = S\psi$, where the unitary matrix S is determined by the Lorentz transformation relating the two frames. The fact that $\gamma^\mu\partial/\partial x^\mu$ may be assumed in effect to be a scalar under Lorentz transformations then leads easily to the proof of covariance of the Dirac equation. Details may be found in standard textbooks.[5]

Plane Waves

Let us review briefly the plane-wave solutions of the free-particle Dirac equation, since these appear when the Dirac field is quantized. The fact that we used $E^2 = \mathbf{p}^2 + m^2$ in the derivation of the Dirac equation implies that there will be negative-energy as well as positive-energy solutions. We denote these by ψ_- and ψ_+, respectively, and write

$$\psi_+ = e^{-iEt}e^{i\mathbf{P}\cdot\mathbf{X}}u(\mathbf{p}) = e^{-ip\cdot x}u(\mathbf{p}), \tag{9.22}$$

$$\psi_- = e^{iEt}e^{-i\mathbf{P}\cdot\mathbf{X}}v(\mathbf{p}) = e^{ip\cdot x}v(\mathbf{p}), \tag{9.23}$$

---

[4] This is what Pauli called the *fundamental theorem* of the gamma matrices. See R. H. Good, Jr., *Rev. Mod. Phys.* **27**, 187 (1955).

[5] See, for instance, Bjorken and Drell (1964).

where u and v are 4-dimensional column vectors ("spinors"). The Dirac equation (9.16) for ψ_+ and ψ_- implies

$$(\not p - m)u(\mathbf{p}) = 0, \quad (\not p + m)v(\mathbf{p}) = 0. \tag{9.24}$$

Consider first a particle at rest, with $\mathbf{p} = 0$ and therefore $p^0 = E = m$. Then $\not p = \gamma^\mu p_\mu = \gamma^0 p_0 = \beta m$ and (9.24) becomes

$$(\beta - 1)u(0) = 0, \quad (\beta + 1)v(0) = 0. \tag{9.25}$$

Each of these equations has two linearly independent solutions:

$$u^1(0) = \begin{pmatrix} 1 \\ 0 \\ 0 \\ 0 \end{pmatrix}, \quad u^2(0) = \begin{pmatrix} 0 \\ 1 \\ 0 \\ 0 \end{pmatrix}, \tag{9.26}$$

$$v^1(0) = \begin{pmatrix} 0 \\ 0 \\ 1 \\ 0 \end{pmatrix}, \quad v^2(0) = \begin{pmatrix} 0 \\ 0 \\ 0 \\ 1 \end{pmatrix}. \tag{9.27}$$

Here, $u^1(0)$ and $u^2(0)$ describe a positive-energy particle with spin "up" and "down," respectively, along the z axis, whereas $v^1(0)$ and $v^2(0)$ describe a negative-energy particle with spin up and down, respectively (Section 9.7).

To obtain plane-wave solutions with $\mathbf{p} \neq 0$, we note that $(\not p - m)(\not p + m) = p^2 - m^2 = 0$, so that $u^i(\mathbf{p}) = (\not p + m)u^i(0)$ and $v^i(\mathbf{p}) = (-\not p + m)v^i(0)$ satisfy (9.24). Explicitly, solutions of (9.24) are

$$u^1(\mathbf{p}) = \frac{A}{E+m}\begin{pmatrix} 1 \\ 0 \\ p_z \\ p_+ \end{pmatrix}, \quad u^2(\mathbf{p}) = \frac{A}{E+m}\begin{pmatrix} 0 \\ 1 \\ p_- \\ p_z \end{pmatrix}, \tag{9.28}$$

and

$$v^1(\mathbf{p}) = \frac{A}{E+m}\begin{pmatrix} p_z \\ p_+ \\ 1 \\ 0 \end{pmatrix}, \quad v^2(\mathbf{p}) = \frac{A}{E+m}\begin{pmatrix} p_- \\ -p_z \\ 0 \\ 1 \end{pmatrix}, \tag{9.29}$$

where $E = \sqrt{\mathbf{p}^2 + m^2}$, A is a normalization factor, and $p_\pm \equiv p_x \pm ip_y$. The choice $|A| = \sqrt{(E+m)/2m}$ gives

$$\bar u^i(\mathbf{p})u^j(\mathbf{p}) = -\bar v^i(\mathbf{p})v^j(\mathbf{p}) = \delta_{ij}. \tag{9.30}$$

Since $\overline{\psi}\psi$ is a scalar under Lorentz transformations, this normalization is Lorentz invariant. Note also that

$$u^i(\mathbf{p})^\dagger u^j(\mathbf{p}) = v^i(\mathbf{p})^\dagger v^j(\mathbf{p}) = \frac{E}{m}\delta_{ij} \ . \tag{9.31}$$

Here, $\psi^\dagger \psi$ transforms as the time component of a four-vector, as indicated by (9.21) and the factor E/m in (9.31). The Lorentz contraction of a volume element is thus compensated by the dilation of $\psi^\dagger \psi$ in the direction of motion, so that $\int d^3x \psi^\dagger \psi$ is a scalar invariant.

Finally we list the following additional spinor properties that will be useful in Chapter 10:

$$\overline{v}^i(\mathbf{p})u^j(\mathbf{p}) = 0, \tag{9.32}$$

$$\sum_i u^i_\alpha(\mathbf{p})\overline{u}^i_\beta(\mathbf{p}) = \left(\frac{\not{p}+m}{2m}\right)_{\alpha\beta}, \tag{9.33}$$

$$\sum_i v^i_\alpha(\mathbf{p})\overline{v}^i_\beta(\mathbf{p}) = \left(\frac{\not{p}-m}{2m}\right)_{\alpha\beta}. \tag{9.34}$$

9.3 Hole Theory: The Dirac Sea

One of the perceived difficulties with the Klein–Gordon equation before 1934 was that the time component of its conserved current j^μ, which was associated with a probability density, is not positive–definite. The corresponding probability density $\psi^\dagger \psi$ for the Dirac equation, by contrast, is of course positive–definite. However, the Dirac equation still admits negative-energy solutions, as we have reviewed in the preceding section. It is well-known that Dirac turned this "vice" into a "virtue" by his interpretation of the negative-energy states in terms of holes.

As noted in connection with the Klein–Gordon equation, negative-energy solutions are required for completeness. In problems such as the Compton scattering of light by electrons, furthermore, negative-energy solutions of the Dirac equation are necessary in order to obtain the Klein–Nishina cross section, which has been accurately tested experimentally. On the other hand, the existence of negative-energy states would seem to imply that matter is unstable against transitions from positive- to negative-energy states, as also noted in connection with the Klein–Gordon equation. Dirac (1930) resolved these difficulties by supposing that all the negative-energy states are filled, so that transitions from positive- to negative-energy states are forbidden by the Pauli exclusion principle for electrons.

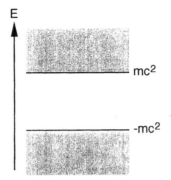

Figure 9.1: Energy spectrum associated with the Dirac equation. Energies between $-mc^2$ and mc^2 are forbidden.

The set of occupied negative-energy states is called the *Dirac sea*. With the Dirac hole concept, the vacuum became more complicated, now consisting of zero-point electromagnetic fields as well as a sea of negative-energy electron states. We shall see in the following chapter that the idea of the Dirac sea is modified by quantum field theory, but the concept still provides a very useful intuitive guide that warrants further discussion here.

Figure 9.1 illustrates the energy spectrum of the Dirac equation (and also the Klein–Gordon equation). For the free-particle Dirac equation we have $E = \pm\sqrt{\mathbf{p}^2 + m^2}$, so that energies between $-mc^2$ and $+mc^2$ are forbidden. The energy interval $(-mc^2, mc^2)$ is analogous to the energy gap between valence and conduction bands in a solid. In the case of a solid, a filled valence band means there will be no current when an external field is applied: as a consequence of the Pauli exclusion principle, the electrons cannot change their states unless they absorb enough energy from the field to be promoted into the conduction band. In particular, a filled valence band at low temperatures means that the solid will be an insulator. If an electron in the valence band is somehow excited into the conduction band, however, it leaves a *hole* in the valence band. If the material is connected to the terminals of a battery, there will be a flow of current associated with the motion of electrons in the conduction band. However, electrons in the valence band are also affected by the potential difference, and can fall into the holes left behind by the electrons that have been promoted into the conduction band. In so doing, the electrons in the valence band drift in the same direction as the electrons in the conduction band *and also contribute to the current*. We can view this situation as one in which electrons in the conduction band move in one direction while the (positively charged) holes in the valence band move in the *opposite* direction.

Such a hole picture was devised by Peierls in 1929 to explain the Hall effect in many metals, where the charge carriers appear to be positively charged. Dirac (1930) independently argued analogously that the absence of a particle in the negative-energy sea corresponds to the presence of a positive-energy, oppositely charged particle.[6] As is well-known, he originally thought that the positively charged states associated with a hole in the negative-energy sea might correspond to protons, but difficulties associated with this interpretation led him to propose later that "A hole, if there were one, would be a new kind of particle, unkown to experimental physics, having the same mass and opposite charge of the electron" (Dirac, 1931). The existence of these hole particles was confirmed with Anderson's discovery of the antielectron e[+], or positron, in 1932.[7]

We noted in Chapter 3 that, in his theory of spontaneous emission based on the quantized radiation field, Dirac in 1927 showed that quantum theory could deal with the creation of particles. In that theory the particles (photons) that can be created or annihilated are massless. With the construction of his relativistic wave equation and the hole theory, Dirac in 1930 took the first step in showing that quantum theory could deal with the creation or annihilation of particles of finite mass. For instance, the absorption of a photon of energy exceeding $2mc^2$ can promote a negative-energy state across the energy gap $2mc^2$ and produce an electron–positron pair. Quantum field theory deals more generally with the creation and annihilation of particles in much the same way as QED deals with the creation and annihilation of photons (Chapter 10).

The theory of the Dirac equation that we have been recalling here is a *single-particle* theory in that ψ is not quantized and there are no annihilation and creation operators for particles in the formalism. The shortcomings of such a single-particle theory are already implicit in the need to resort to the hole theory in order to escape from difficulties with negative-energy states. Nevertheless, hole theory provides a very useful intuitive picture for various relativistic phenomena, and we shall now consider a few examples.

---

[6] For historical analyses of Dirac's hole theory see, for instance, Bromberg (1976) or Moyer (1981).

[7] The existence of the positron was inferred from tracks observed in a cloud chamber exposed to cosmic rays. Evidence for the creation of electron–positron pairs was observed in 1933 by Blackett and Occhialini. The antiproton was discovered by Chamberlain, Segré, Wiegand, and Ypsilantis in 1955.

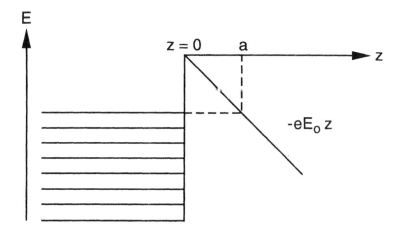

Figure 9.2: Energy diagram for an electron in a square well subject to a uniform electric field E_o; $z = 0$ defines the surface of the metal. An electron of energy E can tunnel across the triangular potential barrier.

9.4   Pair Creation in a Uniform Electric Field

The analogy between the Dirac hole theory and the energy band theory of solids allows a simple approximate treatment of electron–positron pair creation in a uniform electric field. (Itzykson and Zuber, 1980; Aitchison 1985) We consider first a simple model of field emission of electrons from a metal, where each electron is assumed to be confined by a one-dimensional square well potential. The application of an electric field E_o adds a potential energy $V(z) = -eE_o z$, where z is the coordinate normal to the surface of the metal. An electron of energy E can tunnel through this triangular potential barrier (Figure 9.2), and the tunneling probability is given approximately by (Schiff, 1968)

$$P = e^{-2 \int_0^{z_E} dz \sqrt{2m[V(z)-E]/\hbar^2}} \, , \tag{9.35}$$

where z_E is the classical turning point such that $V(z_E) = E$. For the problem of field emission we let $V(z) = -eE_o z$ and $E = -W$, where W is the work function of the metal. In this case $z_E = W/eE_o \equiv a$ and

$$P = e^{-2 \int_0^{a} dz \sqrt{2m(W-eE_o z)/\hbar^2}} = e^{-\frac{4}{3}\sqrt{2mW/\hbar^2}a} \, , \tag{9.36}$$

which is known as the Fowler–Nordheim formula for field emission.

Now in the case of a uniform electric field of strength E_o in the *vacuum*, we might expect intuitively that (9.36) should be applicable to e^+e^- pair production if we let $W = 2mc^2$ and $a = 2mc^2/eE_o$:

$$P_{\text{pair}} = e^{-16m^2c^3/3\hbar eE_o} = e^{-16mc^2/3e\lambda_c E_o} \,, \tag{9.37}$$

where $\lambda_c = \hbar/mc$ is $1/2\pi$ times the Compton wavelength of the electron. The exact calculation for the probability w per unit volume per unit time for pair creation, which was first done by Schwinger (1951), gives

$$w = \frac{e^2 E_o^2}{\pi^2 \hbar^2 c} \sum_{n=1}^{\infty} \frac{1}{n^2} e^{-n\pi m^2 c^3/\hbar eE_o} \,, \tag{9.38}$$

so that the simple calculation based on tunneling and hole theory gives roughly the same exponential dependence on $mc^2/e\lambda_c E_o$.

Note that $mc^2/e\lambda_c \cong 10^{18}$ V/m. For a monochromatic plane wave with this field strength, this corresponds to an intensity $\sim 10^{30}$ W/cm^2. Such enormous electric field strengths are required for pair production that it has never been directly observed. The effect is exceedingly small even for the electric field ($\sim 5 \times 10^{11}$ V/m) binding the electron in the hydrogen atom. In this case $E_o = e^2/r$, with $r = \hbar^2/me^2 = \hbar/\alpha mc = \lambda_c/\alpha$ for the first Bohr orbit. Then $mc^2/e\lambda_c E_o = \alpha^{-3}$ and $P_{\text{pair}} \sim e^{-\alpha^{-3}} \cong e^{-137^3}$. In the field of a nucleus of charge Ze, however, $P_{\text{pair}} \sim e^{-1/Z^3\alpha^3}$, and pair creation might be observable near a nucleus of charge $Z \sim \alpha^{-1} \cong 137$. No such stable nuclei are believed to be possible, but a transient nuclear state of this type might occur in the collision of two stable nuclei of large Z. In this case there is the possibility of a "breakdown" of the Dirac vacuum, with the creation of positrons of well-defined energy (Fulcher, Rafelski, and Klein, 1979; Greiner, Müller, and Rafelski, 1985).

9.5 Vacuum Polarization

Hole theory suggests that a positive-energy electron should electrostatically repel the negative-energy electrons in the Dirac sea, thus in effect polarizing the vacuum in its vicinity. This results in an effective electron charge that is smaller in magnitude than e, i.e., the "bare charge" of the electron is partially screened. This screening is a consequence of virtual pair production in the Coulomb field of the charge, such that the positron tends to be attracted to the electron, while the electron of the virtual pair tends to be repelled. Likewise a positively charged nucleus will have an effective charge smaller than its bare charge. Evidently *observed* charges of all particles are effective charges determined in part by vacuum polarization.

From the energy–time uncertainty relation we know that virtual pairs of particles of mass m can exist for times $\Delta t \sim \hbar/mc^2$, during which time they can separate by a distance $\sim c\Delta t = \hbar/mc = \lambda_c$. Thus at distances smaller than λ_c from a charged particle we expect to "see" the bare charge, whereas at larger distances we can see only the usual observed charge of smaller magnitude.

The effect of vacuum polarization, and in particular the difference between bare charge and effective charge, can be observed indirectly as a contribution to the Lamb shift. At distances close to the nucleus, according to the above argument, the electron in a hydrogen atom should feel a stronger Coulomb attraction than that due to the usual observed charge. Since s state electrons are most likely to be found near the nucleus, this effect should be largest for the s states. As we shall see in Chapter 11, the $2p_{1/2} - 2s_{1/2}$ splitting in hydrogen due to vacuum polarization is about 27 MHz. This is only about 1/40 of the total Lamb shift, but it is essential in the comparison of theory with experiment. In other systems, such as muonic atoms, vacuum polarization can be the *dominant* contribution to the Lamb shift.

9.6   The Klein Paradox

Consider a one-dimensional model of electron scattering by a potential barrier $V_o > 0$ (Figure 9.3). We represent the incident electron of energy E by the spin–up wave function $\psi_{\text{inc}} = e^{ipz}u^1(\mathbf{p})e^{-iEt} = \phi_{\text{inc}}(\mathbf{p})e^{-iEt}$, where

$$
\phi_{\text{inc}}(\mathbf{p}) = a e^{ipz}
\begin{pmatrix}
1 \\
0 \\
\frac{p}{E+m} \\
0
\end{pmatrix}, \quad p = \sqrt{E^2 - m^2} . \tag{9.39}
$$

The wave function in region I is the sum of $\phi_{\text{inc}}(\mathbf{p})$ and a reflected wave of the form (9.39) with $p \to -p$:

$$
\phi_I(\mathbf{p}) = a e^{ipz}
\begin{pmatrix}
1 \\
0 \\
\frac{p}{E+m} \\
0
\end{pmatrix}
+ b e^{-ipz}
\begin{pmatrix}
1 \\
0 \\
\frac{-p}{E+m} \\
0
\end{pmatrix} . \tag{9.40}
$$

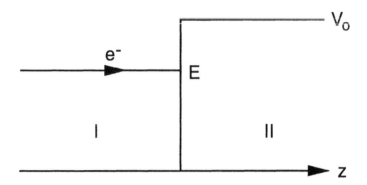

Figure 9.3: An electron of energy E incident on a potential barrier of height V_o.

The transmitted wave in region II has the form of a free-particle plane wave, but with E replaced by $E - V_o$ and p replaced by $p' = \sqrt{(E - V_o)^2 - m^2}$:[8]

$$\phi_{II}(\mathbf{p}) = de^{ip'z} \begin{pmatrix} 1 \\ 0 \\ \frac{p'}{E - V_o + m} \\ 0 \end{pmatrix}. \tag{9.41}$$

Continuity of ψ at $z = 0$ implies that

$$a + b = d \tag{9.42}$$

and $p(a - b)/(E + m) = p'd/(E - V_o + m)$, or

$$a - b = \frac{p'}{p} \frac{E + m}{E - V_o + m} d. \tag{9.43}$$

Thus

$$\frac{b}{a} = \frac{1 - r}{1 + r} \tag{9.44}$$

and

$$\frac{d}{a} = \frac{2}{1 + r}, \tag{9.45}$$

---

[8] There is no need to consider spin–down components in the transmitted and reflected waves: including them and imposing the condition of continuity of the wave function across the boundary at $z = 0$, one easily deduces that their amplitudes must vanish. Thus there is no spin flipping in the scattering process of Figure 9.3.

where

$$r \equiv \frac{p'}{p} \frac{E+m}{E-V_o+m} \; . \tag{9.46}$$

The forms (9.44) and (9.45) are familiar from nonrelativistic theory.

If $|E - V_o| < m$, p' is imaginary and the solution (9.41) in region II is exponentially damped. If $V_o > E + m$, however, p' is real and so the wave function in region II is oscillatory, indicating that there can be a penetration of the barrier if it is *large* enough. In the latter case, according to a common interpretation within the single-particle framework (Bjorken and Drell, 1964), r is real and negative. The incident, transmitted, and reflected current densities $j^3 \equiv j = \psi^\dagger \alpha_3 \psi$ calculated for these solutions satisfy

$$\frac{j_{\text{trans}}}{j_{\text{inc}}} = \frac{4r}{(1+r)^2} \; , \qquad \frac{j_{\text{ref}}}{j_{\text{inc}}} = \frac{(1-r)^2}{(1+r)^2} \; , \tag{9.47}$$

and therefore the transmitted current is *negative* and the reflected current is *greater* than the incident current. These peculiar results for $V_o > E + m$ are referred to as *Klein's paradox*.

To understand the origin of the "paradox," note that for $|E - V_o| < m$ the wave function in region II is exponentially damped over a distance

$$\Delta z \sim \left| \frac{1}{p'} \right| \sim \frac{1}{\sqrt{m^2 - (E-V_o)^2}} \; , \tag{9.48}$$

so that an increase in V_o toward E tends to tighten the localization of the wave function in region II. For $V_o \sim E, \Delta z \sim \hbar/mc$, and the spread in momentum associated with this coordinate localization corresponds to energies large enough for pair creation to be possible. In other words, the single-particle interpretation breaks down, and can lead to apparently paradoxical conclusions, when the potential barrier is so large that particle-antiparticle pairs can be produced. The breakdown of the single-particle interpretation in this regime is the source of the Klein paradox.

We can arrive at a better physical understanding of these results from the perspective of hole theory (Greiner et al., 1985). The potential in region II raises the positive- and negative-energy continua by V_o, as indicated in Figure 9.4, and therefore there is an overlap of the positive-energy continuum in region I with the lower-energy continuum of region II. This allows an electron incident from region I to eject electrons from the occupied lower-energy continuum states in region II. The ejected electrons move to the left in the positive-energy continuum of region I, as shown in Figure 9.4, while the holes (positrons) in region II move to the right. This explains why the reflection coefficient can exceed unity. It also explains why the electron

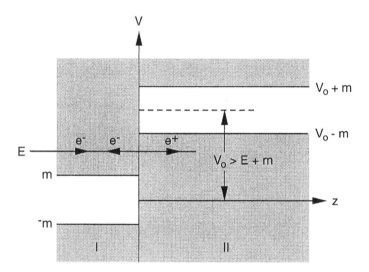

Figure 9.4: Allowed energy continua in the regions I and II of Figure 9.3. An electron e^- incident on the barrier from region I can cause more electrons to appear in region I and positrons e^+ to appear in region II if $V_o > E + m$. After Greiner et al., 1985.

probability current in region II can be negative: a positron current in the positive z direction is equivalent to an electron current in the negative z direction.

9.7 Spin and the Nonrelativistic Limit

The Dirac equation (9.6) has the Hamiltonian form

$$i\frac{\partial\psi}{\partial t} = -i\boldsymbol{\alpha}\cdot\nabla\psi + \beta m\psi = (\boldsymbol{\alpha}\cdot\mathbf{p} + \beta m)\psi \equiv H\psi. \tag{9.49}$$

To obtain a Dirac equation for an electron coupled to a prescribed, external electromagnetic field with vector and scalar potentials $\mathbf{A}$ and Φ, we make the usual substitution $p^\mu \rightarrow p^\mu - eA^\mu$, i.e., $\mathbf{p} \rightarrow \mathbf{p} - e\mathbf{A}$ and $p^0 = i\partial/\partial t \rightarrow i\partial/\partial t - e\Phi$ (see Chapters 4 and 12). Then (9.49) is replaced by

$$i\frac{\partial\psi}{\partial t} = [\boldsymbol{\alpha}\cdot(\mathbf{p} - e\mathbf{A}) + e\Phi + \beta m]\psi \tag{9.50}$$

for an electron in an externally applied field. This identifies the Hamiltonian as

$$H = \boldsymbol{\alpha} \cdot (\mathbf{p} - e\mathbf{A}) + e\Phi + \beta m = \boldsymbol{\alpha} \cdot \mathbf{p} + \beta m + H_{\text{INT}} , \qquad (9.51)$$

where $H_{\text{INT}} = -e\boldsymbol{\alpha} \cdot \mathbf{A} + e\Phi$. Comparison with the classical interaction $-e\mathbf{v} \cdot \mathbf{A} + e\Phi$ suggests the interpretation of $\boldsymbol{\alpha}$ as the operator corresponding to the particle's velocity, i.e., $\boldsymbol{\alpha} = \mathbf{v}/c$. This interpretation is strengthened by the Heisenberg equations of motion that follow straightforwardly from the Hamiltonian (9.51) and the canonical commutation relations:

$$\dot{\mathbf{r}} = \left(\frac{1}{i\hbar}\right) [\mathbf{r}, H] = \boldsymbol{\alpha}, \qquad (9.52)$$

$$\dot{\boldsymbol{\pi}} = \left(\frac{1}{i\hbar}\right) [\boldsymbol{\pi}, H] = e(\mathbf{E} + \boldsymbol{\alpha} \times \mathbf{B}), \qquad (9.53)$$

where $\boldsymbol{\pi} \equiv \mathbf{p} - e\mathbf{A}$ is the "kinetic momentum," $\mathbf{p}$ being the canonical momentum conjugate to $\mathbf{r}$. The interpretation of $\boldsymbol{\alpha}$ as a velocity operator, however, is ambiguous, since $\alpha_i^2 = 1$ would imply that v_i has eigenvalues $\pm c$. We shall return to this point later.

Let us review briefly the nonrelativistic limit of the Dirac equation (9.50). Write

$$\psi = \begin{pmatrix} \tilde{\phi} \\ \tilde{\chi} \end{pmatrix} , \qquad (9.54)$$

where $\tilde{\phi}$ and $\tilde{\chi}$ are each two-component column vectors, so that

$$i\frac{\partial}{\partial t} \begin{pmatrix} \tilde{\phi} \\ \tilde{\chi} \end{pmatrix} = \boldsymbol{\sigma} \cdot \boldsymbol{\pi} \begin{pmatrix} \tilde{\chi} \\ \tilde{\phi} \end{pmatrix} + e\Phi \begin{pmatrix} \tilde{\phi} \\ \tilde{\chi} \end{pmatrix} + m \begin{pmatrix} \tilde{\phi} \\ -\tilde{\chi} \end{pmatrix} . \qquad (9.55)$$

In the nonrelativistic limit the energy mc^2 is large compared with any kinetic or potential energy, and this suggests writing

$$\begin{pmatrix} \tilde{\phi} \\ \tilde{\chi} \end{pmatrix} = e^{-imt} \begin{pmatrix} \phi \\ \chi \end{pmatrix} \qquad (9.56)$$

and assuming that ϕ and χ are slowly varying compared with e^{-imt} in a nonrelativistic approximation. Equation (9.55) becomes

$$i\frac{\partial}{\partial t} \begin{pmatrix} \phi \\ \chi \end{pmatrix} = \boldsymbol{\sigma} \cdot \boldsymbol{\pi} \begin{pmatrix} \chi \\ \phi \end{pmatrix} + e\Phi \begin{pmatrix} \phi \\ \chi \end{pmatrix} - 2m \begin{pmatrix} 0 \\ \chi \end{pmatrix} , \qquad (9.57)$$

and in the nonrelativistic limit the second of the two indicated equations is replaced by

$$\chi \cong \frac{\boldsymbol{\sigma} \cdot \boldsymbol{\pi}}{2m} \phi \, . \tag{9.58}$$

Then the equation for ϕ becomes

$$i\frac{\partial \phi}{\partial t} \cong \left[\frac{(\boldsymbol{\sigma} \cdot \boldsymbol{\pi})^2}{2m} + e\Phi \right] \phi \, . \tag{9.59}$$

This result can be cast in a more familiar form by using the general identity

$$(\boldsymbol{\sigma} \cdot \mathbf{C})(\boldsymbol{\sigma} \cdot \mathbf{D}) = \sigma_i \sigma_j C_i D_j = (\delta_{ij} + i\sigma_k \epsilon_{ijk}) C_i D_j = \mathbf{C} \cdot \mathbf{D} + i\boldsymbol{\sigma} \cdot (\mathbf{C} \times \mathbf{D}), \tag{9.60}$$

where the ϵ_{ijk} are the components of the Levi–Civita pseudotensor. Thus[9]

$$\begin{aligned}
(\boldsymbol{\sigma} \cdot \boldsymbol{\pi})^2 &= \boldsymbol{\pi} \cdot \boldsymbol{\pi} + i\boldsymbol{\sigma} \cdot (\boldsymbol{\pi} \times \boldsymbol{\pi}) \\
&= (\mathbf{p} - e\mathbf{A})^2 + i\boldsymbol{\sigma} \cdot [(\mathbf{p} - e\mathbf{A}) \times (\mathbf{p} - e\mathbf{A})] \\
&= (\mathbf{p} - e\mathbf{A})^2 - ie\boldsymbol{\sigma} \cdot [\mathbf{p} \times \mathbf{A} + \mathbf{A} \times \mathbf{p}] \\
&= (\mathbf{p} - e\mathbf{A})^2 - e\boldsymbol{\sigma} \cdot \mathbf{B},
\end{aligned} \tag{9.61}$$

and we can write (9.59) as the nonrelativistic, two-component Pauli equation,

$$i\hbar\frac{\partial \phi}{\partial t} = \left[\frac{1}{2m}\left(\mathbf{p} - \frac{e}{c}\mathbf{A}\right)^2 - \frac{e\hbar}{2mc}\boldsymbol{\sigma} \cdot \mathbf{B} + e\Phi \right] \phi. \tag{9.62}$$

For a uniform magnetic field $\mathbf{B}$, with $\mathbf{A} = \frac{1}{2}\mathbf{B} \times \mathbf{r}$, this reduces to

$$i\hbar\frac{\partial \phi}{\partial t} \cong \left[\frac{\mathbf{p}^2}{2m} - \frac{e}{2mc}(\mathbf{L} + 2\mathbf{S}) \cdot \mathbf{B} - e\Phi \right] \phi \tag{9.63}$$

if we retain only first-order terms in $\mathbf{B}$. Here $\mathbf{L} = \mathbf{r} \times \mathbf{p}$ is the orbital angular momentum and $\mathbf{S} = (\hbar/2)\boldsymbol{\sigma}$ is the intrinsic (spin) angular momentum. Thus the gyromagnetic ratio of 2 emerges "automatically" from the Dirac equation for an electron in an external field.[10]

By considering the Dirac equation for a free particle or a particle in a central force field, one easily verifies that $[H, \mathbf{J}] = 0$, where $\mathbf{J} \equiv \mathbf{L} + \hbar\boldsymbol{\Sigma}/2$ and

$$\boldsymbol{\Sigma} = \begin{pmatrix} \boldsymbol{\sigma} & 0 \\ 0 & \boldsymbol{\sigma} \end{pmatrix} \, . \tag{9.64}$$

---

[9] We use the fact that the p operator in p × A acts on the product of $\mathbf{A}$ and a scalar function f, and $\mathbf{p} \times (\mathbf{A}f) = -i\nabla \times (\mathbf{A}f) = -if\nabla \times \mathbf{A} + i\mathbf{A} \times \nabla f = -if\mathbf{B} - \mathbf{A} \times \mathbf{p}f$, so that $\mathbf{p} \times \mathbf{A} = -i\mathbf{B} - \mathbf{A} \times \mathbf{p}$.

[10] Recall the discussion in Section 3.13.

Thus the total conserved angular momentum consists of an orbital part and a spin. This is perhaps the simplest way to deduce that $\Sigma/2$ corresponds to a non-orbital, spin angular momentum.

For an electron at rest, the spinors $u^1(0)$ and $v^1(0)$ given in equations (9.26) and (9.27) are eigenvectors of Σ_3 with eigenvalues $+1$; for this reason they are said to have spin "up." Also, $u^2(0)$ and $v^2(0)$ are eigenvectors of Σ_3 with eigenvalues -1, i.e., they have spin "down."

For $\mathbf{p} \neq 0$, the spinors $u^{1,2}(\mathbf{p})$ and $v^{1,2}(\mathbf{p})$ given by (9.28) and (9.29) are not eigenvectors of Σ_3. However, they are eigenvectors of the *helicity* operator $\Sigma \cdot \hat{p} = \Sigma \cdot \mathbf{p}/|\mathbf{p}|$. It is easy to see, in particular, that if we choose axes such that $p_x = p_y = 0$, then $u^1(\mathbf{p})$ and $v^1(\mathbf{p})$ are helicity eigenstates with helicity eigenvalues $+1$, and $u^2(\mathbf{p})$ and $v^2(\mathbf{p})$ are helicity eigenstates of helicity -1. (Note that in nonrelativistic theory, by contrast, any two-component Pauli spinor that is independent of $\mathbf{x}$ and t is an eigenstate of $\boldsymbol{\sigma} \cdot \hat{n}$, with $\hat{n}$ a unit vector in "some" *arbitrary* direction.)

The natural way in which spin emerges from the Dirac theory suggests that it is a relativistic effect. However, we can introduce spin nonrelativistically by writing $\mathbf{p}^2$ equivalently as $(\boldsymbol{\sigma} \cdot \mathbf{p})^2$ in the Schrödinger equation for a free particle, and then making the replacement $\mathbf{p} \to \mathbf{p} - e\mathbf{A}/c$ for electromagnetic coupling. This approach to the derivation of the Pauli equation is less natural than that proceeding from the Dirac equation, but nevertheless provides a logical basis for the introduction of spin in nonrelativistic theory.[11]

It is not difficult to derive relativistic corrections to the Pauli Hamiltonian appearing in (9.62). Retention of the lowest-order corrections gives (Itzykson and Zuber, 1980)

$$
\begin{aligned}
H \;\cong\;& \frac{1}{2m}(\mathbf{p} - \frac{e}{c}\mathbf{A})^2 - \frac{e\hbar}{2mc}\boldsymbol{\sigma} \cdot \mathbf{B} + e\Phi - \frac{1}{8m^3c^2}(\mathbf{p} - \frac{e}{c}\mathbf{A})^4 \\
&+ \frac{e\hbar^2}{4m^2c^2}\boldsymbol{\sigma} \cdot \nabla\Phi \times \mathbf{p} + \frac{e\hbar^2}{8m^2c^2}\nabla^2\Phi \;.
\end{aligned}
\tag{9.65}
$$

The first correction is easily understood from the expansion

$$
[m^2c^4 + (\mathbf{p} - \frac{e}{c}\mathbf{A})^2 c^2]^{1/2} = mc^2 + \frac{1}{2m}(\mathbf{p} - \frac{e}{c}\mathbf{A})^2 - \frac{1}{8m^3c^2}(\mathbf{p} - \frac{e}{c}\mathbf{A})^4 + \dots
\tag{9.66}
$$

---

[11] This approach is often attributed to Feynman, but I have been unable to recall or find where Feynman might have published it. It is also employed by A. Galindo and C. Sanchez del Rio, *Am. J. Phys.* **29**, 582 (1961). I am indebted to Professor Luis de la Peña for bringing this reference to my attention.

The second correction is the usual spin–orbit coupling for $\Phi = \Phi(r)$:

$$\frac{e\hbar^2}{4m^2c^2}\boldsymbol{\sigma}\cdot\nabla\Phi\times\mathbf{p} = \frac{e\hbar^2}{4m^2c^2}\boldsymbol{\sigma}\cdot\frac{1}{r}\frac{d\Phi}{dr}\mathbf{r}\times\mathbf{p} = \frac{e\hbar^2}{4m^2c^2}\frac{1}{r}\frac{d\Phi}{dr}\boldsymbol{\sigma}\cdot\mathbf{L}. \qquad (9.67)$$

The last, so-called Darwin term in (9.65) may be understood, analogously to (3.40), as the correction

$$e\Phi(\mathbf{r}+\Delta\mathbf{r}) - e\Phi(\mathbf{r}) \cong \frac{e}{6}(\Delta\mathbf{r})^2\nabla^2\Phi , \qquad (9.68)$$

with

$$(\Delta\mathbf{r})^2 = \frac{3}{4}\left(\frac{\hbar}{mc}\right)^2 = \frac{3}{4}\lambda_c^2 . \qquad (9.69)$$

Evidently this correction is attributable to fluctuations of the electron's position on a scale on the order of its Compton wavelength. We now turn our attention briefly to this *zitterbewegung* ("trembling motion") of a particle described by the Dirac equation.

9.8   Zitterbewegung

Consider the Heisenberg equation of motion for the "velocity operator" $\boldsymbol{\alpha}$ for a free particle ($H_o = \boldsymbol{\alpha}\cdot\mathbf{p} + \beta m$):

$$i\dot{\boldsymbol{\alpha}} = [\boldsymbol{\alpha}, H_o] = 2\boldsymbol{\alpha}\beta m + 2i\mathbf{p}\times\boldsymbol{\Sigma} = 2\boldsymbol{\alpha}H_o - 2\boldsymbol{\alpha}(\boldsymbol{\alpha}\cdot\mathbf{p}) + 2i\mathbf{p}\times\boldsymbol{\Sigma} , \qquad (9.70)$$

where $\boldsymbol{\Sigma}$ is defined by (9.64) and we have used the commutators

$$[\alpha_i, \alpha_j] = 2i\epsilon_{ijk}\Sigma_k , \qquad [\alpha_i, \beta] = 2\alpha_i\beta . \qquad (9.71)$$

Now

$$\begin{aligned}
i\mathbf{p}\times\boldsymbol{\Sigma} - \boldsymbol{\alpha}(\boldsymbol{\alpha}\cdot\mathbf{p}) &= i\epsilon_{ijk}p_j\Sigma_k - \alpha_i\alpha_j p_j = p_j\left(\frac{1}{2}[\alpha_i,\alpha_j] - \alpha_i\alpha_j\right) \\
&= -\frac{1}{2}p_j\{\alpha_i,\alpha_j\} = -p_i ,
\end{aligned} \qquad (9.72)$$

so that, from (9.70),

$$\dot{\boldsymbol{\alpha}} = 2i(\mathbf{p} - \boldsymbol{\alpha}H_o). \qquad (9.73)$$

This equation has the formal solution

$$\boldsymbol{\alpha}(t) = [\boldsymbol{\alpha}(0) - \mathbf{p}H_o^{-1}]e^{-2iH_o t} + \mathbf{p}H_o^{-1} . \qquad (9.74)$$

Accordingly, since $\boldsymbol{\alpha} = \dot{\mathbf{x}}$, we have

$$\mathbf{x}(t) \quad = \quad \mathbf{a} + \mathbf{p}c^2 H_o^{-1} t - \mathbf{a} e^{-2iH_o t/\hbar} \ , \tag{9.75}$$

$$\mathbf{a} \quad = \quad \mathbf{x}(0) - \frac{i}{2}\hbar c \boldsymbol{\alpha}(0) H_o^{-1} + \frac{i}{2}\hbar c^2 \mathbf{p} H_o^{-2} \ , \tag{9.76}$$

where now we have explicitly included $\hbar$ and c. The first two terms in (9.75) correspond to the familiar motion $\mathbf{a} + \mathbf{v}t$, whereas the last term represents rapid oscillations (zitterbewegung) superimposed on this motion. The frequency of these oscillations exceeds $2mc^2/\hbar \sim 2 \times 10^{21}$ sec^{-1}, and their magnitude is very roughly of order $\hbar c \langle H_o^{-1} \rangle \sim \hbar/mc = \lambda_c$. That is, zitterbewegung is associated with fluctuations of the particle position on the order of a Compton wavelength.

When $\langle \mathbf{x}(t) \rangle$ is evaluated for an arbitrary state of the particle, with this state expressed as a superposition of positive- and negative-energy plane-wave solutions of the Dirac equation, it is found that the zitterbewegung results precisely from the interference of positive- and negative-energy state amplitudes. It is worth noting also that wave packets composed of only positive-energy plane waves are possible only if their spread is comparable to or larger than λ_c (Newton and Wigner, 1949). In other words, zitterbewegung is tied to the localization of a wave packet over distances λ_c and smaller, where the corresponding momentum spread allows energies large enough for pair creation. It implies that even a point particle has an effective linear dimension $\sim \lambda_c$, as noted in Chapter 5. Zitterbewegung can be explained more physically in the language of hole theory. The presence of negative-energy components in the wave function allows for the possibility that a negative-energy electron in the Dirac sea can make a virtual (non-energy-conserving) transition to a positive-energy state. This then allows for the possibility that the original positive-energy electron can make a transition into the unoccupied state (hole) left by the transition, so that in effect the positive- and negative-energy electrons have exchanged roles. From the energy–time uncertainty relation the virtual transition of the negative-energy electron is effective only over a time scale $\Delta t \sim \hbar/2mc^2$, and over such a time is separated from the original electron by a distance $< c\Delta t \sim \lambda_c/2$, i.e., by a distance on the order of the zitterbewegung fluctuation of the electron's position. Zitterbewegung can thus be viewed as a consequence of "exchange scattering" between the positive-energy electron and the negative-energy electrons in the Dirac sea.

9.9   Hydrogen

> I was scared. I was afraid that maybe [bound-state wave functions] would not come out right. Perhaps the whole basis of the idea would have to be abandoned if it should turn out that it was not right to the higher orders and I just could not face that prospect.
> — P. A. M. Dirac (1971)

Let us consider first the Klein–Gordon equation for an electron bound by the Coulomb potential $A^0 = \Phi = -e/r, A^1 = A^2 = A^3 = 0$. The minimal coupling substitution $p^\mu \to p^\mu - eA^\mu$ in the Klein–Gordon equation $(p^\mu p_\mu - m^2)\psi = 0$ yields

$$\left[\left(i\frac{\partial}{\partial t} + \frac{\alpha}{r}\right)^2 + \nabla^2 - m^2\right]\psi(\mathbf{x}, t) = 0, \qquad (9.77)$$

where $\alpha = e^2/\hbar c = e^2$ in natural units. Writing $\psi(\mathbf{x}, t) = \phi(\mathbf{x})e^{-iEt}$, we obtain the energy eigenvalue equation

$$\left[\left(E + \frac{\alpha}{r}\right)^2 + \nabla^2 - m^2\right]\phi(\mathbf{x}) = 0. \qquad (9.78)$$

As in central force problems described by the Schrödinger equation we seek solutions of the form $\phi(\mathbf{x}) = \phi(r)Y_{\ell m}$, where $Y_{\ell m}$ is a spherical harmonic. This results in the radial equation

$$\left[\frac{1}{r^2}\frac{\partial}{\partial r}\left(r^2\frac{\partial}{\partial r}\right) - \frac{\ell(\ell+1) - \alpha^2}{r^2} + \frac{2\alpha E}{r} + E^2 - m^2\right]\phi_\ell(r) = 0. \quad (9.79)$$

Now recall that in the case of the Schrödinger equation the radial equation is

$$\left[\frac{1}{r^2}\frac{\partial}{\partial r}\left(r^2\frac{\partial}{\partial r}\right) - \frac{\ell(\ell+1)}{r^2} + \frac{2m\alpha}{r} + 2mE\right]\phi_\ell(r) = 0. \qquad (9.80)$$

Equation (9.79) has the form (9.80) with the replacements $\ell(\ell+1) \to \ell(\ell+1) - \alpha^2$, $\alpha \to \alpha E/m$, and $E \to (E^2 - m^2)/2m$ in the latter equation. The Bohr energy levels

$$E_n = -\frac{m\alpha^2}{2n^2}, \quad n = 1, 2, 3, \dots \qquad (9.81)$$

deduced from (9.80) are therefore replaced by

$$\frac{E_{n'}^2 - m^2}{2m} = -\frac{m}{2}\frac{\alpha^2 E_{n'}^2}{m^2}\frac{1}{n'^2} \qquad (9.82)$$

or

$$E_{n'} = m \left[1 + \frac{\alpha^2}{n'^2} \right]^{-1/2} .$$ (9.83)

The angular momentum quantum number ℓ' associated with (9.79) is evidently given by $\ell'(\ell' + 1) = \ell(\ell + 1) - \alpha^2$, or in other words[12] $\ell' = -\frac{1}{2} + \sqrt{(\ell + \frac{1}{2})^2 - \alpha^2}$ or $\ell' = \ell - \Delta_\ell$ with $\Delta_\ell \equiv \ell + \frac{1}{2} - \sqrt{(\ell + \frac{1}{2})^2 - \alpha^2}$. Thus ℓ' does not have integer values, whereas in the Schrödinger case ℓ must be a positive integer. However, as in the Schrödinger case $n' - \ell'$ must be an integer, and so if ℓ' is "displaced" from an integer by Δ_ℓ, then so must n'. Thus we take $n' = n - \Delta_\ell$ in (9.83):

$$E_{n'} \to E_{n\ell} = m \left[1 + \left(\frac{\alpha}{n - (\ell + \frac{1}{2}) + \sqrt{(\ell + \frac{1}{2})^2 - \alpha^2}} \right)^2 \right]^{-1/2} .$$ (9.84)

These energies, with $n = 1, 2, 3, \ldots$ and $\ell = 0, 1, 2, \ldots, n-1$ for each n, as in the Schrödinger case, are the allowed energy levels for the Coulomb problem according to the Klein–Gordon equation.

The Klein–Gordon equation is a valid relativistic wave equation and, like the Dirac equation, exhibits relativistic phenomena such as zitterbewegung and the Klein paradox. We shall discuss it further in the following chapter. It is not possible, however, to include the Pauli matrices $\sigma_x, \sigma_y, \sigma_z$ in the Klein–Gordon equation for a particle in an electromagnetic field and still maintain Lorentz covariance. This is a consequence of the fact that $\boldsymbol{\sigma}$ transforms as an ordinary three-vector rather than a four-vector, while ψ itself is a scalar (one-component) wave function (Schiff, 1968). Thus the Klein–Gordon equation is a relativistic wave equation for a *spinless* particle.

As such, the Klein–Gordon equation can be assumed to describe a pionic atom resulting from the capture of a π^- particle by a nucleus, or a mesonic atom resulting from the capture of a K^-. When various corrections including reduced mass and vacuum polarization are accounted for, the energy levels predicted by the Klein–Gordon equation are in excellent agreement with experimental data.

For the Coulomb problem described by the Dirac equation, it is convenient first to put the eigenvalue problem into the form of a second-order differential equation, and then to proceed as in the Klein–Gordon case by analogy to the nonrelativistic theory. Define the projection operators

$$P_1 = \frac{1}{2} \begin{pmatrix} 1 & 1 \\ 1 & 1 \end{pmatrix}, \quad P_2 = \frac{1}{2} \begin{pmatrix} 1 & -1 \\ -1 & 1 \end{pmatrix}$$ (9.85)

---

[12] The $+$ sign is chosen in order to allow a normalizable solution as $\alpha \to 0$.

and the column vectors

$$\psi_1 = P_1\psi = \frac{1}{2}\left(\begin{array}{c} \phi + \chi \\ \phi + \chi \end{array}\right), \qquad (9.86)$$

$$\psi_2 = P_2 = \frac{1}{2}\left(\begin{array}{c} \phi - \chi \\ -\phi + \chi \end{array}\right), \qquad (9.87)$$

where the two-component spinors ϕ and χ are defined by writing

$$\psi = \left(\begin{array}{c} \phi \\ \chi \end{array}\right) = \psi_1 + \psi_2 . \qquad (9.88)$$

Now multiply the Dirac equation $\gamma_\mu \pi^\mu \psi = m\psi$ from the left by P_2:

$$mP_2\psi = m\psi_2 = P_2\gamma_\mu\pi^\mu\psi = \gamma_\mu P_1\pi^\mu\psi = \gamma_\mu\pi^\mu P_1\psi = \gamma_\mu\pi^\mu\psi_1 , \qquad (9.89)$$

since $P_2\gamma_\mu = \gamma_\mu P_1$. Then

$$\psi = \psi_1 + \psi_2 = (1 + \frac{1}{m}\gamma_\mu\pi^\mu)\psi_1 , \qquad (9.90)$$

and the Dirac equation for ψ therefore implies the equation

$$(\gamma_\mu\pi^\mu - m)(\gamma_\mu\pi^\mu + m)\psi_1 = (\gamma_\mu\gamma_\nu\pi^\mu\pi^\nu - m^2)\psi_1 = 0 \qquad (9.91)$$

for ψ_1. From (9.86) it is clear that the eigensolutions of (9.91) will be of the form

$$\psi_1 = \left(\begin{array}{c} \phi_\ell \\ \phi_\ell \end{array}\right) e^{-iEt} . \qquad (9.92)$$

For the Coulomb potential we have $\pi^\mu = p^\mu - eA^\mu$, $A^0 = -e/r$, and $\mathbf{A} = 0$. In this case (9.91) and (9.92) imply, as may be shown by straightforward algebra,[13]

$$\left[\frac{1}{r^2}\frac{\partial}{\partial r}\left(r^2\frac{\partial}{\partial r}\right) - \frac{\ell(\ell+1) - \alpha^2 - i\alpha\boldsymbol{\sigma}\cdot\hat{r}}{r^2} + \frac{2\alpha E}{r} + E^2 - m^2\right]\phi_\ell(\mathbf{r}) = 0 \qquad (9.93)$$

for a state of orbital angular momentum quantum number ℓ.

This result differs from (9.79) only by the presence of the term $i\alpha\boldsymbol{\sigma}\cdot\hat{r}/r^2$, which accounts for the spin of the electron. The effect of this term is to

[13] The term $\boldsymbol{\sigma}\hat{r}/r^2$ results from $\boldsymbol{\sigma}\cdot\nabla\Phi$ with $\Phi = -e/r$.

replace the orbital angular momentum quantum number ℓ in the energy levels (9.84) by the total angular momentum quantum number $j - \ell \pm \frac{1}{2}$:[14]

$$E_{nj} = m \left[1 + \left(\frac{\alpha}{n - (j + \frac{1}{2}) + \sqrt{(j + \frac{1}{2})^2 - \alpha^2}} \right)^2 \right]^{-1/2}, \qquad (9.91)$$

where $n = 1, 2, 3, ..., \ell = 0, 1, 2, ..., n - 1$, and $j = \frac{1}{2}, \frac{3}{2}, ..., n$. Unlike the nonrelativistic energy levels given by the Bohr formula, the levels E_{nj} for the hydrogen atom described by the Dirac equation depend on both n and j. For historical reasons the levels are still labelled $ns_j, np_j, nd_j \ ...$ for $\ell = 0, 1, 2, ...$, as if ℓ were a "good" quantum number. Note that

$$E_{nj} = m - \frac{m\alpha^2}{2n^2} - \frac{m\alpha^4}{2n^3(j + \frac{1}{2})} + \frac{3m\alpha^4}{8n^4} + O(\alpha^6). \qquad (9.95)$$

The first term in this expansion, the electron rest energy, contributes the same energy to each level and is therefore spectroscopically unobservable. The second term gives the Bohr levels of the nonrelativistic (Schrödinger) theory, and the third term accounts for the *fine structure* in the energy levels of hydrogenic atoms. Thus, whereas the nonrelativistic theory predicts that the levels $2p_{1/2}$ and $2p_{3/2}$ should be degenerate, the Dirac theory predicts the difference (see Figure 3.1)

$$E(2p_{3/2}) - E(2p_{1/2}) = -\frac{m\alpha^4}{8}\left(\frac{1}{4} - \frac{1}{2}\right) = \frac{m\alpha^4}{32} = 4.5 \times 10^{-5} \text{eV} = 10.9 \text{ GHz}. \qquad (9.96)$$

This fine structure derives from the term $i\alpha\boldsymbol{\sigma} \cdot \hat{r}/r^2$ in (9.93), which gives rise to the spin–orbit coupling (9.68). The latter gives precisely (9.96).

The accounting for fine structure was historically a major success of the Dirac theory. However, several other effects, not included in the original Dirac theory, must be included in order to arrive at results in good agreement with precise spectroscopic measurements. One of these, of course, is the Lamb shift, which for the $2s_{1/2} - 2p_{1/2}$ splitting is about 1.06 GHz (Figure 3.1). The Lamb shift may be regarded as largely a consequence of the coupling of the electron to the vacuum field, which is not included in the theory outlined earlier (see Chapters 3 and 11). Another effect is the *hyperfine structure* associated with the interaction between the electron spin and the nuclear magnetic field. This results in the splitting of the

---

[14] For details, and for the Coulomb wave functions, we refer the reader to the books in the bibliography, such as Grandy (1991).

$1s_{1/2}$ level into two levels separated by about 1.42 GHz. Nuclear size and recoil effects give further corrections.

Note that the relevant length scale in the hydrogen problem is roughly the size of the atom, i.e., about $\alpha^{-1} \cong 137$ times the Compton radius of the electron. As a consequence, phenomena related to the Klein "paradox" do not appear in the theory of the hydrogen atom.

9.10   The Dirac Vacuum

The requirement that a quantum-mechanical wave equation conform to the requirements of special relativity led Dirac to the almost incredible prediction of antimatter. Among other striking successes, the Dirac equation accounted correctly for the fine structure observed in the hydrogen spectrum and for deviations from the Thomson cross section observed in the Compton scattering of light by electrons.[15]

This chapter has reviewed some topics that are covered in greater detail and depth in many other books. Its purpose has been to provide a bridge between the first part of this book, which has focused on nonrelativistic aspects of the QED vacuum, and the remainder, which focuses on the vacuum in relativistic QED and quantum field theory.

Dirac's hole theory gives us a picture of the vacuum in which, in addition to the zero-point electromagnetic field, there is a sea of filled negative-energy states. Electron–positron pairs can be "kicked out" of this *Dirac vacuum* by a uniform electric field, by a photon of energy $\hbar\omega > 2mc^2$, or by a potential acting to localize an electron to a region of space whose dimensions are comparable to or smaller than the Compton wavelength. The virtual electron–positron pairs of the vacuum lead us to regard the vacuum furthermore as a polarizable medium in which the field of a point charge, for instance, polarizes the vacuum in its vicinity and results in a difference between "bare" and "observed" charge.

As discussed in the following chapter, the existence of antiparticles is closely intertwined with Lorentz invariance, i.e., the requirement that quantum theory be consistent with the theory of special relativity. Quantum theory and special relativity together *require* negative-energy states and the existence of antiparticles. The fact that the number of particles or antiparticles can change as a result of pair creation and annihilation processes suggests a theory involving general particle creation and annihilation operators, analogous to the photon creation and annihilation operators of

---

[15] Compton scattering and the Klein–Nishina formula derived from the Dirac equation are treated in standard texts such as Sakurai (1976).

QED. This is the theory of second quantization, or quantum field theory, to which we now turn our attention.

9.11   Bibliography

Aitchison, I. J. R., "Nothings's Plenty. The Vacuum in Modern Quantum Field Theory," *Contemp. Phys.* **26**, 333 (1985).

Bjorken, J. D. and S. D. Drell, *Relativistic Quantum Mechanics* (McGraw-Hill, New York, 1964).

Bromberg, J., "The Concept of Particle Creation Before and After Quantum Mechanics," in *Historical Studies in the Physical Sciences*, Volume 17, ed. R. McCormmach (Princeton University Press, Princeton, New Jersey, 1976).

Dirac, P. A. M., "The Quantum Theory of the Electron," *Proc. Roy. Soc. Lond.* **A117**, 610 (1928).

Dirac, P. A. M., "A Theory of Electrons and Protons," *Proc. Roy. Soc. Lond.* **A126**, 360 (1930).

Dirac, P. A. M., "Quantised Singularities in the Electromagnetic Field," *Proc. Roy. Soc. Lond.* **A133**, 60 (1931).

Dirac, P. A. M., in *The Development of Quantum Theory* (Gordon and Breach, New York, 1971).

Dirac, P. A. M., in *Directions in Physics*, ed. H. Hora and J. R. Shepanski (Wiley, New York, 1978), p. 16.

Feynman, R. P., *Quantum Electrodynamics* (Benjamin, New York, 1962).

Feynman, R. P., R. B. Leighton, and M. Sands, *The Feynman Lectures on Physics* (Addison–Wesley, Reading, Mass., 1964), Volume 1, Chapter 15.

Fulcher, L. P., J. Rafelski, and A. Klein, "The Decay of the Vacuum," *Sci. Am.* **241**, 150 (December 1979).

Grandy, W. T., Jr., *Relativistic Quantum Mechanics of Leptons and Fields* (Kluwer, Dordrecht, 1991).

Greiner, W., B. Müller, and J. Rafelski, *Quantum Electrodynamics of Strong Fields* (Springer–Verlag, Berlin, 1985).

Itzykson, C. and J.-B. Zuber, *Quantum Field Theory* (McGraw–Hill, New York, 1980).

Moyer, D. F., "Origins of Dirac's Electron, 1925-1928," *Am. J. Phys.* **49**, 944 (1981).

Moyer, D. F., "Evaluations of Dirac's Electron, 1925-1928," *Am. J. Phys.* **49**, 1055 (1981).

Moyer, D. F., "Vindications of Dirac's Electron, 1932-1934," *Am. J. Phys.*

49, 1120 (1981)

Newton, T. D. and E. P. Wigner, "Localized States for Elementary Systems," *Rev. Mod. Phys.* **21**, 400 (1949).

Sakurai, J. J., *Advanced Quantum Mechanics* (Addison–Wesley, Reading, Mass., 1976).

Schiff, L. I., *Quantum Mechanics* (McGraw–Hill, New York, 1968).

Schwinger, J., "On Gauge Invariance and Vacuum Polarization," *Phys. Rev.* **82**, 664 (1951).

Chapter 10

Introduction to Quantum Field Theory

> The inhabitants of the universe [are] conceived to be a set of fields — an electron field, an electromagnetic field ... this point of view ... forms the central dogma of quantum field theory: the essential reality is a set of fields, subject to the rules of special relativity and quantum mechanics; all else is derived as a consequence of the quantum dynamics of these fields. — S. Weinberg (1977)

10.1 Introduction

The quantization of the electromagnetic field as in Chapter 2 allows us to describe the creation and annihilation of photons. Since the creation and annihilation of particles of all kinds is arguably the hallmark of relativistic quantum physics, we now turn to the formulation of quantum mechanics in terms of quantum fields, in a manner analogous to the quantization of the electromagnetic field.

As in the example of the electromagnetic field, all quantum fields have vacuum states and zero-point energies. We begin our discussion in the following section with the simplest example of a quantum field theory, namely, the quantization of the Schrödinger field $\psi(x,t)$ in one spatial dimension x. Next we quantize the Klein–Gordon field in order to emphasize the profound differences between relativistic and nonrelativistic quantum field theories. After considering also the quantization of a charged scalar field in Section 10.4, we turn our attention to the principal goal of this chapter, the quantization of the Dirac field. This will set the stage for a discussion,

mainly in Chapters 11 and 12, of some relativistic QED effects, including effects associated with the vacuum state of the Dirac field. We also
return once again to the quantized electromagnetic field and consider the
"propagators" associated with quantum fields.

10.2 Second Quantization: Nonrelativistic

When we quantize a one-particle, one-dimensional system we take its position x and momentum p to be operators in a Hilbert space, satisfying
$[x, p] = i\hbar$. The "vectors" of the Hilbert space are states $|\psi\rangle$, and the wave
function of the system at time t is the projection $\psi(x, t) = \langle x | \psi(t) \rangle$. The
wave function, of course, is a c-number, not an operator. In *second quantization*, however, we take $\psi(x, t)$ to be an operator, just as the electric and
magnetic fields are operators when the electromagnetic field is quantized.
$\psi(x, t)$ becomes a quantized field, hence the name *quantum field theory*.

To introduce the notion of a quantum field theory as simply as possible, we consider first the second quantization of the Schrödinger equation,
i.e., the quantization of the "classical" c-number field $\psi(x, t)$. We begin
heuristically with the familiar expansion of the wave function in terms of a
complete set of eigenfunctions $\phi_n(x)$ of the Hamiltonian:

$$\psi(x, t) = \sum_n a_n(t) \phi_n(x), \tag{10.1}$$

where $a_n(t) = a_n(0) e^{-iE_n t}$. Let us suppose that, when we second-quantize
the theory and make $\psi(x, t)$ an operator, $a_n(t)$ becomes an operator in
a Hilbert space while $\phi_n(x)$ remains an ordinary, c-number function of
x. Then the Hermitian conjugate of $\psi(x, t)$ will involve the Hermitian
conjugate of $a_n(t)$:

$$\psi^\dagger(x, t) = \sum_n a_n^\dagger(t) \phi_n^*(x) \tag{10.2}$$

and

$$[\psi(x, t), \psi^\dagger(x', t)] = \sum_m \sum_n [a_n(t), a_m^\dagger(t)] \phi_m^*(x') \phi_n(x). \tag{10.3}$$

At this heuristic level there is nothing to tell us what the commutators in
this expression should be. We will show later that the equal-time commutation rules for second quantization can be taken to be

$$[\psi(x, t), \psi^\dagger(x', t)] = \delta(x - x'), \quad [\psi(x, t), \psi(x', t)] = 0. \tag{10.4}$$

These rules, together with the orthonormality relation

$$\int dx \phi_n(x) \phi_m^*(x') = \delta_{mn} ,$$ (10.5)

imply that[1]

$$[a_n(t), a_m^\dagger(t)] = \delta_{mn}, \quad [a_n(t), a_m(t)] = 0.$$ (10.6)

We might guess furthermore that in the second-quantized theory the Hamiltonian can be expressed in the form

$$H = \sum_n E_n a_n^\dagger a_n ,$$ (10.7)

analogous to the Hamiltonian for the quantized electromagnetic field. This, together with the previous commutation rules, suggests that our quantum field theory reduces in essence to a theory of uncoupled harmonic oscillators, with a_n and $a_n^\dagger$ playing the role of annihilation and creation operators for particles in "mode" $\phi_n(x)$.

The vacuum state of our quantum field is the state in which there are no particles in any state, i.e., $a_n|0\rangle = 0$ for every n. Likewise $a_n^\dagger|0\rangle = |1\rangle_n$ is a state in which there is one particle of energy E_n, and $a_n^\dagger a_m^\dagger|0\rangle$ is a two-particle state of total energy $E_n + E_m$. The states obtained by operating on $|0\rangle$ with creation operators in this way are called *Fock states*.[2]

The state

$$\psi^\dagger(x,t)|0\rangle = \sum_n a_n^\dagger(t)|0\rangle \phi_n^*(x) = \sum_n |1\rangle_n \phi_n^*(x) e^{iE_n t}$$ (10.8)

is a one-particle state in which the particle is at x at time t. Evidently the operator $\psi^\dagger(x,t)$ creates a particle at x at time t, and likewise $\psi(x,t)$ annihilates a particle at x.

Before going further with this elementary quantum field theory, let us first provide a more systematic basis for it.

---

[1] Note from (10.1) and orthonormality of the ϕ_n that $a_n(t) = \int dx \psi(x,t)\phi_n^*(x)$ and $a_n^\dagger(t) = \int dx \psi^\dagger(x,t)\phi_n(x)$. The commutators (10.6) then follow from the assumed field commutators (10.4). Note also that we can derive the latter if we first assume (10.6) and use the completeness condition $\sum_n \phi_n^*(x)\phi_n(x') = \delta(x - x')$.

[2] Similarly, states of definite photon number in QED are often referred to as *Fock states*.

Lagrangian, Action, and Hamiltonian

In our quantum field theory we would like the Schrödinger equation

$$i\frac{\partial\psi}{\partial t} = \left[-\frac{1}{2}\frac{\partial^2}{\partial x^2} + V(x)\right]\psi \tag{10.9}$$

to become an operator equation derived from the Heisenberg equation of motion

$$i\dot{\psi} = [\psi, H]. \tag{10.10}$$

Now if we assume the equal-time commutation relations (10.4), a Hamiltonian that produces (10.9) via (10.10) is

$$H = \int dx\,\psi^\dagger(x,t)\left[-\frac{1}{2}\frac{\partial^2}{\partial x^2} + V(x)\right]\psi(x,t). \tag{10.11}$$

This will in fact turn out to be the appropriate Hamiltonian.

In Section 4.2 we reviewed briefly the procedure for constructing a Lagrangian and Hamiltonian for a classical mechanical system. The procedure in classical field theory is similar. In this case the analog of the action $\int_{t_1}^{t_2} dt\, L(q, \dot{q}, t)$ is[3]

$$S[\psi] = \int_{t_1}^{t_2} dt \int dx \mathsf{L}[\psi], \tag{10.12}$$

where L is the Lagrangian density and $\int dx \mathsf{L}$ the Lagrangian. The notation $S[\psi]$ and $\mathsf{L}[\psi]$ is used to denote the fact that S and L are functionals of the field ψ, i.e., they associate according to some rule a number with a function ψ. As in classical mechanics we seek to derive equations of motion by making the action an extremum with respect to variations in ψ, keeping $\psi(x, t_1)$ and $\psi(x, t_2)$ fixed. The variation in the action is

$$\delta S = \int_{t_1}^{t_2} dt \int dx \left[\frac{\partial \mathsf{L}}{\partial \psi}\delta\psi + \frac{\partial \mathsf{L}}{\partial \dot{\psi}}\delta\dot{\psi} + \frac{\partial \mathsf{L}}{\partial(\partial\psi/\partial x)}\delta\left(\frac{\partial\psi}{\partial x}\right)\right], \tag{10.13}$$

where, analogously to the case of classical mechanical theory, we assume that L does not depend on $\ddot{\psi}, \partial^2\psi/\partial x^2$, or higher derivatives. It is easily seen from the definition of the derivatives that

$$\delta\dot{\psi} = \frac{\partial}{\partial t}(\delta\psi), \quad \delta\left(\frac{\partial\psi}{\partial x}\right) = \frac{\partial}{\partial x}(\delta\psi), \tag{10.14}$$

---

[3]The limits on the integral over x in (10.12) may be assumed to be 0 and L, with $\psi(0, t) = \psi(L, t)$ under the assumption of periodic boundary conditions.

so that

$$
\begin{aligned}
\delta S &= \int_{t_1}^{t_2} dt \int dx \left[\frac{\partial L}{\partial \psi} \delta\psi + \frac{\partial L}{\partial \dot\psi} \frac{\partial}{\partial t}(\delta\psi) + \frac{\partial L}{\partial(\partial\psi/\partial x)} \frac{\partial}{\partial x}(\delta\psi) \right] \\
&= \int_{t_1}^{t_2} dt \int dx \left[\frac{\partial L}{\partial \psi} - \frac{\partial}{\partial t}\left(\frac{\partial L}{\partial \dot\psi}\right) - \frac{\partial}{\partial x} \frac{\partial L}{\partial(\partial\psi/\partial x)} \right] \delta\psi. \quad (10.15)
\end{aligned}
$$

The second equality follows by partial integration; the surface terms in the partial integration over x vanish under the assumption of periodic boundary conditions, and those in the partial integration over t vanish because $\delta\psi(x, t_1) = \delta\psi(x, t_2) = 0$ by assumption. Requiring that the variation δS vanish for any arbitrary variation $\delta\psi$ at each x, we obtain

$$
\frac{\partial L}{\partial \psi} - \frac{\partial}{\partial t}\left(\frac{\partial L}{\partial \dot\psi}\right) - \frac{\partial}{\partial x} \frac{\partial L}{\partial(\partial\psi/\partial x)} = 0. \quad (10.16)
$$

Now again analogously to the classical theory of point particle dynamics, we define the *field* momentum conjugate to ψ as

$$
\pi(x, t) = \frac{\partial L}{\partial \dot\psi} . \quad (10.17)
$$

And in analogy to the definition of the Hamiltonian $H = p\dot q - L$ in classical particle dynamics, we define the field Hamiltonian

$$
H = \int dx [\pi(x, t)\dot\psi(x, t) - L]. \quad (10.18)
$$

The Lagrangian density L must be chosen such that (10.16) produces the Schrödinger equation for ψ. The choice

$$
L = i\psi^* \dot\psi - \frac{1}{2} \frac{\partial\psi}{\partial x} \frac{\partial\psi^*}{\partial x} - V\psi^*\psi \quad (10.19)
$$

accomplishes this. With this L the momentum conjugate to ψ is found from (10.17) to be

$$
\pi(x, t) = i\psi^*(x, t) , \quad (10.20)
$$

and the Hamiltonian (10.18) becomes

$$
\begin{aligned}
H &= \int dx \left[i\psi^* \dot\psi - i\psi^* \dot\psi + \frac{1}{2} \frac{\partial\psi}{\partial x} \frac{\partial\psi^*}{\partial x} + V\psi^*\psi \right] \\
&= \int dx \left[-\frac{1}{2} \psi^* \frac{\partial^2\psi}{\partial x^2} + V\psi^*\psi \right] , \quad (10.21)
\end{aligned}
$$

where the second equality follows from a partial integration.

Quantization

In the formalism just outlined $\psi(x,t)$ has been assumed to be a classical field. We now want to quantize the field ψ, i.e., to *second*-quantize the Schrödinger equation. To do this we make ψ an operator, replace ψ^* by $\psi^\dagger$, and impose the equal-time canonical commutation rules, analogous to $[x, p] = i$ for particles,

$$[\psi(x, t), \pi(x', t)] = i\delta(x - x'), \quad [\psi(x, t), \psi(x', t)] = [\pi(x, t), \pi(x', t)] = 0,$$
(10.22)

or, from (10.20),

$$[\psi(x, t), \psi^\dagger(x', t)] = \delta(x - x'), \quad [\psi(x, t), \psi(x', t)] = 0,$$
(10.23)

which is (10.4). Thus the quantized version of the Hamiltonian (10.21),

$$H = \int dx\, \psi^\dagger(x, t)[-\frac{1}{2}\frac{\partial^2}{\partial x^2} + V(x)]\psi(x, t) ,$$
(10.24)

is indeed the Hamiltonian for which the Heisenberg equation of motion for ψ is the Schrödinger equation for the (operator) field ψ.

The expansion (10.1), together with the Heisenberg equation (10.10), implies the equation of motion

$$i\dot{a}_n = [a_n, H]$$
(10.25)

for the annihilation operator a_n; of course this is a special case of the general Heisenberg equation of motion $i\dot{A} = [A, H]$. Use of the expansion (10.1) in (10.24) also implies

$$H = \sum_n E_n a_n^\dagger a_n ,$$
(10.26)

so that $\dot{a}_n = -iE_n a_n$ and $a_n(t) = a_n(0)e^{-iE_n t}$ for the (free) Schrödinger field.

Other operators find a similar expression in terms of ψ and $\psi^\dagger$. For instance, the coordinate operator x is

$$
\begin{aligned}
x &= \int dx\, \psi^\dagger(x, t)x\psi(x, t) = \sum_m \sum_n a_m^\dagger(t)a_n(t)x_{mn} \\
&= \sum_m \sum_n x_{mn}\sigma_{mn}(t),
\end{aligned}
$$
(10.27)

where $x_{mn} \equiv \int dx \phi_m^*(x) x \phi_n(x) = \langle \phi_m | x | \phi_n \rangle$ and $\sigma_{mn} \equiv a_m^\dagger a_n$. Observe that

$$
\begin{aligned}
[\sigma_{mn}, \sigma_{ij}] &= [a_m^\dagger a_n, a_i^\dagger a_j] \\
&= a_m^\dagger (a_i^\dagger a_n + \delta_{in}) a_j - a_i^\dagger (a_m^\dagger a_j + \delta_{jn}) a_n \\
&= \sigma_{mj} \delta_{in} - \sigma_{in} \delta_{jm} ,
\end{aligned} \tag{10.28}
$$

which will be recognized as the commutation relation (4.79). Indeed, our treatment of an atom in Chapter 4 in terms of the operators σ_{mn}, and in particular a "two-state" atom in terms of the Pauli operators σ and σ_z, was a first step toward a quantum field theory of the Schrödinger wave function.

Bosons and Fermions

Recall that $\psi^\dagger(x, t)|0\rangle$ is a one-particle state with the particle at x. A one-particle state $|\Phi_1\rangle$ described by the (c-number) wave function $\phi(x, t)$ may evidently be obtained by multiplying $\psi^\dagger(x, t)|0\rangle$ by the amplitude $\phi(x, t)$ and integrating over all x:

$$
|\Phi_1(t)\rangle = \int dx \phi(x, t) \psi^\dagger(x, t)|0\rangle . \tag{10.29}
$$

Consider now $H|\Phi_1\rangle$, with the Hamiltonian H given by (10.24):

$$
\begin{aligned}
H|\Phi_1\rangle &= \int dx' \psi^\dagger(x', t)[-\frac{1}{2}\frac{\partial^2}{\partial x'^2} + V(x')]\psi(x', t) \int dx \phi(x, t)\psi^\dagger(x, t)|0\rangle \\
&= \int dx' \int dx \psi^\dagger(x', t)[-\frac{1}{2}\frac{\partial^2}{\partial x'^2} + V(x')]\phi(x, t)\psi(x', t)\psi^\dagger(x, t)|0\rangle \\
&= \int dx \psi^\dagger(x, t)[-\frac{1}{2}\frac{\partial^2}{\partial x^2} + V(x)]\phi(x, t)|0\rangle ,
\end{aligned} \tag{10.30}
$$

where we have used (10.23). Thus we have $H|\Phi_1\rangle = E|\Phi_1\rangle$ if

$$
-\frac{1}{2}\frac{\partial^2 \phi}{\partial x^2} + V(x)\phi = E\phi, \tag{10.31}
$$

i.e., if $\phi(x, t) = \phi_n(x)e^{-iE_n t}$ is an eigenstate of the first-quantized Hamiltonian.

Similarly the two-particle state $|\Phi_2\rangle$ described by the wave function $\phi(x_1, x_2, t)$,

$$
|\Phi_2\rangle = \int dx_1 \int dx_2 \phi(x_1, x_2, t)\psi^\dagger(x_1, t)\psi^\dagger(x_2, t)|0\rangle, \tag{10.32}
$$

may be shown to be an eigenstate of H if $\phi(x_1, x_2, t)$ satisfies a two particle first-quantized Schrödinger equation. Note that, since $[\psi^\dagger(x,t), \psi^\dagger(x',t)] = 0$, we can interchange x_1 and x_2 in the integrand in (10.32) without changing the state $|\Phi_2\rangle$. We deduce therefore that

$$\phi(x_1, x_2, t) = \phi(x_2, x_1, t). \tag{10.33}$$

In other words, our quantum field theory describes bosons, and it is not surprising therefore that we obtained the boson commutation relations $[a_n(t), a_m^\dagger(t)] = \delta_{mn}$, $[a_n(t), a_m(t)] = 0$. In order to have a quantum field theory describing fermions, we must evidently have field commutation rules different from (10.23), since these imply bosons.

If we are dealing with fermions we should have

$$\psi^\dagger(x,t)\psi^\dagger(x,t)|0\rangle = 0, \tag{10.34}$$

since we cannot put two particles in the same state. Now if we interchange x_1 and x_2 in (10.32), and use the fact that $\phi(x_2, x_1, t) = -\phi(x_1, x_2, t)$ for fermions, we deduce that

$$\begin{aligned}
|\Phi_2\rangle &= \int dx_1 \int dx_2 \phi(x_1, x_2, t)\psi^\dagger(x_1, t)\psi^\dagger(x_2, t)|0\rangle \\
&= -\int dx_1 \int dx_2 \phi(x_1, x_2, t)\psi^\dagger(x_2, t)\psi^\dagger(x_1, t)|0\rangle , \tag{10.35}
\end{aligned}$$

and consequently that

$$\psi^\dagger(x_2, t)\psi^\dagger(x_1, t) = -\psi^\dagger(x_1, t)\psi^\dagger(x_2, t). \tag{10.36}$$

In other words, the *anticommutator*

$$\psi^\dagger(x_1, t)\psi^\dagger(x_2, t) + \psi^\dagger(x_2, t)\psi^\dagger(x_1, t) \equiv \{\psi^\dagger(x_1, t), \psi^\dagger(x_2, t)\} = 0. \tag{10.37}$$

It follows by Hermitian conjugation that $\{\psi(x_1, t), \psi(x_2, t)\} = 0$. In fact to second-quantize the Schrödinger equation for a fermion system we simply replace commutators by anticommutators. In particular, the boson commutation rules (10.23) are replaced by

$$\{\psi(x,t), \psi^\dagger(x',t)\} = \delta(x-x'), \quad \{\psi(x,t), \psi(x',t)\} = 0. \tag{10.38}$$

Note that these anticommutators, together with (10.1), imply the fermion algebra for annihilation and creation operators:

$$\{a_n(t), a_m^\dagger(t)\} = \delta_{mn}, \quad \{a_n(t), a_m(t)\} = 0. \tag{10.39}$$

The Hamiltonian retains the form (10.26), but with a_n and $a_n^\dagger$ now being annihilation and creation operators for fermions. It follows from (10.39) that

$$(a_n^\dagger a_n)^2 = a_n^\dagger (1 - a_n^\dagger a_n) a_n = a_n^\dagger a_n \ , \tag{10.40}$$

so that the eigenvalues of the number operator $a_n^\dagger a_n$ are 0 and 1, the familiar result for fermions. Denoting the corresponding eigenstates by $|0\rangle$ and $|1\rangle$, we have $a_n^\dagger a_n a_n |1\rangle = 0$ and $a_n^\dagger a_n a_n^\dagger |0\rangle = a_n^\dagger (1 - a_n^\dagger a_n)|0\rangle = a_n^\dagger |0\rangle = |1\rangle$. Thus $a_n |1\rangle$ and $a_n^\dagger |0\rangle$ are $|0\rangle$ and $|1\rangle$, respectively, justifying the terminology of annihilation and creation operators for the fermion operators.

Before proceeding to an example of *relativistic* quantum field theory, we note that, in going from the Hamiltonian (10.21) to the second-quantized version (10.24), we implicitly chose a normal ordering of ψ and $\psi^\dagger$ — creation operators to the left of annihilation operators. Since we can write (10.21) equivalently as

$$H = \int dx \left[\frac{1}{2} \left(-\psi^* \frac{1}{2} \frac{\partial^2 \psi}{\partial x^2} - \psi \frac{1}{2} \frac{\partial^2 \psi^*}{\partial x^2} \right) + \frac{1}{2} (\psi^* V \psi + \psi V \psi^*) \right], \tag{10.41}$$

we can equally well write the second-quantized Hamiltonian as

$$\begin{aligned}
H &= \int dx \frac{1}{2} \psi^\dagger(x,t) \left[-\frac{1}{2} \frac{\partial^2}{\partial x^2} + V(x) \right] \psi(x,t) \\
&\quad + \frac{1}{2} \psi(x,t) \left[-\frac{1}{2} \frac{\partial^2}{\partial x^2} + V(x) \right] \psi^\dagger(x,t) \\
&= \sum_n \frac{1}{2} E_n (a_n^\dagger a_n + a_n a_n^\dagger) \\
&= \sum_n E_n \left(a_n^\dagger a_n + \frac{1}{2} \right) \tag{10.42}
\end{aligned}$$

if we use (10.6). In other words, if we choose a symmetric ordering of annihilation and creation operators in the second-quantization prescription, we obtain the zero-point energy $\sum_n \frac{1}{2} E_n$ for the vacuum state of the second-quantized Schrödinger field. This is discussed further in Section 10.7.

10.3   The Klein–Gordon Field

In Section 9.2 we touched briefly on the Klein–Gordon equation, noting that it admits negative-energy solutions and that there was a difficulty in the single-particle theory with the definition of a probability current.[4] It is easy to define a conserved current for the Klein–Gordon equation; one finds that the spatial component,

$$j_i = -\frac{i}{2m}\left(\psi\frac{\partial\psi^*}{\partial x^i} - \psi^*\frac{\partial\psi}{\partial x^i}\right) = -\frac{i}{2m}(\psi\partial_i\psi^* - \psi^*\partial_i\psi), \qquad (10.43)$$

has the same form as in the case of the nonrelativistic Schrödinger equation, but that the time component

$$j_0 = \rho = \frac{i}{2m}\left(\psi^*\frac{\partial\psi}{\partial t} - \psi\frac{\partial\psi^*}{\partial t}\right) \qquad (10.44)$$

is different, involving derivatives of ψ and ψ^* with respect to t. (The corresponding density ρ in the Schrödinger case is $\psi^*\psi$.) For the plane-wave solution $\psi = e^{-i(Et - \mathbf{k}\cdot\mathbf{x})}$ of the Klein–Gordon equation, we have

$$\rho = \frac{E}{m} = \pm\frac{1}{m}\sqrt{\mathbf{k}^2 + m^2}, \qquad (10.45)$$

which is not a positive–definite probability density because of the negative-energy solutions ($E = -\sqrt{\mathbf{k}^2 + m^2}$). The difficulty with the Klein–Gordon equation, then, lies in the interpretation of negative probability densities. In the case of the Dirac equation, the density $\rho = \psi^\dagger\psi$ is positive–definite irrespective of negative-energy solutions.

We will now consider the second-quantization of the Klein–Gordon field, and show among other things that the difficulty with the *single-particle* Klein–Gordon theory is removed by quantum field theory. (Section 10.4)

The classical action whose variation yields the Klein–Gordon equation is

$$
\begin{aligned}
S[\psi] &= \int dt \int d^3x \frac{1}{2}\left[\left(\frac{\partial\psi}{\partial t}\right)^2 - (\nabla\psi)^2 - m^2\psi^2\right] \\
&= \int d^4x \frac{1}{2}[\partial^\mu\psi(x)\partial_\mu\psi(x) - m^2\psi^2(x)],
\end{aligned}
\qquad (10.46)
$$

where x stands for the space–time point $(\mathbf{x}, t)$ and $d^4x \equiv d^3x\,dt$. The reader may easily check that the generalization of (10.16) to three spatial dimensions, together with the Lagrangian density L implied in (10.46), gives the Klein–Gordon equation. The momentum conjugate to $\psi(x)$ is

$$\pi(x) = \frac{\partial\mathsf{L}}{\partial\dot\psi} = \dot\psi, \qquad (10.47)$$

---

[4] Schrödinger had considered this equation before Klein and Gordon, but abandoned it for these and other reasons.

so that the Hamiltonian is

$$H = \int d^3x[\pi(x)\dot{\psi}(x) - L] = \frac{1}{2}\int d^3x[\dot{\psi}^2(x) + (\nabla\psi(x))^2 + m^2\psi^2(x)]$$
$$= \frac{1}{2}\int d^3x[\pi^2(x) + |\nabla\psi(x)|^2 + m^2\psi^2(x)] \tag{10.48}$$

(We are assuming here that $\psi(x)$ is a real field.) In second quantization we use the generalization of the commutation rules (10.22) to three spatial dimensions,

$$[\psi(\mathbf{x},t), \pi(\mathbf{x}',t)] = i\delta(\mathbf{x} - \mathbf{x}'), \quad [\psi(\mathbf{x},t), \psi(\mathbf{x}',t)] = [\pi(\mathbf{x},t), \pi(\mathbf{x}',t)] = 0. \tag{10.49}$$

Note that this use of equal-time commutators implies a choice of a specific Lorentz frame. However, the theory may nevertheless be shown to be Lorentz–invariant, and in particular these commutators may be generalized to arbitrary (space–like) separations. It is easily verified that the Hamiltonian (10.48), with $\psi(x)$ and $\pi(x)$ operators, gives the Heisenberg equations of motion

$$\dot{\psi} = -i[\psi, H] = \pi, \quad \dot{\pi} = -i[\pi, H] = \nabla^2\psi - m^2\psi, \tag{10.50}$$

i.e., the Klein–Gordon equation for the field operator $\psi(x)$.

We wish to write the analogue of equation (10.1) for the Klein–Gordon field. Now, however, we have to allow for both positive- and negative-energy eigenfunctions, $\psi_{\mathbf{k}}^{(+)}(\mathbf{x})$ and $\psi_{\mathbf{k}}^{(-)}(\mathbf{x})$, respectively:

$$\begin{aligned}\psi(\mathbf{x},t) &= \sum_{\mathbf{k}}[A_{\mathbf{k}}^{(+)}(t)\psi_{\mathbf{k}}^{(+)}(\mathbf{x}) + A_{\mathbf{k}}^{(-)}(t)\psi_{\mathbf{k}}^{(-)}(\mathbf{x})] \\ &= \sum_{\mathbf{k}}[A_{\mathbf{k}}^{(+)}(t)e^{i\mathbf{k}\cdot\mathbf{x}} + A_{\mathbf{k}}^{(-)}(t)e^{-i\mathbf{k}\cdot\mathbf{x}}] \\ &= \sum_{\mathbf{k}}[A_{\mathbf{k}}^{(+)}e^{-i(E_k t - \mathbf{k}\cdot\mathbf{x})} + A_{\mathbf{k}}^{(-)}e^{i(E_k t - \mathbf{k}\cdot\mathbf{x})}] \\ &= \sum_{\mathbf{k}}[A_{\mathbf{k}}^{(+)}e^{-ik\cdot x} + A_{\mathbf{k}}^{(-)}e^{ik\cdot x}]. \end{aligned} \tag{10.51}$$

Since we took $\psi(x)$ classically to be a real field, in second quantization it must be Hermitian. This implies that $A_{\mathbf{k}}^{(-)} = A_{\mathbf{k}}^{\dagger}$, where $A_{\mathbf{k}} \equiv A_{\mathbf{k}}^{(+)}$:

$$\psi(x) = \sum_{\mathbf{k}}[A_{\mathbf{k}}e^{-ik\cdot x} + A_{\mathbf{k}}^{\dagger}e^{ik\cdot x}]. \tag{10.52}$$

Since there is a continuum of possible $\mathbf{k}$ values, we replace (10.52) by

$$\psi(\mathbf{x},t) = \int \frac{d^3k}{(2\pi)^3} \frac{1}{2E_k} [a(\mathbf{k})e^{-ik\cdot x} + a^\dagger(\mathbf{k})e^{ik\cdot x}]. \tag{10.53}$$

The factor $1/(2\pi)^3$ arises as in the replacement $\sum_{\mathbf{k}} \to [V/(2\pi)^3] \int d^3k$ familiar from the electromagnetic case. The factor $1/2E_k$ is included in order to have a Lorentz-invariant measure, as discussed in Appendix G.

Using

$$\pi(\mathbf{x},t) = \dot{\psi}(\mathbf{x},t) = -i \int \frac{d^3k}{(2\pi)^3} \frac{1}{2E_k} E_k [a(\mathbf{k})e^{-ik\cdot x} - a^\dagger(\mathbf{k})e^{ik\cdot x}], \tag{10.54}$$

we can solve for $a(\mathbf{k})$ and $a^\dagger(\mathbf{k})$ in terms of $\psi(\mathbf{x},t)$ and $\pi(\mathbf{x},t)$. Then the commutation rules (10.49) imply

$$[a(\mathbf{k}), a^\dagger(\mathbf{k}')] = (2\pi)^3 2E_k \delta^3(\mathbf{k} - \mathbf{k}') \tag{10.55}$$

and $[a(\mathbf{k}), a(\mathbf{k}')] = 0$, i.e., boson commutators for the annihilation and creation operators. In terms of the annihilation and creation operators the Hamiltonian (10.48) is that for a (continuous) sum of uncoupled harmonic oscillators:

$$H = \frac{1}{2} \int \frac{d^3k}{(2\pi)^3} \frac{1}{2E_k} E_k [a^\dagger(\mathbf{k})a(\mathbf{k}) + a(\mathbf{k})a^\dagger(\mathbf{k})]. \tag{10.56}$$

As in the case of the electromagnetic field, $a(\mathbf{k})$ and $a^\dagger(\mathbf{k})$ are particle annihilation and creation operators. In the case of the Klein–Gordon field, they annihilate and create particles of rest mass m, momentum $\mathbf{k}$, and energy $E_k = \sqrt{\mathbf{k}^2 + m^2}$. The vacuum state $|0\rangle$, as in the electromagnetic case, satisfies $a(\mathbf{k})|0\rangle = 0$, whereas $a^\dagger(\mathbf{k})|0\rangle$ is a state with one particle of energy E_k and momentum $\mathbf{k}$, etc.

The expectation value of H in the vacuum state is

$$\langle 0|H|0\rangle = \frac{1}{2} \int \frac{d^3k}{(2\pi)^3} \frac{1}{2E_k} E_k \langle 0|a(\mathbf{k})a^\dagger(\mathbf{k})|0\rangle = \frac{1}{2} \int d^3k\, E_k \delta^3(0), \tag{10.57}$$

where we have used (10.55).[5] Like the electromagnetic field, therefore, the Klein–Gordon field has an infinite zero-point energy. This infinity can be

---

[5]Note that the state $a^\dagger(\mathbf{k})|0\rangle$ is not normalizable, since it has infinite norm: $\langle 0|a(\mathbf{k})a^\dagger(\mathbf{k})|0\rangle = (2\pi)^3 2E_k \delta^3(0)$. This is because it is a plane-wave state with definite momentum, and so the particle's position probability distribution is uniform within the infinite volume $\delta^3(0)$. This problem is easily remedied (see, for instance, Itzykson and Zuber, 1980, p. 211) and will not concern us.

discarded by simply redefining the zero of the energy scale. Using (10.55) and dropping the zero-point energy, we then have

$$H \rightarrow \int \frac{d^3 k}{(2\pi)^3} \frac{1}{2E_k} E_k a^\dagger(\mathbf{k}) a(\mathbf{k}). \tag{10.58}$$

In other words, we put the annihilation and creation operators in normal order and drop the zero-point term. Normal ordering is often indicated by colons. Thus the Hamiltonian (10.58) is denoted : H :, and similarly : $a(\mathbf{k}) a^\dagger(\mathbf{k})$:= $a^\dagger(\mathbf{k}) a(\mathbf{k})$. That is, within the colons we can freely commute the annihilation and creation operators to obtain normal-ordered products.

Discrete and Continuous Sums over k

When we worked with the quantized electromagnetic field we chose to write the sum over modes in the discrete form $\sum_{\mathbf{k}\lambda}$ rather than as $\int d^3 k \sum_\lambda$. We can do this for the Klein–Gordon field, too, in which case the normally ordered Hamiltonian takes the familiar form

$$: H := \sum_{\mathbf{k}} E_k a_{\mathbf{k}}^\dagger a_{\mathbf{k}} , \tag{10.59}$$

where now

$$[a_{\mathbf{k}}, a_{\mathbf{k}'}^\dagger] = \delta_{\mathbf{k},\mathbf{k}'}^3 . \tag{10.60}$$

The field $\psi(\mathbf{x}, t)$, similarly, takes the form

$$\psi(\mathbf{x}, t) = \sum_{\mathbf{k}} \left(\frac{1}{2E_k V} \right)^{1/2} [a_{\mathbf{k}}(t) e^{i\mathbf{k}\cdot\mathbf{x}} + a_{\mathbf{k}}^\dagger(t) e^{-i\mathbf{k}\cdot\mathbf{x}}], \tag{10.61}$$

with $a_{\mathbf{k}}(t) = a_{\mathbf{k}}(0) e^{-iE_k t}$ for the free field and where V is a quantization volume. This is basically of the same form as the vector potential (2.52) in free space, except that (a) $\psi(\mathbf{x}, t)$ here describes a scalar field, with no polarization index λ, and (b) $E_k = \sqrt{k^2 + m^2}$ replaces $\omega_k = |\mathbf{k}|$ (in units with $\hbar = c = 1$). The "extra" factor of $\sqrt{4\pi}$ in (2.52) is simply a consequence of having chosen Gaussian units for the electromagnetic field, and results in the expression $(\mathbf{E}^2 + \mathbf{B}^2)/8\pi$ rather than $(\mathbf{E}^2 + \mathbf{B}^2)/2$ (in Heaviside–Lorentz units) for the energy density (see Section 10.8).

Causality and Spin Statistics

Acting on Fock states, $\psi(x)$ and $\psi(x')$ change the number of particles at space–time points x and x', respectively. If $(x - x')^2 = (t - t')^2 - (\mathbf{x} -$

$\mathbf{x}')^2 < 0$, i.e., if the separation between x and x' is spacelike, the creation and annihilation events cannot, according to special relativity, affect one another. Thus we should have $[\psi(x), \psi(x')] = 0$ if $(x - x')^2 < 0$. Now from (10.53) and (10.55) we have

$$
\begin{aligned}
[\psi(x), \psi(x')] &= \int \frac{d^3k}{(2\pi)^3} \int \frac{d^3k'}{(2\pi)^3} \frac{1}{4E_k E_{k'}} \left([a(\mathbf{k}), a^\dagger(\mathbf{k}')] e^{-ik\cdot x} e^{ik'\cdot x'} \right. \\
&\quad \left. + [a^\dagger(\mathbf{k}), a(\mathbf{k}')] e^{ik\cdot x} e^{-ik'\cdot x'} \right) \\
&= \int \frac{d^3k}{(2\pi)^3} \frac{1}{2E_k} [e^{-ik\cdot(x-x')} - e^{ik\cdot(x-x')}] \\
&= \int \frac{d^3k}{(2\pi)^3} \frac{1}{2E_k} [e^{-iE_k(t-t')} e^{i\mathbf{k}\cdot(\mathbf{x}-\mathbf{x}')} \\
&\quad - e^{iE_k(t-t')} e^{-i\mathbf{k}\cdot(\mathbf{x}-\mathbf{x}')}] \\
&= -i \int \frac{d^3k}{(2\pi)^3} e^{i\mathbf{k}\cdot(\mathbf{x}-\mathbf{x}')} \frac{1}{E_k} \sin E_k(t - t') \\
&\equiv i\Delta(x - x').
\end{aligned}
\tag{10.62}
$$

Since it involves four-vector dot products and a Lorentz-invariant measure, $\Delta(x - x')$ is Lorentz invariant. Then since $\Delta(x - x') = 0$ for $t - t' = 0$ and is Lorentz invariant, it vanishes for all $(x - x')^2 < 0$, as required. Note also that

$$\Delta(-x) = -\Delta(x), \tag{10.63}$$

$$\dot{\Delta}(x) = -\delta^3(\mathbf{x}) \quad \text{at } t = 0, \tag{10.64}$$

and that $\Delta(x)$ satisfies the Klein–Gordon equation:

$$\left(\frac{\partial^2}{\partial t^2} - \nabla^2 + m^2 \right) \Delta(x) = 0. \tag{10.65}$$

It is crucial, in order to satisfy the causality requirement, that we quantize the Klein–Gordon field using commutators. Had we used anticommutators and replaced the commutator on the left side of (10.62) by an anticommutator, with $a(\mathbf{k})$ and $a^\dagger(\mathbf{k})$ being fermion annihilation and creation operators, we would get a nonvanishing result for spacelike intervals. In fact it is easy to see that we would get a plus sign instead of a minus sign between the two terms in brackets in the second line of (10.62), and the resulting expression does not vanish for spacelike intervals. This is an example of the "spin–statistics theorem," which requires us to use commutators for particles of integer spin and anticommutators for particles of half-integer spin (Section 10.5).

Necessity of Negative-Energy States

An important difference between our second quantization of the nonrelativistic Schrödinger equation and the relativistic Klein–Gordon equation is that in the latter case there are negative-energy eigenfunctions in the field operators (compare (10.1) and (10.52), for instance). The negative-energy states are essential for causality: without negative-energy contributions equation (10.62) is replaced by

$$[\psi(x), \psi(x')] = - \int \frac{d^3k}{(2\pi)^3} \frac{1}{2E_k} e^{-iE_k(t-t')} e^{i\mathbf{k}\cdot(\mathbf{x}-\mathbf{x}')} \;, \qquad (10.66)$$

which does not vanish for spacelike separations.

The second quantization of the Klein–Gordon equation has thus brought out two very important points connected with Lorentz invariance: (1) Lorentz invariance and causality dictate whether the particles associated with the quantized field are bosons or fermions, and (2) Lorentz invariance and causality require the existence of negative-energy states and, therefore, antiparticles.

10.4   Charged Scalar Field

The conserved current j^μ associated with the Klein–Gordon equation vanishes if, as we have assumed, ψ is real or, in the second-quantized theory, the operator ψ is Hermitian [see equations (10.43) and (10.44)]. Furthermore the field theory of the preceding section does not really distinguish between particles and antiparticles. For these reasons we now allow for $\psi(x)$ to be non-Hermitian.

Consider the non-Hermitian field

$$\psi(x) = \frac{1}{\sqrt{2}} [\psi_1(x) + i\psi_2(x)], \qquad (10.67)$$

where $\psi_1(x)$ and $\psi_2(x)$ are Hermitian Klein–Gordon fields. We assume that $\psi_1(x)$ and $\psi_2(x)$ are independent fields in the sense that the action when both fields are present is just the sum of two actions of the form (10.46),

$$S[\psi_1, \psi_2] = \int d^4x \frac{1}{2} [\partial^\mu \psi_1 \partial_\mu \psi_1 + \partial^\mu \psi_2 \partial_\mu \psi_2 - m^2 \psi_1^2(x) - m^2 \psi_2^2(x)], \quad (10.68)$$

and that $\pi_i(x) = \dot{\psi}_i(x), i = 1, 2$, and

$$[\psi_i(\mathbf{x}, t), \pi_j(\mathbf{x}', t)] = i\delta_{ij}\delta^3(\mathbf{x} - \mathbf{x}'), \qquad (10.69)$$

$$[\psi_i(\mathbf{x}, t), \psi_j(\mathbf{x}', t)] - [\pi_i(\mathbf{x}, t), \pi_j(\mathbf{x}', t)] - 0. \qquad (10.70)$$

Once again the use of equal-time commutators implies a particular Lorentz frame.

Using (10.67), we can write the (normally ordered) action (10.68) in terms of $\psi(x)$ and $\psi^\dagger(x)$:

$$S[\psi, \psi^\dagger] = \frac{1}{2} \int d^4x : \partial^\mu \psi^\dagger \partial_\mu \psi - m^2 \psi^\dagger \psi : . \qquad (10.71)$$

This implies $\pi(x) = \dot{\psi}^\dagger(x)$ and the Hamiltonian

$$H = \frac{1}{2} \int d^3x : \pi^\dagger \pi + \nabla \psi^\dagger \cdot \nabla \psi + m^2 \psi^\dagger \psi : . \qquad (10.72)$$

It follows from (10.67) and (10.69) that the equal-time commutator

$$[\psi(\mathbf{x}, t), \pi(\mathbf{x}', t)] = i\delta^3(\mathbf{x} - \mathbf{x}'), \qquad (10.73)$$

whereas the other equal-time commutators, except for the Hermitian conjugate of (10.73), all vanish as usual.

To write $\psi(x)$ and $\psi^\dagger(x)$ in terms of annihilation and creation operators, we note that $\psi_1(x)$ and $\psi_2(x)$ may both be written in the form (10.53):

$$\psi_1(x) = \int \frac{d^3k}{(2\pi)^3} \frac{1}{2E_k} [a_1(\mathbf{k})e^{-ik\cdot x} + a_1^\dagger(\mathbf{k})e^{ik\cdot x}],$$

$$\psi_2(x) = \int \frac{d^3k}{(2\pi)^3} \frac{1}{2E_k} [a_2(\mathbf{k})e^{-ik\cdot x} + a_2^\dagger(\mathbf{k})e^{ik\cdot x}]. \qquad (10.74)$$

A consequence of (10.67) is that

$$\psi(x) = \int \frac{d^3k}{(2\pi)^3} \frac{1}{2E_k} [a(\mathbf{k})e^{-ik\cdot x} + b^\dagger(\mathbf{k})e^{ik\cdot x}],$$

$$\psi^\dagger(x) = \int \frac{d^3k}{(2\pi)^3} \frac{1}{2E_k} [b(\mathbf{k})e^{-ik\cdot x} + a^\dagger(\mathbf{k})e^{ik\cdot x}], \qquad (10.75)$$

where

$$a(\mathbf{k}) = \frac{1}{\sqrt{2}} [a_1(\mathbf{k}) + ia_2(\mathbf{k})], \qquad (10.76)$$

$$b(\mathbf{k}) = \frac{1}{\sqrt{2}} [a_1(\mathbf{k}) - ia_2(\mathbf{k})]. \qquad (10.77)$$

Since $[a_i(\mathbf{k}), a_j^\dagger(\mathbf{k}')] = (2\pi)^3 2E_k \delta_{ij} \delta^3(\mathbf{k} - \mathbf{k}')$, it follows that

$$[a(\mathbf{k}), a^\dagger(\mathbf{k}')] = [b(\mathbf{k}), b^\dagger(\mathbf{k}')] = (2\pi)^3 2E_k \delta^3(\mathbf{k} - \mathbf{k}'), \qquad (10.78)$$

and that $[a(\mathbf{k}), a(\mathbf{k}')]$, and so forth, are zero.

Thus far we have performed trivial manipulations starting from (10.67). To see what all this accomplishes, consider now the operator

$$Q = \int d^3x \, [\psi^\dagger \dot\psi \; \dot\psi^\dagger \psi \;] = \int \frac{d^3k}{(2\pi)^3} \frac{1}{2E_k} [a^\dagger(\mathbf{k})a(\mathbf{k}) \; b^\dagger(\mathbf{k})b(\mathbf{k})] \; N_a \; N_b,$$
(10.79)

where

$$N_a \equiv \int \frac{d^3k}{(2\pi)^3} \frac{1}{2E_k} a^\dagger(\mathbf{k})a(\mathbf{k}),$$
(10.80)

$$N_b \equiv \int \frac{d^3k}{(2\pi)^3} \frac{1}{2E_k} b^\dagger(\mathbf{k})b(\mathbf{k}).$$
(10.81)

N_a and N_b are number operators for a particles and b particles, just as

$$N_1 = \int \frac{d^3k}{(2\pi)^3} \frac{1}{2E_k} a_1^\dagger(\mathbf{k})a_1(\mathbf{k})$$
(10.82)

and

$$N_2 = \int \frac{d^3k}{(2\pi)^3} \frac{1}{2E_k} a_2^\dagger(\mathbf{k})a_2(\mathbf{k})$$
(10.83)

are number operators for particles 1 and 2 associated with the fields ψ_1 and ψ_2. Since $a_1(\mathbf{k})|0\rangle = a_2(\mathbf{k})|0\rangle = 0$ for the vacuum state $|0\rangle$, we have

$$N_1|0\rangle = N_2|0\rangle = N_a|0\rangle = N_b|0\rangle = 0.$$
(10.84)

N_1, N_2, N_a, and N_b all commute with the Hamiltonian, and so are constants of the motion for the free (uncoupled) quantum field under consideration. The Fock space for this field may be taken to be eigenstates $|n_a, n_b\rangle$ of N_a and N_b or eigenstates $|n_1, n_2\rangle$ of N_1 and N_2.

Q corresponds to *charge*. Evidently the a particles have charge $+1$ and the b particles have charge 1. Note from (10.75) that $\psi(x)$ acts to annihilate an a particle and create a b particle, whereas $\psi^\dagger$ annihilates a b particle and creates an a particle. It is also clear that ψ acts to decrease the charge by one unit and $\psi^\dagger$ acts to increase the charge by one unit.

Particles a and b are antiparticles of each other, carrying the same mass and opposite charge. A Fock state $|n_a, n_b\rangle$ with $N_a|n_a, n_b\rangle = n_a|n_a, n_b\rangle$ and $N_b|n_a, n_b\rangle = n_b|n_a, n_b\rangle$ has charge eigenvalue $q = n_a - n_b$, which of course can be positive or negative. What has been accomplished here by introducing the non-Hermitian fields ψ and $\psi^\dagger$ is a reinterpretation of the quantity $\psi^*\dot\psi - \dot\psi^*\psi$ appearing in the conserved Klein–Gordon current, equation (10.44). At first it seemed that the non-positive–definiteness of

ρ signalled a fatal difficulty with a probability interpretation of the Klein–Gordon equation. However, we see now that we can effectively reinterpret the conserved Klein–Gordon current as a *charge* current rather than a probability current. A charge current, of course, can be positive or negative. This reinterpretation of the Klein–Gordon current, following the initial rejection of the Klein–Gordon equation, was proposed by Pauli and Weisskopf in 1934.

It is worth mentioning that the "charge" quantum number here is not necessarily electric charge. More generally charge appears as a coupling constant in quantum field theory. For instance, the mesons K^0 and $\overline{K}^0$ are antiparticles that are electrically uncharged but have different "strangeness" charges $+1$ and -1.

10.5   The Dirac Field

Let us consider now the second quantization of the Dirac equation, $(i\partial\!\!\!/ - m)\psi = 0$. Again we proceed from an action whose variation produces, in this case, the Dirac equation. We express this action in terms of ψ and $\overline{\psi} = \psi^\dagger \beta = \psi^\dagger \gamma^0$:

$$S[\psi, \overline{\psi}] = \int d^4x \mathsf{L} = \int d^4x \overline{\psi}(x)(i\partial\!\!\!/ - m)\psi(x). \qquad (10.85)$$

It is trivial to verify, using again the three-dimensional generalization of (10.16), that the variation of $\overline{\psi}$ or ψ gives the Dirac equation for ψ or $\overline{\psi}$, respectively. From the definition

$$\pi = \frac{\partial \mathsf{L}}{\partial \dot{\psi}} = i\overline{\psi}\gamma^0 = i\psi^\dagger \qquad (10.86)$$

of the momentum conjugate to ψ we obtain the Hamiltonian

$$
\begin{aligned}
H &= \int d^3x [\pi(x)\dot{\psi}(x) - \mathsf{L}] = \int d^3x [i\psi^\dagger \dot{\psi} - \overline{\psi}(i\partial\!\!\!/ - m)\psi] \\
&= \int d^3x [i\psi^\dagger \dot{\psi} - \psi^\dagger \gamma^0 (i\gamma^0 \frac{\partial}{\partial t} + i\boldsymbol{\gamma} \cdot \nabla - m)\psi] \\
&= \int d^3x [-i\psi^\dagger \gamma^0 \boldsymbol{\gamma} \cdot \nabla + m\psi^\dagger \gamma^0 \psi] \\
&= \int d^3x \psi^\dagger [-i\boldsymbol{\alpha} \cdot \nabla + m\beta]\psi, \qquad (10.87)
\end{aligned}
$$

which is consistent with the Hamiltonian form (9.49) of the Dirac equation.

To quantize the field $\psi(x)$ we must impose commutation rules. We saw in Section 10.2 that creation and annihilation operators satisfy fermion commutation rules if anticommutation relations are assumed for the field. Since the Dirac equation describes fermions, we assume equal-time anticommutation relations for the Dirac field:

$$\{\psi_\alpha(\mathbf{x},t), \pi_\beta(\mathbf{x}',t)\} = i\delta_{\alpha\beta}\delta^3(\mathbf{x}-\mathbf{x}') \tag{10.88}$$

or

$$\{\psi_\alpha(\mathbf{x},t), \psi_\beta^\dagger(\mathbf{x}',t)\} = \delta_{\alpha\beta}\delta^3(\mathbf{x}-\mathbf{x}'), \tag{10.89}$$

together with $\{\psi_\alpha(\mathbf{x},t), \psi_\beta(\mathbf{x}',t)\} = 0$ for the four-component field $\psi(x)$.[6]

As in the example of the Klein–Gordon field, we now expand ψ and $\psi^\dagger$ in terms of eigenfunctions multiplied by annihilation and creation operators. For the free Dirac field the appropriate eigenfunctions are the positive- and negative-energy plane waves $u(p)e^{-ip\cdot x}$ and $v(p)e^{ip\cdot x}$ defined in Section 9.2. We write, in analogy to (10.75),

$$\psi(x) = \int \frac{d^3p}{(2\pi)^3}\frac{m}{E}\sum_{i=1}^{2}[b_i(p)u^i(p)e^{-ip\cdot x} + d_i^\dagger(p)v^i(p)e^{ip\cdot x}], \tag{10.90}$$

$$\begin{aligned}
\psi^\dagger(x) &= \int \frac{d^3p}{(2\pi)^3}\frac{m}{E}\sum_{i=1}^{2}[b_i^\dagger(p)u^{i\,\dagger}(p)e^{ip\cdot x} + d_i(p)v^{i\,\dagger}(p)e^{-ip\cdot x}] \\
&= \int \frac{d^3p}{(2\pi)^3}\frac{m}{E}\sum_{i=1}^{2}[b_i^\dagger(p)\overline{u}^i(p)\gamma^0 e^{ip\cdot x} + d_i(p)\overline{v}^i(p)\gamma^0 e^{-ip\cdot x}].
\end{aligned}$$

$$\tag{10.91}$$

Following convention, we write d^3p instead of the d^3k used for the Klein–Gordon field. The $1/E$ factors, with $E = \sqrt{\mathbf{p}^2 + m^2}$, appear as in the Klein–Gordon case in order to have a Lorentz-invariant measure of integration. The factor m and the notation b_i and d_i are, again, conventional.

By Fourier transforming both sides of equations (10.90) and (10.91), we can express the operators $b_i(p)$ and $d_i(p)$ in terms of $\psi(x)$ and $\psi^\dagger(x)$. It

---

[6] The reader may easily verify that these anticommutation relations, together with the Heisenberg equation of motion $i\dot\psi = [\psi, H]$, give the Dirac equation for the operator $\psi(x)$. See Lee (1981), p. 33, for this short calculation. See also the remarks following equation (10.49) regarding the use of equal-time commutators. The same remarks apply here in the case of equal-time *anti*commutators.

then follows from (10.89) and the properties of the spinors $u^i(p)$ and $v^i(p)$ listed at the end of Section 9.2 that[7]

$$\{b_i(p), b_j^\dagger(p')\} = \{d_i(p), d_j^\dagger(p')\} = (2\pi)^3 \frac{E}{m} \delta_{ij} \delta^3(\mathbf{p} - \mathbf{p}'), \qquad (10.92)$$

and that all the other anticommutators, such as $\{b_i(p), b_j(p')\}$, vanish.

Particles and Antiparticles

The anticommutation relations (10.92) mean that $b_i(p)$ and $d_i(p)$ are fermion annihilation operators and that $b_i^\dagger(p)$ and $d_i^\dagger(p)$ are fermion creation operators. [Recall the remarks following equation (10.40).] The Hamiltonian (10.87) can be expressed in the form

$$H = \int \frac{d^3p}{(2\pi)^3} m \sum_{i=1}^{2} [b_i^\dagger(p)b_i(p) - d_i(p)d_i^\dagger(p)] \qquad (10.93)$$

when the expansions (10.90) and (10.91) are used together with the properties of the spinors.

We interpret $b_i(p)$ and $b_i^\dagger(p)$ as annihilation and creation operators for (positive-energy) electrons.[8] With this interpretation $b_i^\dagger(p)b_i(p)$, with eigenvalues 0 and 1 in this (fermion) case, is the number operator for electrons with four-momentum p and helicity index i.

To interpret $d_i(p)$ and $d_i^\dagger(p)$, we use (10.92) to write

$$H = \int \frac{d^3p}{(2\pi)^3} m \sum_{i=1}^{2} [b_i^\dagger(p)b_i(p) + d_i^\dagger(p)d_i(p) - (2\pi)^3 \frac{E}{m} \delta^3(0)]. \qquad (10.94)$$

The lowest possible energy is therefore that for which the number of b particles and d particles is zero. The b particles have already been assumed to be positive-energy electrons. Now recall that in the Dirac hole theory the lowest possible state is that in which there are no positive-energy electrons and all possible negative-energy electron states are filled. This suggests that we should interpret $d_i^\dagger(p)d_i(p)$ as the number operator for *holes* (i.e., unfilled negative-energy electron states) of four-momentum p. If the number

---

[7] Although this is a completely straightforward exercise, the reader is encouraged to do it, if only to check his understanding of the notation.

[8] We assume here that the spin-1/2 particles described by the Dirac equation are electrons.

of positive-energy electrons is zero, and there are no unfilled negative-energy states (holes), the energy of the Dirac vacuum is a minimum.

We therefore interpret $d_i(p)$ and $d_i^\dagger(p)$ as annihilation and creation operators for a hole of four-momentum p, i.e., for a positron.

The charge operator

$$
\begin{aligned}
Q &\equiv \int d^3x \psi^\dagger \psi = \int \frac{d^3p}{(2\pi)^3} \sum_{i=1}^{2} \frac{m}{E}[b_i^\dagger(p)b_i(p) + d_i(p)d_i^\dagger(p)] \\
&= \int \frac{d^3p}{(2\pi)^3} \sum_{i=1}^{2} \frac{m}{E}[b_i^\dagger(p)b_i(p) - d_i^\dagger(p)d_i(p) + (2\pi)^3 \frac{E}{m}\delta^3(0)]
\end{aligned}
$$

(10.95)

reinforces our interpretation of the b and d operators. If the electrons have "charge 1," and their antiparticles (positrons) have "charge -1," then the first two terms in the integrand in the last line of (10.95) represent the charge above and beyond the charge

$$
Q_0 \equiv \int d^3p \sum_{i=1}^{2} \delta^3(0)
$$

(10.96)

of the Dirac sea of filled negative-energy electron states. That is, these terms represent the charge due to positive-energy electrons and positrons.

The infinite $\delta^3(0)$ in (10.95) and (10.96) is a consequence of the anticommutation rule (10.92) with $\mathbf{p} = \mathbf{p}'$. It appears similarly in the zero-point energy (10.57) of the Klein–Gordon field. To interpret this infinity, we observe that, had we chosen to write $\psi(x)$ as a discrete sum over all $\mathbf{p}$ rather than the continuous sum (10.90), the anticommutators (10.92) would be replaced by $[b_{\mathbf{p}i}, b_{\mathbf{p}'j}^\dagger] = [d_{\mathbf{p}i}, d_{\mathbf{p}'j}^\dagger] = \delta_{ij}\delta^3_{\mathbf{p},\mathbf{p}'}$. This would follow closely the quantization of the electromagnetic field using discrete sums over $\mathbf{k}$, leading to $[a_{\mathbf{k}\lambda}, a_{\mathbf{k}'\lambda'}^\dagger] = \delta_{\lambda\lambda'}\delta_{\mathbf{k},\mathbf{k}'}$ [Equation (2.55)]. Had we proceeded in this fashion in our quantization of the Dirac field, we would have obtained

$$
H = \sum_{\mathbf{p}} \sum_{i=1}^{2} E_{\mathbf{p}}[b_{\mathbf{p}i}^\dagger b_{\mathbf{p}i} + d_{\mathbf{p}i}^\dagger d_{\mathbf{p}i} - 1]
$$

(10.97)

and

$$
Q = \sum_{\mathbf{p}} \sum_{i=1}^{2} [b_{\mathbf{p}i}^\dagger(p)b_{\mathbf{p}i} - d_{\mathbf{p}i}^\dagger d_{\mathbf{p}i} + 1]
$$

(10.98)

instead of (10.94) and (10.95). Obviously (10.96) would have been replaced by the infinite sum over states $\sum_\mathbf{p} \sum_i$.

To give a slightly more precise interpretation of the positron operators, consider the linear momentum operator

$$
\begin{aligned}
\mathbf{P} &= -i \int d^3x\, \psi^\dagger \nabla \psi = \int \frac{d^3p}{(2\pi)^3} \frac{m}{E} \mathbf{P} \sum_{i=1}^{2} [b_i^\dagger(p) b_i(p) - d_i(p) d_i^\dagger(p)] \\
&= \int \frac{d^3p}{(2\pi)^3} \frac{m}{E} \mathbf{P} \sum_{i=1}^{2} [b_i^\dagger(p) b_i(p) + d_i^\dagger(p) d_i(p) + (2\pi)^3 \frac{E}{m} \delta_{ij} \delta^3(0)] \\
&= \int \frac{d^3p}{(2\pi)^3} \frac{m}{E} \mathbf{P} \sum_{i=1}^{2} [b_i^\dagger(\mathbf{p}) b_i(\mathbf{p}) + d_i^\dagger(\mathbf{p}) d_i(\mathbf{p})], \quad (10.99)
\end{aligned}
$$

since $\int d^3p\, \mathbf{p} = 0$, $b_i^\dagger(p) b_i(p) = b_i^\dagger(\mathbf{p}) b_i(\mathbf{p})$, and $d_i^\dagger(p) d_i(p) = d_i^\dagger(\mathbf{p}) d_i(\mathbf{p})$. We conclude that $d_i(p)$ and $d_i^\dagger(p)$ are annihilation and creation operators for positrons of four-momentum p, not, as might at first have been guessed from hole theory, $-p$. Similarly the helicity index i on $b_i(p)$ and $d_i(p)$ corresponds to the same helicity for the electrons and positrons.[9]

Causality and Spin Statistics

Consider now the anticommutator $\{\psi_\alpha(x), \overline{\psi}_\beta(x')\}$. By analogy to the argument given in Section 10.3 for the Klein–Gordon field, this *anti*commutator should vanish for spacelike intervals. A straightforward calculation yields

$$
\{\psi_\alpha(x), \overline{\psi}_\beta(x')\} = i(i\not{\partial}_x + im)_{\alpha\beta} \Delta(x - x'), \quad (10.100)
$$

where $\Delta(x - x')$ is the function appearing in equation (10.62) for the commutator of the Klein–Gordon field. Thus the anticommutator for the Dirac field does vanish if $(x - x')$ is a spacelike interval.

As in the Klein–Gordon case we have here an example of Pauli's spin–statistics theorem (1940): *quantum fields for integer spin particles must be quantized according to Bose–Einstein statistics, using commutators, whereas fields for half-integer spin particles must be quantized according to Fermi–Dirac statistics, using anticommutators.* We can attempt to quantize the Dirac field using commutators, and indeed we can obtain the correct Heisenberg equation of motion for $\psi(x)$ by replacing the left side of (10.89) by a commutator, but the commutator $[\psi(x), \overline{\psi}(x')]$ we obtain does not vanish for spacelike intervals. Furthermore, having arrived at an expression of the

---

[9] See Sakurai (1976), p. 152.

form (10.93) with the b's and d's being boson operators, the Hamiltonian we end up with would not be bounded from below, i.e., there would be no stable ground (vacuum) state, since there is no limit to the number of bosons allowed in a given state. *The Dirac vacuum would be unstable.*

10.6   Dirac Vacuum in Field Theory

Let $|0\rangle$ denote the vacuum state of the Dirac field, such that

$$b_i(p)|0\rangle = d_i(p)|0\rangle = 0 \qquad (10.101)$$

for all p and i. $|0\rangle$ is the state with no (positive-energy) electrons or positrons. The state $b_i^\dagger(p)|0\rangle$ has one positive-energy electron of four-momentum p and helicity i, and $d_i^\dagger(p)|0\rangle$ is a state with one positive-energy positron of the same momentum and helicity.

The vacuum expectation value of the Hamiltonian (10.94) [or (10.97)] is

$$\langle 0|H|0\rangle = -\sum_{\mathbf{p}}\sum_{i=1}^{2} 2(\frac{1}{2}E_{\mathbf{p}}) = -\sum_{\mathbf{p}}\sum_{i=1}^{2} 2(\frac{1}{2})\sqrt{\mathbf{p}^2 + m^2} . \qquad (10.102)$$

This corresponds to a zero-point energy $-\frac{1}{2}\sqrt{\mathbf{p}^2 + m^2}$ for each electron state of energy $E_{\mathbf{p}}$ *plus* the zero-point energy of each positron state of the same energy. We can write

$$\langle 0|H|0\rangle = -\sum_{\mathbf{k},i}\frac{1}{2}\hbar\omega_k \text{ (electrons)} - \sum_{\mathbf{k},i}\frac{1}{2}\hbar\omega_k \text{ (positrons)}, \qquad (10.103)$$

where we write $\mathbf{k}$ for $\mathbf{p}$, include $\hbar$, and define $\hbar\omega_k = \sqrt{\mathbf{k}^2 + m^2}$. The zero-point energy associated with electron (positron) states thus has the magnitude $\frac{1}{2}\hbar\omega_k$ familiar from the simple harmonic oscillator or the electromagnetic field, but the *opposite sign*. The minus sign in the zero-point energy of the Dirac field is a consequence of the fermion character of the particles it describes. Thus, whereas the zero-point energy of a boson oscillator arises from the term $1/2$ in

$$aa^\dagger + a^\dagger a = [a, a^\dagger] + a^\dagger a + a^\dagger a = 2a^\dagger a + 1 = 2(a^\dagger a + \frac{1}{2}), \qquad (10.104)$$

that for a fermion oscillator is associated with the -1 in

$$b^\dagger b - dd^\dagger = b^\dagger b - [\{d, d^\dagger\} - d^\dagger d] = b^\dagger b + d^\dagger d - \{d, d^\dagger\} = b^\dagger b + d^\dagger d - 1. \qquad (10.105)$$

This way of writing each term in the discrete-sum form of the Hamiltonian (10.93) suggests that the zero-point energy is associated entirely with the positron anticommutators; however, we can equally well write (10.105) as

$$
\begin{aligned}
\frac{1}{2}(b^\dagger b - bb^\dagger) + \frac{1}{2}(d^\dagger d - dd^\dagger) &= \frac{1}{2}(b^\dagger b + b^\dagger b - 1) + \frac{1}{2}(d^\dagger d + d^\dagger d - 1) \\
&= (b^\dagger b - \frac{1}{2}) + (d^\dagger d - \frac{1}{2}),
\end{aligned} \tag{10.106}
$$

in which case it is naturally attributable to equal zero-point energies of the electron and positron states, as in (10.103). In any event, it should be evident that the negative zero-point energy of the Dirac field, as opposed to the positive zero-point energies of the electromagnetic or Klein–Gordon fields, is a consequence of the fermion character of the Dirac particles.

The negativity of the energy of the Dirac vacuum is, of course, understandable from the hole picture. If all the negative-energy electron states are filled in the Dirac sea, there must be a total energy $\sum_{\mathbf{p},i}(-E_\mathbf{p})$, which is $\langle 0|H|0\rangle$.

The vacuum state $|0\rangle$ is evidently a *physical vacuum* with respect to which energy or charge is measured. We can define a *bare vacuum* as the state in which, in the language of hole theory, the negative-energy electron states are all *empty*. Denoting the bare vacuum state by $|0\rangle_\text{bare}$, we have

$$
|0\rangle = \Pi_\mathbf{p} \Pi_{i=1}^2 d_i(p)|0\rangle_\text{bare} \ . \tag{10.107}
$$

Note that, since $d_i(p)$ annihilates a positron, it in effect creates a negative-energy electron, so that $d_i(p)|0\rangle_\text{bare}$ is a state with one negative-energy electron. Henceforth the term *vacuum* will mean the physical vacuum.

As with the electromagnetic or Klein–Gordon fields, we can get rid of the zero-point energy in the Hamiltonian by normal ordering:

$$
: H := \int \frac{d^3 p}{(2\pi)^3} m \sum_{i=1}^2 [b_i^\dagger(p) b_i(p) + d_i^\dagger d_i(p)]. \tag{10.108}
$$

Similarly the normally ordered charge operator is

$$
: Q := \int \frac{d^3 p}{(2\pi)^3} \frac{m}{E} \sum_{i=1}^2 [b_i^\dagger b_i(p) - d_i^\dagger(p) d_i(p)], \tag{10.109}
$$

and $\langle 0| : H : |0\rangle = \langle 0| : Q : |0\rangle = 0$.

10.7   Casimir Effect for the Dirac Field

We have emphasized that the quantum field theory of massive particles is a natural generalization of the quantum electromagnetic field theory considered in earlier chapters. Whereas the electromagnetic quantum field theory describes the creation and annihilation of photons, the theory of other quantum fields describes the creation and annihilation of other particles that may be fermions or bosons and may or may not have rest mass. The four-component Dirac field describes creation and annihilation processes involving two types of particles, electrons and positrons, with each of which is associated two independent possible helicities.

A characteristic of all quantum fields is zero-point energy. Zero-point energy can be eliminated from the Hamiltonian by normal ordering. As we have seen in the case of the electromagnetic field, however, this does not mean that zero-point energy is without physical consequence. When the field interacts with itself or with something else — another field — there are effects that are attributable to the zero-point energy and fluctuations of the fields. One way to understand this is to work in the Heisenberg picture. In this formalism the solution of a field equation must include a homogeneous, "source-free" part, and this source-free field has nontrivial dynamical properties even if the field state is devoid of any real particles. And of course, as stressed in earlier chapters, the source-free field contributes a zero-point energy to the total energy, irrespective of the subtraction at the outset of zero-point energy from the Hamiltonian.

Conceptually, the simplest electromagnetic vacuum effect is associated with a mere change of boundary conditions from the idealized case of infinite free space. We refer here, of course, to the Casimir effect — a force as a consequence of spatial variations, due to boundaries, of zero-point energy.

If all quantum fields have zero-point energies, then evidently all quantum fields must exhibit Casimir effects of this type. Let us now consider, as an example, a Casimir effect associated with the Dirac field. As discussed later, this example is relevant to the "MIT bag model" for hadrons.

In the case of the electromagnetic Casimir effect, the boundary conditions on the field mode functions are well-known from classical electromagnetic theory. The first question that must be addressed in considering Casimir effects for other quantum fields is: what are the boundary conditions?

We will consider the Dirac field for two parallel plates at $z = 0$ and $z = d$, the geometry assumed in the familiar electromagnetic Casimir effect. The physical boundary condition we impose is that there is no particle current through the walls: $\hat{n} \cdot \mathbf{j}(x) = 0$ at $z = 0$ and $z = d$, where $\hat{n}$ is the unit vector normal to the surface, equal to $\hat{z}$ at $z = 0$ and $-\hat{z}$ at $z = d$. In Lorentz covariant form for an arbitrary reference frame we have

$$n_\mu j^\mu = \overline{\psi} n_\mu \gamma^\mu \psi = \overline{\psi} n \cdot \gamma \psi = 0. \tag{10.110}$$

Now consider $(in \cdot \gamma)^2$:

$$\begin{aligned}
(in \cdot \gamma)^2 &= -n_\mu n_\nu \gamma^\mu \gamma^\nu = -\frac{1}{2} n_\mu n_\nu \{\gamma^\mu, \gamma^\nu\} \\
&= -g_{\mu\nu} n^\mu n^\nu = -[(n^0)^2 - \mathbf{n}^2] = 1, \tag{10.111}
\end{aligned}$$

since $(n^0)^2 - \mathbf{n}^2 = -1$ in the rest frame and is a Lorentz invariant. Thus $in \cdot \gamma$ must have eigenvalues ± 1. Assume

$$in \cdot \gamma \psi(x) = \psi(x) \quad \text{on the surfaces.} \tag{10.112}$$

Then $in_\mu j^\mu = \overline{\psi}\psi$ on the surfaces. But it follows from (10.112) that $i\overline{\psi}\gamma \cdot n = -\overline{\psi}$ on the surfaces, so that $in_\mu j^\mu = i\overline{\psi} n \cdot \gamma \psi = -\overline{\psi}\psi = \overline{\psi}\psi = 0$ on the surfaces, as required. In other words, we can satisfy (10.110) by assuming the boundary condition (10.112) for ψ.

For reasons discussed later, it will suffice for our purposes to consider the Dirac field for *massless* spin-1/2 particles.

The dependence of a single-particle wave function $\psi(\mathbf{x}, t)$ on x and y will not be affected by the plates at $z = 0$ and d. The z-dependence for an otherwise free particle between the plates will involve $e^{\pm ip_3 z}$, as in free space. Consider a positive-energy wave function of the form (Chodos and Thorn, 1974)

$$\psi(\mathbf{x}, t) = e^{-iEt} e^{i(p_1 x + p_2 y)} [e^{ip_3 z} + i\gamma^3 e^{-ip_3 z}] u(\mathbf{p}) = e^{-iEt} \phi(\mathbf{x}), \tag{10.113}$$

where $E = |\mathbf{p}|$ for the massless particles under consideration, and where $u(\mathbf{p})$ is a positive-energy spinor for the massless Dirac equation. It is easy to show that the boundary conditions (10.112),

$$i\gamma^3 \phi(0) = \phi(0), \quad i\gamma^3 \phi(d) = -\phi(d), \tag{10.114}$$

are satisfied by (10.113) if $e^{ip_3 d} = e^{-ip_3 d}$, or

$$p_3 = \frac{n\pi}{2d}, \quad n = 1, 3, 5, \dots. \tag{10.115}$$

The positive energy levels for a massless Dirac particle in the "box" under consideration, subject to the assumed boundary conditions, are therefore

$$E_\mathbf{p} = \sqrt{\mathbf{p}^2 + m^2} = |\mathbf{p}| = \left[p_1^2 + p_2^2 + \frac{n^2 \pi^2}{4d^2} \right]^{1/2}, \quad n = 1, 3, 5, \dots. \tag{10.116}$$

Consider now the total zero-point energy per unit area of the massless Dirac field confined between the two plates:

$$D(d) \quad - \quad -\sum_{\mathbf{p}}\sum_{i=1}^{2} E_{\mathbf{p}} = -2\sum_{\mathbf{p}_\perp}\sum_{n \text{ odd}} \left[\mathbf{p}_\perp^2 + \frac{n^2\pi^2}{4d^2}\right]^{1/2}$$

$$\to \quad -2\sum_{n \text{ odd}} \int \frac{d^2 p_\perp}{(2\pi)^2} \left[\mathbf{p}_\perp^2 + \frac{n^2\pi^2}{4d^2}\right]^{1/2}$$

$$= \quad -2\sum_{n \text{ odd}} \int_0^\infty \frac{2\pi r dr}{(2\pi)^2} \left[r^2 + \frac{n^2\pi^2}{4d^2}\right]^{1/2}$$

$$= \quad -\frac{1}{\pi}\sum_{n \text{ odd}} \int_{n\pi/2d}^\infty dx\, x^2 \,, \tag{10.117}$$

which of course is infinite. We define a "regularized" zero-point energy as

$$E(d) \quad = \quad -\frac{1}{\pi} \lim_{\alpha \to 0} \frac{\partial^2}{\partial\alpha^2} \sum_{n \text{ odd}} \int_{n\pi/2d}^\infty dx\, e^{-\alpha x}$$

$$= \quad -\frac{1}{\pi} \lim_{\alpha \to 0} \frac{\partial^2}{\partial\alpha^2} \frac{1}{\alpha} \sum_{n \text{ odd}} e^{-n\pi\alpha/2d} \,. \tag{10.118}$$

Now

$$\sum_{n \text{ odd}} e^{-n\pi\alpha/2d} \quad = \quad \left[2\sinh\frac{\pi\alpha}{2d}\right]^{-1}$$

$$= \quad \frac{1}{2}\left[\frac{2d}{\pi\alpha} - \frac{1}{6}\left(\frac{\pi\alpha}{2d}\right) + \frac{7}{360}\frac{\pi^3\alpha^3}{8d^3} + O(\alpha^4)\right], \tag{10.119}$$

so that

$$E(d) = -\frac{1}{2\pi} \lim_{\alpha \to 0} \frac{\partial^2}{\partial\alpha^2}\left[\frac{2d}{\pi\alpha^2} - \frac{\pi}{12d} + \frac{7\pi^3}{360}\frac{\alpha^3}{8d^3} + O(\alpha^3)\right]. \tag{10.120}$$

The first term in brackets is an extensive, "volume energy" that may be ignored in a calculation of the force between the plates. Thus (Johnson, 1975)

$$E(d) = -\frac{1}{2\pi}\left(\frac{7\pi^3}{360}\right)\left(\frac{2}{8d^3}\right) = -\frac{7\pi^2}{2880d^3}\,, \tag{10.121}$$

which gives the force per unit area

$$F(d) = -\frac{\partial E}{\partial d} = -\frac{7\pi^2}{960 d^4} \tag{10.122}$$

or

$$F(d) = -\frac{7\pi^2 \hbar c}{960 d^4} \tag{10.123}$$

when the appropriate factors of $\hbar$ and c are restored.

This is 7/4 times the electromagnetic Casimir force per unit area between two parallel conducting plates separated by a distance d. Like the electromagnetic Casimir force, the Casimir force associated with the vacuum Dirac field is attractive. This is a bit surprising, since the electromagnetic zero-point energy is positive, whereas the zero-point energy of the Dirac field is negative. To trace the origin of this result, let us return briefly to the zero-point energy per unit plate area of the electromagnetic field between parallel conducting plates [equation (2.99)]:

$$
\begin{aligned}
E(d) &= \frac{\hbar c}{\pi^2} \sum_{n=0}^{\infty}{}' \frac{1}{4} \int_{-\infty}^{\infty} dk_x \int_{-\infty}^{\infty} dk_y \left[k_x^2 + k_y^2 + \frac{n^2 \pi^2}{d^2} \right]^{1/2} \\
&= \frac{\hbar c}{4\pi^2} \sum_{n=0}^{\infty}{}' \int_0^{\infty} 2\pi r dr \left[r^2 + \frac{n^2 \pi^2}{d^2} \right]^{1/2} \\
&= \frac{\hbar c}{2\pi} \sum_{n=0}^{\infty}{}' \int_{n\pi/d}^{\infty} dx\, x^2 \ , \tag{10.124}
\end{aligned}
$$

where, as discussed in Section 2.7, the prime indicates that a $1/2$ is to be inserted for $n = 0$. Except for a factor $-1/2$, the replacement of d by $2d$, and the fact that the sum is over all the positive integers instead of only the odd ones, this has the same form as (10.117), and can be evaluated similarly:

$$
\begin{aligned}
E(d) &= \frac{\hbar c}{2\pi} \lim_{\alpha \to 0} \frac{\partial^2}{\partial \alpha^2} \frac{1}{\alpha} \sum_{n=0}^{\infty}{}' e^{-n\pi\alpha/d} = \frac{\hbar c}{4\pi} \lim_{\alpha \to 0} \frac{\partial^2}{\partial \alpha^2} \frac{1}{\alpha} \coth \frac{\pi\alpha}{2d} \\
&= \frac{\hbar c}{4\pi} \lim_{\alpha \to 0} \frac{\partial^2}{\partial \alpha^2} \frac{1}{\alpha} \left[\frac{2d}{\pi\alpha} + \frac{\pi\alpha}{2d} - \frac{1}{45} \frac{\pi^3 \alpha^3}{8d^3} + O(\alpha^4) \right] \\
&\to \frac{\hbar c}{4\pi} \left(-\frac{\pi^3}{45 d^3} \right) \left(\frac{2}{8} \right) = -\frac{\pi^2 \hbar c}{720 d^3} \ , \tag{10.125}
\end{aligned}
$$

which is the (electromagnetic) Casimir result, obtained in Chapters 2, 3, and 7 by different methods. It should be clear, by comparing this derivation with that leading to (10.121), that the attractive result in the Dirac case depends critically on the fact that only *odd* integers are summed over in (10.117), i.e., on the choice of the boundary condition (10.112).

We noted in Section 8.10 that the electromagnetic Casimir force on a spherical conducting shell turns out to be repulsive. The extension of the Casimir effect for the Dirac vacuum to the case of a spherical shell likewise turns out to be repulsive. Milton (1983) obtains

$$E(a) \cong \frac{.02}{a} \quad (= \frac{.02\hbar c}{a}) \tag{10.126}$$

for a sphere of radius a.

These results are relevant to the so-called MIT bag model, in which hadrons are regarded as bags containing freely moving quarks. Quark confinement implies that the quark currents through the walls of the bag are zero. If a quark is confined to a region of dimension $a \sim 1$ fm, on the order of the radius of a typical hadron, then the quark mass $m << 1/a$ and can be neglected, as we have done in the foregoing.[10] Assuming a hadron of radius a composed of three quarks, the Casimir energy associated with the confined quark field is three times the energy (10.126):

$$E_{\text{quark}} \cong \frac{.06}{a} \ . \tag{10.127}$$

However, this is not the only Casimir contribution to the bag energy; there is also the Casimir energy associated with the field of eight gluons. The gluon field is mathematically analogous to the electromagnetic field, and contributes a Casimir energy $E(a) \cong .09235/2a$ in the case of a sphere (Section 8.10). For eight gluons, therefore,

$$E_{\text{gluon}} \cong \frac{8(.092)}{2a} \ , \tag{10.128}$$

and the total Casimir energy of the bag is

$$E_{\text{Cas}}(a) = E_{\text{quark}}(a) + E_{\text{gluon}}(a) \cong \frac{0.43}{a} \ . \tag{10.129}$$

For $a = 1$ fm, this is about 85 MeV, i.e., about 9% of the proton mass (Mostepanenko and Trunov, 1988). In practice the bag parameters are adjusted to fit the mass spectra and magnetic moments of hadrons, and quite good agreement can be obtained in this way (Johnson, 1975).

---

[10] If we choose to include the mass in equation (10.116), for instance, then $E_p = \sqrt{p_1^2 + p_2^2 + m^2 + n^2\pi^2/4d^2} \cong \sqrt{p_1^2 + p_2^2 + n^2\pi^2/4d^2}$ for $m << \pi/2d$, i.e., for the Compton radius of the particle much larger than the confinement distance d.

As noted earlier, all quantum fields should exhibit Casimir effects. The zero-point energy (10.42) of the nonrelativistic Schrödinger field implies such a Casimir effect even in the nonrelativistic limit. However, for a nonrelativistic particle in a box, the sum over all n as earlier means the nonrelativistic approximation will break down when $n/d \sim m$, i.e., when n becomes as large as the wall separation d divided by the Compton wavelength of the particle described by the field.

10.8 Maxwell Field Quantization Again

In Chapter 2 we quantized the electromagnetic field in the Coulomb gauge by exploiting the formal equivalence of a field mode to a harmonic oscillator and then quantizing the oscillator. We now return to electromagnetic field quantization from the more general perspective of quantum fields.

First, however, we make another concession to convention (and practicality). In particle physics and quantum field theory it is conventional to employ Heaviside–Lorentz units for the electromagnetic field. In these units the source terms in the Maxwell equations are $\mathbf{J}$ and ρ rather that $4\pi \mathbf{J}$ and $4\pi\rho$, and the energy density of the field is $(\mathbf{E}^2 + \mathbf{B}^2)/2$ rather than $(\mathbf{E}^2 + \mathbf{B}^2)/8\pi$. In order to make it easier for the reader who wishes to consult more advanced treatises such as Itzkyson and Zuber (1980), we now adopt the Heaviside–Lorentz system instead of the Gaussian system employed (somewhat conventionally) in our nonrelativistic theory of Chapters 1–8.

Recall that the vector potential $\mathbf{A}$ and the scalar potential ϕ of the electromagnetic field are defined by writing $\mathbf{B} = \nabla \times \mathbf{A}$, which then implies $\mathbf{E} = -\partial \mathbf{A}/\partial t - \nabla\phi$. Then the Maxwell equation $\nabla \times \mathbf{B} = \partial \mathbf{E}/\partial t + \mathbf{J}$ implies

$$\nabla^2 \mathbf{A} - \frac{\partial^2 \mathbf{A}}{\partial t^2} - \nabla\left(\nabla \cdot \mathbf{A} + \frac{\partial\phi}{\partial t}\right) = -\mathbf{J}, \qquad (10.130)$$

while $\nabla \cdot \mathbf{E} = \rho$ implies

$$\nabla^2 \phi + \frac{\partial}{\partial t}(\nabla \cdot \mathbf{A}) = -\rho . \qquad (10.131)$$

Since $\mathbf{A}$ is introduced only through its curl, we can always add the gradient of a scalar function $\chi(\mathbf{x}, t)$ to $\mathbf{A}$ without changing anything, since $\nabla \times \nabla\chi = 0$. In order to keep $\mathbf{E} = -\partial \mathbf{A}/\partial t - \nabla\phi$ unaffected by this transformation we must add $-\partial\chi/\partial t$ to ϕ. That is, the *gauge transformation* $A^\mu \to A^\mu - \partial^\mu\chi$, or

$$\mathbf{A} \to \mathbf{A}' = \mathbf{A} + \nabla\chi, \qquad (10.132)$$

$$\phi \to \phi' = \phi - \frac{\partial \chi}{\partial t} \qquad (10.133)$$

leaves the physically measurable electric and magnetic fields invariant.

Different *gauges* correspond to different choices of $\mathbf{A}$ and ϕ that leave $\mathbf{E}$ and $\mathbf{B}$ invariant. In the Coulomb gauge, for instance, we choose $\mathbf{A}$ and ϕ such that $\nabla \cdot \mathbf{A} = 0$ and therefore

$$\nabla^2 \phi = -\rho, \qquad (10.134)$$

$$\nabla^2 \mathbf{A} - \frac{\partial^2 \mathbf{A}}{\partial t^2} = -\mathbf{J} + \nabla(\frac{\partial \phi}{\partial t}) . \qquad (10.135)$$

We can also make a gauge transformation from any $\mathbf{A}, \phi$ to $\mathbf{A}', \phi'$ satisfying the Lorentz condition

$$\nabla \cdot \mathbf{A}' + \frac{\partial \phi'}{\partial t} = 0 \qquad (10.136)$$

by choosing χ such that

$$\nabla^2 \chi - \frac{\partial^2 \chi}{\partial t^2} = - \left(\nabla \cdot \mathbf{A} + \frac{\partial \phi}{\partial t} \right). \qquad (10.137)$$

If $\mathbf{A}', \phi'$ satisfy the Lorentz condition we are still free to make a "restricted gauge transformation" of them with a gauge function χ satisfying

$$\nabla^2 \chi - \frac{\partial^2 \chi}{\partial t^2} = 0, \qquad (10.138)$$

so that the transformed potentials still satisfy the Lorentz condition. All potentials related to $\mathbf{A}', \phi'$ by a gauge function satisfying (10.138) are said to belong to the *Lorentz gauge*. In the Lorentz gauge, from (10.130) and (10.131), we have

$$\nabla^2 \phi - \frac{\partial^2 \phi}{\partial t^2} = -\rho, \qquad (10.139)$$

$$\nabla^2 \mathbf{A} - \frac{\partial^2 \mathbf{A}}{\partial t^2} = -\mathbf{J}, \qquad (10.140)$$

so that the scalar and vector potentials are treated in a more symmetric fashion than in the Coulomb gauge.

The charge and current densities ρ and $\mathbf{J}$ satisfy the charge conservation condition $\nabla \cdot \mathbf{J} + \partial \rho / \partial t = 0$ and define a four-vector $J^\mu = (\rho, \mathbf{J})$, in terms of which charge conservation is expressed as

$$\frac{\partial J^\mu}{\partial x^\mu} = \partial_\mu J^\mu = 0. \qquad (10.141)$$

Lorentz covariance implies then that ϕ and $\mathbf{A}$ satisfying (10.139) and (10.140) form a four-vector $A^\mu = (\phi, \mathbf{A})$, in terms of which these equations may be written in the manifestly covariant form

$$\partial_\mu \partial^\mu A^\nu = J^\nu. \qquad (10.142)$$

In classical electromagnetic theory one defines the gauge-invariant, second-rank antisymmetric field tensor

$$F^{\mu\nu} = \frac{\partial A^\nu}{\partial x_\mu} - \frac{\partial A^\mu}{\partial x_\nu} = \partial^\mu A^\nu - \partial^\nu A^\mu$$

$$= \begin{pmatrix} 0 & -E^1 & -E^2 & -E^3 \\ E^1 & 0 & -B^3 & B^2 \\ E^2 & B^3 & 0 & -B^1 \\ E^3 & -B^2 & B^1 & 0 \end{pmatrix}, \qquad (10.143)$$

in terms of which the inhomogeneous Maxwell equations $\nabla \cdot \mathbf{E} = \rho$ and $\nabla \times \mathbf{B} = \partial \mathbf{E}/\partial t + \mathbf{J}$ may be written compactly in the covariant form

$$\partial_\mu F^{\mu\nu} = J^\nu. \qquad (10.144)$$

It is easily shown that

$$F_{\mu\nu} F^{\mu\nu} = -2(\mathbf{E}^2 - \mathbf{B}^2). \qquad (10.145)$$

The classical action for the electromagnetic field is

$$S = -\int d^4x \left[\frac{1}{4} F^{\mu\nu} F_{\mu\nu} + J_\mu A^\mu \right]$$

$$= -\int d^4x \left[\frac{1}{2} \partial^\mu A^\nu \partial_\mu A_\nu - \frac{1}{2} \partial^\mu A^\nu \partial_\nu A_\mu + J_\mu A^\mu \right]$$

$$\equiv \int d^4x \mathsf{L}, \qquad (10.146)$$

i.e., the Euler–Lagrange equation generalizing (10.16),

$$\frac{\partial \mathsf{L}}{\partial A_\mu} - \partial_\nu \left(\frac{\partial \mathsf{L}}{\partial_\nu A_\mu} \right) = 0, \qquad (10.147)$$

yields (10.144), as is easily verified. The latter equation is equivalent to (10.130) and (10.131), from which the set of four Maxwell equations for $\mathbf{E}$ and $\mathbf{B}$ follow from the identifications $\mathbf{B} = \nabla \times \mathbf{A}$ and $\mathbf{E} = -\partial \mathbf{A}/\partial t - \nabla \phi$.

It will be convenient to consider separately the quantization of the field in the Coulomb and Lorentz gauges. Quantization in the Coulomb gauge has already been done in Chapter 2; now we will do it differently.

Coulomb Gauge

We are interested in quantizing the *free* field with Lagrangian density $L = -(1/4)F^{\mu\nu}F_{\mu\nu} = (1/2)(\mathbf{E}^2 - \mathbf{B}^2)$. In the Coulomb gauge, $\nabla \cdot \mathbf{A} = 0$, we have $\mathbf{E} = -\partial \mathbf{A}/\partial t$ and $\mathbf{B} = \nabla \times \mathbf{A}$ for the free field, for which we can take ϕ, satisfying $\nabla^2 \phi = 0$ everywhere, to vanish. Thus

$$L = \frac{1}{2}\left(\frac{\partial \mathbf{A}}{\partial t}\right)^2 - \frac{1}{2}(\nabla \times \mathbf{A})^2, \tag{10.148}$$

and the momentum conjugate to A_i ($i = 1, 2, 3$) is

$$\pi_i = \dot{A}_i = -E_i . \tag{10.149}$$

The Hamiltonian is therefore

$$H = \int d^3x(\pi_i \dot{A}_i - L) = \frac{1}{2}\int d^3x(\mathbf{E}^2 + \mathbf{B}^2). \tag{10.150}$$

Since photons are bosons, it might seem as though we should now impose the equal-time canonical commutation relation

$$[A_i(\mathbf{x}, t), E_j(\mathbf{x}', t)] = -i\delta_{ij}\delta^3(\mathbf{x} - \mathbf{x}'). \tag{10.151}$$

However, this relation is inconsistent with the Coulomb-gauge condition $\partial_i A_i = 0$. In order to satisfy the latter condition we take

$$[A_i(\mathbf{x}, t), E_j(\mathbf{x}', t)] = -i\delta_{ij}^{\perp}(\mathbf{x} - \mathbf{x}') \tag{10.152}$$

instead of (10.151), where $\delta_{ij}^{\perp}(\mathbf{x})$ is the transverse delta function defined in Section 4.4. Equation (10.152) is the same as equation (4.33), except for the trivial factor of 4π associated with the Gaussian units used in obtaining the latter. As discussed in Section 4.4, only the transverse electromagnetic fields are quantized in the Coulomb gauge; the longitudinal field is a c-number whose contribution to the Hamiltonian takes the form of instantaneous Coulomb interactions among charged particles. Although the quantization of the field in the Coulomb gauge is not manifestly Lorentz covariant, the theory is nevertheless Lorentz invariant. In particular, the commutation relations for the $\mathbf{E}$ and $\mathbf{B}$ fields do vanish at spacelike separations, although (10.152) does not [see Section 2.8 and also the discussion following equation (4.59)].

As in Chapter 2 we expand the vector potential in plane waves, now using a continuous rather than discrete sum over plane-wave modes:

$$\mathbf{A}(x) = \int \frac{d^3k}{(2\pi)^3}\frac{1}{2\omega_k}\sum_{\lambda=1}^{2}[a(\mathbf{k}, \lambda)e^{-ik\cdot x} + a^{\dagger}(\mathbf{k}, \lambda)e^{ik\cdot x}]\mathbf{e}(\mathbf{k}, \lambda), \tag{10.153}$$

where $\omega_k - |\mathbf{k}|$ and $e(\mathbf{k}, \lambda)$, $\lambda = 1, 2$, are orthogonal linear polarization unit vectors such that $\mathbf{k} \cdot e(\mathbf{k}, \lambda) = 0$. As in the case of the scalar Klein–Gordon field, the factors $1/2\omega_k$ are introduced in order to have a Lorentz-invariant measure. From $\mathbf{E}(x) = -\dot{\mathbf{A}}(x)$ we obtain

$$\mathbf{E}(x) = i \int \frac{d^3k}{(2\pi)^3} \frac{1}{2} \sum_{\lambda=1}^{2} [a(\mathbf{k}, \lambda)e^{-ik\cdot x} - a^\dagger(\mathbf{k}, \lambda)e^{ik\cdot x}]e(\mathbf{k}, \lambda), \quad (10.154)$$

and (10.152) implies

$$[a(\mathbf{k}, \lambda), a^\dagger(\mathbf{k}', \lambda')] = (2\pi)^3 2\omega_k \delta^3(\mathbf{k} - \mathbf{k}')\delta_{\lambda\lambda'} . \quad (10.155)$$

The Hamiltonian (10.150) becomes

$$: H := \int \frac{d^3k}{(2\pi)^3} \sum_{\lambda} \frac{1}{2\omega_k} \omega_k a^\dagger(\mathbf{k}, \lambda)a(\mathbf{k}, \lambda) \quad (10.156)$$

when normally ordered. Without normal ordering we incur the infinite zero-point energy

$$\begin{aligned}\langle 0|H|0\rangle &= \int \frac{d^3k}{(2\pi)^3} \sum_{\lambda} \frac{1}{2\omega_k} \omega_k \frac{1}{2} (2\pi)^3 2\omega_k \delta^3(0)\\ &= \delta^3(0) \int d^3k \sum_{\lambda} \frac{1}{2}\omega_k , \quad (10.157)\end{aligned}$$

which is the familiar $\sum_{\mathbf{k}\lambda} \frac{1}{2}\hbar\omega_k$ when a discrete sum over modes is adopted.[11]

Lorentz Gauge

The Maxwell equations in the Lorentz gauge are manifestly covariant, i.e., their Lorentz covariance is obvious from their form. This suggests that the quantized field in the Lorentz gauge will involve manifestly covariant equal-time commutators such as

$$[A_\mu(\mathbf{x}, t), \pi_\nu(\mathbf{x}', t)] = ig_{\mu\nu}\delta^3(\mathbf{x} - \mathbf{x}'). \quad (10.158)$$

However, quantization in the Lorentz gauge is a fairly delicate matter, as will be clear from the following observations concerning (10.158).

This commutator involves the temporal component A_0 as well as A_i, $i = 1, 2, 3$. Whereas in the Coulomb gauge the scalar potential A_0 is not

---

[11] $\delta^3(\mathbf{k}) = (1/2\pi)^3 \int d^3x e^{i\mathbf{k}\cdot\mathbf{x}} \to V/8\pi^3$ for $\mathbf{k} \to 0$, and $(V/8\pi^3) \int d^3k \to \sum_{\mathbf{k}}$ when a discrete mode summation is employed as in earlier chapters.

quantized, (10.158) implies that now it *is* quantized, i.e., it is an operator. Evidently we will have "scalar photons" as well as "transverse photons" in the Lorentz gauge. Furthermore, since $\mathbf{A}$ is not generally transverse in the Lorentz gauge, it will have a longitudinal component as well as a transverse part. Thus it appears that we will have "longitudinal photons" in addition to transverse and scalar photons in the Lorentz gauge. Since they correspond to the gauge-dependent part of the vector potential,[12] the longitudinal photons, like the scalar photons, are "unphysical" and should not appear in any measurable quantities. The basic difficulty here is that in the Lorentz gauge we are "over-quantizing," i.e., we are quantizing more parts of the field than are physically required. In particular, the Lorentz condition $\partial A^0/\partial t + \nabla \cdot \mathbf{A} = \partial \cdot A = 0$ means that the scalar and longitudinal parts of the field are not independent, and this constraint must be taken into account.

The Lagrangian $\mathsf{L} = -(1/4)F^{\mu\nu}F_{\mu\nu}$ implies that $\pi_0 = \partial \mathsf{L}/\partial(\partial_t A^0) = 0$, which is inconsistent with (10.158). The standard remedy for this is to consider a different Lagrangian, replacing L by

$$\mathsf{L}' = -\frac{1}{4}F^{\mu\nu}F_{\mu\nu} - \frac{\lambda}{2}(\partial \cdot A)^2. \tag{10.159}$$

For this Lagrangian the momentum conjugate to A^μ is

$$\pi^\mu = \frac{\partial \mathsf{L}}{\partial(\partial_t A_\mu)} = F^{\mu 0} - \lambda g^{\mu 0}(\partial \cdot A), \tag{10.160}$$

so that $\pi^i = -E^i$, as with the original Lagrangian, but

$$\pi^0 = -\lambda(\partial \cdot A) \neq 0. \tag{10.161}$$

The free-space Maxwell equations $\partial_\mu \partial^\mu A^\nu = 0$ in the Lorentz gauge are replaced by

$$\partial_\mu \partial^\mu A_\nu - (1 - \lambda)\partial_\nu(\partial \cdot A) = 0. \tag{10.162}$$

We will take the number $\lambda = 1$, the so-called Feynman gauge for which $\partial_\mu \partial^\mu A_\nu = 0$ and $\pi^0 = -\partial \cdot A$.[13] Now the Lorentz gauge condition on the operator A, $\partial \cdot A = 0$, would make $\pi^0 = 0$, which would be inconsistent

---

[12] Recall the Helmholtz theorem that any vector field may be divided uniquely into longitudinal and transverse parts (Appendix F), and that the transverse part of the vector potential in particular is gauge invariant.

[13] The addition of the term $-(\lambda/2)(\partial \cdot A)^2$ to the field Lagrangian density is not a gauge transformation in the usual sense of a transformation of A^μ that leaves Maxwell's equations invariant. The original theory (based on the original Lagrangian) is recovered by imposing the constraint (10.163), and Maxwell's equations are recovered only in terms of expectation values.

with the desired, manifestly covariant commutator (10.158) and put us right back where we started. We will instead impose the constraint

$$\langle \psi | \partial \cdot A | \psi \rangle = 0 \qquad (10.163)$$

on the field states $|\psi\rangle$. States satisfying this constraint will be considered as "physical" states of the field. This approach to quantization in the Lorentz gauge is the basis for the Gupta–Bleuler method of "indefinite metric quantization," part of which we now describe.[14]

As in earlier chapters we can write $A(x) = A^{(+)}(x) + A^{(-)}(x)$, where $A^{(+)}(x)$ and $A^{(-)}(x)$ are the positive- and negative-frequency parts of the field, involving photon annihilation and creation operators, respectively, for the free field. Now if $\partial \cdot A^{(+)} |\psi\rangle = 0$ then

$$
\begin{aligned}
\langle \psi | \partial \cdot A | \psi \rangle &= \langle \psi | \partial \cdot A^{(+)} | \psi \rangle + \langle \psi | \partial \cdot A^{(-)} | \psi \rangle \\
&= \langle \psi | \partial \cdot A^{(+)} | \psi \rangle + [\partial \cdot A^{(+)} | \psi \rangle]^{\dagger} | \psi \rangle \\
&= 0.
\end{aligned}
\qquad (10.164)
$$

In other words, we can satisfy the constraint (10.163) by taking

$$\partial \cdot A^{(+)} |\psi\rangle = 0. \qquad (10.165)$$

Since $\partial_\nu \partial^\nu A_\mu = 0$, we can again expand the field $A_\mu(x)$ in plane waves:

$$A_\mu(x) = \int \frac{d^3 k}{(2\pi)^3} \frac{1}{2\omega_k} \sum_{\lambda=0}^{3} [a(\mathbf{k}, \lambda) e^{-ik \cdot x} + a^\dagger(\mathbf{k}, \lambda) e^{ik \cdot x}] e_\mu(\mathbf{k}, \lambda). \quad (10.166)$$

The sum over the polarization index[15] λ now involves four terms for each $\mathbf{k}$ instead of two [compare with (10.153)] because we are now accomodating scalar and longitudinal photons as well as transverse photons. We designate the transverse components by $\lambda = 1, 2$, as in the Coulomb gauge. Thus

$$e_0(\mathbf{k}, 1) = e_0(\mathbf{k}, 2) = 0 \qquad (10.167)$$

and

$$k^\mu e_\mu(\mathbf{k}, 1) = k^\mu e_\mu(\mathbf{k}, 2) = 0, \quad \lambda = 1, 2. \qquad (10.168)$$

---

[14] A full employment of the Gupta–Bleuler method involves a redefinition of the scalar product in Hilbert space to allow for "negative probabilities" ("indefinite metric"). This is not the only possible approach to quantization in the Lorentz gauge. See Heitler (1966), p. 89.

[15] λ should not be confused with the parameter λ in (10.160), which we have taken to be unity in the "Feynman gauge."

For any chosen reference frame we can take the four-vectors $e(\mathbf{k}, 0)$ and $e(\mathbf{k}, 3)$ to point along the time axis and the $\mathbf{k}$ direction, respectively. Thus $e_0(\mathbf{k}, 0) = 1$, $e_i(\mathbf{k}, 0) = 0$, $i = 1, 2, 3$, and $e_i(\mathbf{k}, 3) = k_i/|\mathbf{k}|$, $e_0(\mathbf{k}, 3) = 0$. $\lambda = 0$ and $\lambda = 3$ correspond to "scalar" and "longitudinal" photons, respectively. If $\mathbf{k} = |\mathbf{k}|\hat{z}$, for instance, then

$$
e(\mathbf{k}, 0) = \begin{pmatrix} 1 \\ 0 \\ 0 \\ 0 \end{pmatrix}, \qquad e(\mathbf{k}, 1) = \begin{pmatrix} 0 \\ 1 \\ 0 \\ 0 \end{pmatrix},
$$

$$
e(\mathbf{k}, 2) = \begin{pmatrix} 0 \\ 0 \\ 1 \\ 0 \end{pmatrix}, \qquad e(\mathbf{k}, 3) = \begin{pmatrix} 0 \\ 0 \\ 0 \\ 1 \end{pmatrix}. \qquad (10.169)
$$

More generally $e(\mathbf{k}, \lambda) \cdot e(\mathbf{k}, \lambda') = g^{\lambda\lambda'}$.

Now as usual we can determine the commutation relations for the annihilation and creation operators from the commutation relations for the field. It is easily shown that the Lorentz covariant commutation relation (10.158) is satisfied if

$$
[a(\mathbf{k}, \lambda), a^\dagger(\mathbf{k}', \lambda')] = -(2\pi)^3 2\omega_k g^{\lambda\lambda'} \delta^3(\mathbf{k} - \mathbf{k}') \qquad (10.170)
$$

and $[a(\mathbf{k}, \lambda), a(\mathbf{k}', \lambda')] = 0$. The normally ordered Hamiltonian is

$$
\begin{aligned}
: H : \ &= \ \int d^3x : [\pi^\mu \dot{A}^\mu - L'] : \\
&= \ \frac{1}{2} \int d^3x : \left(\sum_{i=1}^3 [A_i^2 + (\nabla A_i)^2] - A_0^2 - (\nabla A_0)^2 \right) : \\
&= \ \int \frac{d^3k}{(2\pi)^3} \frac{1}{2\omega_k} \omega_k \left[\sum_{\lambda=1}^3 a^\dagger(\mathbf{k}, \lambda) a(\mathbf{k}, \lambda) - a^\dagger(\mathbf{k}, 0) a(\mathbf{k}, 0) \right]
\end{aligned}
$$

$$(10.171)$$

The commutator (10.170) has the "wrong" sign for the scalar polarization ($g^{00} = +1$), as does the second term in the last line of (10.171). The latter implies that the energy spectrum is unbounded from below. These results pose no real problem, since we have not yet imposed the constraint (10.165). From (10.166) it follows that

$$
\partial \cdot A^{(+)}(x) = -i \int \frac{d^3k}{(2\pi)^3} \frac{1}{2\omega_k} \sum_{\lambda=0}^3 a(\mathbf{k}, \lambda)[k \cdot e(\mathbf{k}, \lambda)] e^{-ik\cdot x}, \qquad (10.172)
$$

and the constraint (10.165) therefore means that the physical states of the field are those for which

$$\sum_{\lambda=0}^{3}[k \cdot e(\mathbf{k}, \lambda)]a(\mathbf{k}, \lambda)|\psi\rangle = 0. \tag{10.173}$$

Now $k \cdot e(\mathbf{k}, \lambda) = 0$ for the transverse polarizations $\lambda = 1, 2$ [equation (10.168)]. Furthermore $k \cdot e(\mathbf{k}, 0) = -k \cdot e(\mathbf{k}, 3)$ [see (10.169)], so that (10.173) reduces to

$$[a(\mathbf{k}, 3) - a(\mathbf{k}, 0)]|\psi\rangle = 0. \tag{10.174}$$

Thus the physical states of the field are those for which the unphysical scalar and longitudinal photons cancel each other's effects in the sense of (10.174).

Ghosts

The procedure just described allows Maxwell's equations in the Lorentz gauge to be satisfied in terms of expectation values. The Lorentz gauge condition $\partial \cdot A = 0$ is itself satisfied only as an expectation value, as indicated by (10.164). In addition to the physical states $|\psi\rangle$ constrained by (10.174), the Hilbert space for the field contains *ghost states*. Consider, for instance, the state with one scalar photon (Itzykson and Zuber, 1980):

$$|\phi_1\rangle = \int \frac{d^3k}{(2\pi)^3}\frac{1}{2\omega_k}f(k)a^\dagger(\mathbf{k}, 0)|0\rangle . \tag{10.175}$$

Here $|0\rangle$ is the bare vacuum state such that $a(\mathbf{k}, \lambda)|0\rangle = 0$, and a distribution $f(k)$ is introduced in order that $|\phi_1\rangle$ have a finite norm. This norm, however, is negative as a consequence of (10.170):

$$\begin{aligned}\langle\phi_1|\phi_1\rangle &= \int \frac{d^3k}{(2\pi)^6}\left(\frac{1}{2\omega_k}\right)^2 |f(k)|^2\langle 0|a(\mathbf{k}, 0)a^\dagger(\mathbf{k}, 0)|0\rangle \\ &= -\int \frac{d^3k}{(2\pi)^3}\frac{1}{2\omega_k}|f(k)|^2 . \end{aligned} \tag{10.176}$$

The negative norm is a reflection of the unphysical nature of the scalar photons and the ghost state $|\phi_1\rangle$.

The ghost states, however, pose no problems so long as we remember to impose the constraint (10.174). Suppose we write the physical states $|\psi\rangle$ of the field in the form $|\psi_T\rangle|0\rangle$, where $|\psi_T\rangle$ involves only physical, transverse photons and $|\phi\rangle$ involves unphysical, scalar and longitudinal photons. (This can always be done because the constraint is linear.) Then (10.174) implies

$[a(\mathbf{k}, 3) - a(\mathbf{k}, 0)]|\phi\rangle = 0$. For the expectation value of the normally ordered Hamiltonian (10.171) we have

$$\langle \psi | : H : |\psi\rangle = \int \frac{d^3 k}{(2\pi)^3} \frac{1}{2\omega_k} \omega_k \sum_{\lambda = 0}^{3} \langle \psi_T | a^\dagger(\mathbf{k}, \lambda) a(\mathbf{k}, \lambda) |\psi_T\rangle \, , \qquad (10.177)$$

where the sum over $\lambda = 3$ and 4 vanishes due to (10.174). This expectation value is positive–definite, as required, as is the norm of all physical states $|\psi\rangle$. The norm of the ghost state $|\phi_1\rangle$, by contrast, is negative because it has no longitudinal photon to balance the scalar photon, i.e., it does not satisfy the constraint (10.174).

From a physical standpoint, of course, there is no difference between the Coulomb and Lorentz gauges. Nature cares little about scalar and longitudinal photons, ghost states, or different gauges — although it does appear to insist that our theories be gauge–invariant. For the physicist there is, however, an advantage to working in the Lorentz gauge: the equations are manifestly covariant and, in particular, the Lorentz-gauge propagator is neater in form than the Coulomb-gauge propagator, as we shall see in the following section.

10.9 Propagators

We have already noted that the state $\psi^\dagger(x)|0\rangle$ describes a particle at $x = (\mathbf{x}, t)$. The probability amplitude for the particle to propagate from x to x' is therefore the scalar product of $\psi^\dagger(x)|0\rangle$ and $\psi^\dagger(x')|0\rangle$: $(\psi^\dagger(x')|0\rangle)^\dagger \times \psi^\dagger(x)|0\rangle = \langle 0|\psi(x')\psi^\dagger(x)|0\rangle$. This can be interpreted as the creation of a particle at x out of the vacuum, followed by the annihilation of the particle at x'. Causality demands that $t' > t$ in order for this interpretation to make sense. Thus we replace $\langle 0|\psi(x')\psi^\dagger(x)|0\rangle$ by $\theta(t' - t)\langle 0|\psi(x')\psi^\dagger(x)|0\rangle$, where $\theta(t' - t)$ is the unit step function, and interpret this as the probability amplitude for a particle described by the quantized field $\psi(x)$ to propagate from x to x'. Except for a possible phase factor of modulus unity, we call this amplitude a *propagator*.

Consider first the example of the nonrelativistic Schrödinger field. From (10.1), with the simple generalization of the formalism of Section 10.2 to three spatial dimensions, we have

$$\theta(t' - t)\langle 0|\psi(\mathbf{x}', t')\psi^\dagger(\mathbf{x}, t)|0\rangle$$

$$= \theta(t' - t) \sum_m \sum_n \langle 0|a_m a_n^\dagger|0\rangle \phi_m(\mathbf{x}')\phi_n^*(\mathbf{x}) e^{i(E_n t - E_m t')}$$

$$= \theta(t' - t) \sum_n \phi_n(\mathbf{x}')\phi_n^*(\mathbf{x})e^{-iE_n(t'-t)} \qquad (10.178)$$

Since $\dot{\theta}(t) = \delta(t)$, we obtain by differentiation

$$\frac{\partial}{\partial t'}\left[\theta(t' - t)\langle 0|\psi(\mathbf{x}',t')\psi^\dagger(\mathbf{x},t)|0\rangle\right]$$

$$= \delta(t' - t)\langle 0|\psi(\mathbf{x}',t')\psi^\dagger(\mathbf{x},t)|0\rangle - i\theta(t' - t)\sum_n E_n\phi_n(\mathbf{x}')\phi_n^*(\mathbf{x})e^{-iE_n(t'-t)}$$

$$= \delta(t' - t)\sum_n \phi_n(\mathbf{x}')\phi_n^*(\mathbf{x}) - iH(\mathbf{x}')\theta(t' - t) \times \sum_n \phi_n(\mathbf{x}')\phi_n^*(\mathbf{x})e^{-iE_n(t'-t)} ,$$

$$(10.179)$$

or

$$\left[\frac{\partial}{\partial t'} + iH(\mathbf{x}')\right]\theta(t'-t)\langle 0|\psi(\mathbf{x}',t')\psi^\dagger(\mathbf{x},t)|0\rangle = \delta^3(\mathbf{x}'-\mathbf{x})\delta(t'-t), \quad (10.180)$$

where we have employed the completeness relation $\sum_n \phi_n(\mathbf{x}')\phi_n^*(\mathbf{x}) = \delta^3(\mathbf{x}'-\mathbf{x})$ for the eigenfunctions of the Schrödinger equation.[16] We define the propagator

$$G(\mathbf{x}',t';\mathbf{x},t) \equiv -i\theta(t' - t)\langle 0|\psi(\mathbf{x}',t')\psi^\dagger(\mathbf{x},t)|0\rangle , \qquad (10.181)$$

so that

$$\left[i\frac{\partial}{\partial t'} - H(\mathbf{x}')\right]G(\mathbf{x}',t';\mathbf{x},t) = \delta^3(\mathbf{x}' - \mathbf{x})\delta(t' - t) . \qquad (10.182)$$

In other words, the propagator is a Green function for the (time-dependent) Schrödinger equation, subject to the boundary condition that $G(\mathbf{x}',t';\mathbf{x},t) = 0$ for $t' < t$.

For a free particle with $H(\mathbf{x}') = -\nabla^2_{\mathbf{x}'}$, the propagator $G_o(\mathbf{x}',t';\mathbf{x},t)$ may be evaluated explicitly using

$$\delta^3(\mathbf{x}' - \mathbf{x})\delta(t' - t) = \left(\frac{1}{2\pi}\right)^3 \int d^3p\, e^{i\mathbf{P}\cdot(\mathbf{x}'-\mathbf{x})} \left(\frac{1}{2\pi}\right)\int_{-\infty}^{\infty} d\omega\, e^{-i\omega(t'-t)}$$

$$(10.183)$$

---

[16]Completeness of the ϕ_n implies $f(\mathbf{x}) = \sum_n c_n\phi_n(\mathbf{x})$ for any function $f(\mathbf{x})$. Then $c_n = \int d^3x' f(\mathbf{x}')\phi_n^*(\mathbf{x}')$ and $f(\mathbf{x}) = \int d^3x' f(\mathbf{x})\sum_n \phi_n^*(\mathbf{x}')\phi_n(\mathbf{x})$, which implies $\sum_n \phi_n^*(\mathbf{x}')\phi_n(\mathbf{x}) = \delta^3(\mathbf{x}' - \mathbf{x})$.

and writing $G_o(\mathbf{x}', t'; \mathbf{x}, t) = (1/2\pi)^4 \int d^3p \int_{-\infty}^{\infty} d\omega g_o(\mathbf{p}, \omega) e^{i\mathbf{p} \cdot (\mathbf{x}' - \mathbf{x})} \times e^{-i\omega(t'-t)}$. Then it follows from (10.182) and a contour integration that

$$G_o(\mathbf{x}', t'; \mathbf{x}, t) = -i \left[\frac{m}{2\pi i(t'-t)} \right]^{3/2} e^{im|\mathbf{x}'-\mathbf{x}|^2/2(t'-t)} \theta(t'-t) \quad (10.184)$$

for the case of a free particle.

The Schrödinger equation for a particle subject to an interaction potential $V(\mathbf{x}, t)$ is $[i\partial/\partial t + \nabla_{\mathbf{x}}^2]\psi(\mathbf{x}, t) = V(\mathbf{x}, t)\psi(\mathbf{x}, t)$. From (10.182) it follows that

$$\psi(\mathbf{x}', t') = \psi_o(\mathbf{x}', t') + \int d^3x \int_{-\infty}^{\infty} dt G_o(\mathbf{x}', t'; \mathbf{x}, t) V(\mathbf{x}, t)\psi(\mathbf{x}, t), \quad (10.185)$$

where $\psi_o(\mathbf{x}', t')$ is a solution of the Schrödinger equation with $V = 0$, i.e., in the "remote past," before the interaction is "on." The formal solution (10.185) can be iterated to develop a perturbation expansion for $\psi(\mathbf{x}', t')$:

$$\begin{aligned}
\psi(\mathbf{x}', t') = {} & \psi_o(\mathbf{x}', t') + \int d^3x \int dt G_o(\mathbf{x}', t'; \mathbf{x}, t) V(\mathbf{x}, t)\psi_o(\mathbf{x}, t) \\
& + \int d^3x_1 \int d^3x_2 \int dt_1 \int dt_2 G_o(\mathbf{x}', t'; \mathbf{x}, t) V(\mathbf{x}_1, t_1) \\
& \times G_o(\mathbf{x}_1, t_1; \mathbf{x}_2, t_2) V(\mathbf{x}_2, t_2)\psi_o(\mathbf{x}_2, t_2) + \dots. \quad (10.186)
\end{aligned}$$

Of course all this is well-known, and we do not need quantum field theory, starting from the amplitude $\langle 0|\psi(\mathbf{x}', t')\psi^\dagger(\mathbf{x}, t)|0\rangle$, to derive (10.184)–(10.186). Things become more interesting when we consider the propagator for a *relativistic* quantum field. We now turn to the propagators for the Klein–Gordon, Dirac, and electromagnetic fields. Our discussion will be rather formal and terse, as our principal aim is to evaluate these propagators for use later on.

Klein–Gordon Propagator

Based on (10.181), it would seem natural to define the Klein–Gordon propagator as $G(x', x) = -i\theta(t'-t)\langle 0|\psi(x')\psi^\dagger(x)|0\rangle$, where $\psi(x)$ is the quantized Klein Gordon field and $|0\rangle$ is its vacuum state. However, $\langle 0|\psi(x')\psi^\dagger(x)|0\rangle$ is not the total amplitude by which a unit charge, say, can be propagated from x to x'. For in addition to creating a particle at x out of the vacuum and annihilating it at x', we can accomplish the same transport of charge from x to x' by creating an antiparticle at x' and annihilating it at x. The amplitude for the latter process is $\langle 0|\psi^\dagger(x)\psi(x')|0\rangle$. This amplitude, multiplied by $\theta(t-t')$ to ensure that the antiparticle at x' is created before it is annihilated at x, must be added to $-i\theta(t'-t)\langle 0|\psi(x')\psi^\dagger(x)|0\rangle$ in order to get the total probability amplitude for transporting a unit of charge from x to x'. Thus

$$iG(x',x) = \theta(t'-t)\langle 0|\psi(x')\psi^\dagger(x)|0\rangle + \theta(t-t')\langle 0|\psi^\dagger(x)\psi(x')|0\rangle. \quad (10.187)$$

We define the time-ordering operator T by

$$
\begin{aligned}
T\psi(x)\psi^\dagger(x') &= \psi(x)\psi^\dagger(x') \quad \text{if } t' < t \\
&= \psi^\dagger(x')\psi(x) \quad \text{if } t' > t.
\end{aligned} \quad (10.188)
$$

T simply orders the operators according to the order of their time arguments, later times appearing to the left of earlier times. Using T, we can write (10.187) more compactly as

$$iG(x',x) = \langle 0|T\psi^\dagger(x)\psi(x')|0\rangle. \quad (10.189)$$

To evaluate $G(x',x)$ we use the plane-wave expansions (10.75) for the charged scalar field, together with the commutators (10.78):[17]

$$
\begin{aligned}
iG(x',x) &= \int \frac{d^3k}{(2\pi)^3} \frac{1}{2E_k} \left[\theta(t'-t)e^{-ik\cdot(x'-x)} + \theta(t-t')e^{ik\cdot(x'-x)} \right] \\
&= \frac{1}{2\pi i} \int \frac{d^3k}{(2\pi)^3} \frac{1}{2E_k} \left[\int_{-\infty}^{\infty} \frac{d\omega}{\omega - i\epsilon} e^{ik\cdot(x'-x)} e^{-i(E_k-\omega)(t'-t)} \right. \\
&\quad \left. + \int_{-\infty}^{\infty} \frac{d\omega}{\omega - i\epsilon} e^{-ik\cdot(x'-x)} e^{i(E_k-\omega)(t'-t)} \right] \\
&= \frac{1}{2\pi i} \int \frac{d^3k}{(2\pi)^3} \frac{1}{2E_k} \left[\int_{-\infty}^{\infty} dy \frac{e^{-iy(t'-t)} e^{ik\cdot(x'-x)}}{E_k - y - i\epsilon} \right. \\
&\quad \left. + \int_{-\infty}^{\infty} dy \frac{e^{-iy(t'-t)} e^{ik\cdot(x'-x)}}{E_k + y - i\epsilon} \right] \\
&= -i \int \frac{d^3k}{(2\pi)^3} \int \frac{dk_0}{2\pi} e^{-i[k_0(t'-t)-k\cdot(x'-x)]} \frac{1}{2E_k} \\
&\quad \times \left[\frac{1}{E_k - k_0 - i\epsilon} + \frac{1}{E_k + k_0 - i\epsilon} \right]
\end{aligned}
$$

---

[17] We use the representation $\theta(t) = (1/2\pi i)\lim_{\epsilon\to 0+} \int_{-\infty}^{\infty} d\omega e^{i\omega t}(\omega - i\epsilon)^{-1}$ of the unit step function.

$$
= -i \int \frac{d^4 k}{(2\pi)^4} e^{-ik\cdot(x'-x)} \frac{1}{E_k^2 - k_0^2 - i\epsilon}
$$

$$
= -i \int \frac{d^4 k}{(2\pi)^4} \frac{e^{-ik\cdot(x'-x)}}{m^2 - k^2 - i\epsilon} , \tag{10.190}
$$

since $E_k^2 - k_0^2 = \mathbf{k}^2 + m^2 - k_0^2 = m^2 - k \cdot k = m^2 - k^2$; here the limit $\epsilon \to 0^+$ is implicit. Thus the Klein–Gordon propagator has the manifestly covariant form

$$
G(x', x) = G(x' - x) = \int \frac{d^4 k}{(2\pi)^4} \frac{e^{-ik\cdot(x'-x)}}{k^2 - m^2 + i\epsilon} . \tag{10.191}
$$

It is obvious from this form that $G(x' - x)$ is a Green function for the Klein–Gordon equation:

$$
\begin{aligned}
(\partial_{\mu'}\partial^{\mu'} + m^2)G(x' - x) &= -\int \frac{d^4 k}{(2\pi)^4} e^{-ik\cdot(x'-x)} = -\delta^3(\mathbf{x'} - \mathbf{x})\delta(t' - t) \\
&= -\delta^4(x' - x). \tag{10.192}
\end{aligned}
$$

The particular Green function (10.191) satisfies (10.192) and the boundary condition that it is the amplitude for a particle to propagate from x' to x when $t' > t$, and for an antiparticle to propagate from x' to x when $t' < t$.

Dirac Propagator

The Dirac propagator $S_F(x', x)$ is defined by

$$
iS_F(x', x) = \langle 0|T\psi(x')\overline{\psi}(x)|0\rangle , \tag{10.193}
$$

where the time-ordering operator T in the case of fermions is defined by

$$
\begin{aligned}
T\psi(x')\psi(x) &= \psi(x')\psi(x) \quad \text{if } t' > t \\
&= -\psi(x)\psi(x') \quad \text{if } t' < t. \tag{10.194}
\end{aligned}
$$

The minus sign appears because fermion amplitudes are antisymmetric under an interchange of coordinates. Without the minus sign we would have $T[\psi(x')\psi(x) + \psi(x)\psi(x')] = 2\psi(x')\psi(x)\theta(t' - t) + 2\psi(x)\psi(x')\theta(t - t')$, which would be inconsistent with the anticommutation relation $\psi(x')\psi(x) + \psi(x)\psi(x') = \{\psi(x'), \psi(x)\} = 0$. Thus

$$
iS_F(x', x) = \theta(t' - t)\langle 0|\psi(x')\overline{\psi}(x)|0\rangle - \theta(t - t')\langle 0|\overline{\psi}(x)\psi(x')|0\rangle. \tag{10.195}
$$

This is quite analogous to (10.187) for the spin–0 field: the amplitude to transport a charge from x to x' is the amplitude to create a charged particle at x and annihilate it at x', *minus* the amplitude to create an (oppositely charged) antiparticle at x' and annihilate it at x, with the appropriate boundary conditions on the time ordering for these events. Note that $S_F(x',x)$ is a 4 × 4 matrix; we can write (10.193) as

$$iS_F(x',x)_{\alpha\beta} = \langle 0|T\psi_\alpha(x')\overline{\psi}_\beta(x)|0\rangle. \tag{10.196}$$

$S_F(x',x)$ can be evaluated in basically the same way as the Klein–Gordon propagator. From the plane-wave expansions (10.90) and (10.91), and the spinor identities (9.33) and (9.34), we obtain straightforwardly the result

$$iS_F(x'-x) \;=\; \int \frac{d^3p}{(2\pi)^3}\frac{1}{2E}\left[\theta(t'-t)(\not{p}+m)e^{-ip\cdot(x'-x)}\right.$$
$$\left. -\;\theta(t-t')(\not{p}-m)e^{ip\cdot(x'-x)}\right]. \tag{10.197}$$

Using again the integral representation of the unit step function θ, and a change of integration variable as in (10.190), we get the manifestly covariant result

$$S_F(x'-x) \;=\; \int \frac{d^4p}{(2\pi)^4}e^{-ip\cdot(x'-x)}\frac{\not{p}+m}{p^2-m^2+i\epsilon}$$
$$=\; \int \frac{d^4p}{(2\pi)^4}e^{-ip\cdot(x'-x)}S_F(p), \tag{10.198}$$

where the Fourier transform

$$S_F(p) = \frac{\not{p}+m}{p^2-m^2+i\epsilon} \equiv \frac{1}{\not{p}-m+i\epsilon}. \tag{10.199}$$

The latter form allows us to see easily that

$$(i\not{\partial}'-m)S_F(x'-x) = \delta^4(x'-x), \tag{10.200}$$

i.e., $S_F(x'-x)$ is a Green function for the Dirac equation. The subscript F stands for Feynman, and S_F is often called the *Feynman propagator* or the *electron propagator*. Sakurai (1976) refers to S_F as "one of the most important functions in twentieth-century physics." We shall defer further discussion of this propagator to the following chapter.

Photon Propagator: Coulomb Gauge

It should by now be obvious how to define the propagator for the Maxwell four-potential field $A(x)$:[18]

$$D_{\mu\nu}(x', x) = -i\langle 0|T A_\mu(x') A_\nu(x)|0\rangle \qquad (10.201)$$

$D_{\mu\nu}(x', x)$ will depend on the choice of gauge. In the Coulomb gauge, where the vector potential is transverse ($\nabla \cdot \mathbf{A} = 0$), we denote $D_{\mu\nu}(x', x)$ by $D^C_{\mu\nu}(x', x)$. The Coulomb-gauge photon propagator may be evaluated following the same procedure as for the Klein–Gordon and Dirac propagators. Using the plane-wave expansion (10.153) for the Coulomb-gauge vector potential $A_i, i = 1, 2, 3$, we easily calculate

$$D^C_{ij}(x', x) = \int \frac{d^4k}{(2\pi)^4} \frac{e^{-ik\cdot(x'-x)}}{k^2 + i\epsilon} \sum_{\lambda=1}^{2} e_i(\mathbf{k}, \lambda) e_j(\mathbf{k}, \lambda) \qquad (10.202)$$

for $i, j = 1, 2, 3$. To evaluate the sum over λ, note that, since $\mathbf{k} \cdot \mathbf{e}(\mathbf{k}, \lambda) = 0$, we can write any vector $\mathbf{a}$ as

$$\mathbf{a} = (\mathbf{a} \cdot \mathbf{k})\mathbf{k}/\mathbf{k}^2 + \sum_{\lambda=1}^{2}[\mathbf{a} \cdot \mathbf{e}(\mathbf{k}, \lambda)]\mathbf{e}(\mathbf{k}, \lambda) \qquad (10.203)$$

or

$$
\begin{aligned}
a_i &= \sum_{j=1}^{3} a_j k_i k_j / \mathbf{k}^2 + \sum_{j=1}^{3}\sum_{\lambda=1}^{2} a_j e_j(\mathbf{k}, \lambda) e_i(\mathbf{k}, \lambda) \\
&= \sum_{j=1}^{3} a_j \left[\frac{k_i k_j}{\mathbf{k}^2} + \sum_{\lambda=1}^{2} e_i(\mathbf{k}, \lambda) e_j(\mathbf{k}, \lambda)\right],
\end{aligned} \qquad (10.204)
$$

from which

$$\sum_{\lambda=1}^{2} e_i(\mathbf{k}, \lambda) e_j(\mathbf{k}, \lambda) = \delta_{ij} - \frac{k_i k_j}{\mathbf{k}^2} \qquad (10.205)$$

and

$$
\begin{aligned}
D^C_{ij}(x' - x) &= \int \frac{d^4k}{(2\pi)^4} \frac{e^{-ik\cdot(x'-x)}}{k^2 + i\epsilon} \left(\delta_{ij} - \frac{k_i k_j}{\mathbf{k}^2}\right) \\
&= \int \frac{d^4k}{(2\pi)^4} e^{-ik\cdot(x'-x)} D^C_{ij}(k) ,
\end{aligned} \qquad (10.206)
$$

---

[18] The use of the symbol D for the photon propagator, like S_F for the electron propagator, is common but not universal in the literature. Often the photon propagator is denoted D_F and called the *Feynman propagator for electromagnetic radiation* (Bjorken and Drell, 1964).

where

$$D_{ij}^C(k) = \frac{1}{k^2 + i\epsilon} \left(\delta_{ij} - \frac{k_i k_j}{k^2} \right). \qquad (10.207)$$

It follows from (10.206) that

$$
\begin{aligned}
\left(\frac{\partial^2}{\partial t'^2} - \nabla'^2 \right) D_{ij}^C(x' - x) &= \partial'_\mu \partial'^\mu D_{ij}^C(x' - x) \\
&= -\int \frac{d^4 k}{(2\pi)^4} \left(\delta_{ij} - \frac{k_i k_j}{k^2} \right) e^{-ik \cdot (x'-x)} \\
&= -\delta(t' - t) \int \frac{d^3 k}{(2\pi)^3} \left(\delta_{ij} - \frac{k_i k_j}{k^2} \right) e^{i\mathbf{k} \cdot (\mathbf{x}' - \mathbf{x})} \\
&= -\delta(t' - t) \delta_{ij}^\perp(\mathbf{x}' - \mathbf{x}), \qquad (10.208)
\end{aligned}
$$

where $\delta_{ij}^\perp(\mathbf{x}' - \mathbf{x})$ is the transverse delta function defined by equation (4.34). Thus $D_{ij}^C(x' - x)$ is a Green function for the transverse part of the vector potential. To see this explicitly, consider the equation (10.135) for the Coulomb-gauge vector potential. From the continuity equation $\nabla \cdot \mathbf{J} + \partial\rho/\partial t = 0$ that follows from the Maxwell equations, and the solution of equation (10.134) for the scalar potential in the Coulomb gauge, we have

$$
\begin{aligned}
-\mathbf{J} + \nabla \left(\frac{\partial \phi}{\partial t} \right) &= -\mathbf{J} + \nabla \frac{1}{4\pi} \int d^3 x' \frac{\partial\rho(\mathbf{x}', t)/\partial t}{|\mathbf{x} - \mathbf{x}'|} \\
&= -\mathbf{J} - \frac{1}{4\pi} \nabla \int d^3 x' \frac{\nabla' \cdot \mathbf{J}(\mathbf{x}', t)}{|\mathbf{x} - \mathbf{x}'|} \\
&= -\mathbf{J} + \mathbf{J}^\| = -\mathbf{J}^\perp, \qquad (10.209)
\end{aligned}
$$

so that $\mathbf{A}(\mathbf{x}, t)$ satisfies

$$\nabla^2 \mathbf{A} - \frac{\partial^2 \mathbf{A}}{\partial t^2} = -\mathbf{J}^\perp \qquad (10.210)$$

and the Green function for this equation satisfies (10.208).

The simplest way to calculate $D_{00}^C(x' - x)$ is to use the fact that it is a Green function for the Poisson equation satisfied by $A_0(x) = \phi(x)$ in the Coulomb gauge. Thus $\nabla'^2 D_{00}^C(x' - x) = -\delta^3(\mathbf{x}' - \mathbf{x})$, so that

$$D_{00}^C(k) = \frac{1}{k^2} \qquad (10.211)$$

and

$$D_{00}^C(x' - x) = \int \frac{d^4 k}{(2\pi)^4} \frac{e^{-ik \cdot (x'-x)}}{k^2} = \delta(t' - t) \int \frac{d^3 k}{(2\pi)^3} \frac{e^{i\mathbf{k} \cdot (\mathbf{x}' - \mathbf{x})}}{k^2}$$

$$
= \delta(t'-t)\frac{2\pi}{8\pi^3}\int_0^\infty dk\, k^2 \int_0^\pi d\theta \sin\theta \frac{e^{ikr\cos\theta}}{k^2}
$$

$$
= \delta(t'-t)\frac{1}{2\pi^2 r}\int_0^\infty dk\frac{\sin kr}{k}
$$

$$
= \frac{\delta(t'-t)}{4\pi r} = \frac{\delta(t'-t)}{4\pi|\mathbf{x}'-\mathbf{x}|} . \tag{10.212}
$$

It is easily seen, from the definition of the propagator, that the remaining components $D_{0j}^C(x',x)$ and $D_{i0}^C(x',x)$, $i = 1,2,3$, all vanish.

The part of the photon propagator associated with the scalar potential, $D_{00}^C(x'-x)$, obviously corresponds to nonretarded, instantaneous propagation from x to x'. Instantaneous interactions associated with the scalar potential in the Coulomb gauge are cancelled by interactions associated with the transverse vector potential, as we showed in Section 4.6 for the example of the dipole–dipole interaction.

Photon Propagator: Lorentz Gauge

Using the plane-wave expansion (10.166) for $A_\mu(x)$ in the Lorentz gauge, we obtain from (10.201), in the now familiar way, the manifestly covariant expression

$$
\begin{aligned}
D_{\mu\nu}(x'-x) &= -g_{\mu\nu}\int \frac{d^4k}{(2\pi)^4}\frac{e^{-ik\cdot(x'-x)}}{k^2+i\epsilon} \\
&= \int \frac{d^4k}{(2\pi)^4}e^{-ik\cdot(x'-x)}D_{\mu\nu}(k),
\end{aligned} \tag{10.213}
$$

with

$$
D_{\mu\nu}(k) = -\frac{g_{\mu\nu}}{k^2+i\epsilon} . \tag{10.214}
$$

Note that $D_{\mu\nu}(x'-x)$ is $-g_{\mu\nu}$ times the $m \to 0$ limit of the Klein–Gordon propagator.

10.10   Remarks

We noted in Chapters 2 and 3 that, although the electromagnetic field was first quantized by Born, Heisenberg, and Jordan in 1926, the first important application of quantum field theory — to spontaneous emission — was made by Dirac in 1927. Dirac showed that the quantum theory of the electromagnetic field could deal with the creation and annihilation of photons. Soon after that Jordan and Wigner (1928), and then Fermi (1929) and Heisenberg and Pauli (1929, 1930), showed that particles of finite mass could also be understood as field quanta, just as photons are quanta of the electromagnetic field. In other words, by around 1930 the idea of quantized fields beyond electromagnetism had already been established. Indeed Fermi (1934) employed ideas of quantum field theory in his theory of the beta decay of nuclei: the electron associated with beta decay, according to Fermi, is *created* through the interaction of fields, analogous to the creation of a photon via the interaction between an atom and the quantized electromagnetic field.

However, the belief that quantum fields are the "essential reality" declined during the 1930s and 1940s, although Pauli's spin–statistics theorem (1940) was certainly a major triumph of quantum field theory. This decline in popularity began with the calculation of Oppenheimer (1930) in which the interaction of an atomic electron with the quantized electromagnetic field produced an *infinite* shift in the electron's energy. The Lamb–Retherford experiments in the late 1940s compelled physicists to finally confront such infinities, and the calculation of the Lamb shift and other QED effects was successfully done by the method of renormalization, as described in Chapter 3 in connection with the nonrelativistic theory and in the following chapter for the correctly relativistic theory. With this extremely successful calculational tool, quantum fields could again be claimed to be the "essential reality" of the universe.

In this chapter we have shown, in relatively very simple terms, how the theory of quantized fields leads to creation and annihilation operators for material particles, in much the same way that QED involves creation and annihilation operators for photons. We have emphasized that *all* quantum fields have zero-point energy and nontrivial vacuum states. We showed, for instance, that there is a Casimir effect associated with the Dirac field of which electrons and positrons are the quanta. According to quantum field theory the vacuum is a complicated state of affairs involving virtual photons, electrons, positrons, quarks, gluons, ... ; there is a quantized field, with vacuum fluctuations, associated with every kind of fundamental particle. In calculations we must take into account the myriad possibilities for the creation of particles out of the vacuum and their annihilation. In the following two chapters, and particularly in Chapter 12, we discuss how such calculations are done in the "best theory we have," QED. In particular, we will show how the sorts of calculations done in previous chapters are modified in *relativistic* QED.

10.11   Bibliography

The reprint volume edited by Schwinger (1958) is a very useful collection of important papers on QED and quantum field field theory.

Bjorken, J. D. and S. D. Drell, *Relativistic Quantum Mechanics* (McGraw-Hill, New York, 1964).

Chodos, A. and C. B. Thorn, "Electromagnetic Self-Energy of Confined Massless Quarks," *Phys. Lett.* B53, 359 (1974).

Fermi, E., "Sopra l'elettrodinamica quantistica," *Lincei Acc. Naz. Rend.* 9, 881 (1929).

Fermi, E., "Versuch einer Theorie der β-Strahlen," *Z. Phys.* 88, 161 (1934).

Heisenberg, W. and W. Pauli, "Zur Quantendynamik der Wellenfelder," *Z. Phys.* 56, 1 (1929).

Heisenberg, W. and W. Pauli, "Zur Quantendynamik der Wellenfelder. II," *Z. Phys.* 59, 168 (1930).

Heitler, W., *The Quantum Theory of Radiation*, 3rd ed. (Oxford University Press, London, 1966).

Itzykson, C. and J.-B. Zuber, *Quantum Field Theory* (McGraw-Hill, New York, 1980).

Johnson, K., "The MIT Bag Model," *Acta Phys. Polonica* B6, 865 (1975).

Johnson, K. A., "The Bag Model of Quark Confinement," *Sci. Am.* 241, 112 (January 1979).

Jordan, P. and E. Wigner, "Über das Paulische Äquivalenzverbot," *Z. Phys.* 47, 631 (1928).

Lee, T. D., *Particle Physics and Introduction to Field Theory* (Harwood Academic Publishers, Chur, Switzerland, 1981).

Milton, K. A., "Fermionic Casimir Stress on a Spherical Bag," *Ann. Phys.* (New York) 150, 432 (1983).

Mostepanenko, V. M. and N. N. Trunov, "The Casimir Effect and Its Applications," *Sov. Phys. Usp.* 31, 965 (1988).

Oppenheimer, J. R., "Note on the Theory of the Interaction of Field and Matter," *Phys. Rev.* 35, 461 (1930).

Pauli, W. and V. Weisskopf, "Über die Quantisierung der skalaren relativistischen Wellengleichung," *Helv. Phys. Acta* 7, 709 (1934).

Pauli, W., "The Connection between Spin and Statistics," *Phys. Rev.* 58, 716 (1940).

Sakurai, J. J., *Advanced Quantum Mechanics* (Addison-Wesley, Reading, Mass., 1976).

Schiff, L. I., *Quantum Mechanics* (McGraw-Hill, New York, 1968).

Schwinger, J. (editor), *Selected Papers on Quantum Electrodynamics*

(Dover Books, New York, 1958).

Weinberg, S., "The Search for Unity: Notes for a History of Quantum Field Theory," *Daedalus* **106**, 17 (1977).

Chapter 11

Self-Energies and Renormalization

> No progress was made for 20 years. Then a development came,
> initiated by Lamb's discovery and explanation of the Lamb shift,
> which fundamentally changed the character of theoretical physics. It
> involved setting up rules for discarding ... infinities ...
> — P. A. M. Dirac (1989)

11.1 Introduction

The interactions of electrons and positrons with the electromagnetic field
are treated in quantum field theory through the coupling of the quantized
Dirac and Maxwell fields. The resulting theory of relativistic quantum electrodynamics is the most accurately tested theory in the history of physics.

For nearly half a century the most convenient and widespread formulation of QED has involved covariant perturbation theory in which terms in
a perturbation expansion correspond to Feynman diagrams. This formulation is the subject of the following chapter. In this chapter we take an
old-fashioned approach to some fundamentally important aspects of QED:
the electron self-energy, renormalization, the Lamb shift, and vacuum polarization. There are two compelling reasons for such an approach. First,
the results obtained in this way lend themselves to a more direct physical interpretation than is possible without some working familiarity with
the more powerful methods of covariant perturbation theory. Second, the
power of the methods described in the following chapter can be better appreciated when they are compared with the older, noncovariant methods.

There is yet another, historical reason for devoting some attention to the older ways: the first correct results for the Lamb shift, for instance, were obtained by old-fashioned perturbation theory. The discussion here in fact relies heavily on those early efforts (Kroll and Lamb, 1949; French and Weisskopf, 1949).

11.2 Coupled Dirac and Maxwell Fields

The Lagrangian density for the Dirac field is given by equation (10.85) as

$$\mathsf{L_D} = \overline{\psi}(x)(i\slashed{\partial} - m)\psi(x). \tag{11.1}$$

For the Maxwell field we have, from (10.146),

$$\mathsf{L_M} = -\frac{1}{4}F^{\mu\nu}F_{\mu\nu} = \frac{1}{2}(\mathbf{E}^2 - \mathbf{B}^2). \tag{11.2}$$

The contribution to the Lagrangian density arising from the coupling of the electromagnetic field to a current density j^μ can also be read from (10.146):

$$\mathsf{L_{DM}}(x) = -ej_\mu(x)A^\mu(x). \tag{11.3}$$

Using the current density $j^\mu = e\overline{\psi}\gamma^\mu\psi$ for the Dirac field, we can write this as

$$\mathsf{L_{DM}}(x) = -e\overline{\psi}(x)\gamma_\mu\psi(x)A^\mu(x), \tag{11.4}$$

so that the QED Lagrangian density for the coupled Dirac and Maxwell fields is

$$
\begin{aligned}
\mathsf{L}(x) &= \mathsf{L_D}(x) + \mathsf{L_M}(x) + \mathsf{L_{DM}}(x) \\
&= \overline{\psi}(i\slashed{\partial} - m)\psi + \frac{1}{2}(\mathbf{E}^2 - \mathbf{B}^2) - e\overline{\psi}\gamma_\mu\psi A^\mu \ .
\end{aligned} \tag{11.5}
$$

In the Coulomb gauge the vector potential $\mathbf{A}$ is transverse, i.e., $\nabla \cdot \mathbf{A} = 0$. The time component $A^0(x)$ of $A^\mu(x)$ in this gauge is nondynamical and can be eliminated by solving for it in terms of the charge density $\rho(\mathbf{x}, t)$ as in equation (4.15):

$$A^0(\mathbf{x}, t) = \phi(\mathbf{x}, t) = \frac{e}{4\pi}\int d^3x' \frac{\rho(\mathbf{x}', t)}{|\mathbf{x} - \mathbf{x}'|} = \frac{e}{4\pi}\int d^3x' \frac{\psi^\dagger(\mathbf{x}', t)\psi(\mathbf{x}', t)}{|\mathbf{x} - \mathbf{x}'|} \tag{11.6}$$

in Heaviside–Lorentz units. The fact that $A^0(\mathbf{x}, t)$ is not retarded in the Coulomb gauge is not a problem because, as discussed in Section 4.6, it is only the total field, longitudinal plus transverse, that must be retarded.

In the Coulomb gauge the longitudinal and transverse parts of the electric field are given respectively by

$$\mathbf{E}^{\parallel}(\mathbf{x}, t) = -\nabla A^0(\mathbf{x}, t) \tag{11.7}$$

and

$$\mathbf{E}^{\perp}(\mathbf{x}, t) = -\dot{\mathbf{A}}(\mathbf{x}, t), \tag{11.8}$$

and the latter has an unretarded part that exactly cancels the unretarded contribution $\mathbf{E}^{\parallel}$ to the complete electric field $\mathbf{E} = \mathbf{E}^{\parallel} + \mathbf{E}^{\perp}$ (Section 4.6).

From (11.5) we obtain the momenta conjugate to the fields $\mathbf{A}$ and ψ in the Coulomb gauge:

$$\frac{\partial \mathsf{L}}{\partial(\partial_t A_i)} = -E_i = -\nabla_i A_0 + \dot{A}_i, \tag{11.9}$$

$$\frac{\partial \mathsf{L}}{\partial \dot{\psi}} = i\psi^\dagger. \tag{11.10}$$

The Hamiltonian following from the Lagrangian density (11.5) is therefore found from the prescription given in Chapter 10 to be

$$\begin{aligned} H &= \int d^3x [\psi^\dagger(-i\boldsymbol{\alpha} \cdot \nabla + \beta m)\psi + \frac{1}{2}(\mathbf{E}^2 + \mathbf{B}^2) \\ &\quad + \mathbf{E} \cdot \nabla A_0 + e\overline{\psi}\gamma_\mu\psi A^\mu]. \end{aligned} \tag{11.11}$$

Now by partial integration

$$\int d^3x \mathbf{E} \cdot \nabla A_0 = -\int d^3x (\nabla \cdot \mathbf{E})A_0 = -e \int d^3x \rho(\mathbf{x}, t)A_0(\mathbf{x}, t), \tag{11.12}$$

which exactly cancels the term $e\overline{\psi}\gamma_0\psi A^0 = e\psi^\dagger\psi A_0 = e\rho A_0$ in the integrand of (11.11). Thus

$$H = \int d^3x \left[\psi^\dagger(-i\boldsymbol{\alpha} \cdot \nabla + \beta m)\psi + \frac{1}{2}(\mathbf{E}^2 + \mathbf{B}^2) + e\overline{\psi}\gamma_i\psi A^i\right], \tag{11.13}$$

where the repeated index i in the last term implies a summation over the spatial components $i = 1, 2, 3$.

We can further simplify H using

$$\int d^3x \mathbf{E}^2 = \int d^3x (\mathbf{E}^{\parallel\,2} + \mathbf{E}^{\perp\,2} + 2\mathbf{E}^{\parallel} \cdot \mathbf{E}^{\perp}). \tag{11.14}$$

Note first that

$$\int d^3x \mathbf{E}^{\parallel} \cdot \mathbf{E}^{\perp} = \int d^3x \nabla A_0 \cdot \dot{\mathbf{A}} = -\int d^3x A_0(\nabla \cdot \dot{\mathbf{A}}) = 0. \tag{11.15}$$

Furthermore

$$
\int d^3x \mathbf{E}^{\|2} = \int d^3x \nabla A_0 \cdot \nabla A_0 = -\int d^3x A_0 \nabla^2 A_0
$$
$$
= e \int d^3x A_0 \rho = \frac{e^2}{4\pi} \int d^3x \int d^3x' \frac{\rho(\mathbf{x},t)\rho(\mathbf{x}',t)}{|\mathbf{x}-\mathbf{x}'|}
$$
(11.16)

from (11.6). Therefore

$$
H = \int d^3x \left[\psi^\dagger(-i\boldsymbol{\alpha}\cdot\nabla + \beta m)\psi + \frac{1}{2}(\mathbf{E}^{\perp 2} + \mathbf{B}^2) + e\overline{\psi}\gamma_i\psi A^i \right]
$$
$$
+ \frac{e^2}{8\pi} \int d^3x \int d^3x' \frac{\rho(\mathbf{x},t)\rho(\mathbf{x}',t)}{|\mathbf{x}-\mathbf{x}'|} .
$$
(11.17)

We rewrite this, using $\overline{\psi}\gamma_i A^i \psi = -\psi^\dagger \beta\boldsymbol{\gamma}\cdot\mathbf{A}\psi = -\psi^\dagger\boldsymbol{\alpha}\cdot\mathbf{A}\psi$, as

$$
H = \int d^3x \left[\psi^\dagger(-i\boldsymbol{\alpha}\cdot\nabla + \beta m)\psi + \frac{1}{2}(\mathbf{E}^2 + \mathbf{B}^2) - e\psi^\dagger\boldsymbol{\alpha}\cdot\mathbf{A}\psi \right]
$$
$$
+ \frac{e^2}{8\pi} \int d^3x \int d^3x' \frac{\rho(\mathbf{x},t)\rho(\mathbf{x}',t)}{|\mathbf{x}-\mathbf{x}'|} ,
$$
(11.18)

with the understanding that $\mathbf{E}$ and $\mathbf{A}$ are *transverse* field operators. They satisfy the equal-time commutation relations (Section 10.8)

$$
[A_i(\mathbf{x},t), E_j(\mathbf{x}',t)] = -i\delta_{ij}^\perp(\mathbf{x}-\mathbf{x}'),
$$
(11.19)

$$
[A_i(\mathbf{x},t), A_j(\mathbf{x}',t)] = [E_i(\mathbf{x},t), E_j(\mathbf{x}',t)] = 0,
$$
(11.20)

whereas the Dirac field operators satisfy the equal-time anticommutation relations

$$
\{\psi_\alpha(\mathbf{x},t), \psi_\beta(\mathbf{x}',t)\} = \delta_{\alpha\beta}\delta^3(\mathbf{x}-\mathbf{x}'),
$$
(11.21)
$$
\{\psi_\alpha(\mathbf{x},t), \psi_\beta(\mathbf{x}',t)\} = 0.
$$
(11.22)

Commutators of Dirac field operators ψ_α with Maxwell field operators A_i and $\dot{A}_i$ vanish, e.g.,

$$
[\psi_\alpha(\mathbf{x},t), A_i(\mathbf{x}',t)] = 0.
$$
(11.23)

It follows from (11.6) and (11.21), however, that

$$
\{\psi_\alpha(\mathbf{x},t), A_0(\mathbf{x}',t)\} = \frac{e}{4\pi}\frac{\psi_\alpha(\mathbf{x},t)}{|\mathbf{x}-\mathbf{x}'|} .
$$
(11.24)

The Hamiltonian (11.18) is the relativistic generalization of the nonrelativistic Hamiltonian (4.17) in the Coulomb gauge. The part of the Hamiltonian attributable to the transverse field, involving the energy density proportional to $\mathbf{E}^{\perp 2} + \mathbf{B}^2$, has the same form, as does the part associated with the instantaneous Coulomb interactions among the charged particles. (In the nonrelativistic Hamiltonian (4.17) the Coulomb interactions are written explicitly in the form appropriate for classical point particles.) The main difference between the relativistic and nonrelativistic Hamiltonians lies in the treatment of the fermion particle dynamics. The nonrelativistic kinetic energy operator $\mathbf{p}^2/2m$ is replaced in the second-quantized relativistic theory by a Hamiltonian density $\overline{\psi}(\boldsymbol{\alpha} \cdot \mathbf{p} + \beta m)\psi$, while the interaction term $-(e/mc)\mathbf{A} \cdot \mathbf{p} + (e^2/2mc^2)\mathbf{A}^2$ is replaced by the density $e\overline{\psi}\slashed{A}\psi$ in relativistic units and notation. As in classical theory — or actually as a *result* of what we know classically — we are using different particle dynamics in the relativistic case, whereas the electromagnetic field is treated in essentially the same way as in nonrelativistic theory. Of course this difference may be traced back to purely classical theory, where the Maxwell equations have the correct transformation properties under Lorentz transformations but the Newton equations do not.

In obtaining the Hamiltonian H given by (11.18) we have assumed, in (11.16), that A_0 is the electrostatic field due to the charge density $e\rho = e\psi^\dagger \psi$. More generally there will be an *external* electrostatic field A_0^{ext} giving an additional contribution $e\overline{\psi}\gamma^0\psi A_0^{\text{ext}} = e\psi^\dagger\psi A_0^{\text{ext}} \equiv \psi^\dagger V\psi$ to the Hamiltonian density. Then

$$
\begin{aligned}
H &= \int d^3x \left[\psi^\dagger(\boldsymbol{\alpha} \cdot \mathbf{p} + \beta m + V)\psi + \frac{1}{2}(\mathbf{E}^2 + \mathbf{B}^2) - e\psi^\dagger\boldsymbol{\alpha} \cdot \mathbf{A}\psi \right] \\
&\quad + \frac{e^2}{8\pi} \int d^3x \int d^3x' \frac{\rho(\mathbf{x},t)\rho(\mathbf{x}',t)}{|\mathbf{x} - \mathbf{x}'|} .
\end{aligned}
\tag{11.25}
$$

For a bound electron, for instance, V would be the binding potential, such as the Coulomb potential due to the proton in the hydrogen atom.

11.3 Self-Energy: The Old-Fashioned Way

We consider now, based on the second-quantized relativistic Hamiltonian (11.25), the energy associated with the coupling of an electron to the vacuum electromagnetic field. We refer to this energy as the "self-energy" of the electron.

First we write the Dirac field $\psi(\mathbf{x}, t)$ a little differently. The expression (10.90) gives $\psi(\mathbf{x}, t)$ as a continuous sum of free-electron plane-wave states. More generally we will want to express $\psi(\mathbf{x}, t)$ in terms of one-particle eigenfunctions that differ from $u^i(p)$ and $v^i(p)$, and which may correspond to discrete as well as continuous spectra. Thus we now replace (10.90) and (10.91) by

$$\psi(\mathbf{x}, t) = \sum_k \left[b_k \phi_{k+}(\mathbf{x}) e^{-iE_k t} + d_k^\dagger \phi_{k-}(\mathbf{x}) e^{iE_k t} \right], \tag{11.26}$$

$$\psi^\dagger(\mathbf{x}, t) = \sum_k \left[b_k^\dagger \phi_{k+}^*(\mathbf{x}) e^{iE_k t} + d_k \phi_{k-}^*(\mathbf{x}) e^{-iE_k t} \right]. \tag{11.27}$$

Here $\phi_{k+}(\mathbf{x})$ and $\phi_{k-}(\mathbf{x})$ are four-component positive- and negative-energy eigenfunctions replacing the plane-wave eigenfunctions $u(p) e^{i\mathbf{p}\cdot\mathbf{x}}$ and $v(p)$ $\times e^{-i\mathbf{p}\cdot\mathbf{x}}$, respectively. We choose here, in order to follow the older literature more closely, to omit the covariant normalization factors m/E_k included in equations (10.90) and (10.91). Thus the normalization of the Dirac spinors will differ from the choice made in Chapter 9, a minor difference that will be noted later, when it is important. The formulas (11.26) and (11.27) generalize the plane-wave expansions (10.90) and (10.91), and in writing them we allow for the possibility that at least some of the $\phi_{k+}(\mathbf{x})$ and $\phi_{k-}(\mathbf{x})$ are bound-state eigenfunctions, as already noted.

We consider first the expectation value of the last term in the Hamiltonian (11.18) for the state in which there is one electron described by the eigenfunction $\phi_a(\mathbf{x})$. From the fact that $\rho(\mathbf{x}, t) = \psi^\dagger(\mathbf{x}, t)\psi(\mathbf{x}, t)$ we see that the expectation value of this operator will involve expectation values of products of four annihilation and creation operators such as $b_k^\dagger b_\ell b_r^\dagger b_m$, $b_k^\dagger b_\ell d_r d_m^\dagger$, etc. For the state of interest the only nonvanishing contributions come from $\langle d_\sigma d_\rho^\dagger d_\rho d_\sigma^\dagger \rangle = \langle b_a^\dagger b_a d_\sigma d_\sigma^\dagger \rangle = \langle d_\sigma b_r b_r^\dagger d_\sigma^\dagger \rangle = \langle b_a^\dagger b_r b_r^\dagger b_a \rangle = \langle d_\sigma d_\sigma^\dagger b_a^\dagger b_a \rangle = 1$, where σ, ρ designate negative-energy eigenfunctions [e.g., $\phi_{\sigma-}(\mathbf{x})$] and r designates a positive-energy eigenfunction. Thus

$$
\begin{aligned}
\langle a|H_S|a \rangle &\equiv \langle a| \frac{e^2}{8\pi} \int d^3x \int d^3x' \frac{\rho(\mathbf{x}, t)\rho(\mathbf{x}', t)}{|\mathbf{x} - \mathbf{x}'|} |a \rangle \\
&= \sum_\sigma \sum_\rho A_{\sigma\sigma\rho\rho} + \sum_\sigma A_{aa\sigma\sigma} + \sum_\sigma \sum_{r \neq a} A_{\sigma r r \sigma} \\
&\quad + \sum_r A_{arra} + \sum_\sigma A_{\sigma\sigma aa} \\
&= \sum_\sigma \sum_\rho A_{\sigma\sigma\rho\rho} + 2\sum_\sigma A_{aa\sigma\sigma}
\end{aligned}
$$

$$+ \sum_{\sigma} \sum_{r \neq a} A_{\sigma r r \sigma} + \sum_{r} A_{arra} , \qquad (11.28)$$

where

$$A_{k\ell mn} \equiv \frac{e^2}{8\pi} \int d^3 x \int d^3 x' |\mathbf{x} - \mathbf{x}'|^{-1} \phi_k^*(\mathbf{x}) \phi_\ell(\mathbf{x}) \phi_m^*(\mathbf{x}') \phi_n(\mathbf{x}'). \qquad (11.29)$$

We argue that the physically significant electrostatic energy is the difference between $\langle a|H_S|a \rangle$ and $\langle \text{vac}|H_S|\text{vac} \rangle$, where $|\text{vac}\rangle$ is the vacuum state of no electrons, positrons, or photons. The latter expectation value is easily seen to be given by

$$\langle \text{vac}|H_S|\text{vac} \rangle = \sum_{\sigma} \sum_{\rho} A_{\sigma\sigma\rho\rho} + \sum_{\sigma} \sum_{r} A_{\sigma r r \sigma} , \qquad (11.30)$$

so that

$$\begin{aligned} W_S &\equiv \langle a|H_S|a \rangle - \langle \text{vac}|H_S|\text{vac} \rangle \\ &= 2 \sum_{\sigma} A_{aa\sigma\sigma} + \sum_{r} A_{arra} - \sum_{\sigma} A_{\sigma a a \sigma} \\ &= 2 \sum_{\sigma} A_{aa\sigma\sigma} + \sum_{n} \pm A_{anna} , \end{aligned} \qquad (11.31)$$

where the upper or lower sign in the second term is used depending on whether n is a positive- or negative-energy state, respectively.

It is convenient to define $\delta_j = E_j/|E_j| = +1$ for positive-energy states and -1 for negative-energy states. In this notation,

$$W_S = \sum_{j} (1 - \delta_j) A_{aajj} + \sum_{j} \delta_j A_{ajja} , \qquad (11.32)$$

where j denotes any state, regardless of the sign of the energy.

The interaction $-e \int d^3 x \psi^\dagger \boldsymbol{\alpha} \cdot \mathbf{A} \psi$ in (11.18) has expectation value zero for the state in which there is an electron in state a and there are no positrons or photons; in fact its expectation value vanishes for any state of definite photon number. However, this term contributes to an energy shift when taken to second order in perturbation theory. We denote this second order "dynamical" contribution by ΔE_a. To evaluate ΔE_a we first write out the interaction more explicitly:

$$H_{\text{int}} = -e \int d^3 x \psi^\dagger \boldsymbol{\alpha} \cdot \mathbf{A} \psi$$

$$\rightarrow \ -e \int \frac{d^3k}{(2\pi)^3} \frac{1}{2\omega_k} \sum_{\lambda=1}^{2} \sum_m \sum_\ell \Big[C^\lambda_{m+\ell+}(\mathbf{k}) b^\dagger_m b_\ell$$

$$+ C^\lambda_{m+\ell-}(\mathbf{k}) b^\dagger_m d^\dagger_\ell + C^\lambda_{m-\ell+}(\mathbf{k}) d_m d_\ell$$

$$+ C^\lambda_{m-\ell-}(\mathbf{k}) d_m d^\dagger_\ell \Big] a^\dagger(\mathbf{k}, \lambda), \qquad (11.33)$$

where

$$C^\lambda_{m+\ell+}(\mathbf{k}) \equiv \int d^3x\, \phi^*_{m+}(\mathbf{x}) \boldsymbol{\alpha} \cdot \mathbf{e}_{\mathbf{k}\lambda} \phi_{\ell+}(\mathbf{x}) e^{-i\mathbf{k}\cdot\mathbf{x}} \quad \text{etc.,} \qquad (11.34)$$

and we have used (11.26), (11.27), and the expansion (10.153) for the Coulomb-gauge vector potential. In writing (11.33) we have kept explicitly the photon creation operators $a^\dagger(\mathbf{k}, \lambda)$ but not the annihilation operators $a(\mathbf{k}, \lambda)$. The reason for this is that the photon creation part produces a nonnull state vector when H_{int} acts on the initial state $|i\rangle = |1_a; 0; 0_{\mathbf{k}\lambda}\rangle$ in which there is an electron in state a, no positrons, and no photons. Thus only the photon creation part contributes to $\langle I|H_{\text{int}}|i\rangle$ in the expression

$$\Delta E_i = \sum_{I \neq i} \frac{\langle i|H_{\text{int}}|I\rangle\langle I|H_{\text{int}}|i\rangle}{E_i - E_I} \qquad (11.35)$$

for the second-order energy shift of state i. Each intermediate state $|I\rangle$ will have one photon in the field. The intermediate states generated by $H_{\text{int}}|a\rangle$ are easily seen to be of the following types:

$$|1_a; 0; 1_{\mathbf{k}\lambda}\rangle \quad (\ell = m = a)$$
$$|1_m; 0; 1_{\mathbf{k}\lambda}\rangle \quad (\ell = a)$$
$$|1_m 1_a; 1_\ell; 1_{\mathbf{k}\lambda}\rangle \quad (m \neq a)$$
$$|1_a; 0; 1_{\mathbf{k}\lambda}\rangle \quad (\ell - m)$$

with $E_I = E_a + \omega_k, E_m + \omega_k, E_m + E_a + E_\ell + \omega_k$. These states are eigenstates of the unperturbed Hamiltonian

$$H_0 = \int d^3x\, \psi^\dagger[\boldsymbol{\alpha} \cdot \mathbf{p} + \beta m + V]\psi + \frac{1}{2}\int d^3x (\mathbf{E}^2 + \mathbf{B}^2)$$

$$= \sum_m E_m[b^\dagger_m b_m + d^\dagger_m d_m] + \sum_{\mathbf{k}\lambda} \omega_k a^\dagger_{\mathbf{k}\lambda} a_{\mathbf{k}\lambda}\,. \qquad (11.36)$$

With these intermediate states it follows from (11.35) that

$$\Delta E_a = e^2 \int \frac{d^3k}{(2\pi)^3} \frac{1}{2\omega_k} \sum_{\lambda=1}^{2} \left[\sum_m \frac{C^\lambda_{a+m+}(-\mathbf{k}) C^\lambda_{m+a+}(\mathbf{k})}{E_a - E_m - \omega_k} \right.$$

$$+ \sum_\ell \sum_{m \neq a} \frac{C^\lambda_{\ell-m+}(-\mathbf{k})C^\lambda_{m+\ell-}(\mathbf{k})}{-E_a - E_\ell - \omega_k}$$

$$+ \sum_\ell \frac{C^\lambda_{\ell-\ell-}(-\mathbf{k})C^\lambda_{\ell-\ell-}(\mathbf{k})}{-\omega_k}$$

$$+ 2 \sum_\ell \frac{C^\lambda_{a+a+}(-\mathbf{k})C^\lambda_{\ell-\ell-}(\mathbf{k})}{-\omega_k} \Bigg] . \qquad (11.37)$$

We assume once again that the physically significant energy shift due to H_{int} of the state a is the difference between ΔE_a and ΔE_{vac}, where ΔE_{vac} is the shift due to H_{int} of the vacuum state $|\text{vac}\rangle = |0; 0; 0_{\mathbf{k}\lambda}\rangle$. For the latter the intermediate states $H_{\text{int}}|\text{vac}\rangle$ are of the type

$$|1_m; 1_\ell; 1_{\mathbf{k}\lambda}\rangle$$
$$|0; 0; 1_{\mathbf{k}\lambda}\rangle \quad (\ell = m)$$

with $E_I = E_m + E_\ell + \omega_k$ and $E_I = \omega_k$, respectively, and $E_i = 0$. Then

$$\Delta E_{\text{vac}} = e^2 \int \frac{d^3k}{(2\pi)^3} \frac{1}{2\omega_k} \sum_{\lambda=1}^{2} \Bigg[\sum_\ell \sum_m \frac{C^\lambda_{\ell-m+}(-\mathbf{k})C^\lambda_{m+\ell-}(\mathbf{k})}{-E_m - E_\ell - \omega_k}$$

$$+ \sum_\ell \frac{C^\lambda_{\ell-\ell-}(-\mathbf{k})C^\lambda_{\ell-\ell-}(\mathbf{k})}{-\omega_k} \Bigg] , \qquad (11.38)$$

and consequently the physically significant "dynamical" contribution to the energy shift is

$$W_D \equiv \Delta E - \Delta E_{\text{vac}}$$

$$= e^2 \int \frac{d^3k}{(2\pi)^3} \frac{1}{2\omega_k} \sum_{\lambda=1}^{2} \Bigg[\sum_m \frac{C^\lambda_{a+m+}(-\mathbf{k})C^\lambda_{m+a+}(\mathbf{k})}{E_a - E_m - \omega_k}$$

$$+ \sum_\ell \frac{C^\lambda_{\ell-a+}(-\mathbf{k})C^\lambda_{a+\ell-}(\mathbf{k})}{E_a - E_\ell + \omega_k} + 2 \sum_\ell \frac{C^\lambda_{a+a+}(-\mathbf{k})C^\lambda_{\ell-\ell-}(\mathbf{k})}{-\omega_k} \Bigg]$$

$$= e^2 \int \frac{d^3k}{(2\pi)^3} \frac{1}{2\omega_k} \sum_{\lambda=1}^{2} \sum_j \frac{C^\lambda_{aj}(-\mathbf{k})C^\lambda_{ja}(\mathbf{k})}{E_a - E_j - \omega_k \delta_j}$$

$$- e^2 \int \frac{d^3k}{(2\pi)^3} \frac{1}{2\omega_k} \sum_{\lambda=1}^{2} \sum_j (1 - \delta_j) C^\lambda_{aa}(-\mathbf{k})C^\lambda_{jj}(\mathbf{k}) . \qquad (11.39)$$

It is interesting to compare this with the result of nonrelativistic QED. In the latter there are no negative-energy states, so that the second term in the last expression for W_D above is absent, as are the negative-energy contributions to the first term:

$$W_D \to e^2 \int \frac{d^3k}{(2\pi)^3} \frac{1}{2\omega_k} \sum_{\lambda=1}^{2} \sum_n \frac{C_{an}^\lambda(-\mathbf{k})C_{na}^\lambda(\mathbf{k})}{E_a - E_n - \omega_k} . \qquad (11.40)$$

In nonrelativistic theory, furthermore, the "velocity" $\boldsymbol{\alpha} \to \mathbf{p}/m$ and therefore

$$C_{na}^\lambda(\mathbf{k}) \to \frac{1}{m} \int d^3x \phi_n^*(\mathbf{x})\mathbf{p} \cdot \mathbf{e}_{\mathbf{k}\lambda}\phi_a(\mathbf{x})e^{-i\mathbf{k}\cdot\mathbf{x}} = \frac{1}{m}\langle n|\mathbf{p} \cdot \mathbf{e}_{\mathbf{k}\lambda}e^{-i\mathbf{k}\cdot\mathbf{x}}|a\rangle \qquad (11.41)$$

and

$$
\begin{aligned}
W_D \quad &\to \quad \frac{e^2}{m^2} \int \frac{d^3k}{(2\pi)^3} \frac{1}{2\omega_k} \sum_{\lambda=1}^{2} \sum_n \\
&\quad \times \frac{\langle a|\mathbf{p} \cdot \mathbf{e}_{\mathbf{k}\lambda}e^{i\mathbf{k}\cdot\mathbf{x}}|n\rangle\langle n|\mathbf{p} \cdot \mathbf{e}_{\mathbf{k}\lambda}e^{-i\mathbf{k}\cdot\mathbf{x}}|a\rangle}{\omega_{an} - \omega_k} \\
&= \quad \frac{2\pi\alpha}{m^2}\frac{1}{V} \sum_{\mathbf{k}\lambda} \frac{1}{\omega_k} \sum_n \frac{\langle a|\mathbf{p} \cdot \mathbf{e}_{\mathbf{k}\lambda}e^{i\mathbf{k}\cdot\mathbf{x}}|n\rangle\langle n|\mathbf{p} \cdot \mathbf{e}_{\mathbf{k}\lambda}e^{-i\mathbf{k}\cdot\mathbf{x}}|a\rangle}{\omega_{an} - \omega_k} ,
\end{aligned}
\qquad (11.42)
$$

where α is the fine structure constant ($\alpha = e^2/4\pi\hbar c$ in Heaviside–Lorentz units) and we have reverted to a discrete mode summation with the quantization volume V. This is indeed the nonrelativistic QED expression for the level shift of state a due to the coupling of a bound electron to the vacuum field, as given by equation (3.54).

Exchange and Nonexchange Contributions

Both W_S [equation (11.32)] and W_D [equation (11.39)] may be divided into "exchange" and "nonexchange" parts according to the way the intermediate states j appear:

$$W_S = W_S^X + W_S^{NX} , \qquad (11.43)$$

$$W_S^X = \sum_j \delta_j A_{ajja} , \qquad (11.44)$$

$$W_S^{NX} = \sum_j (1 - \delta_j)A_{aajj} ; \qquad (11.45)$$

$$W_D = W_D^X + W_D^{NX} , \qquad (11.46)$$

$$W_D^X = \frac{\alpha}{4\pi^2} \int \frac{d^3k}{k} \sum_{\lambda=1}^{2} \sum_j \frac{C_{aj}^\lambda(-\mathbf{k})C_{ja}^\lambda(\mathbf{k})}{E_a - E_j - k\delta_j} , \qquad (11.47)$$

$$W_D^{NX} = -\frac{\alpha}{4\pi^2} \int \frac{d^3k}{k^2} \sum_{\lambda=1}^{2} \sum_j (1 - \delta_j)C_{aa}^\lambda(-\mathbf{k})C_{jj}^\lambda(\mathbf{k}), \quad (11.48)$$

where now we write k instead of ω_k for the photon energy.

It will be convenient to work with the full exchange and nonexchange energies $W^X = W_S^X + W_D^X$ and $W^{NX} = W_S^{NX} + W_D^{NX}$. For this purpose we rewrite W_S^X and W_S^{NX} in forms more closely resembling W_D^X and W_D^{NX}. We note first that

$$
\begin{aligned}
A_{k\ell mn} &= \frac{e^2}{8\pi} \int d^3x \int d^3x'\, \phi_k^*(\mathbf{x})\phi_\ell(\mathbf{x})\phi_m^*(\mathbf{x}')\phi_n(\mathbf{x}') \\
&\quad \times \frac{1}{2\pi^2} \int \frac{d^3k}{k^2} e^{i\mathbf{k}\cdot(\mathbf{x}-\mathbf{x}')} \\
&= \frac{\alpha}{4\pi^2} \int \frac{d^3k}{k^2} \int d^3x\, \phi_k^*(\mathbf{x})e^{i\mathbf{k}\cdot\mathbf{x}}\phi_\ell(\mathbf{x}) \\
&\quad \times \int d^3x'\, \phi_m^*(\mathbf{x}')e^{-i\mathbf{k}\cdot\mathbf{x}'}\phi_n(\mathbf{x}'), \qquad (11.49)
\end{aligned}
$$

where we have used the Fourier representation of $1/|\mathbf{x} - \mathbf{x}'|$. Furthermore

$$[\mathsf{H}, e^{\pm i\mathbf{k}\cdot\mathbf{x}}] = \pm\boldsymbol{\alpha}\cdot\mathbf{k}e^{\pm i\mathbf{k}\cdot\mathbf{x}} \qquad (11.50)$$

and

$$\langle m|[\mathsf{H}, e^{\pm i\mathbf{k}\cdot\mathbf{x}}]|n\rangle = (E_m - E_n)\langle m|e^{\pm i\mathbf{k}\cdot\mathbf{x}}|n\rangle , \qquad (11.51)$$

where $\mathsf{H} = \boldsymbol{\alpha}\cdot\mathbf{p}+\beta m+V(\mathbf{x})$ is the first-quantized, one-particle Hamiltonian, so that

$$(E_m - E_n)\langle m|e^{\pm i\mathbf{k}\cdot\mathbf{x}}|n\rangle = \pm k\langle m|\alpha_k e^{\pm i\mathbf{k}\cdot\mathbf{x}}|n\rangle , \qquad (11.52)$$

where $\alpha_k \equiv \boldsymbol{\alpha}\cdot\mathbf{k}/k$. After some straightforward algebra we obtain from these results

$$
\begin{aligned}
W_S^X &= \frac{\alpha}{4\pi^2} \int \frac{d^3k}{k} \sum_j \frac{1}{E_a - E_j - k\delta_j} \\
&\quad \times [C_{aj}^3(-\mathbf{k})C_{ja}^3(\mathbf{k}) - C_{aj}^0(-\mathbf{k})C_{ja}^0(\mathbf{k})] , \qquad (11.53)
\end{aligned}
$$

where we define, following (11.34),

$$C_{aj}^3(\mathbf{k}) \equiv \frac{1}{k} \int d^3x\, \phi_a^*(\mathbf{x})\boldsymbol{\alpha}\cdot\mathbf{k}\phi_j(\mathbf{x})e^{-i\mathbf{k}\cdot\mathbf{x}} , \qquad (11.54)$$

$$C_{aj}^0(\mathbf{k}) \equiv \int d^3x\, \phi_a^*(\mathbf{x})\phi_j(\mathbf{x})e^{-i\mathbf{k}\cdot\mathbf{x}} . \qquad (11.55)$$

Thus

$$
\begin{aligned}
W^X &= W_D^X + W_S^X \\
&= \frac{\alpha}{4\pi^2} \int \frac{d^3k}{k} \sum_j \frac{1}{E_a - E_j - k\delta_j} \\
&\quad \times \left[C_{aj}^1(-\mathbf{k})C_{ja}^1(\mathbf{k}) + C_{aj}^2(-\mathbf{k})C_{ja}^2(\mathbf{k}) \right. \\
&\quad \left. + C_{aj}^3(-\mathbf{k})C_{ja}^3(\mathbf{k}) - C_{aj}^0(-\mathbf{k})C_{ja}^0(\mathbf{k}) \right] \\
&\equiv \frac{\alpha}{4\pi^2} \int \frac{d^3k}{k} \sum_j \sum_{\lambda=0}^{3} {}' \frac{C_{aj}^\lambda(-\mathbf{k})C_{ja}^\lambda(\mathbf{k})}{E_a - E_j - k\delta_j} ,
\end{aligned}
\tag{11.56}
$$

where

$$
\sum_{\lambda=0}^{3} {}' f(\lambda) \equiv f(1) + f(2) + f(3) - f(0).
\tag{11.57}
$$

We have similarly

$$
\begin{aligned}
W_S^{NX} &= \sum_j (1 - \delta_j) A_{aajj} \\
&= \frac{\alpha}{4\pi^2} \int \frac{d^3k}{k} \sum_j (1 - \delta_j) C_{aa}^0(-\mathbf{k})C_{jj}^0(\mathbf{k})
\end{aligned}
\tag{11.58}
$$

and therefore

$$
\begin{aligned}
W^{NX} &= W_D^{NX} + W_S^{NX} \\
&= -\frac{\alpha}{4\pi^2} \int \frac{d^3k}{k^2} \sum_j (1 - \delta_j) \left[C_{aa}^1(-\mathbf{k})C_{jj}^1(\mathbf{k}) \right. \\
&\quad \left. + C_{aa}^2(-\mathbf{k})C_{jj}^2(\mathbf{k}) - C_{aa}^0(-\mathbf{k})C_{jj}^0(\mathbf{k}) \right].
\end{aligned}
\tag{11.59}
$$

Similarly

$$
A_{aajj} = \frac{\alpha}{4\pi^2} \int \frac{d^3k}{k^2} C_{aa}^0(-\mathbf{k})C_{jj}^0(\mathbf{k})
\tag{11.60}
$$

and $C_{jj}^3(\mathbf{k}) = 0$ from (11.52), and so

$$
W_S^{NX} = \frac{\alpha}{4\pi^2} \int \frac{d^3k}{k^2} \sum_j (1 - \delta_j) \left[C_{aa}^0(-\mathbf{k})C_{jj}^0(\mathbf{k}) - C_{aa}^3(-\mathbf{k})C_{jj}^3(\mathbf{k}) \right]
$$

$$
\tag{11.61}
$$

and

$$
\begin{aligned}
W^{NX} &= W_S^{NX} + W_D^{NX} \\
&= \frac{\alpha}{4\pi^2} \int \frac{d^3 k}{k^2} \sum_j (1 - \delta_j) \left[C_{aa}^0(-\mathbf{k}) C_{jj}^0(\mathbf{k}) \right. \\
&\quad \left. - \sum_{\lambda=1}^{3} C_{aa}^\lambda(-\mathbf{k}) C_{jj}^\lambda(\mathbf{k}) \right] .
\end{aligned}
\tag{11.62}
$$

Now

$$
\begin{aligned}
\int \frac{d^3 k}{k^2} C_{aa}^0(-\mathbf{k}) C_{jj}^0(\mathbf{k}) &= 2\pi^2 \int d^3 x \int d^3 x' \frac{\phi_a^*(\mathbf{x})\phi_a(\mathbf{x})\phi_j^*(\mathbf{x}')\phi_j(\mathbf{x}')}{|\mathbf{x} - \mathbf{x}'|} \\
&= \frac{2\pi^2}{e^2} \int d^3 x \int d^3 x' \frac{J_a^0(\mathbf{x}) J_j^0(\mathbf{x}')}{|\mathbf{x} - \mathbf{x}'|}
\end{aligned}
\tag{11.63}
$$

and

$$
\begin{aligned}
\int \frac{d^3 k}{k^2} \sum_{\lambda=1}^{3} C_{aa}^\lambda(-\mathbf{k}) C_{jj}^\lambda(\mathbf{k}) &= \frac{2\pi^2}{e^2} \sum_{\lambda=1}^{3} \int d^3 x \int d^3 x' \\
&\quad \times \frac{\phi_a^*(\mathbf{x})\alpha_\lambda\phi_a(\mathbf{x})\phi_j^*(\mathbf{x}')\alpha_\lambda\phi_j(\mathbf{x}')}{|\mathbf{x} - \mathbf{x}'|} \\
&= \frac{2\pi^2}{e^2} \int d^3 x \int d^3 x' \frac{\mathbf{J}_a(\mathbf{x}) \cdot \mathbf{J}_j(\mathbf{x}')}{|\mathbf{x} - \mathbf{x}'|} ,
\end{aligned}
\tag{11.64}
$$

where $\alpha_0 = 1; \alpha_\lambda = \boldsymbol{\alpha} \cdot \mathbf{e}_{\mathbf{k}\lambda}, \lambda = 1, 2; \alpha_3 = \mathbf{k} \cdot \boldsymbol{\alpha}/k$; and $J_a^\mu(\mathbf{x})$ and $J_j^\mu(\mathbf{x})$ are the current densities associated with states a and j. Then

$$
\begin{aligned}
W^{NX} &= \frac{1}{8\pi} \int d^3 x \int d^3 x' \sum_j (1 - \delta_j) \frac{J_{a\mu}(\mathbf{x}) J_j^\mu(\mathbf{x}')}{|\mathbf{x} - \mathbf{x}'|} \\
&= \frac{1}{4\pi} \int d^3 x \int d^3 x' \frac{J_{a\mu}(\mathbf{x}) J_{\text{neg}}^\mu(\mathbf{x}')}{|\mathbf{x} - \mathbf{x}'|} ,
\end{aligned}
\tag{11.65}
$$

where $J_{\text{neg}}^\mu(\mathbf{x})$ is the current density associated with negative-energy states.

The shift of level a is given to second order in the fine structure constant by the sum of W^X [equation (11.56)] and W^{NX} [equation (11.62) or (11.65)]. As the form (11.65) suggests, W^{NX} is associated with vacuum polarization. This term has no analog in nonrelativistic QED. W^X reduces in the nonrelativistic limit to the shift in level a due to the coupling of the electron to the vacuum electromagnetic field or alternatively, as discussed in Chapter 4, the coupling of the electron to its radiation reaction field. We will now consider a few applications of these expressions for W^X and W^{NX}.

11.4   Self-Energy of a Free Electron

We will first apply the results of the preceding section to the case of a free electron.

In this case $\mathbf{J}_{\text{neg}}(\mathbf{x}') = 0$, since in the absence of any external potential V there is no preferential direction for current associated with all the negative-energy states. Then, according to (11.65),

$$
\begin{aligned}
W^{NX} &= \frac{1}{4\pi} \int d^3x \int d^3x' \frac{J_{a0}(\mathbf{x}) J_{\text{neg}}^0(\mathbf{x}')}{|\mathbf{x} - \mathbf{x}'|} \\
&= \frac{e^2}{4\pi} \int d^3x \int d^3x' \frac{\rho_a(\mathbf{x}) \rho_{\text{neg}}(\mathbf{x}')}{|\mathbf{x} - \mathbf{x}'|} ,
\end{aligned}
\tag{11.66}
$$

which, in the language of hole theory, is just the Coulomb interaction of the free electron in state a with all the electrons of the filled Dirac sea of negative-energy states. For a free electron described by any of the plane-wave states of Section 9.2, W^{NX} is a (divergent) constant, independent of the state (momentum) of the free electron.[1] Thus we can take this unmeasurable, physically uninteresting energy of a free electron to be zero. (In the presence of an external potential, however, W^{NX} gives rise to an additional, vacuum polarization energy, as discussed in Section 11.8.)

Consider next the energy W_S^X for a free electron. Defining $\mathbf{r}$ and $\mathbf{u}$ by writing $\mathbf{x} = \mathbf{r} - \mathbf{u}/2$ and $\mathbf{x}' = \mathbf{r} + \mathbf{u}/2$ in A_{ajja}, we have

$$
W_S^X = \frac{e^2}{8\pi} \int d^3u \frac{G(\mathbf{u})}{u} ,
\tag{11.67}
$$

where

$$
G(\mathbf{u}) \equiv \sum_j \delta_j \int d^3r \phi_a^*(\mathbf{r} - \tfrac{1}{2}\mathbf{u}) \phi_j(\mathbf{r} - \tfrac{1}{2}\mathbf{u}) \phi_j^*(\mathbf{r} + \tfrac{1}{2}\mathbf{u}) \phi_a(\mathbf{r} + \tfrac{1}{2}\mathbf{u}).
\tag{11.68}
$$

---

[1] To calculate $\rho(\mathbf{x})$, for instance, it must be remembered that (11.66) was derived without the covariant normalization factor m/E_k in (11.26) and (11.27). Thus the appropriate plane-wave normalization factor is not $|A| = \sqrt{(E+m)/2m}$, as in Section 9.2, but $\sqrt{(E+m)/2E}$. Then $\rho_a(\mathbf{x}) = \text{constant}$. Alternatively, of course, we can include the covariant normalization factor in (11.26) and (11.27), use $|A| = \sqrt{(E+m)/2m}$, and obtain the same result.

For an electron at rest with spin up in the z direction we have (Section 9.2)

$$\psi_a(\mathbf{x}) = \sqrt{\frac{1}{V}} \begin{pmatrix} 1 \\ 0 \\ 0 \\ 0 \end{pmatrix}. \qquad (11.69)$$

For the positive-energy intermediate states the wave functions are given by

$$\phi_j(\mathbf{x}) = \sqrt{\frac{1}{V}} \sqrt{\frac{E_j + m}{2E_j}} \frac{1}{E_j + m} \begin{pmatrix} E_j + m \\ 0 \\ p_{jz} \\ p_{j+} \end{pmatrix} e^{i\mathbf{p}_j \cdot \mathbf{x}}, \qquad (11.70)$$

$$\phi_j(\mathbf{x}) = \sqrt{\frac{1}{V}} \sqrt{\frac{E_j + m}{2E_j}} \frac{1}{E_j + m} \begin{pmatrix} 0 \\ E_j + m \\ p_{j-} \\ -p_{jz} \end{pmatrix} e^{i\mathbf{p}_j \cdot \mathbf{x}} \qquad (11.71)$$

for spin up and down, respectively, while for the negative-energy intermediate states the corresponding wave functions are[2]

$$\phi_j(\mathbf{x}) = \sqrt{\frac{1}{V}} \sqrt{\frac{E_j + m}{2E_j}} \frac{1}{E_j + m} \begin{pmatrix} p_{jz} \\ p_{j+} \\ E_j + m \\ 0 \end{pmatrix} e^{-i\mathbf{p}_j \cdot \mathbf{x}}, \qquad (11.72)$$

$$\phi_j(\mathbf{x}) = \sqrt{\frac{1}{V}} \sqrt{\frac{E_j + m}{2E_j}} \frac{1}{E_j + m} \begin{pmatrix} p_{j-} \\ -p_{jz} \\ 0 \\ E_j + m \end{pmatrix} e^{-i\mathbf{p}_j \cdot \mathbf{x}}. \qquad (11.73)$$

We have introduced the factor $\sqrt{1/V}$ in order to have the normalization $\int d^3x \, \phi^*(\mathbf{x})\phi(\mathbf{x}) = 1$ for $\phi_a(\mathbf{x})$ and $\phi_j(\mathbf{x})$. Using these results, we calculate

$$\phi_a^*(\mathbf{r} - \tfrac{1}{2}\mathbf{u})\phi_j(\mathbf{r} - \tfrac{1}{2}\mathbf{u}) = \frac{1}{V} \sqrt{\frac{E_j + m}{2E_j}} e^{i\mathbf{p}_j \cdot (\mathbf{r} - \mathbf{u}/2)} \qquad (11.74)$$

for positive-energy states j with spin up and $\phi_a^*(\mathbf{r} - \tfrac{1}{2}\mathbf{u})\phi_j(\mathbf{r} - \tfrac{1}{2}\mathbf{u}) = 0$ for positive-energy states with spin down. For negative-energy states j with

---

[2] Recall that $E_j = +\sqrt{\mathbf{p}_j^2 + m^2}$ in the definitions of these spinors.

spin up and down we obtain respectively

$$
\phi_a^*(\mathbf{r} - \frac{1}{2}\mathbf{u})\phi_j(\mathbf{r} - \frac{1}{2}\mathbf{u}) = \frac{1}{V}\sqrt{\frac{E_j + m}{2E_j}}\frac{p_{jz}}{E_j + m}e^{-i\mathbf{p}_j\cdot(\mathbf{r}-\mathbf{u}/2)} \quad (11.75)
$$

and

$$
\phi_a^*(\mathbf{r} - \frac{1}{2}\mathbf{u})\phi_j(\mathbf{r} - \frac{1}{2}\mathbf{u}) = \frac{1}{V}\sqrt{\frac{E_j + m}{2E_j}}\frac{p_{j-}}{E_j + m}e^{-i\mathbf{p}_j\cdot(\mathbf{r}-\mathbf{u}/2)} \ . \quad (11.76)
$$

Then

$$
\begin{aligned}
G(\mathbf{u}) &= \frac{1}{V^2}\int d^3r \sum_j{}' \frac{E_j + m}{2E_j}\left[e^{-i\mathbf{p}_j\cdot\mathbf{u}} - \frac{\mathbf{p}_j^2}{(E_j + m)^2}e^{i\mathbf{p}_j\cdot\mathbf{u}}\right] \\
&= \frac{1}{V}\sum_j{}' \frac{E_j + m}{2E_j}\left[1 - \frac{E_j - m}{E_j + m}\right]e^{i\mathbf{p}_j\cdot\mathbf{u}} \\
&= \frac{1}{V}\sum_j{}' \frac{m}{E_j}e^{i\mathbf{p}_j\cdot\mathbf{u}} \ , \quad (11.77)
\end{aligned}
$$

where the prime on the summation symbol implies that the sum over spin states has already been performed.

We now go to the continuum limit and replace the sum over momenta $\mathbf{p}_j$ in (11.77) by an integral, so that $\sum_j{}' \to (V/8\pi^3 \int d^3p$, the replacement familiar from the continuous summation over electromagnetic field modes:

$$
G(\mathbf{u}) = \frac{m}{8\pi^3}\int d^3p\frac{e^{i\mathbf{p}\cdot\mathbf{u}}}{E(\mathbf{p})} = \frac{m}{8\pi^3}\int d^3p\frac{e^{i\mathbf{p}\cdot\mathbf{u}}}{\sqrt{p^2 + m^2}} \ . \quad (11.78)
$$

Then, from (11.67),

$$
\begin{aligned}
W_S^X &= \frac{e^2}{8\pi}\int \frac{d^3u}{u}\frac{m}{8\pi^3}\int d^3p\frac{e^{i\mathbf{p}\cdot\mathbf{u}}}{\sqrt{p^2 + m^2}} \\
&= \frac{me^2}{64\pi^4}\int \frac{d^3p}{\sqrt{p^2 + m^2}}\int d^3u\frac{e^{i\mathbf{p}\cdot\mathbf{u}}}{u} \\
&= \frac{me^2}{16\pi^3}\int \frac{d^3p}{p^2\sqrt{p^2 + m^2}} = \frac{me^2}{4\pi^2}\int_0^\infty \frac{dp}{\sqrt{p^2 + m^2}} \ . \quad (11.79)
\end{aligned}
$$

Thus W_S^X is logarithmically divergent (Weisskopf, 1939). Replacing the upper limit by Λ ($\Lambda \to \infty$), we write

$$
W_S^X = \frac{me^2}{4\pi^2}\log\frac{\Lambda + \sqrt{\Lambda^2 + m^2}}{m} \to \frac{m\alpha}{\pi}\log\frac{\Lambda}{m} \quad (\Lambda \to \infty) \quad (11.80)
$$

In nonrelativistic theory the negative-energy contributions to (11.77) are absent, so that

$$G(u) \quad \frac{1}{V} \sum_j' \frac{E_j + m}{2E_j} e^{i\mathbf{p}_j \cdot \mathbf{u}} \simeq \frac{1}{V} \sum_j' e^{i\mathbf{p} \cdot \mathbf{u}} - \frac{1}{8\pi^3} \int d^3p\, e^{i\mathbf{p} \cdot \mathbf{u}} - \delta^3(u)$$

(11.81)

and

$$W_S^X \to \frac{e^2}{8\pi} \int d^3u \frac{\delta^3(\mathbf{u})}{u} ,$$

(11.82)

which, when expressed in Gaussian units, is identical to the electrostatic energy (5.16) calculated nonrelativistically for a point charge with charge density $\rho(\mathbf{x}) = e\delta^3(\mathbf{x})$. Evidently relativistic QED effectively "spreads out" the electron charge compared with the nonrelativistic distribution $e\delta^3(\mathbf{x})$. This is discussed in the following section. The result is that, whereas the electrostatic energy of the electron diverges linearly in nonrelativistic QED, the divergence in relativistic QED is "only" logarithmic.

Let us now turn our attention to the "dynamical" contribution to the energy of a free electron at rest. From (11.47) with $E_a = m$ for an electron at rest,

$$
\begin{aligned}
W_D^X &= \frac{\alpha}{4\pi^2} \int \frac{d^3k}{k} \sum_{\lambda=1}^2 \sum_j \frac{\langle a|\alpha_\lambda e^{i\mathbf{k}\cdot\mathbf{x}}|j\rangle \langle j|\alpha_\lambda e^{-i\mathbf{k}\mathbf{x}}|a\rangle}{m - E_j - k\delta_j} \\
&= \frac{\alpha}{4\pi^2} \frac{V}{8\pi^3} \int \frac{d^3k}{k} \int d^3p \sum_{\lambda=1}^2 \sum_{s=1}^2 \\
&\quad \times \left[\frac{\langle a|\alpha_\lambda e^{i\mathbf{k}\cdot\mathbf{x}}|\mathbf{p},+,s\rangle \langle \mathbf{p},+,s|\alpha_\lambda e^{-i\mathbf{k}\cdot\mathbf{x}}|a\rangle}{m - E_p - k} \right. \\
&\quad \left. + \frac{\langle a|\alpha_\lambda e^{i\mathbf{k}\cdot\mathbf{x}}|\mathbf{p},-,s\rangle \langle \mathbf{p},-,s|\alpha_\lambda e^{-i\mathbf{k}\cdot\mathbf{x}}|a\rangle}{m + E_p + k} \right],
\end{aligned}
$$

(11.83)

where $E_p = +\sqrt{\mathbf{p}^2 + m^2}$ and $|\mathbf{p}, \pm, s\rangle$ is a free-electron state of energy $\pm E_p$, momentum $\mathbf{p}$, and spin s "up" or "down" along the direction of $\mathbf{p}$. Since

$$|a\rangle = \sqrt{\frac{1}{V}} \begin{pmatrix} 1 \\ 0 \\ 0 \\ 0 \end{pmatrix} \equiv \sqrt{\frac{1}{V}} u_a$$

(11.84)

and

$$|\mathbf{p}, \pm, s\rangle = \sqrt{\frac{1}{V}} \sqrt{\frac{1}{2E_p(E_p + m)}} w_{\mathbf{p},\pm}^{(s)} e^{\pm i\mathbf{p}\cdot\mathbf{x}} , \qquad (11.85)$$

where the column vectors $w_{\mathbf{p},\pm}^{(s)}$ are those appearing in (11.70)–(11.73), it follows that

$$\langle a|\alpha_\lambda e^{i\mathbf{k}\cdot\mathbf{x}}|\mathbf{p}, \pm, s\rangle = \frac{1}{V}\sqrt{\frac{1}{2E_p(E_p + m)}}\{u_a^\dagger \alpha_\lambda w_{\mathbf{p},\pm}^{(s)}\} \times \int d^3 x e^{i(\mathbf{k}\pm\mathbf{p})\cdot\mathbf{x}}$$

$$= \frac{8\pi^3}{V}\sqrt{\frac{1}{2E_p(E_p + m)}}\{u_a^\dagger \alpha_\lambda w_{\mathbf{p},\pm}^{(s)}\}\delta^3(\mathbf{k}\pm\mathbf{p}), \qquad (11.86)$$

and

$$\langle a|\alpha_\lambda e^{i\mathbf{k}\cdot\mathbf{x}}|\mathbf{p}, \pm, s\rangle\langle\mathbf{p}, \pm, s|\alpha_\lambda e^{-i\mathbf{k}\cdot\mathbf{x}}|a\rangle = \frac{8\pi^3}{V}\left[\frac{1}{2E_p(E_p + m)}\right]$$

$$= |u_a^\dagger \alpha_\lambda w_{\mathbf{p},\pm}^{(s)}|^2 \delta^3(\mathbf{k}\pm\mathbf{p})\int d^3 x e^{-i(\mathbf{k}\pm\mathbf{p})\cdot\mathbf{x}}$$

$$= \frac{8\pi^3}{V}\left[\frac{1}{2E_p(E_p + m)}\right]|u_a^\dagger \alpha_\lambda w_{\mathbf{p},\pm}^{(s)}|^2 \delta^3(\mathbf{k}\pm\mathbf{p}).$$

$$(11.87)$$

Then

$$W_D^X = -\frac{\alpha}{4\pi^2}\int\frac{d^3k}{k}\frac{1}{2E_k(E_k + m)}\sum_{\lambda=1}^{2}\sum_{s=1}^{2}$$

$$\times\left[\frac{|u_a^\dagger \alpha_\lambda w_{-\mathbf{k},+}^{(s)}|^2}{E_k + k - m} - \frac{|u_a^\dagger \alpha_\lambda w_{\mathbf{k},-}^{(s)}|^2}{E_k + k + m}\right], \qquad (11.88)$$

with $E_k = \sqrt{k^2 + m^2}$. Using equations (11.70)–(11.73) for the column vectors $w_{\mathbf{p},\pm}^{(s)}$, we calculate by simple matrix algebra the spin and polarization sums

$$\sum_{\lambda=1}^{2}\sum_{s=1}^{2}|u_a^\dagger \alpha_\lambda w_{-\mathbf{k},+}^{(s)}|^2 = 2k^2 , \qquad (11.89)$$

$$\sum_{\lambda=1}^{2}\sum_{s=1}^{2}|u_a^\dagger \alpha_\lambda w_{\mathbf{k},-}^{(s)}|^2 = 2(E_k + m)^2 , \qquad (11.90)$$

and consequently

$$
\begin{aligned}
W_D^X &= -\frac{\alpha}{4\pi^2} \int \frac{d^3k}{k} \frac{1}{E_k(E_k+m)} \left[\frac{k^2}{E_k+k-m} - \frac{(E_k+m)^2}{E_k+k+m} \right] \\
&\quad - \frac{\alpha}{4\pi^2} \int \frac{d^3k}{k} \frac{1}{E_k(E_k+m)} \left[\frac{2km(E_k+m)}{2k(E_k+k)} \right] \\
&= \frac{m\alpha}{4\pi^2} \int_0^\infty \frac{4\pi k^2 dk}{k} \frac{1}{E_k(E_k+m)} \\
&= \frac{m\alpha}{\pi} \int_0^\infty dk\, k\frac{1}{k} \left(\frac{1}{E_k} - \frac{1}{E_k+k} \right) \\
&= \frac{m\alpha}{\pi} \left[\int_0^\infty \frac{dk}{\sqrt{k^2+m^2}} - \int_0^\infty \frac{dk}{k+\sqrt{k^2+m^2}} \right] \\
&\to \frac{m\alpha}{2\pi} \log\frac{\Lambda}{m} \quad (\Lambda \to \infty),
\end{aligned}
\tag{11.91}
$$

where, as in the calculation of W_S^X, we have cut off the upper integration limits at $k = \Lambda$.

We showed in the preceding section how W_D^X reduces in the appropriate limit to the nonrelativistic self-energy discussed in Chapter 3. In the case of a free electron the nonrelativistic self-energy, including retardation, is given by (3.59):

$$
\Delta E^{\text{free}} = \delta m = \frac{16\alpha}{3} \int_0^\infty \frac{dk\, k}{k+k^2/2m} \to \frac{32}{3} m\alpha \log\frac{\Lambda}{m} \quad (\Lambda \to \infty) \tag{11.92}
$$

in the present units ($\hbar = c = 1$, $\alpha = e^2/4\pi$) and with the momentum cutoff Λ.

In relativistic theory the complete self-energy is given by the sum of (11.80) and (11.91):

$$
\Delta E^{\text{free}} = \delta m = W_S^X + W_D^X = \frac{3m\alpha}{2\pi} \log\frac{\Lambda}{m} . \tag{11.93}
$$

This result was first reported by Weisskopf (1939). In the following chapter we will derive it again using the methods of covariant perturbation theory and Feynman diagrams.

Note that for both W_S^X and W_D^X the divergence would be linear rather than logarithmic if the negative-energy states were excluded.[3] For both

---

[3] This was noted following (11.82) for W_S^X. For W_D^X the omission of negative-energy states is equivalent to dropping the second term in brackets in the first line of (11.91) and is easily seen to cause a linear divergence.

terms, furthermore, an addition of positive- and negative-energy contributions, rather than the correct subtraction, makes the divergence linear
rather than logarithmic. A linear divergence is traditionally associated with
nonrelativistic theory.[4] In this connection it is perhaps of interest to recall
some remarks of Weisskopf (1989):

> Pauli asked me to calculate the self-energy of the electron on the
> basis of the positron theory to see if this energy is less divergent in
> that theory. I found that it diverges equally badly and I published
> this result. A few weeks after the publication I received a letter from
> Wendell Furry, who worked with Oppenheimer at the time, informing
> me that I had made a simple mistake of a sign in my calculation. If
> it is done correctly, the divergence is only logarithmic. The positron
> theory improved things considerably, in contradiction to my paper. I
> was down and depressed to have made and published a silly mistake
> in such a fundamental problem! I went to Pauli and said that I
> wanted to give up physics, that I would never survive this blemish.
> Pauli tried to console me: He said, "Don't take it too seriously, many
> people published wrong papers; I never did!"
>
> What followed shows how decent the relations between physicists
> were at that time. I asked Furry by letter to publish his result under
> his name or at least to coauthor a paper correcting the mistake.
> But Furry was a gentleman. He answered, no, I should publish a
> correction in my name only and mention him as the person who drew
> my attention to the error. Since then, the logarithmic divergence of
> the self-energy of the electron goes with my name and not Furry's
> ...

11.5   How Big Is an Electron?

The free electron electrostatic energy $W_S = W_S^X$ calculated in the preceding
section derives from the expression [equation (11.28)]

$$
\begin{aligned}
\langle a|H_S|a\rangle &= \frac{e^2}{8\pi}\int d^3x \int d^3x' \frac{1}{|\mathbf{x}-\mathbf{x}'|}\langle a|\rho(\mathbf{x})\rho(\mathbf{x}')|a\rangle \\
&= \frac{e^2}{8\pi}\int \frac{d^3u}{u}\int d^3r\langle a|\rho(\mathbf{r}-\tfrac{1}{2}\mathbf{u})\rho(\mathbf{r}+\tfrac{1}{2}\mathbf{u})|a\rangle \\
&= \frac{e^2}{8\pi}\int \frac{d^3u}{u}\int d^3r\langle a|\rho(\mathbf{r})\rho(\mathbf{r}+\mathbf{u})|a\rangle .
\end{aligned}
\tag{11.94}
$$

---

[4] *Nonrelativistic theory* here means that retardation is neglected. See the penultimate
paragraph of Section 3.9.

Comparison with (11.67) suggests the interpretation of $G(\mathbf{u})$ as the probability for finding charge at two points separated by the distance $|\mathbf{u}|$ within a charge distribution described by the charge density $e\rho(\mathbf{r})$.[5] Let us accept this interpretation and see what it says about the charge "distribution" for the electron.

$G(\mathbf{u})$ is given by (11.78), and can be evaluated as follows:

$$
\begin{aligned}
G(\mathbf{u}) &= \frac{m}{8\pi^3} \int d^3p \frac{e^{i\mathbf{p}\cdot\mathbf{u}}}{\sqrt{p^2 + m^2}} \\
&= \frac{m}{8\pi^3} 2\pi \int_0^\infty \frac{dp\, p^2}{\sqrt{p^2 + m^2}} \int_0^\pi d\theta \sin\theta e^{ipu\cos\theta} \\
&= \frac{m}{2\pi^2}\frac{1}{u} \int_0^\infty \frac{dp\, p\sin pu}{\sqrt{p^2 + m^2}} = -\frac{m}{2\pi^2}\frac{1}{u}\frac{\partial}{\partial u} \int_0^\infty \frac{dp \cos pu}{\sqrt{p^2 + m^2}} \\
&= \frac{m}{4\pi i}\frac{1}{u}\frac{\partial}{\partial u} H_0^{(1)}(imu), \qquad\qquad\qquad\qquad (11.95)
\end{aligned}
$$

where $H_0^{(1)}$ is the zero-order Hankel function of the first kind. For $mu \ll 1$, i.e., for distances u small compared with the Compton wavelength m^{-1} $(= \hbar/mc)$,

$$
G(\mathbf{u}) \cong \frac{m}{2\pi^2}\frac{1}{u^2} \quad (u \ll m^{-1}). \qquad\qquad (11.96)
$$

For $mu \gg 1$,

$$
G(\mathbf{u}) \cong \left(\frac{m}{2\pi u}\right)^{3/2} e^{-mu} \quad (mu \gg 1). \qquad\qquad (11.97)
$$

We can go a bit further and construct an effective charge density $\rho_{\text{eff}}(\mathbf{r})$ such that [cf. (11.94) and (11.67)]

$$
\int d^3r \rho_{\text{eff}}(\mathbf{r})\rho_{\text{eff}}(\mathbf{r}+\mathbf{u}) = G(\mathbf{u}) = \frac{m}{8\pi^3} \int d^3p \frac{e^{i\mathbf{p}\cdot\mathbf{u}}}{E(\mathbf{p})}, \qquad (11.98)
$$

Writing $\rho_{\text{eff}}(\mathbf{r}) = \int d^3\hat\rho(\mathbf{p})e^{i\mathbf{p}\cdot\mathbf{r}}$, and solving (11.98) for $\hat\rho(\mathbf{p})$, we find

$$
\begin{aligned}
\rho_{\text{eff}}(\mathbf{r}) &= \frac{m^{1/2}}{8\pi^3} \int d^3p \frac{e^{i\mathbf{p}\cdot\mathbf{r}}}{\sqrt{E(\mathbf{p})}} = -\frac{m}{2\pi^2}\frac{1}{r}\frac{\partial}{\partial r} \int_0^\infty \frac{dp \cos pr}{[p^2 + m^2]^{1/4}} \\
&\cong 2^{-5/2}\pi^{-3/2}mr^{-5/2} \qquad\qquad\qquad\qquad\qquad\qquad (11.99)
\end{aligned}
$$

for $r \ll m^{-1}$; for $r \gg m^{-1}$, $\rho_{\text{eff}}(\mathbf{r})$ decreases exponentially as a function of r (Weisskopf, 1939).

---

[5] Of course ρ in (11.94) is an operator.

These results indicate that relativistic effects act to "spread out" the electron: whereas the nonrelativistic theory, with negative-energy contributions absent and $E(\mathbf{p}) \cong m$, gives $G(\mathbf{u}) = \delta^3(\mathbf{u})$ [equation (11.81)], the relativistic theory introduces a natural length scale m^{-1}, the Compton wavelength, such that the effective electron charge distribution decreases exponentially with mr. Like the nonrelativistic theory, relativistic QED gives an infinite charge density at $r = 0$. However, the divergence is weaker in the relativistic theory with $G(\mathbf{u}) \propto u^{-2}$ near $u = 0$ rather than the behavior $G(\mathbf{u}) = \delta^3(\mathbf{u})$ predicted nonrelativistically. The divergence of $G(\mathbf{u})$ at $\mathbf{u} = 0$ is responsible for the divergence of W_S^X:

$$
\begin{aligned}
W_S^X &= \frac{e^2}{8\pi} \int \frac{d^3u}{u} G(\mathbf{u}) \\
&= \frac{e^2}{8\pi} 4\pi \left[\int_0^{u_1} \frac{du\, u^2}{u} G(\mathbf{u}) + \int_{u_1}^{\infty} \frac{du\, u^2}{u} G(\mathbf{u}) \right] \\
&= \frac{e^2}{2} \left[\int_0^{u_1} \frac{du\, u^2}{u} \frac{m}{2\pi^2} \frac{1}{u^2} + \int_{u_1}^{\infty} \frac{du\, u^2}{u} G(\mathbf{u}) \right], \quad (11.100)
\end{aligned}
$$

where u_1 is small compared with m^{-1} but otherwise arbitrary. The second term in (11.100) converges, whereas the first term diverges owing to the lower integration limit $u = 0$. Let us replace the lower integration limit in the first term by some very small length $u_m = \Lambda^{-1}$. Neglecting the convergent second term, and choosing $u_1 = am^{-1}$, where $a \sim 1$, we have

$$
W_S^X \rightarrow \frac{me^2}{4\pi^2} \int_{\Lambda^{-1}}^{am^{-1}} \frac{du}{u} = \frac{me^2}{4\pi^2} \log \frac{am^{-1}}{\Lambda^{-1}} = \frac{m\alpha}{\pi} \log \frac{\Lambda}{m} \quad (\Lambda \rightarrow \infty).
$$
$$(11.101)$$

This is indeed the same as (11.80) and shows that the singularity of $G(\mathbf{u})$ at $u = 0$ is responsible for the divergence of the electrostatic self-energy of the electron. The point is that, although relativistic QED does in effect give a spread-out electron, with the length m^{-1} as the characteristic spread, the function $G(\mathbf{u})$ still diverges badly enough at $u = 0$ to make W_S^X infinite.

We can infer from results such as (11.97) that, as a consequence of vacuum fluctuations, an electron in some respects behaves as though it is spread out over a distance on the order of its Compton wavelength, $m^{-1} = 3.86 \times 10^{-11}$ cm. This distance is about 137 times larger than the classical electron radius r_0 that arises from classical considerations when one supposes that the observed electron mass is entirely electromagnetic (Chapter 5). As we have emphasized in earlier chapters, relativistic QED makes such classical considerations largely irrelevant, for any attempt to localize the electron to within a distance $\sim r_0$ introduces vacuum fluctuations associated with electron–positron pair creation, making it impossible to probe distances $r_0 \ll m^{-1}$.

The spread of the electron charge distribution just considered is associated with electrostatic self-energy and involves the electron–positron Dirac vacuum. The interaction of the electron with the electromagnetic vacuum field is also modified by relativistic effects. In connection with Welton's interpretation of the Lamb shift (Section 3.6) we calculated the mean-square displacement

$$\langle (\Delta \mathbf{r})^2 \rangle = \frac{2\alpha}{\pi m^2} \int_0^\infty \frac{dk}{k} \qquad (11.102)$$

due to the coupling of an unbound electron to the vacuum field. A relativistic calculation modifies the integrand of (11.102) by a factor $(m/k)^2$ for $k \gg 2mc^2$:[6]

$$\langle (\Delta \mathbf{r})^2 \rangle \rightarrow \frac{2\alpha}{\pi m^2} \int_0^{k_m} \frac{dk}{k} + \frac{2\alpha m}{\pi} \int_{k_m}^\infty \frac{dk}{k^3} \rightarrow \frac{2\alpha}{\pi m^2} \log \frac{k_m}{k_0} , \qquad (11.103)$$

where $k_m \sim m$ and k_0 is a low-energy cutoff on the order of the binding energy of a bound electron.[7] Thus relativistic QED gives a convergent result for $\langle (\Delta \mathbf{r})^2 \rangle$, with root-mean-square displacement $\langle (\Delta \mathbf{r})^2 \rangle^{1/2} \sim m^{-1}$ if we take $k_m \sim m = 0.51$ MeV and $k_0 \sim 13.6$ eV.

It should be emphasized that the "spread" of the electron associated with relativistic effects does not alter the fact that the electron in QED is regarded as a pure point particle. The "spread" is associated with quantum fluctuations in the position of the point electron: the electron jiggles around as a consequence of vacuum fluctuations.

High-energy scattering experiments probing small distances indicate that the electron, if it is not a point, is certainly no larger than about 10^{-15} cm. A simple application of the uncertainty principle then suggests that any constituent particles within the electron, analogous to the quark constituents of hadrons, would have to have kinetic energies $> p = 10^{15}$ cm$^{-1} \sim 20$ GeV, which is about four orders of magnitude greater than the electron rest energy. It seems improbable, then, that the electron has any "structure."

11.6 Mass Renormalization

Because we considered specifically an electron at rest, we can regard ΔE^{tree} in (11.93) as the electromagnetic mass δm, as of course we have already

---

[6] The corresponding "reduction factor" compared with nonrelativistic theory in (11.91) is m/k for $k \gg m$.

[7] See the remark following equation (3.41).

done. The observed mass is evidently

$$m = m_o + \delta m = m_o + \frac{3m_o\alpha}{2\pi} \log \frac{\Lambda}{m_o} , \qquad (11.104)$$

where m_o is the bare mass (Sections 3.5 and 5.3). That is, δm is the part of the electron mass attributable to its interaction with the electromagnetic field, whereas m_o is of nonelectromagnetic origin. We have replaced m by m_o in the expression for the electromagnetic mass, recognizing now that the mass in the Hamiltonian *without* coupling to the electromagnetic field should properly have been taken to be the bare mass rather than the actual, observed electron mass m.

Equation (11.104) presents a difficulty: the observed mass seems to depend on the arbitrary choice of the cutoff Λ. However, since the bare mass, like the electromagnetic mass, is not by itself observable, we can argue that it can depend on Λ, and in such a way that m is in fact independent of Λ. Thus we suppose that $m_o = m_o(\Lambda)$ and require that $dm/d\Lambda = 0$. Equation (11.104) then yields

$$\frac{dm_o}{d\Lambda} \cong -\frac{3\alpha}{2\pi} \frac{m_o}{\Lambda} \qquad (11.105)$$

to order α. The solution of this equation to order α, subject to the "boundary condition" (11.104), is

$$m_o(\Lambda) \cong m \left[1 - \frac{3\alpha}{2\pi} \log \frac{\Lambda}{m} \right] \qquad (11.106)$$

to order α.

To see what this accomplishes, consider now the mass term in the original Hamiltonian without coupling to the field:

$$\int d^3x \psi^\dagger \beta m_o \psi \cong \int d^3x \psi^\dagger \beta m\psi - \frac{3m\alpha}{2\pi} \int d^3x \psi^\dagger \beta \log \frac{\Lambda}{m} \psi$$

$$\equiv \int d^3x \psi^\dagger \beta m\psi + H_{\text{counter}} . \qquad (11.107)$$

This means that we can use the *observed* mass m in the Hamiltonian, provided that we include an additional "counter term,"

$$H_{\text{counter}} = -\frac{3m\alpha}{2\pi} \int d^3x \psi^\dagger \beta \log \frac{\Lambda}{m} \psi = -\delta m \int d^3x \psi^\dagger \beta \psi , \qquad (11.108)$$

in the Hamiltonian. And now when we calculate ΔE^{free} to order α as before, using the observed mass in the uncoupled Hamiltonian, we obtain an energy shift given by the last term in (11.104) with $m_o \to m$, plus the shift due to the counter term:

$$\Delta E^{\text{free}} \to \frac{3m\alpha}{2\pi} \log \frac{\Lambda}{m} - \frac{3m\alpha}{2\pi} \log \frac{\Lambda}{m} = 0. \qquad (11.109)$$

Thus, to order α, we can eliminate the infinity associated with electromagnetic mass by an appropriate reconstruction of the Hamiltonian.

Obviously this is just a way to "hide" an infinity, a different way of performing mass renormalization. We are still left with the fact that $m_o \to -\infty$ in the limit $\Lambda \to \infty$ of no cutoff. However, we cannot say for sure whether this is a real problem. When we go to higher orders in α in perturbation theory, we find that the electron charge, like the mass, must be renormalized, that the bare charge e_o must depend on Λ, and that $e_o \to \infty$ as $\Lambda \to \infty$. In particular, the fine structure constant $\alpha = e_o^2/4\pi$ must itself be renormalized, and as a function of Λ it can exceed unity — at which point the whole perturbation procedure breaks down and we cannot sensibly compute m_o in the manner described previously.

It seems fair to say, though, that there is nothing wrong with the renormalization procedure per se: we will have to renormalize (or use a counter term in the Hamiltonian) even if m_o and δm are found in a future theory to be finite (Section 3.5). In a sense, then, the present-day formalism of renormalization is a way to obtain correct answers without having to know the bare masses and charges, be they finite or infinite.

11.7 The Lamb Shift

We have already given nonrelativistic descriptions of the $2s_{1/2} - 2p_{1/2}$ Lamb shift in hydrogen in Chapter 3. We now consider, using old-fashioned perturbation theory, the relativistic QED theory of the Lamb shift (Kroll and Lamb, 1949; French and Weisskopf, 1949). The expressions derived in Section 11.3 allow us, in principle, to calculate the energy shift of any electron state $|a\rangle$ to the lowest order in perturbation theory:

$$W(a) = W^X(a) + W^{NX}(a) = W^X_S(a) + W^X_D(a) + W^{NX}(a). \qquad (11.110)$$

Consider first

$$
\begin{aligned}
W^X_S(a) &= \sum_j \delta_j A_{ajja} = \frac{\alpha}{4\pi} \int \frac{d^3k}{k^2} \sum_j \delta_j \langle a|e^{i\mathbf{k}\cdot\mathbf{x}}|j\rangle\langle j|e^{i\mathbf{k}\cdot\mathbf{x}}|a\rangle \\
&= \frac{\alpha}{4\pi^2} \int \frac{d^3k}{k^2} \sum_j \delta_j \langle a|e^{-i\mathbf{k}\cdot\mathbf{x}}|j\rangle\langle j|e^{i\mathbf{k}\cdot\mathbf{x}}|a\rangle ,
\end{aligned}
\qquad (11.111)
$$

where we have written A_{ajja} as defined by (11.49) in bra/ket notation. Now

$$
\begin{aligned}
\sum_j \delta_j \langle a|e^{-i\mathbf{k}\cdot\mathbf{x}}|j\rangle\langle j|e^{i\mathbf{k}\cdot\mathbf{x}}|a\rangle
&= \sum_j \langle a|e^{-i\mathbf{k}\cdot\mathbf{x}}\frac{H}{|H|}|j\rangle\langle j|e^{i\mathbf{k}\cdot\mathbf{x}}|a\rangle \\
&= \langle a|e^{-i\mathbf{k}\cdot\mathbf{x}}\frac{H}{|H|}\left(\sum_j |j\rangle\langle j|\right) e^{i\mathbf{k}\cdot\mathbf{x}}|a\rangle \\
&= \langle a|e^{-i\mathbf{k}\cdot\mathbf{x}}\frac{H}{|H|}e^{i\mathbf{k}\cdot\mathbf{x}}|a\rangle .
\end{aligned}
\tag{11.112}
$$

Here we have used the completeness relation $\sum_j |j\rangle\langle j| = 1$ and have defined $|H|$ as the operator having the same energy spectrum as the unperturbed electron Hamiltonian H, except that all its eigenvalues are taken to be positive. Thus we can define $|H| = \sqrt{H^2}$. Using the general operator identity

$$
e^{-i\mathbf{k}\cdot\mathbf{x}}F(\mathbf{p}, V)e^{i\mathbf{k}\cdot\mathbf{x}} = F(\mathbf{p}+\mathbf{k}, V)
\tag{11.113}
$$

for any function (such as $H/|H|$) having a series expansion in $\mathbf{p}$ and V, we may write

$$
W_S^X(a) = \frac{\alpha}{4\pi^2}\int \frac{d^3k}{k^2}\langle a|\left[\frac{H}{|H|}\right]_{\mathbf{p}+\mathbf{k}}|a\rangle ,
\tag{11.114}
$$

where $[F(\mathbf{p}, V)]_{\mathbf{p}+\mathbf{k}} \equiv F(\mathbf{p}+\mathbf{k}, V)$. Thus $[H]_{\mathbf{p}+\mathbf{k}}$, for instance, is simply the original unperturbed Hamiltonian $\boldsymbol{\alpha}\cdot\mathbf{p} + \beta m + V$ with the *c-number* $\mathbf{k}$ added to the operator $\mathbf{p}$. The evaluation of (11.114), then, reduces essentially to the evaluation of $1/|H|_{\mathbf{p}+\mathbf{k}}$ in terms of operators whose expectation values in the state $|a\rangle$ can be calculated.

To this end we follow Kroll and Lamb and write

$$
\frac{1}{|H|_{\mathbf{p}+\mathbf{k}}} = \frac{1}{E_k + |H|_{\mathbf{p}+\mathbf{k}} - E_k} \equiv \frac{1}{E_k + \Delta_k} ,
\tag{11.115}
$$

with $E_k \equiv (k^2 + m^2)^{1/2}$ and $\Delta_k \equiv |H|_{\mathbf{p}+\mathbf{k}} - E_k$ and assume the validity (see later) of the expansion

$$
\frac{1}{|H|_{\mathbf{p}+\mathbf{k}}} = \frac{1}{E_k} - \frac{\Delta_k}{E_k^2} + \frac{\Delta_k^2}{E_k^3} - \cdots .
\tag{11.116}
$$

To obtain an expression for the operator Δ_k, we note first that

$$
|H|_{\mathbf{p}+\mathbf{k}} = (H^2)^{1/2}_{\mathbf{p}+\mathbf{k}} = [(\boldsymbol{\alpha}\cdot\mathbf{p} + \beta m + V)^2]^{1/2}_{\mathbf{p}+\mathbf{k}}
$$

$$
\begin{aligned}
&= [\mathbf{p}^2 + \mathbf{k}^2 + m^2 + 2\mathbf{k}\cdot\mathbf{p} + V^2 + \boldsymbol{\alpha}\cdot(\mathbf{p}+\mathbf{k})V \\
&\quad + V\boldsymbol{\alpha}\cdot(\mathbf{p}+\mathbf{k}) + 2\beta mV]^{1/2} \\
&= [E_k^2 + \mathbf{p}^2 + 2\mathbf{k}\cdot\mathbf{p} + V^2 + 2V(\boldsymbol{\alpha}\cdot\mathbf{p} + \boldsymbol{\alpha}\cdot\mathbf{k} + \beta m) \\
&\quad + \boldsymbol{\alpha}\cdot\boldsymbol{\pi}V]^{1/2} , \tag{11.117}
\end{aligned}
$$

where $\boldsymbol{\alpha}\cdot\boldsymbol{\pi}V \equiv \boldsymbol{\alpha}\cdot\mathbf{p}V - V\boldsymbol{\alpha}\cdot\mathbf{p}$, i.e., $\boldsymbol{\pi}$ denotes the operation of $\mathbf{p}$ on the function immediately to its right. Thus

$$
\Delta_k = |H|_{\mathbf{p}+\mathbf{k}} - E_k = (E_k^2 + \delta_k)^{1/2} - E_k = \frac{\delta_k}{2E_k} - \frac{\delta_k^2}{8E_k^3} + \frac{\delta_k^3}{16E_k^5} - \dots , \tag{11.118}
$$

with

$$
\delta_k \equiv \mathbf{p}^2 + 2\mathbf{k}\cdot\mathbf{p} + V^2 + 2V(\boldsymbol{\alpha}\cdot\mathbf{p} + \boldsymbol{\alpha}\cdot\mathbf{k} + \beta m) + \boldsymbol{\alpha}\cdot\boldsymbol{\pi}V. \tag{11.119}
$$

Equations (11.116), (11.118), and (11.119) allow us to evaluate $W_S^X(a)$ in terms of expectation values in state $|a\rangle$ of operators $\mathbf{p}^2, V^2, V\boldsymbol{\alpha}\cdot\mathbf{p}, \beta m, \boldsymbol{\alpha}\cdot\boldsymbol{\pi}V, \dots$, each such expectation value being multiplied by an integral over k as indicated in (11.114). A similar calculation can be made for $W_D^X(a)$. First we write $W_D^X(a)$ in a form resembling (11.114):

$$
\begin{aligned}
W_D^X(a) &= \frac{\alpha}{4\pi^2} \int \frac{d^3k}{k} \sum_{\lambda=1}^{2} \sum_{j} \frac{\langle a|\alpha_\lambda e^{-i\mathbf{k}\cdot\mathbf{x}}|j\rangle\langle j|\alpha_\lambda e^{i\mathbf{k}\cdot\mathbf{x}}|a\rangle}{E_a - E_j - k\delta_j} \\
&= -\frac{\alpha}{4\pi^2} \int \frac{d^3k}{k} \sum_{\lambda=1}^{2} \left[\sum_{j+} \frac{\langle a|\alpha_\lambda e^{-i\mathbf{k}\cdot\mathbf{x}}|j\rangle\langle j|\alpha_\lambda e^{i\mathbf{k}\cdot\mathbf{x}}|a\rangle}{|E_j| + k - E_a} \right. \\
&\quad \left. - \sum_{j-} \frac{\langle a|\alpha_\lambda e^{-i\mathbf{k}\cdot\mathbf{x}}|j\rangle\langle j|\alpha_\lambda e^{i\mathbf{k}\cdot\mathbf{x}}|a\rangle}{|E_j| + k + E_a} \right] \\
&= -\frac{\alpha}{8\pi^2} \int \frac{d^3k}{k} \sum_{\lambda=1}^{2} \langle a|\alpha_\lambda \\
&\quad \times \left[\frac{H/|H| + 1}{|H| + k - E_a} + \frac{H/|H| - 1}{|H| + k + E_a} \right]_{\mathbf{p}+\mathbf{k}} \alpha_\lambda|a\rangle. \tag{11.120}
\end{aligned}
$$

Then we expand the operators in the denominators analogously to (11.116):

$$
\frac{1}{[|H| + k \pm E_a]_{\mathbf{p}+\mathbf{k}}} - \frac{1}{D_k^{\mp} + \Delta_k \pm w_a} = \frac{1}{D_k^{\mp}} - \frac{(\Delta_k + w_a)}{(D_k^{\mp})^2} + \dots , \tag{11.121}
$$

where $D_k^{\mp} \equiv E_k + k \pm m$, $w_a \equiv E_a - m$, and Δ_k is defined as in (11.118). The calculation of $W_D^X(a)$, like $W_S^X(a)$, now requires the evaluation of expectation values of various operators $\mathbf{p}^2, V^2, \ldots$ in state $|a\rangle$, multiplied by integrals over k. For $W_D^X(a)$ we must also sum over the polarization states $\lambda = 1, 2$.

The calculation of $W_S^X(a) + W_D^X(a)$ is thus a straightforward but tedious exercise, even when the expansions above are truncated at some low order.[8] Kroll and Lamb (1949) retained all terms effectively up to fourth and lower order in v/c, and were still faced with the evaluation of "a sum of expectation values of various operators ... each multiplied by a combination of some 50 elementary integrals over k." They found to this order that[9]

$$
\begin{aligned}
[W_S^X(a) + W_D^X(a)]' &= \langle a|\beta \delta m|a\rangle + \frac{\alpha}{6\pi}\langle a|\boldsymbol{\alpha} \cdot \mathbf{p}|a\rangle \\
&+ \frac{2\alpha}{3\pi}\frac{k_i}{m}\langle a|\boldsymbol{\alpha} \cdot \mathbf{p}|a\rangle + \frac{\alpha}{4\pi m}\langle a|\beta\boldsymbol{\alpha} \cdot \boldsymbol{\pi}V|a\rangle \\
&- \frac{\alpha}{3\pi m^2}\left(\log \frac{m}{2k_i} + \frac{11}{24}\right)\langle a|\boldsymbol{\pi}^2 V|a\rangle. \quad (11.122)
\end{aligned}
$$

To explain the meaning of k_i, and the prime on the left side of this equation, we consider briefly now the validity of the expansion procedure used in obtaining this result.

The expansion (11.116) can be expected to be valid for all values of k if $\mathbf{p}^2/m^2$ and V^2/m^2 can in effect be regarded as numbers less than unity, since $E_k \to 1$ as $k \to 0$. Although $V(\mathbf{x})$ can in fact be large and even infinite for small $\mathbf{x}$, this occurs for the Coulomb potential only for a small region of $\mathbf{x}$, and does not substantially alter expectation values involving V. Similarly $\mathbf{p}^2\phi_a(\mathbf{x})/\phi_a(\mathbf{x})$ can be large for small $\mathbf{x}$, but this does not invalidate the expansion procedure if the expansion is carried out to fourth and lower order in v/c. Similar considerations apply to the expansion (11.121) *except* that $D_k^+ = E_k + k - m \to 0$ as $k \to 0$, and so $1/D_k^+$ cannot be regarded as small compared with unity for all values of k. Its integral over k can be carried out only down to some k_i, and for $0 \le k \le k_i$ this term must be treated separately.

This separate treatment of the $1/D_k^+$ term defines the energy k_i in (11.122). The prime on the left side indicates that the low-k contribution from the $1/D_k^+$ term has not yet been taken into account.

The latter, low-k contribution, however, is essentially nonrelativistic, and can be evaluated as the level shift calculated nonrelativistically in

---

[8] The calculations are sufficiently tedious that the author has not himself attempted to verify them in detail. The same result as Kroll and Lamb, obtained independently in essentially the same manner, was reported by French and Weisskopf (1949).

[9] We have written this in a form slightly different from that of Kroll and Lamb, and in particular we have written explicitly the electromagnetic mass δm.

Chapter 3. Prior to the nonrelativistic mass renormalization, that shift is given by

$$\Delta E_a = \Delta E_a^{\text{obs}} + \Delta E_a^{\text{free}} \tag{11.123}$$

in the notation of Section 3.5. Using the present units, and introducing a high-energy cutoff k_i in (3.23), we write

$$\Delta E_a^{\text{free}} = -\frac{2\alpha}{3\pi m^2} \sum_m |\mathbf{p}_{ma}|^2 \int_0^{k_i} dk = -\frac{2\alpha}{3\pi m^2} k_i \langle a|\mathbf{p}^2|a\rangle . \tag{11.124}$$

From (3.26) and (3.27), using now k_i instead of m for the energy cutoff in (3.25), we get

$$\Delta E_a^{\text{obs}} = \frac{2\alpha}{3\pi m^2} \log\frac{k_i}{\epsilon}\frac{1}{2}\langle a|\nabla^2|a\rangle = -\frac{\alpha}{3\pi m^2}\log\frac{k_i}{\epsilon}\langle a|\boldsymbol{\pi}^2 V|a\rangle, \tag{11.125}$$

where ϵ is Bethe's "average excitation energy." Thus, adding (11.124) and (11.125) to (11.122), one finds

$$
\begin{aligned}
W_S^X(a) + W_D^X(a) &= \langle a|\beta\delta m|a\rangle + \frac{\alpha}{6\pi}\langle a|\boldsymbol{\alpha}\cdot\mathbf{p}|a\rangle \\
&\quad + \frac{\alpha}{4\pi m}\langle a|\beta\boldsymbol{\alpha}\cdot\boldsymbol{\pi}V|a\rangle \\
&\quad + \frac{2\alpha}{3\pi}\frac{k_i}{m}\left[\langle a|\boldsymbol{\alpha}\cdot\mathbf{p}|a\rangle - \frac{1}{m}\langle a|\mathbf{p}^2|a\rangle\right] \\
&\quad - \frac{\alpha}{3\pi m^2}\left(\log\frac{m}{2k_i} + \log\frac{k_i}{\epsilon} + \frac{11}{24}\right)\langle a|\boldsymbol{\pi}^2 V|a\rangle.
\end{aligned}
\tag{11.126}
$$

Now writing $\langle a| = (\phi_a^\dagger\ \xi_a^\dagger)$, we have $\langle a|\boldsymbol{\alpha}\cdot\mathbf{p}|a\rangle = 2\text{Re}\phi_a^\dagger\boldsymbol{\sigma}\cdot\mathbf{p}\xi_a$. Using the nonrelativistic approximation $\xi_a \cong (\boldsymbol{\sigma}\cdot\mathbf{p})\phi_a/2m$ for the "small component," as in (9.58), we have furthermore $\langle a|\boldsymbol{\alpha}\cdot\mathbf{p}|a\rangle \cong \text{Re}[\phi_a(\boldsymbol{\sigma}\cdot\mathbf{p})^2\phi_a]/m \cong \langle a|(\boldsymbol{\sigma}\cdot\mathbf{p})^2|a\rangle/m = \langle a|(\mathbf{p}^2/m)|a\rangle$. Then the term proportional to k_i in (11.126) vanishes to lowest order in v/c. That is, the nonrelativistic mass renormalization term (11.124) cancels the third term on the right side of (11.122). We also see that the result obtained by joining the low-k contribution to (11.122) is independent of the "joining energy" k_i:

$$
\begin{aligned}
W_S^X(a) + W_D^X(a) &= \langle a|\beta\delta m|a\rangle + \frac{\alpha}{6\pi}\langle a|\boldsymbol{\alpha}\cdot\mathbf{p}|a\rangle \\
&\quad + \frac{\alpha}{4\pi m}\langle a|\beta\boldsymbol{\alpha}\cdot\boldsymbol{\pi}V|a\rangle \\
&\quad - \frac{\alpha}{3\pi m^2}\left(\log\frac{m}{\epsilon} - \log 2 + \frac{11}{24}\right)\langle a|\boldsymbol{\pi}^2 V|a\rangle.
\end{aligned}
\tag{11.127}
$$

There is one more contribution to the energy shift of the electron in state $|a\rangle$, namely $W^{NX}(a)$. As discussed in the next section, the evaluation of this term yields

$$W^{NX}(a) = -\frac{2\alpha}{3\pi}\int_0^\infty \frac{dk\,k^2}{E_k^3}\langle a|V|a\rangle - \frac{\alpha}{15\pi m^2}\langle a|\nabla^2 V|a\rangle . \tag{11.128}$$

The level shift of state $|a\rangle$ is therefore

$$\begin{aligned}
W(a) &= W_S^X(a) + W_D^X(a) + W^{NX}(a) \\
&= \langle a|\beta|a\rangle\delta m - \langle a|V|a\rangle\frac{2\alpha}{3\pi}\int_0^\infty \frac{dk\,k^2}{E_k^3} + \frac{\alpha}{6\pi}\langle a|\boldsymbol{\alpha}\cdot\mathbf{p}|a\rangle \\
&\quad - \frac{i\alpha}{4\pi m}\langle a|\beta\boldsymbol{\alpha}\cdot\nabla V|a\rangle \\
&\quad + \frac{\alpha}{3\pi m^2}\left(\log\frac{m}{\epsilon} - \log 2 + \frac{11}{24} - \frac{1}{5}\right)\langle a|\nabla^2 V|a\rangle. \tag{11.129}
\end{aligned}$$

The term[10]

$$\frac{\alpha}{3\pi m^2}\log\frac{m}{\epsilon}\langle a|\nabla^2 V|a\rangle = \frac{\alpha}{3\pi m^2}\log\frac{m}{\epsilon}\sum_m |\mathbf{p}_{ma}|^2(E_m - E_a) \tag{11.130}$$

is identical to Bethe's nonrelativistic result (3.26) for the Lamb shift. In the result (11.129) of the relativistic calculation, however, the mass in the factor $\log(m/\epsilon)$ appears naturally, from an integral over k, not as a high-energy cutoff. The cutoff k_i corresponding to Bethe's nonrelativistic cutoff, as already noted, cancels out in the relativistic calculation.

Energy Difference of the $2s_{1/2}$ and $2p_{1/2}$ States

For the $2s_{1/2}$ and $2p_{1/2}$ states of the hydrogen atom, with equal unperturbed energy levels according to the solution of the Dirac equation, the expectation values of β and V are equal, and it therefore follows that the expectation values of $\boldsymbol{\alpha}\cdot\mathbf{p} = H - \beta m - V$ are also equal. The remaining expectation values in (11.129) have the following values:

$$\langle 2s_{1/2}|\nabla^2 V|2s_{1/2}\rangle = m^2\alpha^2\frac{e^2}{2a_o} , \tag{11.131}$$

$$\langle 2p_{1/2}|\nabla^2 V|2p_{1/2}\rangle = 0, \tag{11.132}$$

$$\langle 2s_{1/2}|\beta\boldsymbol{\alpha}\cdot\nabla V|2s_{1/2}\rangle = \frac{im}{2}\alpha^2\frac{e^2}{2a_o} , \tag{11.133}$$

$$\langle 2p_{1/2}|\beta\boldsymbol{\alpha}\cdot\nabla V|2p_{1/2}\rangle = -\frac{im}{6}\alpha^2\frac{e^2}{2a_o} . \tag{11.134}$$

---

[10] See equation (3.27).

Then

$$
\begin{aligned}
W(2s_{1/2}) - W(2p_{1/2}) &= -\frac{i\alpha}{4\pi}\left[\frac{i}{2} + \frac{i}{6}\right]\alpha^2 \frac{e^2}{2a_o} \\
&\quad + \frac{\alpha}{3\pi}\left(\log\frac{m}{\epsilon} - \log 2 + \frac{11}{24} - \frac{1}{5}\right)\alpha^2 \frac{e^2}{2a_o} \\
&= \frac{\alpha^3}{3\pi}\frac{e^2}{2a_o}\left(\log\frac{m}{\epsilon} - \log 2 + \frac{23}{24} - \frac{1}{5}\right).
\end{aligned}
$$

$$(11.135)$$

Assuming the average excitation energy $\epsilon = 17.8\ R_\infty$ used in Section 3.5, we calculate 1046 MHz for this energy difference. Kroll and Lamb assumed Bethe's revised values $\log(m/\epsilon) = 7.7169 - .0293$ and $(\alpha^3/3\pi)R_\infty = 135.580$ MHz and calculated

$$W(2s_{1/2}) - W(2p_{1/2}) = 1051\ \text{MHz}. \qquad (11.136)$$

This result, besides being in excellent accord with experiment, is convergent and independent of any arbitrary cutoff parameter.

Shifts of Individual Levels

In the preceding calculation of the difference $W(2s_{1/2}) - W(2p_{1/2})$ the first three terms in (11.129) drop out because they are the same for the two levels. It is also of interest to consider the individual shifts $W(2s_{1/2})$ and $W(2p_{1/2})$, where these terms must be dealt with.

The (divergent) term $\langle a|\beta\delta m|a\rangle$ in (11.129) is an electromagnetic mass contribution and could have been eliminated at the outset by adding a counter term to the Hamiltonian, as discussed in Section 11.6. The term proportional to $\langle a|V|a\rangle$ is also logarithmically divergent, and will be seen in the following section to correspond to a charge renormalization, analogous to mass renormalization.

The term $(\alpha/6\pi)\langle a|\boldsymbol{\alpha}\cdot\mathbf{p}|a\rangle$, unlike the remaining contributions to (11.129), is independent of V. For the case of a free electron of momentum $\mathbf{p}$ we find, using the free-electron wave function (11.70), for instance, that $(\alpha/6\pi)\langle a|\boldsymbol{\alpha}\cdot\mathbf{p}|a\rangle = (\alpha/6\pi)\mathbf{p}^2/\sqrt{\mathbf{p}^2 + m^2}$. Note, however, that the correction to the self-energy of a free electron of momentum $\mathbf{p}$ is given, to first order in δm, by[11]

$$\sqrt{\mathbf{p}^2 + (m + \delta m)^2} - \sqrt{\mathbf{p}^2 + m^2} \cong \frac{m\delta m}{\sqrt{\mathbf{p}^2 + m^2}}, \qquad (11.137)$$

---

[11] This assumes that a correct (but not yet available) calculation of δm would give $\delta m/m \ll 1$.

so that $(\alpha/6\pi)\langle a|\boldsymbol{\alpha}\cdot\mathbf{p}|a\rangle$ does not represent a p-dependent addition to the self-energy δm of a free electron. In particular, it is not covariant, as such a correction would have to be.

To handle this term, Kroll and Lamb (1949) argue that free electron operators can be subtracted from the operators appearing in the self-energy expression in such a way that, in the resulting expression, the self-energy of a free electron is zero. The resulting expectation value is then the observable difference in self-energy between the electron in state $|a\rangle$ and a free electron. Without the mass and charge renormalization terms in (11.129), the level shift

$$W(a) \quad \rightarrow \quad \frac{\alpha}{6\pi}\langle a|\boldsymbol{\alpha}\cdot\mathbf{p}|a\rangle - \frac{i\alpha}{4\pi m}\langle a|\beta\boldsymbol{\alpha}\cdot\nabla V|a\rangle$$
$$+ \frac{\alpha}{3\pi m^2}\left(\log\frac{m}{\epsilon} - \log 2 + \frac{11}{24} - \frac{1}{5}\right)\langle a|\nabla^2 V|a\rangle ,$$
$$(11.138)$$

and it would appear that the first, V-independent term should be dropped in order to have $W(a) \rightarrow 0$ for a free electron. However, there are other free electron operators of fourth and lower order in v/c, whose expectation values vanish for a free electron, that can be subtracted as well in this approach, so that the simple deletion of $(\alpha/6\pi)\langle a|\boldsymbol{\alpha}\cdot\mathbf{p}|a\rangle$ is not a unique prescription. Kroll and Lamb show, by consideration of the (seven) other free electron operators up to fourth order in v/c, that the effect of these other operators would be to add a term proportional to the second term in (11.138). In other words, the lack of uniqueness as to which free electron operators should be subtracted is equivalent to an indeterminacy with respect to the coefficient multiplying $(-i/2m)\langle a|\beta\boldsymbol{\alpha}\cdot\nabla V|a\rangle$. This indeterminacy, however, may be removed by observing that the coefficient $(\alpha/2\pi)$ in (11.138) corresponds precisely (to order α) to the magnetic moment correction for the electron, as shown later. Therefore $(\alpha/6\pi)\langle a|\boldsymbol{\alpha}\cdot\mathbf{p}|a\rangle$ can in effect be simply dropped from (11.138):

$$W(a) \quad \rightarrow \quad -\frac{i\alpha}{4\pi m}\langle a|\beta\boldsymbol{\alpha}\cdot\nabla V|a\rangle$$
$$+ \frac{\alpha}{3\pi m^2}\left(\log\frac{m}{\epsilon} - \log 2 + \frac{11}{24} - \frac{1}{5}\right)\langle a|\nabla^2 V|a\rangle. \quad (11.139)$$

Then, from (11.130)–(11.133),

$$W(2s_{1/2}) \quad \rightarrow \quad \frac{\alpha^3}{3\pi}\frac{e^2}{2a_o}\left(\log\frac{m}{\epsilon} - \log 2 + \frac{11}{24} - \frac{1}{5} + \frac{3}{8}\right)$$

$$= \ 1034 \text{ MHz}, \tag{11.140}$$

$$W(2p_{1/2}) \ \rightarrow \ -\frac{\alpha^3}{3\pi}\frac{e^2}{2a_o}\left(\frac{1}{8}\right) = -17 \text{ MHz}. \tag{11.141}$$

Effect of Anomalous Magnetic Moment

The argument just given for dropping $(\alpha/6\pi)\langle a|\boldsymbol{\alpha} \cdot \mathbf{p}|a\rangle$ rests on the attribution of the second term in (11.138) to the anomalous magnetic moment of the electron (Section 3.13). For this purpose consider the Dirac equation $(\not{p} - m - e\not{A})\psi = 0$ multiplied on both sides by $(\not{p} + m - e\not{A})$:

$$[(\not{p} - e\not{A})^2 - m^2]\psi = 0. \tag{11.142}$$

Using $\gamma^\mu\gamma^\nu = g^{\mu\nu} + \frac{1}{2}(\gamma^\mu\gamma^\nu + \gamma^\nu\gamma^\mu) - \gamma^\nu\gamma^\mu = g^{\mu\nu} + \frac{1}{2}[\gamma^\mu, \gamma^\nu] \equiv g^{\mu\nu} - i\sigma^{\mu\nu}$, and $p = i\partial$, we find that (11.142) is equivalent to

$$\left[(p - eA)^2 - \frac{e}{2}\sigma^{\mu\nu}F_{\mu\nu} - m^2\right]\psi = 0, \tag{11.143}$$

where $F_{\mu\nu}$ is the electromagnetic field tensor defined by (10.143). Some simple manipulations show that

$$\frac{1}{2}\sigma^{\mu\nu}F_{\mu\nu} = i\boldsymbol{\alpha} \cdot \mathbf{E} - \boldsymbol{\Sigma} \cdot \mathbf{B}, \tag{11.144}$$

where $\boldsymbol{\Sigma}$ is defined by (9.64). Now since $-(e/2m)\boldsymbol{\Sigma} \cdot \mathbf{B} = -(g/2)(e/2m)$ $\times \boldsymbol{\Sigma} \cdot \mathbf{B}$ corresponds in the Pauli equation (9.62) to the energy of a spin in a magnetic field, we surmise that $(g/2)(e/2m)i\boldsymbol{\alpha} \cdot \mathbf{E}$ corresponds to the energy of the spin in an electric field. Actually, of course, this is not quite true, as $i\boldsymbol{\alpha}$ is not Hermitian; more detailed considerations show that the operator corresponding to the energy of a spin in an electric field is $(g/2)(e/2m)i\beta\boldsymbol{\alpha} \cdot$ $\mathbf{E}$.[12] Writing

$$\frac{g}{2}\left(\frac{e}{2m}\right)i\beta\boldsymbol{\alpha} \cdot \mathbf{E} = -\frac{g}{2}\left(\frac{1}{2m}\right)i\beta\boldsymbol{\alpha} \cdot \nabla V, \tag{11.145}$$

we see that the second term in (11.138) is associated with an additional magnetic moment such that the change in g is (Section 3.13)

$$\frac{\Delta g}{2} = \frac{g-2}{2} = \frac{\alpha}{2\pi} . \tag{11.146}$$

This justifies the assertion preceding (11.139).

A different approach to the ambiguity associated with (11.138) was taken by French and Weisskopf (1949). They calculated the energy shift in both magnetic and electrostatic fields and found in both cases that the additional magnetic moment is $\alpha/2\pi$ Bohr magnetons to lowest order, thus removing an ambiguity in their mass renormalization procedure.

---

[12]See Grandy (1991a), pp. 72–73.

11.8   Vacuum Polarization

We consider now the nonexchange term $W^{NX}(a)$. As (11.65) indicates, this term may be interpreted as an interaction between the current density $J_a^\mu(\mathbf{x})$ associated with the electron in state $|a\rangle$ and the current density $J_{\text{neg}}^\mu(\mathbf{x})$ associated with the negative-energy states of the Dirac sea. It can be expected that, in the case of an electrostatic external field giving rise to V, the nonexchange energy reduces to an interaction between charge densities $J_a^0(\mathbf{x})$ and $J_{\text{neg}}^0(\mathbf{x})$:

$$W^{NX}(a) = \frac{1}{4\pi} \int d^3x\, J_{a0}(\mathbf{x}) \int \frac{d^3x'\, J_{\text{neg}}^0(\mathbf{x}')}{|\mathbf{x} - \mathbf{x}'|} \,, \tag{11.147}$$

where

$$J_{\text{neg}}^0(\mathbf{x}) = \frac{e}{2} \sum_j (1 - \delta_j)\phi_j^*(\mathbf{x})\phi_j(\mathbf{x}). \tag{11.148}$$

Because of the completeness relation $\sum_j \phi_j^*(\mathbf{x})\phi_j(\mathbf{x}') = \delta^3(\mathbf{x} - \mathbf{x}')$, $J_{\text{neg}}^0(\mathbf{x})$ consists of an infinite, physically uninteresting V-independent charge density plus

$$J_{\text{neg}}^{\prime 0}(\mathbf{x}) = -\frac{e}{2} \sum_j \delta_j\, \phi_j^*(\mathbf{x})\phi_j(\mathbf{x}). \tag{11.149}$$

The latter will depend on V through the dependence of the eigenfunctions $\phi_j(\mathbf{x})$ on V. Consider

$$\begin{aligned}
\rho(\mathbf{x}, \mathbf{x}') &\equiv \sum_j \delta_j \phi_j^*(\mathbf{x})\phi_j(\mathbf{x}') = \sum_j \phi_j^*(\mathbf{x})(H/|H|)\phi_j(\mathbf{x}') \\
&= \sum_j \sum_{\mu=1}^4 \sum_{\nu=1}^4 \phi_{j\mu}^*(\mathbf{x})(H/|H|)_{\mu\nu}\phi_{j\nu}(\mathbf{x}').
\end{aligned} \tag{11.150}$$

The completeness relation

$$\sum_j \phi_{j\mu}^*(\mathbf{x})\phi_{j\nu}(\mathbf{x}') = \delta_{\mu\nu}\delta^3(\mathbf{x}-\mathbf{x}') = \delta_{\mu\nu}\left(\frac{1}{2\pi}\right)^3 \int d^3k\, e^{i\mathbf{k}\cdot(\mathbf{x}-\mathbf{x}')} \tag{11.151}$$

gives

$$\begin{aligned}
\rho(\mathbf{x}, \mathbf{x}') &= \left(\frac{1}{2\pi}\right)^3 \int d^3k\, e^{i\mathbf{k}\cdot\mathbf{x}}\text{Tr}(H/|H|)e^{-i\mathbf{k}\cdot\mathbf{x}'} \\
&= \left(\frac{1}{2\pi}\right)^3 \int d^3k\, e^{i\mathbf{k}\cdot(\mathbf{x}-\mathbf{x}')}[\text{Tr}(H/|H|)]_{\mathbf{p}+\mathbf{k}} \tag{11.152}
\end{aligned}$$

and therefore

$$J'^0_{\text{neg}}(\mathbf{x}) = -\frac{e}{2}\rho(\mathbf{x},\mathbf{x}) = -\frac{e}{16\pi^3}\int d^3k[\text{Tr}(H/|H|)]_{\mathbf{p}+\mathbf{k}} . \tag{11.153}$$

The trace in the integrand can be evaluated following the expansion procedure beginning with equation (11.115). The result, to the lowest order required, is (Kroll and Lamb, 1949)

$$J'^0_{\text{neg}}(\mathbf{x}) = \left[\frac{e}{16\pi^2}\int\frac{dk\,k^2}{E_k^3}\right]\nabla^2 V + \frac{e}{60\pi^2 m^2}\nabla^4 V. \tag{11.154}$$

From (11.147), then,

$$W^{NX}(a) = W_1^{NX}(a) + W_2^{NX}(a) , \tag{11.155}$$

$$W_1^{NX}(a) \equiv \frac{1}{4\pi}\frac{e^2}{6\pi^2}\int\frac{dk\,k^2}{E_k^3}\int d^3x\,\phi_a^*(\mathbf{x})\phi_a(\mathbf{x})\int\frac{d^3x'\,\nabla'^2 V(\mathbf{x}')}{|\mathbf{x}-\mathbf{x}'|}$$

$$= \frac{1}{4\pi}\frac{e^2}{6\pi^2}\int\frac{dk\,k^2}{E_k^3}\int d^3x\,\phi_a^*(\mathbf{x})\phi_a(\mathbf{x})$$

$$\times \int d^3x'\,V(\mathbf{x}')\nabla'^2\frac{1}{|\mathbf{x}-\mathbf{x}'|}$$

$$= -\frac{e^2}{6\pi^2}\int\frac{dk\,k^2}{E_k^3}\langle a|V|a\rangle = -\frac{2\alpha}{3\pi}\int\frac{dk\,k^2}{E_k^3}\langle a|V|a\rangle$$

$$\equiv -C\langle a|V|a\rangle , \tag{11.156}$$

$$W_2^{NX}(a) \equiv \frac{\alpha}{60\pi^2 m^2}\int d^3x\,\phi_a^*(\mathbf{x})\phi_a(\mathbf{x})\int\frac{d^3x'\,\nabla'^4 V(\mathbf{x}')}{|\mathbf{x}-\mathbf{x}'|}$$

$$= -\frac{\alpha}{15\pi m^2}\int d^3x\,\phi_a^*(\mathbf{x})\phi_a(\mathbf{x})\nabla^2 V(\mathbf{x})$$

$$= -\frac{\alpha}{15\pi m^2}\langle a|\nabla^2 V|a\rangle . \tag{11.157}$$

These are the results used in writing (11.128).

As remarked at the end of Section 11.3, $W^{NX}(a)$ is attributable to the polarization of the vacuum by the electrostatic potential V (Section 9.5). From (11.138) and (11.131) we see that the effect of vacuum polarization on the $2s_{1/2}$ state of hydrogen is to shift its energy by

$$W_2^{NX}(2s_{1/2}) = -\frac{\alpha}{15\pi m^2}m^2\alpha^2\frac{e^2}{2a_o} \cong -27.1 \text{ MHz}. \tag{11.158}$$

Measurements of the Lamb shift, of course, are far more accurate than this, and so one concludes that the Lamb shift provides strong support for the reality of vacuum polarization.[13]

The vacuum polarization contribution $W_2^{NX}(a)$ is seen from (11.147), (11.63), and (11.58) to derive from the last, electrostatic contribution to the Hamiltonian (11.25). As such, it has nothing directly to do with the quantization of the transverse electromagnetic field. In fact, its original calculation by Uehling (1935) made no reference to electromagnetic field quantization. Vacuum polarization is, however, a relativistic effect involving electron–positron pairs, as the hole-theoretic interpretation assumes: an electrostatic field causes a redistribution of charge in the Dirac sea and thus polarizes the vacuum. A single charged particle, in particular, will polarize the vacuum near it, so that its observed charge is actually smaller than its "bare charge." A proton, for instance, will attract electrons and repel positrons of the Dirac sea, resulting in a partial screening of its bare charge and a modification of the Coulomb potential in the hydrogen atom.[14] According to (11.157) the Coulomb interaction is changed from $V(\mathbf{x}) = -e^2/4\pi|\mathbf{x} - \mathbf{x}'|$ to

$$
\begin{aligned}
V_{\text{eff}}(\mathbf{x}) \; &= \; -\frac{1}{4\pi}\frac{e^2}{|\mathbf{x} - \mathbf{x}'|} - \frac{\alpha}{15\pi m^2}\nabla^2\frac{e^2}{4\pi|\mathbf{x} - \mathbf{x}'|} \\
&= \; -\frac{1}{4\pi}\frac{e^2}{|\mathbf{x} - \mathbf{x}'|} + \frac{\alpha e^2}{15\pi m^2}\delta^3(\mathbf{x}) \; .
\end{aligned}
\tag{11.159}
$$

This is an approximation to a more general modification of the Coulomb potential. For two identical point charges (Serber 1935; Uehling 1935)

$$
\begin{aligned}
V_{\text{eff}}(\mathbf{x}) \; &\cong \; \frac{1}{4\pi}\frac{e^2}{|\mathbf{x}|}\left[\frac{2\alpha}{3\pi}\log\frac{\lambda_c}{|\mathbf{x}|} - \gamma + \frac{1}{6}\right] \quad (|\mathbf{x}| << \lambda_c), \\
&\cong \; \frac{1}{4\pi}\frac{e^2}{|\mathbf{x}|}\left[1 + \frac{\alpha}{2\sqrt{\pi}}\left(\frac{\lambda_c}{|\mathbf{x}|}\right)^{3/2}e^{-2|\mathbf{x}|/\lambda_c}\right] \quad (|\mathbf{x}| >> \lambda_c),
\end{aligned}
\tag{11.160}
$$

where $\lambda_c = 1/m \; (= \hbar/mc)$ and $\gamma = 0.5772$ is the Euler–Mascheroni constant. Not surprisingly, the scale length determining the modification of the Coulomb potential resulting from the polarization of the vacuum by the charges is the Compton wavelength, λ_c.

---

[13] We refer the reader to Drake (1982) for a review of Lamb-shift theory in few-electron atoms. It is interesting that a primary motivation for the original Lamb–Retherford experiments was to investigate the contribution of vacuum polarization to the energy levels of hydrogen.

[14] The physical arguments for the vacuum polarization contribution to the Lamb shift are summarized in Section 9.5.

The charges and masses in (11.157) and (11.158) are the *observed* values. Mass renormalization was discussed in Section 11.6. To appreciate the charge renormalization associated with vacuum polarization, consider now the term $W_1^{NX}(a)$. Since $V \propto e^2$, we can evidently interpret $W_1^{NX}(a)$ in terms of a change in e^2. That is, the interaction without vacuum polarization would be $-e^2/4\pi|\mathbf{x} - \mathbf{x}'|$, say, and with vacuum polarization it is changed by $-CV = Ce^2/4\pi|\mathbf{x} - \mathbf{x}'|$ (to lowest order), and so equivalently e^2 changes to $(1-C)e^2$ because of vacuum polarization. Since the observed charge must include the effect of vacuum polarization, we deduce that

$$e_{\text{obs}}^2 = (1 - C)e_{\text{bare}}^2 , \tag{11.161}$$

where

$$C = \frac{2\alpha}{3\pi} \int_0^\infty \frac{dk\,k^2}{E_k^3} \to \frac{2\alpha}{3\pi} \log \frac{\Lambda}{m} \tag{11.162}$$

when we introduce a cutoff Λ as in the case of mass renormalization.

The methodology of charge renormalization is basically the same as that of mass renormalization. Renormalization of e and m gives us finite, cutoff-independent results involving *observed* charges and masses. We do not know where the bare masses and charges come from or how to calculate them. As the discussion in Section 9.5 suggests, we might expect to "see" the bare charge at distances smaller than λ_c from an electron. But the correction to e_{bare} is small even at extremely high energies (and therefore extremely short distances): choosing Λ in (11.160) to be 50 BeV, we calculate $C \cong$ .018 and $e_{\text{bare}}/e_{\text{obs}} \cong 1.009$. Thus, an understanding of e_{bare} requires an understanding of how QED itself ought to be modified at extremely short distances.[15] Nevertheless, it is a very interesting possibility that in the limit of infinite energies the interactions of charged particles are determined by their bare charges.[16] The vacuum polarization contribution to the Lamb shift, for instance, is a consequence of the fact that the electron in the $2s_{1/2}$ state spends more time near the nucleus than the $2p_{1/2}$ electron, and consequently sees "more" of the bare charge of the nucleus and a stronger Coulomb potential.

Although vacuum polarization is a 1% effect in the $2s_{1/2} - 2p_{1/2}$ Lamb shift of hydrogen, it contributes about 90% of the analogous Lamb shift in muonic helium (Petermann and Yamaguchi, 1959; Glauber, Rarita, and Schwed, 1960). This can be understood from the smaller Bohr radii of muonic atoms, which in turn is a consequence of the fact that the muon

---

[15] There are also vacuum polarization contributions from heavier virtual particle pairs (muons, pions, etc.), but these are smaller than the electron–positron contributions because of the mass dependence of expressions such as (11.157).

[16] See, for instance, Gell-Mann and Low (1954).

mass is about 207 times the mass of the electron. Vacuum polarization also manifests itself in proton–proton scattering as a result of its modification of the Coulomb potential (Foldy and Eriksen, 1954).

11.9   Radiation Reaction and the Vacuum Field

Throughout this book much has been made of the fact that vacuum fluctuations constitute the "fluctuation side" of the fluctuation–dissipation relation between the vacuum and radiation reaction fields. The radiation reaction and vacuum fields are two aspects of the same thing when it comes to physical interpretations of various QED processes including the Lamb shift, van der Waals forces, and Casimir effects. The alternative physical interpretations afforded by the fluctuation–dissipation connection have been discussed thus far only in the context of nonrelativistic theory. We now address these interpretations in relativistic QED.

As in the nonrelativistic theory, physical interpretations are facilitated by working in the Heisenberg picture. We begin by writing the Hamiltonian (11.18) in terms of annihilation and creation operators for the Maxwell and Dirac fields, using (10.153), (11.26), and (11.27) for these fields:

$$
\begin{aligned}
H = & \sum_n E_n[b_n^\dagger b_n - d_n d_n^\dagger] + \sum_{\mathbf{k}\lambda} k a_{\mathbf{k}\lambda}^\dagger a_{\mathbf{k}\lambda} - e \sum_m \sum_n \sum_{\mathbf{k}\lambda} \frac{1}{\sqrt{2k}} \\
& \times \Big[C_{m+n+}^\lambda(\mathbf{k}) b_m^\dagger b_n a_{\mathbf{k}\lambda} + C_{m+n-}^\lambda(\mathbf{k}) b_m^\dagger d_n^\dagger a_{\mathbf{k}\lambda} \\
& + C_{m+n+}^\lambda(-\mathbf{k}) a_{\mathbf{k}\lambda}^\dagger b_m^\dagger b_n + C_{m+n-}^\lambda(-\mathbf{k}) a_{\mathbf{k}\lambda}^\dagger b_m^\dagger d_n^\dagger \\
& + C_{m-n+}^\lambda(\mathbf{k}) b_n d_m a_{\mathbf{k}\lambda} + C_{m-n-}^\lambda(\mathbf{k}) d_m d_n^\dagger a_{\mathbf{k}\lambda} \\
& + C_{m-n+}^\lambda(-\mathbf{k}) a_{\mathbf{k}\lambda}^\dagger b_n d_m \\
& + C_{m-n-}^\lambda(-\mathbf{k}) a_{\mathbf{k}\lambda}^\dagger d_m d_n^\dagger \Big] + H_{\text{el}} .
\end{aligned}
\tag{11.163}
$$

The first two terms correspond to the unperturbed Dirac and Maxwell fields, respectively. In order to compare with the nonrelativistic theory of Chapter 4, we use a discrete summation over modes for the Maxwell field. The remaining terms, except for H_{el}, form the interaction $-e \int d^3x \psi^\dagger \boldsymbol{\alpha} \cdot \mathbf{A} \psi$ between the Dirac and transverse electromagnetic fields, with the coupling constants $C_{m+n+}^\lambda(\mathbf{k})$, etc. defined by (11.34). H_{el} is the electrostatic interaction given by the last term of (11.25). This term, of course, can be written in a form involving b's, d's, and their adjoints, but for our purposes here it will not be necessary to do so.

Using the fermion anticommutation relations $\{b_m, b_n^\dagger\} = \delta_{mn}$, $\{b_m, b_n\} = 0$, together with the fact that the b's and $b^\dagger$'s commute with the equal-time operators $a_{\mathbf{k}\lambda}, a_{\mathbf{k}\lambda}^\dagger, d_m$, and $d_m^\dagger$, we obtain straightforwardly the following Heisenberg equation of motion for $b_\ell(t)$:

$$
\begin{aligned}
\dot{b}_\ell &= -iE_\ell b_\ell + ie \sum_n \sum_{\mathbf{k}\lambda} \frac{1}{\sqrt{2k}} \left[C_{\ell+n+}^\lambda(\mathbf{k}) b_n a_{\mathbf{k}\lambda} + C_{\ell+n+}^\lambda(-\mathbf{k}) a_{\mathbf{k}\lambda}^\dagger b_n \right. \\
&\quad + C_{n-\ell+}^\lambda(\mathbf{k}) d_n a_{\mathbf{k}\lambda} + C_{n-\ell+}^\lambda(-\mathbf{k}) a_{\mathbf{k}\lambda}^\dagger d_n + C_{\ell+n-}^\lambda a_{\mathbf{k}\lambda}^\dagger d_n^\dagger \\
&\quad \left. + C_{\ell+n-}^\lambda(\mathbf{k}) d_n^\dagger a_{\mathbf{k}\lambda} \right] - 2ie \sum_{m,n} \sum_{\mathbf{k}\lambda} \frac{1}{\sqrt{2k}} \left[C_{m+n-}^\lambda(\mathbf{k}) d_n^\dagger b_m^\dagger b_\ell a_{\mathbf{k}\lambda} \right. \\
&\quad + C_{m+n-}^\lambda(-\mathbf{k}) a_{\mathbf{k}\lambda}^\dagger d_n^\dagger b_m^\dagger b_\ell + C_{m-n+}^\lambda(\mathbf{k}) b_m b_\ell d_m a_{\mathbf{k}\lambda} \\
&\quad \left. + C_{m-n+}^\lambda(-\mathbf{k}) a_{\mathbf{k}\lambda}^\dagger b_n b_\ell d_m \right] - i[b_\ell, H_{\text{el}}].
\end{aligned}
\tag{11.164}
$$

Similarly

$$
\begin{aligned}
\dot{a}_{\mathbf{k}\lambda} &= -ika_{\mathbf{k}\lambda} + \frac{ie}{\sqrt{2k}} \sum_m \sum_n \left[C_{m+n+}^\lambda(-\mathbf{k}) b_m^\dagger b_n \right. \\
&\quad \left. + C_{m+n-}^\lambda(-\mathbf{k}) b_m^\dagger d_n^\dagger + C_{m-n+}^\lambda(-\mathbf{k}) b_n d_m + C_{m-n-}^\lambda(-\mathbf{k}) d_m d_n^\dagger \right].
\end{aligned}
\tag{11.165}
$$

As in the nonrelativistic formulation of Chapter 4, we use the formal solution of (11.165),

$$
\begin{aligned}
a_{\mathbf{k}\lambda}(t) &= a_{\mathbf{k}\lambda}(0)e^{-ikt} + \frac{ie}{\sqrt{2k}} \sum_m \sum_n \\
&\quad \times \left[C_{m+n+}^\lambda(-\mathbf{k}) \int_0^t dt' b_m^\dagger(t') b_n(t') e^{ik(t'-t)} \right. \\
&\quad + C_{m+n-}^\lambda(-\mathbf{k}) \int_0^t dt' b_m^\dagger(t') d_n^\dagger(t') e^{ik(t'-t)} \\
&\quad + C_{m-n+}^\lambda(-\mathbf{k}) \int_0^t dt' b_n(t') d_m(t') e^{ik(t'-t)} \\
&\quad \left. + C_{m-n-}^\lambda(-\mathbf{k}) \int_0^t d_m(t') d_n^\dagger(t') e^{ik(t'-t)} \right]
\end{aligned}
$$

$$= a_{\mathbf{k}\lambda}^{(o)}(t) + a_{\mathbf{k}\lambda}^{(s)}(t), \tag{11.166}$$

$$a_{\mathbf{k}\lambda}^{(o)}(t) \equiv a_{\mathbf{k}\lambda}(0)e^{-ikt}, \tag{11.167}$$

in (11.164). When we then take expectation values for an initial state $|\psi\rangle$ in which there are no photons, so that $a_{\mathbf{k}\lambda}^{(o)}(t)|\psi\rangle = \langle\psi|a_{\mathbf{k}\lambda}^{(o)\dagger}(t) = 0$, we find of course that there are contributions only from the source part, $a_{\mathbf{k}\lambda}^{(s)}(t)$, of the field. We obtain, analogously to the nonrelativistic theory,

$$
\begin{aligned}
\langle \dot{b}_\ell(t)\rangle &= -iE_\ell\langle b_\ell(t)\rangle - e^2 \sum_p \sum_m \sum_n \sum_{\mathbf{k}\lambda} \frac{1}{2k}[C_{\ell+p+}^\lambda(\mathbf{k})C_{m+n+}^\lambda(-\mathbf{k}) \\
&\quad \times \int_0^t dt' \langle b_p(t)b_m^\dagger(t')b_n(t')\rangle e^{ik(t'-t)} \\
&\quad + C_{\ell+p+}^\lambda(\mathbf{k})C_{m-n-}^\lambda(-\mathbf{k})\int_0^t dt'\langle b_p(t)d_m(t')d_n^\dagger(t')\rangle e^{ik(t'-t)} \\
&\quad + C_{\ell+p+}^\lambda(-\mathbf{k})C_{-m-n-}^{\lambda*}(-\mathbf{k})\int_0^t dt'\langle d_n(t')d_m^\dagger(t')b_p(t)\rangle e^{-ik(t'-t)} \\
&\quad + C_{\ell+p-}^\lambda(-\mathbf{k})C_{m+n-}^{\lambda*}(-\mathbf{k})\int_0^t dt'\langle b_m(t)d_p(t')d_n^\dagger(t')\rangle e^{-ik(t'-t)} \\
&\quad + \ \dots],
\end{aligned}
\tag{11.168}
$$

where we write explicitly only the first few terms.

We can now apply the Markovian approximation in essentially the same way as in Chapter 4. For instance, we make the approximation

$$
\begin{aligned}
\int_0^t dt'\langle b_p(t)b_m^\dagger(t')b_n(t')\rangle e^{-ik(t'-t)} &\cong \langle b_p(t)b_m^\dagger(t)b_n(t)\rangle \\
&\quad \times \int_0^t dt' e^{i(E_m-E_n+k)(t'-t)} \\
&\rightarrow \langle b_p(t)b_m^\dagger(t)b_n(t)\rangle \left[\pi\delta(k-E_n+E_m) - iP\frac{1}{k-E_n+E_m}\right] \\
&= \left[\delta_{pm}\langle b_n(t)\rangle - \langle b_m^\dagger(t)b_p(t)b_n(t)\rangle\right] \\
&\quad \times \left[\pi\delta(k-E_n+E_m) - iP\frac{1}{k-E_n+E_m}\right],
\end{aligned}
\tag{11.169}
$$

where we have used (4.98). The first term within the first bracket contributes

$$- e^2 \sum_p \sum_m \sum_n \sum_{\mathbf{k}\lambda} \frac{1}{2k} C^\lambda_{\ell+p+}(\mathbf{k}) C^\lambda_{m+n+}(-\mathbf{k}) \delta_{pm} \langle b_n(t) \rangle$$

$$\times \left[\pi \delta(k - E_n + E_m) - iP \frac{1}{k - E_n + E_m} \right]$$

$$= - e^2 \sum_m \sum_n \sum_{\mathbf{k}\lambda} \frac{1}{2k} C^\lambda_{\ell+m+}(\mathbf{k}) C^\lambda_{m+n+}(-\mathbf{k}) \langle b_n(t) \rangle$$

$$\times \left[\pi \delta(k - E_n + E_m) - iP \frac{1}{k - E_n + E_m} \right] \tag{11.170}$$

to the second term on the right side of (11.168). Part of this term, the one with $n = \ell$, contributes

$$- e^2 \sum_m \sum_{\mathbf{k}\lambda} \frac{1}{2k} C^\lambda_{\ell+m+}(\mathbf{k}) C_{m+\ell+}(-\mathbf{k}) [\pi \delta(k - E_\ell + E_m)$$

$$- iP \frac{1}{k - E_\ell + E_m}] \langle b_\ell(t) \rangle$$

$$\rightarrow \quad - \frac{e^2}{16\pi^3} \sum_m \int \frac{d^3k}{k} \sum_{\lambda=1}^{2} C^\lambda_{\ell+m+}(\mathbf{k}) C^\lambda_{m+\ell+}(-\mathbf{k}) [\pi \delta(k - E_\ell + E_m)$$

$$- iP \frac{1}{k - E_\ell + E_m}] \langle b_\ell(t) \rangle \tag{11.171}$$

in the mode continuum limit. In particular, the principal part term contributes[17]

$$- \frac{i\alpha}{4\pi^2} \sum_m \int \frac{d^3k}{k} \sum_{\lambda=1}^{2} \frac{C^\lambda_{\ell+m+}(\mathbf{k}) C^\lambda_{m+\ell+}(-\mathbf{k})}{E_\ell - E_m - k} \langle b_\ell(t) \rangle \tag{11.172}$$

to (11.168), i.e.,

$$\langle \dot{b}_\ell(t) \rangle \cong - i \left[E_\ell + \frac{\alpha}{4\pi^2} \sum_m \int \frac{d^3k}{k} \sum_{\lambda=1}^{2} \frac{C^\lambda_{\ell+m+}(\mathbf{k}) C^\lambda_{m+\ell+}(-\mathbf{k})}{E_\ell - E_m - k} \right]$$

$$\times \langle b_\ell(t) \rangle \ | \ \dots . \tag{11.173}$$

The second term in brackets represents a shift in the energy E_ℓ. Comparison with (11.47) shows that this shift is just the contribution to $W^X_D(\ell)$ from

---

[17] As usual the integrals are to be understood as Cauchy principal parts.

positive-energy intermediate states. The delta function part of (11.171) gives similarly a contribution to the spontaneous emission rate from state $|\ell\rangle$.

To get the full level shifts and spontaneous decay rates we must, of course, show that most of the terms not explicitly written in (11.172) either cancel or are negligible.[18] This is straightforward but cumbersome. The fact that we obtain the correct radiative shifts and decay rates, when we collect all the terms, shows that the Markovian (or Weisskopf–Wigner) approximation gives the same level shifts and widths as second-order perturbation theory, just as in the nonrelativistic theory of Chapter 4.

What is of interest here, however, is that the contribution to the level shift and width from the interaction $-e \int d^3x\, \psi^\dagger \boldsymbol{\alpha} \cdot \mathbf{A} \psi$ with the transverse electromagnetic field comes entirely from the *source* part of the field. Normal ordering of the field operators, in the calculation just outlined, removes all explicit contributions from the vacuum electromagnetic field. Thus the Lamb shift, apart from the electrostatic vacuum polarization contribution, is attributable to radiation reaction, just as in the nonrelativistic theory.

However, we could equally well have started by writing the Heisenberg equation (11.164) in the antinormally ordered form

$$
\dot{b}_\ell = \quad - \quad iE_\ell b_\ell + ie \sum_n \sum_{\mathbf{k}\lambda} \frac{1}{\sqrt{2k}} \left[C_{\ell+n+}^\lambda(\mathbf{k}) a_{\mathbf{k}\lambda} b_n + C_{\ell+n+}^\lambda(-\mathbf{k}) b_n a_{\mathbf{k}\lambda}^\dagger \right.
$$
$$
+ \quad C_{n-\ell+}^\lambda(\mathbf{k}) a_{\mathbf{k}\lambda} d_n + \ \dots \ \Big] - i[b_\ell, H_{\mathrm{el}}] \tag{11.174}
$$

in which the photon annihilation operators are at the extreme left and the photon creation operators are at the extreme right. Using the formal solution of this equation, and the corresponding Heisenberg equation for d_n, we can evaluate the righthand side of (11.174) in the Markovian approximation and then take expectation values for a state of no photons. When this is done we obtain explicit "vacuum field" contributions $\langle a_{\mathbf{k}\lambda}(0) a_{\mathbf{k}\lambda}^\dagger(0) \rangle = 1$. The entire procedure is exactly analogous to the nonrelativistic calculation of Section 4.11, and we draw the same conclusion: the Lamb shift, for instance, is attributable to either the source (radiation reaction) or vacuum field, or some combination of the two, depending on the ordering chosen for the commuting (equal-time) electron, positron, and photon annihilation and creation operators.[19]

---

[18] The "negligible" terms are those that oscillate at frequencies different from E_ℓ and are ignorable in a rotating-wave approximation. They are essentially the "Landau terms" mentioned briefly in Section 4.16.

[19] In the relativistic theory, of course, we have also the electrostatic vacuum polarization contribution attributable to the Dirac vacuum rather than the electromagnetic vacuum.

11.10   Discussion

In the final section of the preceding chapter it was noted that Oppenheimer in 1930 found that the coupling of an atomic electron to the quantized electromagnetic field led to an infinite energy shift. A similar result was found at about the same time by Weisskopf and Wigner in their work on the natural lineshape. Weisskopf (1972) recalls that he

> ... tried to convince Wigner that the integral could be made to vanish. Wigner said: "No, no, it is infinite." I didn't believe him, but he was right, of course. This paper, part of which became my thesis, was the first paper in which divergent integrals appeared. They have not yet been resolved; they are still there after 40 years. One ought to be ashamed of it.

The infinities are still with us after 60 years now.

Although the infinite self-energy calculated for a bound (or free) electron may be considered shameful, the extraction of finite numbers in fantastic agreement with experiment is also one of the great triumphs of twentieth-century science. Consider, for instance, the degree of agreement between theory and experiment for the anomalous magnetic moment of the electron (Section 3.13).

Whatever one thinks of the divergences in quantum field theory, renormalization seems logically necessary in order to avoid "double counting" of electromagnetic mass, for instance: if the electron mass in the Hamiltonian is the observed mass, then the electromagnetic mass calculated when the electron is coupled to the electromagnetic field had better be subtracted away, for it is already part of the observed mass. (See the discussion in Section 3.5 concerning the nonrelativistic theory of electromagnetic mass, and in Section 11.6 for the relativistic theory.) In other words, renormalization would be required *even if the electromagnetic mass were finite.* Indeed renormalization is required even in the *classical* theory of radiation reaction. The need for renormalization means in effect that a knowledge of the actual (presumably finite!) electromagnetic mass is not required for the calculation of observable phenomena such as the Lamb shift. Moreover, as noted near the end of Section 11.6, it is not altogether clear that the divergence of the electromagnetic mass calculated in low-order perturbation theory is in fact a real difficulty of QED.

In this chapter we have demonstrated a result that was already presumed to be true in earlier, nonrelativistic descriptions: as a consequence

---

In other words, the "Lamb shift" in this statement refers to the shift without the vacuum polarization contribution.

of the negative-energy states of relativistic theory, an electron is effectively spread out over a distance on the order of its Compton wavelength. This leads to a weaker divergence — logarithmic rather than linear - of the electron self-energy in relativistic QED. It might at first be thought that this weaker divergence is simply a consequence of the retardation factors $e^{\pm i \mathbf{k} \cdot \mathbf{x}}$ in the relativistic theory; retardation in the nonrelativistic theory, after all, also leads to a reduced divergence (Section 3.9). However, this is not the case. The Green function analogous to G_n (Section 3.9) in relativistic theory is $(E_n - \boldsymbol{\alpha} \cdot \mathbf{p} - \beta m - V - \omega_k)$ as opposed to $(E_n - \mathbf{p}^2/2m - V - \omega_k)$. The effect of retardation, then, is not to give an extra factor of k^2 in a denominator, as in (3.57), but only another term linear in k. The logarithmic divergence of δm in relativistic theory is attributable instead to virtual electron–positron pairs causing an effective spreading of the charge distribution of the point electron.

In the calculation of the Lamb shift it is also necessary to renormalize the electron charge, an effect that has little to do with electromagnetic field quantization. Charge renormalization is a consequence of the polarization of the Dirac vacuum by the electron. As noted in Section 9.4, an extremely strong electrostatic field can produce real electron–positron pairs. The treatment of vacuum polarization in this chapter implicitly assumes that the field is small compared with the critical value $E_{\mathrm{cr}} = m^2 c^3/e\hbar$ for pair production[20] (Section 9.4).

Virtual creation of electron–positron pairs results in effect in a coupling of the electromagnetic field to itself. For arbitrarily strong but slowly varying fields, and in particular for frequencies $\omega \ll m$, the self-interaction energy can be calculated to lowest order in α, as first done by Heisenberg and Euler (1936). The Heisenberg–Euler Lagrangian density is (Schwinger 1951)

$$\mathsf{L} = \frac{1}{2}(\mathbf{E}^2 - \mathbf{B}^2) + \frac{2\alpha}{45}\frac{1}{E_{\mathrm{cr}}^2}\left[(\mathbf{E}^2 - \mathbf{B}^2)^2 + 7(\mathbf{E} \cdot \mathbf{B})^2 + \ldots\right], \qquad (11.175)$$

where we write the first two terms of an expansion involving α/E_{cr}^2. Such a Lagrangian describes various QED phenomena associated with an effective nonlinear refractive index of the vacuum, including photon–photon scattering and "photon splitting." Because of the extremely high field strengths required for such phenomena, however, there is still no prospect of ever observing them (Bialynicka-Birula and Bialynicki-Birula, 1970).

---

[20] Even at a distance of a Compton wavelength from an electron the electric field strength is about 1/137 times this critical field. Recall also the discussion at the end of Section 9.4.

The Heisenberg–Euler Lagrangian can also be used to describe the effect
on an external field of the modification of the vacuum electromagnetic field
by parallel mirrors (Barton, 1990). Any physical effects of this modification
are too small to be measurable.

In this chapter we have shown, mostly following Kroll and Lamb (1949)
and French and Weisskopf (1949), how a finite level shift can be obtained in
spite of the sort of divergence first obtained by Weisskopf and Wigner and
Oppenheimer. The kinds of calculations summarized in this chapter are
seldom done now; they have long since been replaced mainly by covariant
perturbation theory and Feynman diagrams, as described in the following
chapter. However, the diagrammatic methods essentially just reproduce
the results obtained here in the old-fashioned way. In other words, *the
diagrammatic techniques do not represent a new physical theory, but "just"
a more convenient way of doing the calculations and renormalizing.* We
now turn to these methods.

11.11   Bibliography

Barton, G., "Faster-than-*c* Light Between Parallel Mirrors. The
  Scharnhorst Effect Rederived," *Phys. Lett.* **B237**, 559 (1990).
Bialynicka-Birula, Z. and I. Bialynicki-Birula, "Nonlinear Effects in
  Quantum Electrodynamics. Photon Propagation and Photon
  Splitting in an External Field," *Phys. Rev.* **D2**, 2341 (1970).
Dirac, P. A. M., in *From a Life of Physics*, ed. A. Salam et al.
  (World Scientific, Singapore, 1989).
Drake, G. W. F., "Quantum Electrodynamic Effects in Few-Electron
  Atomic Systems," *Adv. At. Mol. Phys.* **18**, 399 (1982).
Foldy, L. L. and E. Eriksen, "Some Physical Consequences of Vacuum
  Polarization," *Phys. Rev.* **95**, 1048 (1954).
French, J. B. and V. F. Weisskopf, "The Electromagnetic Shift of Energy
  Levels," *Phys. Rev.* **75**, 1240 (1949).
Gell-Mann, M. and F. Low, "Quantum Electrodynamics at Small
  Distances," *Phys. Rev.* **95**, 1300 (1954).
Glauber, R., W. Rarita, and P. Schwed, "Vacuum Polarization Effects
  on Energy Levels in μ-Mesonic Atoms," *Phys. Rev.* **120**, 609 (1960).
Grandy, W. T., Jr., *Relativistic Quantum Mechanics of Leptons and Fields*
  (Kluwer, Dordrecht, 1991a)
Grandy, W. T., Jr., "The Explicit Nonlinearity of Quantum Electrodynam-
  ics," in *The Electron*, ed. D. Hestenes and A. Weingartshofer
  (Kluwer, Dordrecht, 1991b).
Heisenberg, W. and H. Euler, "Folgerungen aus der Diracschen Theorie

des Positrons," *Z. Physik* **98**, 714 (1936).

Heitler, W., *The Quantum Theory of Radiation*, 3rd ed. (Oxford University Press, London, 1966).

Itzykson, C. and J.-B. Zuber, *Quantum Field Theory* (McGraw- Hill, New York, 1980).

Kroll, N. M. and W. E. Lamb, Jr., "On the Self-Energy of a Bound Electron," *Phys. Rev.* **75**, 388 (1949).

Oppenheimer, J. R., "Note on the Theory of the Interaction of Field and Matter," *Phys. Rev.* **35**, 461 (1930).

Petermann, A. and Y. Yamaguchi, "Corrections to the 3D-2P Transitions in μ-Mesonic Phosphorous and the Mass of the Muon," *Phys. Rev. Lett.* **2**, 359 (1959).

Schwinger, J., "On Gauge Invariance and Vacuum Polarization," *Phys. Rev.* **82**, 664 (1951).

Serber, R., "Linear Modifications in the Maxwell Field Equations," *Phys. Rev.* **48**, 49 (1935).

Uehling, E. A., "Polarization Effects in the Positron Theory," *Phys. Rev.* **48**, 55 (1935).

Weisskopf, V. F., "On the Self-Energy and the Electromagnetic Field of the Electron," *Phys. Rev.* **56**, 72 (1939).

Weisskopf, V. F., *Physics in the Twentieth Century: Selected Essays* (MIT Press, Cambridge, Mass., 1972), p. 4.

Weisskopf, V. F., *The Privilege of Being a Physicist* (W. H. Freeman and Company, New York, 1989), p. 161.

Chapter 12

Feynman Diagrams

> The theories about the rest of physics are very similar to the theory of quantum electrodynamics: they all involve the interaction of spin 1/2 objects (like electrons and quarks) with spin 1 objects (like photons, gluons, or W's) within a framework of amplitudes by which the probability of an event is the square of the length of an arrow. Why are all the theories of physics so similar in their structure?
> — Richard P. Feynman (1985)

12.1 Introduction

Following the Second World War, relativistically covariant formulations of quantum electrodynamics were developed by Tomonaga, Schwinger, Feynman, and Dyson. These formulations involved techniques for the subtraction of divergences in relativistically covariant ways, and the subtractions were shown to be equivalent to mass and charge renormalization. In particular, the diagrammatic methods developed by Feynman (1949a,b, 1950), and elucidated by Dyson, led to a simplification in the theory for higher order processes and a proof of renormalizability. The equivalence of the different approaches was shown by Dyson (1949a,b).[1]

Feynman's way was intuitive and did not require quantum field theory. After Dyson demonstrated the equivalence of Feynman's approach to the methods of quantum field theory, Wick (1950) showed how Feynman's diagrams could be obtained from an S-matrix expansion by operator algebraic methods.

---

[1] The reprint volume edited by Schwinger (1958) includes some of the historic papers of QED.

In this chapter we show how the diagrammatic methods of QED can be deduced from quantum field theory. We also describe the original and much simpler approach of Feynman's. We begin by reviewing first the interaction picture in general quantum theory, and then the S matrix and its iterative expansion in perturbation theory.

12.2   The Interaction Picture

The Schrödinger equation,

$$i\frac{\partial}{\partial t}|\Psi(t)\rangle = H|\Psi(t)\rangle = (H_0 + H_I)|\Psi(t)\rangle, \qquad (12.1)$$

describes the evolution in time of the state vector $|\Psi(t)\rangle$ of a system with Hamiltonian H, which for systems of interest to us is an "unperturbed" part H_0 plus a perturbed or "interaction" part H_I. The (unitary) time evolution operator $U(t, t_0)$ is defined such that

$$|\Psi(t)\rangle = U(t, t_0)|\Psi(t_0)\rangle, \qquad (12.2)$$

where $|\Psi(t_0)\rangle$ is the state vector at some initial time t_0. It follows from the Schrödinger equation that

$$i\frac{\partial}{\partial t}U(t, t_0) = HU(t, t_0), \qquad (12.3)$$

and, from (12.2), $U(t_0, t_0) = 1$, the unit operator.

In earlier chapters we found it useful to work in the Heisenberg picture, where the state vector stays fixed at $|\Psi(t_0)\rangle$ while operators evolve in time. A (time-independent) Schrödinger operator A_S with matrix elements $\langle\Psi_1(t)|A_S|\Psi_2(t)\rangle$ will have matrix elements $\langle\Psi_1(t_0)|A_H(t)|\Psi_2(t_0)\rangle$ in the Heisenberg picture. In order for these to be equal,

$$\begin{aligned}
\langle\Psi_1(t)|A_S|\Psi_2(t)\rangle &= \langle\Psi_1(t_0)|U^\dagger(t, t_0)A_S U(t, t_0)|\Psi_2(t_0)\rangle \\
&= \langle\Psi_1(t_0)|A_H(t)|\Psi_2(t_0)\rangle, \qquad (12.4)
\end{aligned}$$

or $A_H(t) = U^\dagger(t, t_0)A_S U(t, t_0)$. Equation (12.3) then implies the Heisenberg equation of motion

$$i\frac{d}{dt}A_H(t) = [A_H(t), H] \qquad (12.5)$$

for operators in the Heisenberg picture.

For reasons described later, it will be convenient in this chapter to work in the *interaction picture*. In this representation of quantum theory, both the state vectors and the operators evolve in time. To introduce this picture we write the time evolution operator $U(t, t_0)$ defined by (12.2) and (12.3) in the form

$$U(t, t_0) = U_0(t, t_0)u(t, t_0), \tag{12.6}$$

where $U_0(t, t_0)$, defined by

$$i\frac{\partial}{\partial t}U_0(t, t_0) = H_0 U_0(t, t_0) \tag{12.7}$$

and $U_0(t, t_0) = 1$, is the time evolution operator governed by the unperturbed Hamiltonian:

$$U_0(t, t_0) = e^{-iH_0(t-t_0)} . \tag{12.8}$$

From equations (12.3), (12.6), and (12.7),

$$i\frac{\partial U_0}{\partial t}u + iU_0\frac{\partial u}{\partial t} = H_0 U_0 u + iU_0\frac{\partial u}{\partial t} = (H_0 + H_I)U_0 u \tag{12.9}$$

or

$$i\frac{\partial u}{\partial t} = U_0^\dagger(t, t_0)H_I U_0(t, t_0)u(t) \equiv h_I(t)u, \tag{12.10}$$

with $u(t_0, t_0) = 1$. The interaction Hamiltonian in the interaction picture is

$$h_I(t) = U_0^\dagger(t, t_0)H_I U_0(t, t_0). \tag{12.11}$$

We define interaction-picture operators more generally by

$$A_I(t) = U_0^\dagger(t, t_0)A_S U_0(t, t_0). \tag{12.12}$$

Thus a matrix element in the Schrödinger picture can be written

$$
\begin{aligned}
\langle\Psi_1(t)|A_S|\Psi_2(t)\rangle &= \langle\Psi_1(t_0)|u^\dagger(t, t_0)U_0^\dagger(t, t_0)A_S U_0(t, t_0) \\
&\quad \times u(t, t_0)|\Psi_2(t_0)\rangle \\
&= \langle\Psi_1(t_0)|u^\dagger(t, t_0)A_I(t)u(t, t_0)|\Psi_2(t_0)\rangle \\
&= \langle\psi_1(t)|A_I(t)|\psi_2(t)\rangle,
\end{aligned} \tag{12.13}
$$

where the state vector in the interaction picture is defined by

$$|\psi(t)\rangle = u(t, t_0)|\Psi(t_0)\rangle = U_0^\dagger(t, t_0)|\Psi(t)\rangle. \tag{12.14}$$

Note that a state vector in the interaction picture evolves in time according to the equation

$$i\frac{\partial}{\partial t}|\psi(t)\rangle = i\frac{\partial u}{\partial t}|\Psi(t_0)\rangle = h_I(t)u(t,t_0)|\Psi(t_0)\rangle = h_I(t)|\psi(t)\rangle. \qquad (12.15)$$

Equation (12.10) has the formal solution

$$
\begin{aligned}
u(t,t_0) &= 1 - i\int_{t_0}^{t} dt' h_I(t')u(t',t_0)\\
&= 1 + (-i)\int_{t_0}^{t} dt_1 h_I(t_1) + (-i)^2 \int_{t_0}^{t} dt_1 h_I(t_1)\int_{t_0}^{t_1} dt_2 h_I(t_2)\\
&\quad + (-i)^3 \int_{t_0}^{t} dt_1 h_I(t_1)\int_{t_0}^{t_1} dt_2 h_I(t_2)\int_{t_0}^{t_2} dt_3 h_I(t_3) + \cdots,
\end{aligned}
$$

$$ (12.16) $$

which is sometimes called the *Dyson expansion*. Before continuing with our formal discussion, let us pause to recall a simple but important application of this expansion. Suppose that at $t = 0$ we have a system in the initial state $|\Psi\rangle = |i\rangle$, an eigenstate of H_0 with eigenvalue E_i: $H_0|i\rangle = E_i|i\rangle$. What is the probability, if a periodic perturbation $H_I = 2V\cos\omega t$ is applied, that the system will be found in the state $|f\rangle$ at the time t? The probability amplitude for the transition is

$$
\begin{aligned}
a_{fi}(t) &\equiv \langle f|\Psi(t)\rangle = \langle f|U(t,0)|i\rangle = \langle f|U_0(t,0)u(t,0)|i\rangle\\
&= e^{-iE_f t}\langle f|u(t,0)|i\rangle, \qquad (12.17)
\end{aligned}
$$

and we approximate it in lowest order perturbation theory by retaining only the first two terms of (12.16), assuming $\langle f|i\rangle = 0$:

$$
\begin{aligned}
a_{fi}(t) &\cong -ie^{-iE_f t}\int_0^t dt_1 \langle f|h_I(t_1)|i\rangle\\
&= -2ie^{-iE_f t}\int_0^t dt_1 \langle f|U_0^\dagger(t_1,0)VU_0(t_1,0)|i\rangle\cos\omega t_1\\
&= -2ie^{-iE_f t}\int_0^t dt_1 \langle f|e^{iH_0 t_1}Ve^{-iH_0 t_1}|i\rangle\cos\omega t_1\\
&= -2ie^{-iE_f t}\langle f|V|i\rangle\int_0^t dt_1 e^{i(E_f - E_i)t_1}\cos\omega t_1\\
&= -ie^{-iE_f t}\langle f|V|i\rangle\left[e^{i(E_f - E_i + \omega)t/2}\frac{\sin\frac{1}{2}(E_f - E_i + \omega)t}{\frac{1}{2}(E_f - E_i + \omega)}\right.\\
&\quad\left. + e^{i(E_f - E_i - \omega)t/2}\frac{\sin\frac{1}{2}(E_f - E_i - \omega)t}{\frac{1}{2}(E_f - E_i - \omega)}\right].
\end{aligned}
$$

$$ (12.18) $$

For upward, absorptive transitions, $E_f > E_i$ and $\sin[\frac{1}{2}(E_f - E_i + \omega)t]/[E_f - E_i + \omega]$ has many small-amplitude oscillations over a time $t >> E_f - E_i + \omega$ and may be ignored compared with $\sin[\frac{1}{2}(E_f - E_i - \omega)t]/[E_f - E_i - \omega]$, which in fact will grow with t if the resonance condition $E_f - E_i = \omega$ is satisfied. We therefore approximate the transition probability by

$$|a_{fi}(t)|^2 \cong |\langle f|V|i\rangle|^2 \frac{\sin^2 \frac{1}{2}(E_f - E_i - \omega)t}{[\frac{1}{2}(E_f - E_i - \omega)]^2} . \qquad (12.19)$$

This leads in the familiar way to the Fermi golden rule for the transition *rate* when we suppose there is a continuum of possible final states with density $\rho(E_f)$. Thus, from the delta-function property of $\sin^2 xt/x^2$,

$$\begin{aligned}
|a_{fi}(t)|^2 &\rightarrow \int dE_f \rho(E_f)|a_{fi}(t)|^2 \\
&\cong |\langle f|V|i\rangle|^2 \int dE_f \rho(E_f) \frac{\sin^2 \frac{1}{2}(E_f - E_i - \omega)t}{[\frac{1}{2}(E_f - E_i - \omega)]^2} \\
&\rightarrow |\langle f|V|i\rangle|^2 \rho(E_f = E_i + \omega)2\pi t , \qquad (12.20)
\end{aligned}$$

or

$$R_{fi} = \frac{d}{dt}|a_{fi}(t)|^2 = \frac{2\pi}{\hbar}|\langle f|V|i\rangle|^2 \rho(E_f = E_i + \hbar\omega) , \qquad (12.21)$$

where we reinstate $\hbar$ in order to recover the conventional expression of Fermi's golden rule for the transition rate.

There is a small point worth mentioning about this result, simple and well-known though it is. Whereas the energy conservation condition $E_f = E_i + \hbar\omega$ is naturally interpretable in terms of the absorption of a photon of energy $\hbar\omega$, when the perturbation results from the coupling of our system with the electromagnetic field, we have derived it *without quantizing the field*. In fact many phenomena conventionally associated with photons, such as the photoelectric effect, can be adequately explained without invoking photons, i.e., without field quantization (Scully and Sargent, 1972). We can similarly derive "n-photon" absorption (and stimulated emission) rates without field quantization.[2] Of course the classical treatment of the field will ultimately fail even for absorption and stimulated emission processes. For instance, it predicts that a field of exactly one photon has a nonvanishing probability of exciting *two* detectors.[3]

---

[2] That the stimulated emission rate can be obtained without field quantization was noted by P. A. M. Dirac, *Proc. Camb. Phil. Soc.* **26**, 361 (1930) Spontaneous emission, however, demands field quantization for an explanation of some of its essential features. See Section 4.15 and Milonni (1976). Expressions for n-photon absorption rates are derived in many places; see, for instance, P. W. Milonni and B. Sundaram, *Progress in Optics*, Vol. 31, ed. E. Wolf (Elsevier, Amsterdam, 1993).

[3] See, for instance, Milonni (1984).

12.3   The S Matrix: Perturbation Theory

We observe that each term in the series (12.16) is symmetric in $t_1, t_2, ..., t_n$ and write $u(t) \equiv u(t, -\infty)$ in the equivalent form[4]

$$u(t) = \sum_{n=0}^{\infty} \frac{(-i)^n}{n!} \int_{-\infty}^{t} dt_1 \int_{-\infty}^{t} dt_2 ... \int_{-\infty}^{t} dt_n P h_I(t_n)...h_i(t_2)h_I(t_1).$$

(12.22)

Here $t_0 \to -\infty$ formally and the "chronological product" P involves arrangement of the operators in the order of increasing time from right to left, i.e., $t_n > t_{n-1} > ... > t_2 > t_1$.

Probability amplitudes for states of the system at $t = \infty$ are determined by the S matrix, which is defined as $u(\infty)$. Replacing $h_I(t_i)$ in (12.22) by the interaction Hamiltonian density $h_I(t_i)$, we write

$$S = \sum_{n=0}^{\infty} \frac{(-i)^n}{n!} \int d^4x_1 \int d^4x_2 ... \int d^4x_n P h_I(x_n)...h_I(x_2)h_I(x_1), \quad (12.23)$$

where the integrations extend over all space and time.

The S matrix element $S_{fi} = \langle f|u(\infty)|i\rangle$ is presumably the probability amplitude for the transition from an initial, noninteracting "bare" state $|i\rangle$ to a final "bare" state $|f\rangle$. However, things are hardly quite so simple, as even a "noninteracting" electron, say, still has a self-interaction associated with radiation reaction and electromagnetic mass, and the electron charge is itself due in part to the polarization of the vacuum by the electron. In other words, the electron is always interacting with the vacuum electromagnetic and Dirac fields, if nothing else. We will ignore such subtleties, which in fact pose no real difficulties, and proceed with our goal of seeing where the Feynman diagrams come from and why they are useful.

Equation (12.23) is a perturbation series in the interaction h_I. The $n = 0$ term, $S^{(0)} = 1$, contributes $\langle f|i\rangle = 0$ to the transition amplitude for orthogonal initial and final states. The first-order term,

$$S^{(1)} = -i \int d^4x\, h_I(x), \tag{12.24}$$

contributes

$$S_{fi}^{(1)} = -i \int d^4x \langle f|h_I(x)|i\rangle . \tag{12.25}$$

---

[4] This is easily shown for the $n = 2$ term, for example, by integrating by parts and keeping track of the order of the operators $h_I(t_1)$ and $h_I(t_2)$.

The interaction Hamiltonian density appropriate to QED can be read off from (11.5): in the Lorentz gauge,

$$h_I(x) = e\overline{\psi}(x)\gamma_\mu\psi(x)A^\mu(x) = e\overline{\psi}(x)\rlap{/}A(x)\psi(x), \qquad (12.26)$$

whereas in the Coulomb gauge, from (11.25),

$$
\begin{aligned}
h_I(x) &= -e\psi^\dagger(x)\boldsymbol{\alpha}\cdot\mathbf{A}\psi(x) + \frac{e^2}{8\pi}\int d^3y\,\frac{\rho(\mathbf{x},t)\rho(\mathbf{y},t)}{|\mathbf{x}-\mathbf{y}|} \\
&= -e\overline{\psi}(x)\rlap{/}A(x)\psi(x) + \frac{e^2}{8\pi}\int d^3y\,\frac{\rho(\mathbf{x},t)\rho(\mathbf{y},t)}{|\mathbf{x}-\mathbf{y}|}, \qquad (12.27)
\end{aligned}
$$

where $\mathbf{A}$ is transverse. We will work in the Lorentz gauge.

Consider the example of electron scattering from the Coulomb potential $A^0(x) = -Ze/4\pi|\mathbf{x}|$, in which case (12.25) becomes

$$S_{fi}^{(1)} = iZ\alpha\int d^4x\langle f|\overline{\psi}(x)\gamma^0\psi(x)|i\rangle\frac{1}{|\mathbf{x}|}. \qquad (12.28)$$

The initial and final states are both states of one electron and no positrons, and so the nonvanishing contributions to (12.28) involve an electron annihilation operator from $\psi(x)$ and an electron creation operator from $\overline{\psi}(x)$, where $\psi(x)$ and $\overline{\psi}(x) = \psi^\dagger(x)\gamma^0$ are defined by (10.90) and (10.91). Denoting the initial electron momentum and spin by $\mathbf{p}_i$ and s_i, we have[5]

$$
\begin{aligned}
\psi(x)|i\rangle &= \psi(x)|\mathbf{p}_i,s_i\rangle = C_i\psi(x)b_i^\dagger(p_i)|0\rangle \\
&= C_i\int\frac{d^3p}{(2\pi)^3}\frac{m}{E}\sum_{j=1}^2 b_j(p)u^j(p)e^{-ip\cdot x}b_i^\dagger(p_i)|0\rangle \\
&= C_i\int\frac{d^3p}{(2\pi)^3}\frac{m}{E}\sum_{j=1}^2 u^j(p)e^{-ip\cdot x}\{b_j(p),b_i^\dagger(p_i)\}|0\rangle \\
&= C_i u^i(p_i)e^{-ip_i\cdot x}, \qquad (12.29)
\end{aligned}
$$

where we have used the fermion anticommutator (10.92). C_i is a normalization constant determined by $\langle i|i\rangle = 1$:

$$
\begin{aligned}
|C_i|^2\langle 0|b_i(p_i)b_i^\dagger(p_i)|0\rangle &= |C_i|^2\langle 0|\{b_i(p_i),b_i^\dagger(p_i)\}|0\rangle \\
&= |C_i|^2(2\pi)^3\frac{E_i}{m}\delta^3(0) = |C_i|^2\frac{E_i}{m}V, \qquad (12.30)
\end{aligned}
$$

where V is a volume defined by taking $\mathbf{p}\to 0$ in the representation $\delta^3(\mathbf{p}) = (1/2\pi)^3\int d^3x e^{i\mathbf{p}\cdot\mathbf{x}}$ of the delta function. Thus we take $C_i = \sqrt{m/E_iV}$. Similarly, for a final state of momentum $\mathbf{p}_f$ and spin s_f (Figure 12.1),

---

[5] The spin is labelled by the subscript i on $b_i^\dagger$ and by the superscript i on u^i.

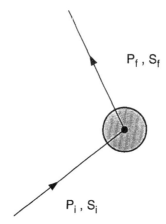

Figure 12.1: Scattering of an electron by a Coulomb potential.

$$\langle f|\overline{\psi}(x) = C_f \overline{u}^f(p_f)e^{ip_f \cdot x} = \sqrt{\frac{m}{E_f V}}\overline{u}^f(p_f)e^{ip_f \cdot x} \ . \qquad (12.31)$$

Then

$$
\begin{aligned}
S_{fi}^{(1)} &= iZ\alpha\sqrt{\frac{m^2}{E_i E_f}}\frac{1}{V}\int d^4 x\, \overline{u}^f(p_f)\gamma^0 u^i(p_i)e^{i(p_f - p_i)\cdot x}\frac{1}{|\mathbf{x}|}\\
&= iZ\alpha\sqrt{\frac{m^2}{E_i E_f}}\frac{1}{V}\overline{u}^f(p_f)\gamma^0 u^i(p_i)\int d^3 x\frac{e^{-i(\mathbf{p}_f - \mathbf{p}_i)\cdot \mathbf{x}}}{|\mathbf{x}|}\\
&\quad \times \int dt\, e^{i(E_f - E_i)t}\\
&= 4\pi iZ\alpha\sqrt{\frac{m^2}{E_i E_f}}\frac{1}{V}\overline{u}^f(p_f)\gamma^0 u^i(p_i)\frac{1}{|\mathbf{q}|^2}\int dt\, e^{i(E_f - E_i)t} \ ,
\end{aligned}
$$
$$(12.32)$$

where $\mathbf{q} \equiv \mathbf{p}_f - \mathbf{p}_i$, and

$$\left|S_{fi}^{(1)}\right|^2 = (4\pi Z\alpha)^2\frac{m^2}{E_i E_f V^2}\frac{|\overline{u}^f(p_f)\gamma^0 u^i(p_i)|^2}{|\mathbf{q}|^4}\left|\int dt\, e^{i(E_f - E_i)t}\right|^2 . \quad (12.33)$$

The integration over t extends formally from $-\infty$ to ∞. We can write instead

$$\left|\int dt e^{i(E_f - E_i)t}\right|^2 = \left|\int_{-T/2}^{T/2} dt e^{i(E_f - E_i)t}\right|^2 = \frac{\sin^2 \frac{1}{2}(E_f - E_i)T}{[\frac{1}{2}(E_f - E_i)]^2}$$
$$\longrightarrow 2\pi T\delta(E_f - E_i) \tag{12.34}$$

for large T. The transition rate $\left|S_{fi}^{(1)}\right|^2 /T$ is therefore

$$R_{fi}^{(1)} = 2\pi(4\pi Z\alpha)^2 \frac{m^2}{E_i E_f} \frac{1}{V^2} \frac{|\bar{u}^f(p_f)\gamma^0 u^i(p_i)|^2}{|\mathbf{q}|^4}\delta(E_f - E_i), \tag{12.35}$$

or, when we sum over possible final momentum states,

$$R \equiv \int \frac{V d^3 p_f}{(2\pi)^3} R_{fi}^{(1)} = V \int \frac{p_f^2 dp_f d\Omega_f}{(2\pi)^3} R_{fi}^{(1)}$$
$$= 4Z^2\alpha^2 m^2 \int dp_f d\Omega_f p_f^2 \frac{1}{E_i E_f} \frac{1}{V} \frac{|\bar{u}^f(p_f)\gamma^0 u^i(p_i)|^2}{|\mathbf{q}|^4}\delta(E_f - E_i). \tag{12.36}$$

The incident flux is $|\mathbf{v}_i|/V = |\mathbf{p}_i|/E_i V$, which leads us to the definition of the differential scattering cross section:

$$\frac{d\sigma}{d\Omega} \equiv 4Z^2\alpha^2 m^2 \int \frac{dp_f p_f^2}{E_i E_f} \frac{1}{V} \left(\frac{|\mathbf{p}_i|}{E_i V}\right)^{-1} \frac{|\bar{u}^f(p_f)\gamma^0 u^i(p_i)|^2}{|\mathbf{q}|^4}\delta(E_f - E_i)$$
$$= \frac{4Z^2\alpha^2 m^2}{|\mathbf{q}|^4}|\bar{u}^f(p_f)\gamma^0 u^i(p_i)|^2, \tag{12.37}$$

where we have used the fact that $dp_f p_f = dE_f E_f$, which follows trivially from $E_f = \sqrt{p_f^2 + m^2}$.

If the polarization of the final state is not observed, and the incident electron is equally likely to have spin up or down, then the experimentally relevant cross section is obtained as usual by averaging over initial states and summing over final states:

$$\frac{d\sigma}{d\Omega} \longrightarrow \frac{4Z^2\alpha^2 m^2}{2|\mathbf{q}|^4} \sum_{s_i, s_f} |u^f(p_f)\gamma^0 u^i(p_i)|^2. \tag{12.38}$$

The evaluation of the sum is a straightforward exercise facilitated by the use of "trace theorems." We simply record the result (Bjorken and Drell,

1064):

$$\frac{d\sigma}{d\Omega} = \frac{Z^2\alpha^2}{4\beta^2|\mathbf{p}_i|^2\sin^4(\theta/2)}\left(1 - \beta^2\sin^2\frac{\theta}{2}\right), \qquad (12.39)$$

with $|\mathbf{q}|^2 = 4|\mathbf{p}|^2\sin^2(\theta/2)$ and $\beta = |\mathbf{p}_i|/E_i$. This is the well-known Mott cross section for Rutherford scattering, which in the nonrelativistic limit $\beta \ll 1$ reduces to the (classical) Rutherford formula,

$$\frac{d\sigma}{d\Omega} \cong \frac{Z^2\alpha^2}{4\sin^4(\theta/2)}\frac{E_i^2}{|\mathbf{p}_i|^4} \cong \frac{Z^2\alpha^2}{4m^2v_i^4\sin^4(\theta/2)}. \qquad (12.40)$$

12.4   Second Order

The second-order term in the expansion (12.23) is

$$
\begin{aligned}
S^{(2)} &= -\frac{1}{2}\int d^4x_1\int d^4x_2 P\mathrm{h}_I(x_2)\mathrm{h}_I(x_1)\\
&= -\frac{e^2}{2}\int d^4x_1\int d^4x_2 P\overline{\psi}(x_2)A(x_2)\psi(x_2)\overline{\psi}(x_1)A(x_1)\psi(x_1).
\end{aligned}
$$
$$(12.41)$$

Now the advantage of working in the interaction picture becomes clear.[6] In the interaction picture the operators $\psi(x)$ and $A(x)$ evolve in time as unperturbed operators satisfying equations of motion for *free* (unperturbed) fields. In particular, they can be expanded in terms of bare (free-field) annihilation and creation operators acting on bare states, with time dependence of the type $a_k(t) = a_k(0)e^{-iE_k t}$. The Dirac operators commute for all times with Maxwell field operators. Thus

$$S^{(2)} = -\frac{e^2}{2}\int d^4x_1\int d^4x_2 P\overline{\psi}(x_2)\gamma^\mu\psi(x_2)\overline{\psi}(x_1)\gamma^\nu\psi(x_1)A_\mu(x_2)A_\nu(x_1). \qquad (12.42)$$

It will be convenient to replace the chronological product P by the time-ordering operator T introduced in Section 10.9. For the (boson) electromagnetic field operators $A(x)$ the two are equivalent, whereas for the (fermion) operators $\psi(x)$ there is at most a sign difference. However, since the $\psi(x)$ operators occur in pairs in (12.42), no sign difference arises:

$$
\begin{aligned}
S^{(2)} &= -\frac{e^2}{2}\int d^4x_1\int d^4x_2 T\overline{\psi}(x_2)\gamma^\mu\psi(x_2)\overline{\psi}(x_1)\gamma^\nu\psi(x_1)\\
&\quad \times A_\mu(x_2)A_\nu(x_1).
\end{aligned}
$$
$$(12.43)$$

---

[6] The advantages of the interaction representation were exploited by Tomonaga and Schwinger in their formulations of QED.

At this point we can plug in the expansions of the ψ and A operators in terms of annihilation and creation operators and evaluate matrix elements for any initial and final states of the Dirac and Maxwell fields. A great simplification can be realized, however, by taking a different approach, the one that leads to Feynman diagrams.

The basic idea is to use normally ordered field operators, where creation operators appear to the left of annihilation operators. The vacuum expectation value of a normally ordered product vanishes, since $a|0\rangle = \langle 0|a^\dagger = 0$, where a and $a^\dagger$ are annihilation and creation operators for either bosons or fermions. Furthermore a matrix element of a normally ordered product does not involve emission and reabsorption of "intermediate" particles that arise from a term like $aa^\dagger$ in a nonnormal product. For these reasons, normal ordering of annihilation and creation operators simplifies the evaluation of matrix elements.

Therefore, rather than evaluating the matrix elements of (12.43) directly, taking the operators in the order in which they appear, we now want to take advantage of the properties of normally ordered products. That is, we want to express the time-ordered products in (12.43) in terms of normally ordered products.

Consider first the time-ordered product $TA_\mu(x_2)A_\nu(x_1)$. For $t_2 > t_1$ this is, by definition, $A_\mu(x_2)A_\nu(x_1)$. Now

$$A(x) = A^{(+)}(x) + A^{(-)}(x), \tag{12.44}$$

where $A^{(+)}(x)$ and $A^{(-)}(x)$ are respectively the annihilation and creation parts of $A(x)$. (Sections 3.11, 7.4, and 10.8) Then

$$
\begin{aligned}
A_\mu(x_2)A_\nu(x_1) &= A_\mu^{(+)}(x_2)A_\nu^{(+)}(x_1) + A_\mu^{(+)}(x_2)A_\nu^{(-)}(x_1) \\
&\quad + A_\mu^{(-)}(x_2)A_\nu^{(+)}(x_1) + A_\mu^{(-)}(x_2)A_\nu^{(-)}(x_1) \\
&= A_\mu^{(+)}(x_2)A_\nu^{(+)}(x_1) + A_\mu^{(-)}(x_2)A_\nu^{(-)}(x_1) \\
&\quad + A_\nu^{(-)}(x_1)A_\mu^{(+)}(x_2) + A_\mu^{(-)}(x_2)A_\nu^{(+)}(x_1) \\
&\quad + [A_\mu^{(+)}(x_2), A_\nu^{(-)}(x_1)] \\
&= N[A_\mu(x_2)A_\nu(x_1)] + [A_\mu^{(+)}(x_2), A_\nu^{(-)}(x_1)] \tag{12.45}
\end{aligned}
$$

where $N[A_\mu(x_2)A_\nu(x_1)]$ is the normally ordered product of $A_\mu(x_2)$ and $A_\nu(x_1)$. Since $A_\mu^{(+)}(x_\lambda)|0\rangle = 0$, we can write the commutator equivalently as a vacuum expectation value:

$$
\begin{aligned}
[A_\mu^{(+)}(x_2), A_\nu^{(-)}(x_1)] &= \langle 0|A_\mu^{(+)}(x_2)A_\nu^{(-)}(x_1)|0\rangle \\
&= \langle 0|A_\mu(x_2)A_\nu(x_1)|0\rangle , \tag{12.46}
\end{aligned}
$$

which is a simple generalization of $[a, a^\dagger] - \langle 0|aa^\dagger|0\rangle = \langle 0|(a + a^\dagger)(a + a^\dagger)|0\rangle$. Therefore

$$A_\mu(x_2)A_\nu(x_1) = N[A_\mu(x_2)A_\nu(x_1)] + \langle 0|A_\mu(x_2)A_\nu(x_1)|0\rangle \qquad (12.47)$$

for $t_2 > t_1$, and similarly, for $t_1 > t_2$, the time-ordered product is

$$A_\mu(x_1)A_\nu(x_2) = N[A_\mu(x_2)A_\nu(x_1)] + \langle 0|A_\nu(x_1)A_\mu(x_2)|0\rangle . \qquad (12.48)$$

These two equations may be combined to give

$$TA_\mu(x_2)A_\nu(x_1) = N[A_\mu(x_2)A_\nu(x_1)] + \langle 0|TA_\mu(x_2)A_\nu(x_1)|0\rangle . \qquad (12.49)$$

But the vacuum expectation value on the right is just i times $D_{\mu\nu}(x_2, x_1)$, the photon propagator of Section 10.9, from which

$$TA_\mu(x_2)A_\nu(x_1) = N[A_\mu(x_2)A_\nu(x_1)] + iD_{\mu\nu}(x_2, x_1). \qquad (12.50)$$

For the Dirac field we write, analogously to (12.44),

$$\psi(x) = \psi^{(+)}(x) + \psi^{(-)}(x), \qquad (12.51)$$

where the annihilation and creation parts are respectively

$$\psi^{(+)}(x) = \int \frac{d^3p}{(2\pi)^3} \frac{m}{E} \sum_{i=1}^{2} b_i(p)u^i(p)e^{-ip\cdot x} , \qquad (12.52)$$

$$\psi^{(-)}(x) = \int \frac{d^3p}{(2\pi)^3} \frac{m}{E} \sum_{i=1}^{2} d_i^\dagger(p)v^i(p)e^{ip\cdot x} . \qquad (12.53)$$

Similarly [equation (10.91)]

$$\overline{\psi}(x) = \overline{\psi}^{(+)}(x) + \overline{\psi}^{(-)}(x), \qquad (12.54)$$

$$\overline{\psi}^{(+)}(x) = \int \frac{d^3p}{(2\pi)^3} \frac{m}{E} \sum_{i=1}^{2} d_i(p)\overline{v}^i(p)e^{-ip\cdot x} , \qquad (12.55)$$

$$\overline{\psi}^{(-)}(x) = \int \frac{d^3p}{(2\pi)^3} \frac{m}{E} \sum_{i=1}^{2} b_i^\dagger(p)\overline{u}^i(p)e^{ip\cdot x} , \qquad (12.56)$$

and, from the fermion algebra of the electron annihilation and creation operators,

$$\{\psi_\alpha^{(+)}(x), \overline{\psi}_\beta^{(-)}(x')\} = \langle 0|\psi_\alpha^{(+)}(x)\overline{\psi}_\beta^{(-)}(x')|0\rangle = \langle 0|\psi_\alpha(x)\overline{\psi}_\beta(x')|0\rangle, \qquad (12.57)$$

$$\{\overline{\psi}_\alpha^{(+)}(x), \psi_\beta^{(-)}(x')\} = \langle 0|\overline{\psi}_\alpha^{(+)}(x)\psi_\beta^{(-)}(x')|0\rangle = \langle 0|\overline{\psi}_\alpha(x)\psi_\beta(x')|0\rangle. \quad (12.58)$$

All other anticommutators of the fields (12.52), (12.53), (12.55), and (12.56) vanish. From these properties it follows easily that

$$\begin{aligned}
T\psi_\alpha(x_2)\overline{\psi}_\beta(x_1) &- N[\psi_\alpha(x_2)\overline{\psi}_\beta(x_1)] + \langle 0|T\psi_\alpha(x_2)\overline{\psi}_\beta(x_1)|0\rangle \\
&= N[\psi_\alpha(x_2)\overline{\psi}_\beta(x_1)] + iS_F(x_2, x_1)_{\alpha\beta}, \quad (12.59)
\end{aligned}$$

where $S_F(x_2, x_1)$ is the Feynman propagator defined by (10.193). Here

$$\begin{aligned}
N[\psi_\alpha(x_2)\overline{\psi}_\beta(x_1)] &= \psi_\alpha^{(+)}(x_2)\overline{\psi}_\beta^{(+)}(x_1) + \psi_\alpha^{(-)}(x_2)\overline{\psi}_\beta^{(-)}(x_1) \\
&\quad + \psi_\alpha^{(-)}(x_2)\overline{\psi}_\beta^{(+)}(x_1) - \overline{\psi}_\beta^{(-)}(x_1)\psi_\alpha^{(+)}(x_2). \\
&\quad\quad\quad\quad\quad\quad\quad\quad\quad\quad\quad\quad\quad (12.60)
\end{aligned}$$

Note the similarity between (12.50) and (12.59). Each involves a field propagator, the photon propagator in (12.50) and the Feynman (electron) propagator in (12.59). Note also the minus sign in the last term of (12.60), which is a consequence of the fermion character of the Dirac field, i.e.,
$$N[\psi_\alpha^{(+)}(x_2)\overline{\psi}_\beta^{(-)}(x_1)] = -N[\overline{\psi}_\beta^{(-)}(x_1)\psi_\alpha^{(+)}(x_2)] = -\overline{\psi}_\beta^{(-)}(x_1)\psi_\alpha^{(+)}(x_2).$$

Writing (12.43) in a form in which the components of ψ and $\overline{\psi}$ are indicated,

$$\begin{aligned}
S_{fi}^{(2)} &= -\frac{e^2}{2}\gamma_{\alpha\beta}^\mu\gamma_{\alpha'\beta'}^\nu \int d^4x_1 \int d^4x_2 \langle f|T\overline{\psi}_\alpha(x_2)\psi_\beta(x_2) \\
&\quad \times \overline{\psi}_{\alpha'}(x_1)\psi_{\beta'}(x_1)A_\mu(x_2)A_\nu(x_1)|i\rangle, \quad (12.61)
\end{aligned}$$

it is obvious that the normal ordering of the ψ and $\overline{\psi}$ operators will lead to the appearance of the Feynman propagator, as in (12.59). Obvious too is the fact that converting the time-ordered product to a form involving normally ordered products is messy even in second order. Before showing how to reduce the labor required to write time-ordered products in terms of normal-ordered products, let us consider two such normal-ordered constituents of $S_{fi}^{(2)}$.

12.5 Example: Compton Scattering

In the course of rewriting the time-ordered product in (12.61) in a form involving normal-ordered products, we will pick up a term

$$2i\overline{\psi}_\alpha^{(-)}(x_2)S_F(x_2, x_1)_{\beta\alpha'}\psi_{\beta'}^{(+)}(x_1)N[A_\mu(x_2)A_\nu(x_1)], \quad (12.62)$$

resulting from the first term of (12.49a) and the second term of (12.59). The factor 2 arises because we can "contract" either $\psi_\beta(x_2)\psi_{\alpha'}(x_1)$ or $\overline{\psi}_\alpha(x_2)\psi_{\beta'}(x_1)$ in (12.61), each contraction giving rise to a Feynman propagator S_F, and their contributions to (12.61) are equal. This results in the contribution

$$S_{fi}^{(2)}(C) \equiv -ie^2\gamma_{\alpha\beta}^\mu\gamma_{\alpha'\beta'}^\nu \int d^4x_1 \int d^4x_2 S_F(x_2, x_1)_{\beta\alpha'}$$

$$\times \langle f|\overline{\psi}_\alpha^{(-)}(x_2)\psi_{\beta'}^{(+)}(x_1)N[A_\mu(x_2)A_\nu(x_1)]|i\rangle \quad (12.63)$$

to the second-order S-matrix element between initial and final states $|i\rangle$ and $|f\rangle$. $S_{fi}^{(2)}(C)$ describes a process in which an electron is annihilated at x_1 and an electron is created at x_2. The operator $N[A_\mu(x_2)A_\nu(x_1)]$ will annihilate or create a photon at x_1 and create or annihilate a photon at x_2. In other words, $S_{fi}^{(2)}(C)$ describes Compton scattering, where

$$|i\rangle = |\mathbf{p}_i, s_i; \mathbf{k}_i, \lambda_i\rangle \quad (12.64)$$

is a state with an electron of momentum $\mathbf{p}_i$ and spin s_i and a photon occupying the mode $(\mathbf{k}_i, \lambda_i)$, and similarly

$$|f\rangle = |\mathbf{p}_f, s_f; \mathbf{k}_f, \lambda_f\rangle . \quad (12.65)$$

The nonvanishing matrix elements of $N[A_\mu(x_2)A_\nu(x_1)]$ result from $A_\mu^{(-)}(x_2)A_\nu^{(+)}(x_1)$ and $A_\nu^{(-)}(x_1)A_\mu^{(+)}(x_2)$, so that

$$S_{fi}^{(2)}(C) = -ie^2 \int d^4x_1 \int d^4x_2 S_F(x_2, x_1)$$

$$\times \left[\langle f|A_\mu^{(-)}(x_2)\overline{\psi}^{(-)}(x_2)\gamma^\mu S_F(x_2, x_1)\gamma^\nu \psi^{(+)}(x_1)A_\nu^{(+)}(x_1)|i\rangle \right.$$

$$+ \left. \langle f|A_\nu^{(-)}(x_1)\overline{\psi}^{(-)}(x_2)\gamma^\mu S_F(x_2, x_1)\gamma^\nu \psi^{(+)}(x_1)A_\mu^{(+)}(x_2)|i\rangle \right]$$

$$\equiv S_{fi}^{(2)}(C1) + S_{fi}^{(2)}(C2). \quad (12.66)$$

We can represent these two amplitudes by the Feynman diagrams shown in Figure 12.2. The diagram $C1$ represents the process in which the incident electron (solid line) and photon (wavy line) are annihilated at x_1, and the final electron and photon are both created at x_2. Between x_1 and x_2 the electron propagates with amplitude determined by the propagator $S_F(x_2, x_1)$. The diagram $C2$ represents the process in which the incident photon is annihilated at x_2 and the final photon is created at x_1.

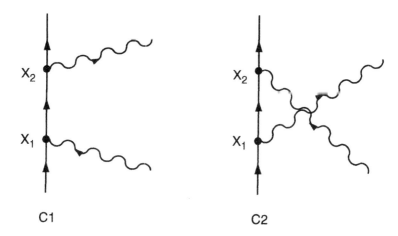

Figure 12.2: Feynman diagrams for the two amplitudes $C1$ and $C2$ contributing to Compton scattering.

Consider in a bit more detail the amplitude for the process $C1$. For the initial state $|i\rangle$ given by (12.64),

$$\psi^{(+)}(x_1)A_\nu^{(+)}(x_1)|i\rangle = \int \frac{d^3p}{(2\pi)^3}\frac{m}{E}\sum_{j=1}^{2} b(\mathbf{p},s_j)u(\mathbf{p},s_j)e^{-ip\cdot x_1}$$

$$\times \int \frac{d^3k}{(2\pi)^3}\frac{1}{2\omega_k}\sum_{\lambda=0}^{3} a(\mathbf{k},\lambda)e^{-ik\cdot x_1}e_\nu(\mathbf{k},\lambda)|\mathbf{p}_i,s_i;\mathbf{k}_i,\lambda_i\rangle$$

$$= C_i \int \frac{d^3p}{(2\pi)^3}\frac{m}{E}\sum_{j=1}^{2} u(\mathbf{p},s_j)e^{-ip\cdot x_1}\int \frac{d^3k}{(2\pi)^3}\frac{1}{2\omega_k}\sum_{\lambda=0}^{3} e_\nu(\mathbf{k},\lambda)$$

$$\times e^{-ik\cdot x_1}b(\mathbf{p},s_j)b^\dagger(\mathbf{p}_i,s_i)a(\mathbf{k},\lambda)a^\dagger(\mathbf{k}_i,\lambda_i)|0\rangle$$

$$= C_i \int \frac{d^3p}{(2\pi)^3}\frac{m}{E}\sum_{j=1}^{2} u(\mathbf{p},s_j)e^{-ip\cdot x_1}\int \frac{d^3k}{(2\pi)^3}\frac{1}{2\omega_k}\sum_{\lambda=0}^{3} e_\nu(\mathbf{k},\lambda)$$

$$\times e^{-ik\cdot x_1}\{b(\mathbf{p},s_j),b^\dagger(\mathbf{p}_i,s_i)\}[a(\mathbf{k},\lambda),a^\dagger(\mathbf{k}_i,\lambda_i)]|0\rangle$$

$$= C_i u(\mathbf{p}_i,s_i)e_\nu(\mathbf{k}_i,\lambda_i)e^{-i(p_i+k_i)\cdot x_1}, \tag{12.67}$$

where we now write $u(\mathbf{p},s_j)$ and $b(\mathbf{p},s_j)$ instead of $u^j(\mathbf{p})$ and $b_j(\mathbf{p})$. C_i is a normalization factor chosen so that $\langle i|i\rangle = 1$:

$$|C_i|^2\langle 0|\{b(\mathbf{p}_i,s_i),b^\dagger(\mathbf{p}_i,s_i)\}[a(\mathbf{k}_i,\lambda_i),a^\dagger(\mathbf{k}_i,\lambda_i)]|0\rangle$$

$$- |C_i|^2 (2\pi)^3 \frac{E_i}{m} \delta^3(0)(2\pi)^3 2k_i \delta^3(0) - |C_i|^2 \frac{E_i}{m} 2k_i V^2 \qquad (12.68)$$

or $C_i = (m/2k_i E_i)^{1/2}/V$, analogous to (12.30). Similarly

$$\langle f|A_\mu^{(-)}(x_2)\overline{\psi}^{(-)}(x_2) = C_f \overline{u}(\mathbf{p}_f, s_f) e_\mu(\mathbf{k}_f, \lambda_f) e^{i(p_f + k_f)\cdot x_2} , \qquad (12.69)$$

with $C_f = (m/2k_f E_f)^{1/2}/V$. Then

$$\begin{aligned}
S_{fi}^{(2)}(C1) &= \frac{-ie^2}{V^2} \left[\frac{m^2}{4k_i k_f E_i E_f}\right]^{1/2} \overline{u}(\mathbf{p}_f, s_f)\not{e}(\mathbf{k}_f, \lambda_f) \\
&\quad \times \left[\int d^4x_1 \int d^4x_2 S_F(x_2, x_1) e^{i(p_f + k_f)\cdot x_2} e^{-i(p_i + k_i)\cdot x_1}\right] \\
&\quad \times \not{e}(\mathbf{k}_i, \lambda_i) u(\mathbf{p}_i, s_i) .
\end{aligned} \qquad (12.70)$$

The integration over x_1 and x_2 is easily carried out using the expression (10.198) for $S_F(x_2, x_1)$:

$$\begin{aligned}
&\int d^4x_1 \int d^4x_2 e^{i(p_f + k_f)\cdot x_2} e^{-i(p_i + k_i)\cdot x_1} S_F(x_2, x_1) \\
&= \int d^4x_1 \int d^4x_2 e^{i(p_f + k_f)\cdot x_2} e^{-i(p_i + k_i)\cdot x_1} \int \frac{d^4p}{(2\pi)^4} e^{-ip\cdot(x_2 - x_1)} S_F(p) \\
&= (2\pi)^4 \int d^4p \, \delta^4(p_f + k_f - p)\delta^4(p_i + k_i - p) S_F(p) \\
&= (2\pi)^4 \delta^4(p_i + k_i - p_f - k_f) S_F(p_i + k_i) \\
&= (2\pi)^4 \delta^4(p_i + k_i - p_f - k_f) \frac{1}{\not{p}_i + \not{k}_i - m + i\epsilon} ,
\end{aligned} \qquad (12.71)$$

so that

$$\begin{aligned}
S_{fi}^{(2)}(C1) &= -i(2\pi)^4 \delta^4(p_i + k_i - p_f - k_f) \frac{e^2}{V^2} \left[\frac{m^2}{4k_i k_f E_i E_f}\right]^{1/2} \\
&\quad \times \overline{u}(\mathbf{p}_f, s_f) \frac{\not{e}(\mathbf{k}_f, \lambda_f)\not{e}(\mathbf{k}_i, \lambda_i)}{\not{p}_i + \not{k}_i - m + i\epsilon} u(\mathbf{p}_i, s_i).
\end{aligned} \qquad (12.72)$$

The amplitude $S_{fi}^{(2)}(C2)$ has a similar expression, except that the exponentials in (12.70) are replaced by $e^{i(p_f - k_i)\cdot x_2}$ and $e^{-i(p_i - k_f)\cdot x_1}$. This leads to the same delta function as in (12.72), but with $\not{p}_i + \not{k}_i$ replaced by $\not{p}_i - \not{k}_f$:

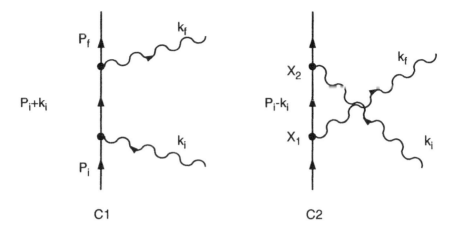

Figure 12.3: Momentum–space Feynman diagrams for the two amplitudes $C1$ and $C2$ contributing to Compton scattering.

$$S_{fi}^{(2)}(C2) = -i(2\pi)^4\delta^4(p_i + k_i - p_f - k_f)\frac{e^2}{V^2}\left[\frac{m^2}{4k_i k_f E_i E_f}\right]^{1/2}$$

$$\times \bar{u}(\mathbf{p}_f, s_f)\frac{\not{e}(\mathbf{k}_f, \lambda_f)\not{e}(\mathbf{k}_i, \lambda_i)}{\not{p}_i - \not{k}_f - m + i\epsilon}u(\mathbf{p}_i, s_i). \tag{12.73}$$

The Fourier transformation (12.71) converts the space–time propagator $S_F(x_2, x_1)$ to the momentum–space propagator $S_F(p)$, and we can draw momentum–space Feynman diagrams representing (12.72) and (12.73) (Figure 12.3). The calculation of the differential scattering cross section for Compton scattering from the absolute square of the sum of (12.72) and (12.73) is given in Bjorken and Drell (1964), for instance, and leads to the Klein–Nishina formula

$$\frac{d\sigma}{d\Omega} = \frac{\alpha^2}{2m^2}\left(\frac{k_f}{k_i}\right)^2\left[\frac{k_f}{k_i} + \frac{k_i}{k_f} - \sin^2\theta\right] \tag{12.74}$$

when we sum over final electron spins and photon polarizations and average over initial spins and polarizations. Here θ is the scattering angle and

$$k_f = \frac{k_i}{1 + (k_i/m)(1 - \cos\theta)}, \tag{12.75}$$

which implies the familiar Compton shift $\Delta\lambda = (h/mc)(1 - \cos\theta)$ in the wavelength of the scattered radiation. When integrated over all solid angles to give the total scattering cross section, (12.74) yields the Thomson cross section $(8\pi/3)\alpha^2/m^2$ in the low-energy limit in which $k_i/m \to 0$, and $(\pi\alpha^2/k_i m)\log(2k_i/m) + 1/2$ in the high-energy limit $k_i/m \gg 1$ of very short wavelengths.

12.6   Electron Self-Energy

We have considered only one contribution to (12.43) arising from the normal ordering of the time-ordered product, namely the term (12.62) for Compton scattering. Another contribution appears from the replacement in (12.62) of the normal product $N[A_\mu(x_2)A_\nu(x_1)]$ by the second term of (12.50), the photon propagator $iD_{\mu\nu}(x_2, x_1)$:

$$2i\overline{\psi}_\alpha^{(-)}(x_2)S_F(x_2, x_1)_{\beta\alpha'}\psi_{\beta'}^{(+)}(x_1)iD_{\mu\nu}(x_2, x_1). \tag{12.76}$$

This gives the S-matrix element

$$
\begin{aligned}
S_{fi}^{(2)}(E) &= e^2\gamma_{\alpha\beta}^\mu\gamma_{\alpha'\beta'}^\nu \int d^4x_1 \int d^4x_2 S_F(x_2, x_1)_{\beta\alpha'} \\
&\quad \times \langle f|\overline{\psi}_\alpha^{(-)}(x_2)\psi_{\beta'}^{(+)}(x_1)D_{\mu\nu}(x_2, x_1)|i\rangle \\
&= e^2 \int d^4x_1 \int d^4x_2 \langle f|\overline{\psi}^{(-)}(x_2)\gamma^\mu S_F(x_2, x_1)\gamma^\nu \psi^{(+)}(x_1)|i\rangle \\
&\quad \times D_{\mu\nu}(x_2, x_1) .
\end{aligned}
\tag{12.77}
$$

This amplitude vanishes if the initial and final states have different numbers of photons. We consider therefore the initial and final states $|i\rangle = |\mathbf{p}_i, s_i\rangle$ and $|f\rangle = |\mathbf{p}_f, s_f\rangle$ indicated by the Feynman diagram of Figure 12.4. For these states we have, as in (12.29) and (12.31),

$$\psi^{(+)}(x_1)|i\rangle = \sqrt{\frac{m}{E_i V}}u(\mathbf{p}_i, s_i)e^{-ip_i \cdot x_1} , \tag{12.78}$$

$$\langle f|\overline{\psi}^{(-)}(x_2) = \sqrt{\frac{m}{E_f V}}\overline{u}(\mathbf{p}_f, s_f)e^{ip_f \cdot x_2} . \tag{12.79}$$

Then, using the expressions (10.198) and (10.214) for the electron and photon propagators and integrating over x_1 and x_2 in (12.77), in the manner of (12.71), we obtain straightforwardly

$$
\begin{aligned}
S_{fi}^{(2)}(E) &= e^2\sqrt{\frac{m^2}{E_i E_f}}\frac{1}{V}\delta^4(p_f - p_i)\int d^4k \overline{u}(\mathbf{p}_f, s_f)\gamma^\mu S_F(p_i - k)\gamma^\nu \\
&\quad \times u(\mathbf{p}_i, s_i)D_{\mu\nu}(k)
\end{aligned}
$$

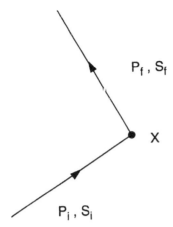

Figure 12.4: Feynman diagram for the electron transition amplitude $|\mathbf{p}_i, s_i\rangle \rightarrow |\mathbf{p}_f, s_f\rangle$.

$$= -i(2\pi)^4 \delta^4(p_f - p_i)\sqrt{\frac{m^2}{E_i E_f}}\frac{1}{V}\bar{u}(\mathbf{p}_f, s_f)$$

$$\times \left[-ie^2 \int \frac{d^4k}{(2\pi)^4} \frac{g_{\mu\nu}}{k^2 + i\epsilon}\gamma^\mu \frac{1}{\not{p}_i - \not{k} - m + i\epsilon}\gamma^\nu\right] u(\mathbf{p}_i, s_i)$$

$$\equiv -i(2\pi)^4 \delta^4(p_f - p_i)\sqrt{\frac{m^2}{E_i E_f}}\frac{1}{V}\bar{u}(\mathbf{p}_f, s_f)\Sigma(p_i)u(\mathbf{p}_i, s_i).$$

$$(12.80)$$

The photon propagator, from the definition (10.201), involves the creation and then the annihilation of a (virtual) photon from the vacuum. We therefore associate with the amplitude (12.80) the momentum–space Feynman diagram of Figure 12.5: an electron propagates along, emits, then reabsorbs a virtual photon indicated by the wavy line, and then propagates again as a free particle. This is evidently a *self-energy diagram* associated in this example with electromagnetic mass. Before concerning ourselves with the physical interpretation, however, let us proceed to evaluate $\Sigma(p)$, which we can write equivalently as

$$\Sigma(p) = -ie^2 \int \frac{d^4k}{(2\pi)^4} \frac{1}{k^2}\gamma_\nu \frac{\not{p} - \not{k} + m}{(p - k)^2 - m^2}\gamma^\nu, \qquad (12.81)$$

where the $i\epsilon$ in the denominator is now implicit. We have multiplied the

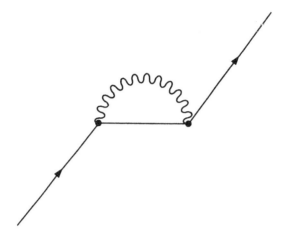

Figure 12.5: Self-energy diagram associated with the amplitude (12.80).

numerator and denominator by $\not{p} - \not{k} + m$, as in (10.199). Now we use the general identities $\gamma_\nu \gamma^\nu = 4$ and $\gamma_\nu \not{p} \gamma^\nu = -2\not{p}$:

$$\Sigma(p) = -2ie^2 \int \frac{d^4k}{(2\pi)^4} \frac{1}{k^2} \frac{\not{k} - \not{p} + 2m}{(p-k)^2 - m^2} \cdot \qquad (12.82)$$

We consider a Taylor series expansion about $\not{p} = m$. If $\not{p} - m = 0$ we have $(\not{p} - m)(\not{p} + m) = p^2 - m^2 = 0$, or $m^2 = p^2$. Thus

$$\frac{\not{k} - \not{p} + 2m}{(p-k)^2 - m^2} = \frac{\not{k} + m}{k^2 - 2kp} + (\not{p} - m)F(k, \not{p} - m) \qquad (12.83)$$

or

$$\begin{aligned}
\Sigma(p) &= -2ie^2 \int \frac{d^4k}{(2\pi)^4} \frac{1}{k^2} \frac{\not{k} + m}{k^2 - 2kp} \\
&\quad - 2ie^2(\not{p} - m) \int \frac{d^4k}{(2\pi)^4} \frac{1}{k^2} F(k, \not{p} - m) \\
&\equiv A + B(\not{p} - m).
\end{aligned} \qquad (12.84)$$

Since $(\not{p}_i - m)u(\mathbf{p}_i, s_i) = 0$, the second term makes no contribution to (12.80). Thus we need consider only

$$A = -2ie^2 \int \frac{d^4k}{(2\pi)^4} \frac{\not{k} + m}{k^2(k^2 - 2kp)} \cdot \qquad (12.85)$$

To do the integral we use a trick of Feynman's: the identity

$$\frac{1}{ab} = \int_0^1 dt \frac{1}{[at + b(1-t)]^2} , \qquad (12.86)$$

with $a - k^2 - 2kp$ and $b - k^2$, gives

$$A = -2ie^2 \int_0^1 dt \int \frac{d^4k}{(2\pi)^4} \frac{\not{k} + m}{(k^2 - 2kpt)^2} \qquad (12.87)$$

or, introducing the new integration variable $q = k - pt$,

$$A = -2ie^2 \int_0^1 dt \int \frac{d^4q}{(2\pi)^4} \frac{\not{q} + tm + m}{(q^2 - m^2t^2)^2} . \qquad (12.88)$$

Now $q = (q^0, \mathbf{q})$, and because the integral over q^0 extends from $-\infty$ to ∞, and the integral over $\mathbf{q}$ is over all solid angles, the term with $\not{q}$ in the numerator makes no contribution to the integral. Thus

$$A = -2ime^2 \int_0^1 dt(t+1) \int \frac{d^4q}{(2\pi)^4} \frac{1}{(q^2 - m^2t^2)^2} . \qquad (12.89)$$

The integral over q is[7]

$$\left(\frac{1}{2\pi}\right)^4 \int d^3q \int_{-\infty}^{\infty} \frac{dq_0}{(q_0^2 - \mathbf{q}^2 - m^2t^2)^2}$$

$$= \left(\frac{1}{2\pi}\right)^4 \frac{i\pi}{2} \int d^3q \frac{1}{(\mathbf{q}^2 + m^2t^2)^{3/2}}$$

$$= \frac{i}{32\pi^3} 4\pi \int \frac{dq\, q^2}{(q^2 + m^2t^2)^{3/2}}$$

$$\rightarrow \frac{i}{8\pi^2} \int_0^{\Lambda/mt} \frac{dx\, x^2}{(x^2 + 1)^{3/2}} \cong \frac{i}{8\pi^2} \log \frac{\Lambda}{mt} \qquad (12.90)$$

when we introduce a high-energy cutoff $\Lambda \gg m$. Therefore

$$A \cong \frac{me^2}{4\pi^2} \int_0^1 dt(t+1) \left(\log \frac{\Lambda}{m} - \log t\right)$$

$$\rightarrow \frac{3m\alpha}{2\pi} \log \frac{\Lambda}{m} \quad (\Lambda \gg m). \qquad (12.91)$$

[7] The integral over q_0 is of the form given in I. S. Gradshteyn and I. M. Ryzhik, *Table of Integrals, Series, and Products* (Academic Press, New York, 1980), Section 3.241, No. 4.

Thus $A = \delta m$, the electromagnetic mass (11.93) calculated previously by "old-fashioned" methods.

In Section 11.6 we showed that the electromagnetic mass term resulting from the coupling of the electron to the field can be cancelled by an appropriate counter term in the Hamiltonian, so that the mass in the Hamiltonian is in fact the renormalized, *observed* mass. The Hamiltonian density associated with the counter term, according to equation (11.108), is

$$\mathsf{h}_{\text{counter}} = -\delta m \psi^\dagger \beta \psi = -\delta m \overline{\psi} \psi. \tag{12.92}$$

Consider the first-order contribution of this term to the S matrix. From (12.25), (12.78), and (12.79),

$$
\begin{aligned}
S_{fi}^{(1)} &= -i \int d^4 x \langle f | \mathsf{h}_{\text{counter}}(x) | i \rangle \\
&= i\delta m \sqrt{\frac{m^2}{E_i E_f}} \frac{1}{V} \overline{u}(\mathbf{p}_f, s_f) u(\mathbf{p}_i, s_i) \int d^4 x\, e^{i(p_f - p_i)\cdot x} \\
&= i\delta m (2\pi)^4 \delta^4(p_f - p_i) \sqrt{\frac{m^2}{E_i E_f}} \frac{1}{V} \overline{u}(\mathbf{p}_f, s_f) u(\mathbf{p}_i, s_i). \tag{12.93}
\end{aligned}
$$

This cancels the S-matrix element (12.80). In other words (12.80), with its diagram shown in Figure 12.5, is the term responsible for electromagnetic mass in the relativistic S-matrix theory.

12.7   Remarks

From the viewpoint of quantum field theory, Feynman diagrams are associated with the normal ordering of time-ordered products, each normal-ordered constituent corresponding to one or more diagrams. The normal ordering of the time-ordered product in (12.61), for instance, has given us the Compton scattering term (12.62) and the associated diagram of Figure 12.2 (or 12.3), and also the self-energy term (12.76) and its diagram, Figure 12.5. Wick's theorem, as described in the next section, provides a systematic way of writing a time-ordered product as a sum of normal-ordered constituents. Alternatively, as discussed in Section 12.10, we can deduce the diagrams from the more intuitive Feynman approach, without relying on quantum field theory or Wick's theorem.

Whichever way one chooses to get to the diagrams, it should be emphasized that we are dealing with the same old physical theory of QED, but with a far more convenient formulation of that theory (Weinberg, 1977):

> ... the theory of Schwinger, Tomonaga, Feynman, and Dyson
> was not really a new physical theory. It was simply the old quantum
> field theory of Heisenberg, Pauli, Fermi, Oppenheimer, Furry, and
> Weisskopf, but cast in a form far more convenient for calculation,
> and equipped with a more realistic definition of physical parameters
> like masses and charges.

The simplification achieved by the Feynman formulation, in particular,
is already evident in the self-energy calculation of the preceding section.
The starting point of that calculation is (12.77);[8] the calculation itself proceeds straightforwardly to (12.91). Things were much more involved in
the old-fashioned calculation of the same result in Chapter 11. There we
separated the static and dynamic contributions resulting from the use of
the Coulomb gauge. Then we had to subtract the contributions (11.30)
and (11.38) associated with the pure vacuum state. Finally we arrived at
(11.56) and (11.65), from which the electron self-energy was calculated in
Section 11.4.

The result of the old-fashioned and covariant perturbation theories, of
course, is the same: the electromagnetic mass δm is logarithmically divergent. However, the "modern" formulation is far more manageable in higher
orders of perturbation theory. It furthermore provides an unambiguous
renormalization procedure and allows the renormalizability of the theory
to be demonstrated. This is a very important point, for a tenet of modern
physics is that *any fundamental theory must satisfy three highly restrictive
criteria*: it must be Lorentz-invariant, gauge–invariant, and renormalizable.

Before embarking on the intuitive path of Feynman in Section 12.10, we
now discuss Wick's theorem for converting a time-ordered product to a sum
of normal-ordered constituents and how this leads to Feynman diagrams.

12.8   Wick's Theorem

Consider the generalization of (12.50) to the case of three fields. Proceeding
in exactly the same fashion as in the derivation of (12.50), we obtain

$$
\begin{aligned}
TA_\mu(x_3)A_\nu(x_2)A_\sigma(x_1) = \ & N[A_\mu(x_3)A_\nu(x_2)A_\sigma(x_1)] \\
& + iD_{\mu\nu}(x_3, x_2)A_\sigma(x_1) \\
& + iD_{\mu\sigma}(x_3, x_1)A_\nu(x_2) \\
& + iD_{\nu\sigma}(x_2, x_1)A_\mu(x_3) \ .
\end{aligned}
\tag{12.94}
$$

---

[8] In Section 12.10 we shall see that (12.77) follows almost trivially in Feynman's
approach.

We can generalize by induction to the product of n fields. The result is called *Wick's theorem*. In simplified notation, it reads

$$TA_n A_{n-1}...A_2 A_1 = N[A_n A_{n-1}...A_2 A_1]$$

$$+ \sum_p iD(x_{p(n)} - x_{p(n-1)})N[A_{p(n-2)}...A_{p(1)}]$$

$$+ \sum_p iD(x_{p(n)} - x_{p(n-1)})iD(x_{p(n-2)} - x_{p(n-3)})N[A_{p(n-4)}...A_{p(1)}]$$

$$+ ... + L_n , \tag{12.95}$$

where the sums are over permutations p of the coordinates x_j, $p(j) > p(j-1)$, and

$$L_n = \sum_p iD(x_{p(n)} - x_{p(n-1)})...iD(x_{p(3)} - x_{p(2)})A(x_{p(1)}) \tag{12.96}$$

for n odd and

$$L_n = \sum_p iD(x_{p(n)} - x_{p(n-1)})...iD(x_{p(2)} - x_{p(1)}) \tag{12.97}$$

for n even. Equations (12.50) and (12.94) are, of course, special cases of Wick's theorem. Note that, because of the normal orderings in (12.95), we have the vacuum expectation values

$$\langle 0|TA_n A_{n-1}...A_2 A_1|0\rangle = \langle 0|L_n|0\rangle , \tag{12.98}$$

which vanish when n is odd.

In the case of fermion fields the generalization of (12.59) is similar, except that we must keep track of signs arising from the algebra of fermion annihilation and creation operators. Thus, for instance,

$$
\begin{aligned}
T\overline{\psi}(4)\psi(3)\overline{\psi}(2)\psi(1) &= N[\overline{\psi}(4)\psi(3)\overline{\psi}(2)\psi(1)] + iS_F(3,2)N[\overline{\psi}(4)\psi(1)]\\
&\quad - iS_F(3,4)N[\overline{\psi}(2)\psi(1)] - iS_F(1,2)N[\overline{\psi}(4)\psi(3)]\\
&\quad - iS_F(1,4)N[\psi(3)\overline{\psi}(2)] + [-iS_F(3,4)][-iS_F(1,2)]\\
&\quad + [iS_F(3,2)][-iS_F(1,4)] .
\end{aligned} \tag{12.99}
$$

The minus signs in front of the propagators S_F appear when an interchange of a $\overline{\psi}$ and a ψ is necessary to put them in the order $\psi\overline{\psi}$ appropriate to the definition (10.193) of the propagator. Except for these signs, Wick's theorem has the same form for fermion fields as for boson fields (Wick, 1950; Itzykson and Zuber, 1980).

The second-order S-matrix element (12.43) involves

$$T\overline{\psi}(x_2)\gamma^\mu\psi(x_2)\overline{\psi}(x_1)\gamma^\nu\psi(x_1)A_\mu(x_2)A_\nu(x_1)$$
$$= Tj^\mu(x_2)j^\nu(x_1)A_\mu(x_2)A_\nu(x_1) \tag{12.100}$$

when we use the definition $j^\mu(x) = \overline{\psi}(x)\gamma^\mu\psi(x)$ of the current density. Actually, however, this expression should be antisymmetrized to read (Pauli, 1941; Dyson 1949a)

$$j^\mu(x) = \frac{e}{2}[\overline{\psi}_\alpha(x), \psi_\beta(x)]\gamma^\mu_{\alpha\beta}, \tag{12.101}$$

since $\overline{\psi}(x)$ and $\psi(x)$ are anticommuting *operators* in quantum field theory. The consequence of this antisymmetrization for our purposes is simply that the vacuum expectation value of $j^\mu(x)$ vanishes, and so we can in effect drop some terms in (12.99), for instance, when $x_4 = x_3$ and $x_2 = x_1$:

$$T[\overline{\psi}(2)\psi(2)\overline{\psi}(1)\psi(1)] \rightarrow N[\overline{\psi}(2)\psi(2)\overline{\psi}(1)\psi(1)] + iS_F(2,1)N[\overline{\psi}(2)\psi(1)]$$
$$- iS_F(1,2)N[\psi(2)\overline{\psi}(1)]$$
$$+ [iS_F(2,1)][-iS_F(1,2)]. \tag{12.102}$$

Then

$$T\overline{\psi}_\alpha(x_2)\psi_\beta(x_2)\overline{\psi}_{\alpha'}(x_1)\psi_{\beta'}(x_1)A_\mu(x_2)A_\nu(x_1)$$

$$= \{N[\overline{\psi}_\alpha(x_2)\psi_\beta(x_2)\overline{\psi}_{\alpha'}(x_1)\psi_{\beta'}(x_1)] + iS_F(x_2,x_1)_{\beta\alpha'}N[\overline{\psi}_\alpha(x_2)\psi_{\beta'}(x_1)]$$
$$- iS_F(x_1,x_2)_{\beta'\alpha}N[\psi_\beta(x_2)\overline{\psi}_{\alpha'}(x_1)] + [iS_F(x_2,x_1)_{\beta\alpha'}][-iS_F(x_1,x_2)_{\beta'\alpha}]\}$$
$$\times \{N[A_\mu(x_2)A_\nu(x_1)] + iD_{\mu\nu}(x_2,x_1)\}. \tag{12.103}$$

Since we integrate over all x_1 and x_2 in calculating transition amplitudes, we can interchange x_1 and x_2 in the third term within the first set of curly brackets, and replace this term by $+iS_F(x_2,x_1)_{\beta'\alpha}N[\overline{\psi}_{\alpha'}(x_2)\psi_\beta(x_1)]$. This then combines with the second term to give a factor of 2 times either (identical) term when we sum over α, β, α', and β' as in (12.61). Thus, multiplying by $\gamma^\mu_{\alpha\beta}\gamma^\nu_{\alpha'\beta'}$ as in (12.61), and using the fact that μ and ν are dummy indices that are summed over, we obtain for the second-order S matrix the expression

$$S^{(2)} = -\frac{e^2}{2}\int d^4x_1 \int d^4x_2 N[\overline{\psi}(x_2)\slashed{A}(x_2)\psi(x_2)\overline{\psi}(x_1)\slashed{A}(x_1)\psi(x_1)]$$
$$\tag{12.104}$$

$$- \quad e^2 \int d^4x_1 \int d^4x_2 N[\overline{\psi}(x_2)\not{A}(x_2)iS_F(x_2,x_1)\not{A}(x_1)\psi(x_1)]$$

(12.105)

$$- \quad \frac{e^2}{2} \int d^4x_1 \int d^4x_2 N[\overline{\psi}(x_2)\gamma^\mu \psi(x_2)iD_{\mu\nu}(x_2,x_1)\overline{\psi}(x_1)\gamma^\nu \psi(x_1)]$$

(12.106)

$$- \quad e^2 \int d^4x_1 \int d^4x_2 N[\overline{\psi}(x_2)\gamma^\mu iS_F(x_2,x_1)iD_{\mu\nu}(x_2,x_1)\gamma^\nu \psi(x_1)]$$

(12.107)

$$+ \quad \frac{e^2}{2} \int d^4x_1 \int d^4x_2 N[\gamma^\mu iS_F(x_2,x_1)\gamma^\nu iS_F(x_1,x_2)A_\mu(x_2)A_\nu(x_1)$$

(12.108)

$$+ \quad \frac{e^2}{2} \int d^4x_1 \int d^4x_2 \gamma^\mu iS_F(x_2,x_1)\gamma^\nu iS_F(x_1,x_2)iD_{\mu\nu}(x_2,x_1) \ .$$

(12.109)

This expression can, of course, be derived without Wick's theorem. The normal ordering of higher order contributions to the S matrix, however, is greatly facilitated by Wick's theorem.

Each of the normal constituents (12.104)–(12.109) of the second-order S matrix may be associated now with Feynman diagrams, and more generally the normal constituents of high-order terms in the S matrix may similarly be represented diagrammatically. We now discuss some conventions and examples for constructing the diagrams.

12.9   Diagrams

The interaction

$$h_I = e\overline{\psi}\not{A}\psi = e[\overline{\psi}^{(+)} + \overline{\psi}^{(-)}]\gamma^\mu[A_\mu^{(+)} + A_\mu^{(-)}][\psi^{(+)} + \psi^{(-)}] \qquad (12.110)$$

involves electron annihilation $[\psi^{(+)}]$ and creation $[\overline{\psi}^{(-)}]$ operators, positron annihilation $[\overline{\psi}^{(+)}]$ and creation $[\psi^{(-)}]$ operators, and photon annihilation $[A_\mu^{(+)}]$ and creation $[A_\mu^{(-)}]$ operators [recall equations (12.44), (12.51)–(12.53), and (12.54)–(12.56)]. The part $e\overline{\psi}^{(-)}(x)\gamma^\mu A_\mu^{(+)}(x)\psi^{(+)}(x)$ of h_I, for instance, corresponds to a process in which an electron is annihilated, a photon is annihilated, and an electron is created. Such a process is represented in Figure 12.6, where the solid and wavy lines represent the electron and the photon, respectively. The arrows point toward the *vertex*

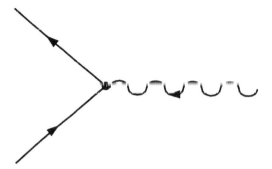

Figure 12.6: Feynman diagram corresponding to $e\overline{\psi}^{(-)}(x)A_\mu^{(+)}(x)\psi^{(+)}(x)$.

at x for electrons and photons that are annihilated, and away from the vertex for electrons and photons that are created. For processes involving positrons, an arrow pointing toward a vertex will designate the creation of a positron, and an arrow pointing away from the vertex the annihilation of a positron. Thus electron and positron lines point in opposite senses. (The rationale for this convention is given in the following section.) The interaction $e\overline{\psi}^{(+)}(x)A_\mu(x)\psi^{(-)}(x)$, in which a positron is created, a photon is annihilated or created $[A_\mu^{(+)}$ or $A_\mu^{(-)}]$, and a positron is annihilated, is therefore represented by the diagram of Figure 12.7, in which the directions of the fermion lines are reversed from Figure 12.6. The interactions $e\overline{\psi}^{(+)}(x)A_\mu(x)\psi^{(+)}(x)$ and $e\overline{\psi}^{(-)}(x)A_\mu(x)\psi^{(-)}(x)$ likewise correspond to (electron–positron) pair annihilation and creation, respectively, and are represented by the diagrams of Figure 12.8.

The diagrams we draw do not necessarily correspond to allowed processes. The one-photon pair annihilation and creation processes of Figure 12.8, for instance, are forbidden by energy–momentum conservation. (Section 12.10)

Because h_I involves $\overline{\psi}$, ψ, and $A\!\!\!/$, these first-order processes correspond to diagrams with two fermion lines and one photon line. We call these lines *external* because none of the operators has been "contracted" to form the propagators S_F or $D_{\mu\nu}$. Similarly the fermion and boson operators in the first-order S matrix element $S^{(1)}$ may be called *external* fields in this sense.

Consider now the normal constituents (12.104)–(12.109) of $S^{(2)}$. The diagrams corresponding to these terms will all involve *two* vertices (x_1 and x_2).

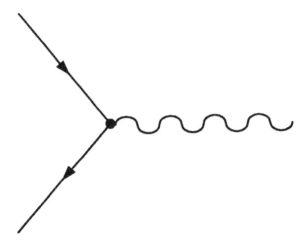

Figure 12.7: Feynman diagram corresponding to $e\overline{\psi}^{(+)}(x)A_\mu^{(\pm)}(x)\psi^{(-)}(x)$.

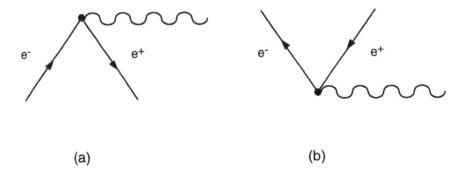

Figure 12.8: Pair annihilation (a) and creation (b) corresponding to the interactions $e\overline{\psi}^{(+)}(x)A_\mu(x)\psi^{(+)}(x)$ and $e\overline{\psi}^{(-)}(x)A_\mu(x)\psi^{(-)}(x)$, respectively.

Compton Scattering

The term (12.105), for instance, describes various processes, depending on how the annihilation and creation operators effect transitions. The part $\overline{\psi}^{(-)}(x_2)A_\mu^{(-)}(x_2)iS_F(x_2,x_1)A_\nu^{(+)}(x_1)\psi^{(+)}(x_1)$ is represented by a diagram with two external electron lines and two external photon lines, together with an "internal" electron line [resulting from the contraction of $\psi(x_2)\overline{\psi}(x_1)$ to form the propagator $S_F(x_2,x_1)$] joining the vertices x_1 and x_2. This part corresponds to the Compton scattering diagram $C1$ of Figure 12.2.

Now since

$$iS_F(x_2,x_1) = \langle 0|T\psi(x_2)\overline{\psi}(x_1)|0\rangle = \langle 0|\psi(x_2)\overline{\psi}(x_1)|0\rangle$$
$$= \langle 0|\psi^{(+)}(x_2)\overline{\psi}^{(-)}(x_1)|0\rangle \tag{12.111}$$

for $x_{20} > x_{10}$, we can think of

$$\overline{\psi}^{(-)}(x_2)A_\mu^{(-)}(x_2)iS_F(x_2,x_1)A_\nu^{(+)}(x_1)\psi^{(+)}(x_1) \tag{12.112}$$

as effecting a transition in which a real (initial) electron and photon are annihilated at x_1, a virtual electron is created at x_1 and annihilated at x_2, and a real (final) electron and photon are created at x_2. In other words, the internal line connecting the vertices x_1 and x_2 in the diagram $C1$ is associated with the creation and annihilation of a virtual electron.

For $x_{10} > x_{20}$ we have

$$iS_F(x_2,x_1) = \langle 0|\overline{\psi}(x_1)\psi(x_2)|0\rangle = \langle 0|\overline{\psi}^{(+)}(x_1)\psi^{(-)}(x_2)|0\rangle. \tag{12.113}$$

For $x_{10} > x_{20}$, therefore, the diagram $C1$ can be thought of as representing the creation of an electron, a positron, and a photon at x_2, followed by the annihilation of the positron and the creation of the (initial) electron and photon at x_1. Thus $iS_F(x_2,x_1)$ describes the creation and annihilation of a virtual electron or a virtual positron, depending on whether $x_{20} > x_{10}$ or $x_{10} > x_{20}$.

Of course (12.105) includes other processes besides electron Compton scattering: two-photon pair creation, described by

$$\overline{\psi}^{(-)}(x_2)A_\nu^{(+)}(x_2)iS_F(x_2,x_1)A_\mu^{(+)}(x_1)\psi^{(-)}(x_1) \tag{12.114}$$

and the Feynman diagram in Figure 12.9; two-photon pair annihilation described by

$$\overline{\psi}^{(+)}(x_2)A_\nu^{(-)}(x_2)iS_F(x_2,x_1)A_\mu^{(-)}(x_1)\psi^{(+)}(x_1) \tag{12.115}$$

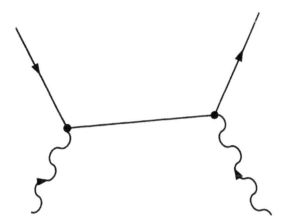

Figure 12.9: Feynman diagram for two-photon pair creation.

and the diagram in Figure 12.10; and positron Compton scattering described by (Figure 12.11)

$$\psi^{(-)}(x_2)A_\mu^{(-)}(x_2)iS_F(x_2,x_1)A_\nu^{(+)}(x_1)\overline{\psi}^{(+)}(x_1) \qquad (12.116)$$

and

$$\psi^{(-)}(x_2)A_\mu^{(+)}(x_2)iS_F(x_2,x_1)A_\nu^{(-)}(x_1)\overline{\psi}^{(+)}(x_1) \qquad (12.117)$$

As the reader may easily show, there are no other processes associated with (12.105) that are consistent with energy conservation.

Electron Self-Energy

The term (12.107) is identical to the operator $S^{(2)}(E)$ appearing in (12.77); i.e., it corresponds to the electron self-energy (electromagnetic mass) and the diagram in Figure 12.5. Since it involves both the electron propagator and the photon propagator, it can be interpreted as effecting the emission and absorption of a virtual electron or positron and the emission and absorption of a virtual photon (Figure 12.5). That is, the photon propagator defined by $iD_{\mu\nu}(x_2,x_1) = \langle 0|TA_\mu(x_2)A_\nu(x_1)|0\rangle = \langle 0|A_\mu(x_2)A_\nu(x_1)|0\rangle = \langle 0|A_\mu^{(+)}(x_2)A_\nu^{(-)}(x_1)|0\rangle$ for $x_{20} > x_{10}$, for instance, corresponds to the emission of a photon $[A_\nu^{(-)}(x_1)]$ followed by the absorption of the photon $[A_\mu^{(+)}(x_2)]$, just as $iS_F(x_2,x_1)$ corresponds to the emission and reabsorption of a virtual electron or positron.

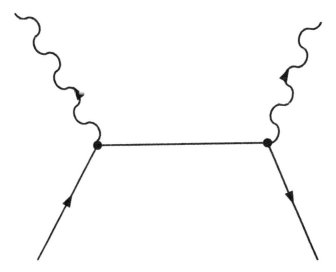

Figure 12.10: Feynman diagram for two-photon pair annihilation.

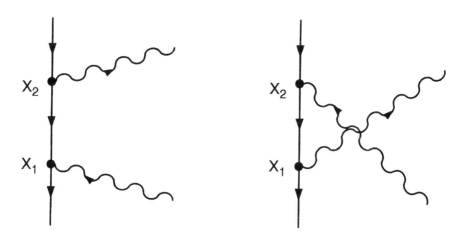

Figure 12.11: Positron Compton scattering.

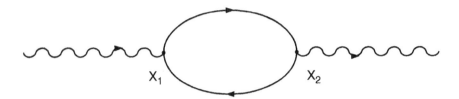

Figure 12.12: Vacuum polarization diagram.

Photon Self-Energy (Vacuum Polarization)

The term (12.108) involves two external photon lines and two internal fermion lines, and corresponds to the Feynman diagram in Figure 12.12. We have already noted that $iS_F(x_2, x_1)$ may be associated with the creation and annihilation of a virtual electron or positron. The product $iS_F(x_2, x_1)iS_F(x_1, x_2)$ is easily seen by similar arguments to correspond to the creation of a virtual electron–positron pair followed by the annihilation of the pair (Figure 12.12). As the "closed loop" in the Feynman diagram of Figure 12.12 suggests, (12.108) describes the modification of the propagating photon by virtual electron–positron pairs, i.e., it describes vacuum polarization ("photon self-energy"). We have already encountered vacuum polarization in connection with the Lamb shift, and in Section 11.8 we outlined an old-fashioned calculation of vacuum polarization. The covariant calculation based on (12.108) may be found, for instance, in Bjorken and Drell (1964).

Other Second-Order Processes (Diagrams)

The term (12.106) involves four external fermion lines and one internal photon line, the latter associated with the propagator $D_{\mu\nu}(x_2, x_1)$, i.e., the emission and absorption of a virtual photon. The part $\overline{\psi}^{(-)}(x_2)\psi^{(+)}(x_2) \times iD_{\mu\nu}(x_2, x_1)\overline{\psi}^{(-)}(x_1)\psi^{(+)}(x_1)$, for instance, describes electron-electron scat-

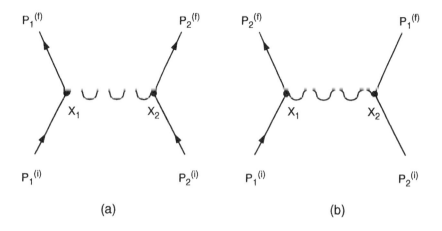

Figure 12.13: Direct (a) and exchange (b) amplitudes for electron-electron scattering.

tering. Because $\overline{\psi}^{(-)}(x_1)$ and $\overline{\psi}^{(-)}(x_2)$ can create either final electron, we have the "direct" and "exchange" amplitudes shown in Figure 12.13. As a consequence of the fermion algebra (i.e., the Pauli principle), these two amplitudes add with opposite signs in the total electron-electron scattering amplitude. Equation (12.106) also describes positron-positron scattering and electron–positron pair annihilation in the field of an electron or positron (Figure 12.14).

The term (12.109) has only internal lines (Figure 12.15) and therefore does not effect any transitions. The term (12.104), finally, has only external lines, and corresponds simply to two first-order processes, of the type shown in Figure 12.6, proceeding independent of each other.

Lines and Vertices

It should be clear by now that the basic building blocks of Feynman diagrams are vertices and propagators. For a given normal-ordered constituent in the nth order contribution to the S-matrix, we have n vertices $x_1, x_2, \dots x_n$. For each pair of Dirac operators $\psi(x_i)$ and $\overline{\psi}(x_j)$ that combine to give a propagator $S_F(x_i, x_j)$, we draw a solid (fermion) line from x_i to x_j. For each pair of Maxwell field operators $A_\mu(x_i)$ and $A_\nu(x_j)$ that combine to give a propagator $D_{\mu\nu}(x_i, x_j)$, likewise, we draw a wavy (photon) line from x_i to x_j. Unpaired operators $\overline{\psi}(x_i)$ and $\psi(x_i)$ correspond to external fermion lines drawn from x_i to the outer part of the diagram

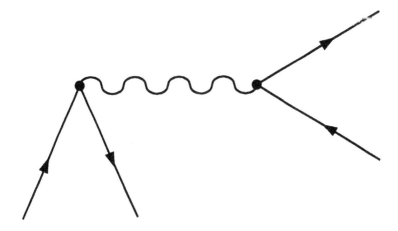

Figure 12.14: Electron-positron pair annihilation in the field of an electron.

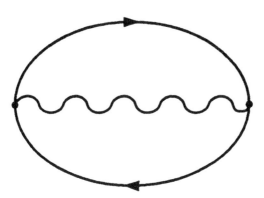

Figure 12.15: The "vacuum diagram" corresponding to (12.109).

in the case of $\overline{\psi}(x_i)$, and to x_i in the case of $\psi(x_i)$. Unpaired operators $A(x_i)$ and $A(x_i)$ correspond to external photon lines, drawn from or to x_i in the case of $A^{(-)}(x_i)$ and $A^{(+)}(x_i)$, respectively. The external lines are associated with the annihilation of "initial" particles and the creation of the "final" particles. With experience it becomes easier in many instances to start from the diagrams than from the normal-ordered constituents of the nth-order S matrix $S^{(n)}$.

Having completed our survey of Feynman diagrams associated with the second-order S matrix, we refer the reader to Bjorken and Drell (1964) or Sakurai (1976) for detailed calculations of the amplitudes for these diagrams. Let us consider now Feynman's way, where the diagrams (amplitudes) are deduced without quantization of the Dirac or Maxwell fields.

12.10   Feynman's Way

The Dirac equation (9.50) for a spin–1/2 particle in a field $A^\mu = (\Phi, \mathbf{A})$ can be written as

$$(\not{p} - m)\psi(x) = (i\not{\partial} - m)\psi(x) = e\not{A}(x)\psi(x). \tag{12.118}$$

We can attempt to solve this equation by introducing the Green function $G(x', x)$ satisfying

$$(\not{p} - m)G(x', x) = \delta^4(x' - x), \tag{12.119}$$

so that a solution of (12.111) is

$$\psi(x') = \psi_0(x') + e \int d^4x\, G(x', x)\not{A}(x)\psi(x), \tag{12.120}$$

where $\psi_0(x)$ is a solution of the free-particle Dirac equation, $(\not{p} - m)\psi_0(x) = 0$. Once $G(x', x)$ is found, we can solve (12.120) iteratively as a perturbation series:

$$\begin{aligned}
\psi(x') &= \psi_0(x') + e \int d^4x\, G(x', x)\not{A}(x)\psi_0(x) \\
&\quad + e^2 \int d^4x \int d^4x''\, G(x', x)\not{A}(x)G(x, x'')\not{A}(x'')\psi_0(x'') \\
&\quad + \dots .
\end{aligned} \tag{12.121}$$

To find $G(x', x)$ we write

$$G(x', x) = \left(\frac{1}{2\pi}\right)^4 \int d^4p\, G(p)e^{-ip\cdot(x'-x)}, \tag{12.122}$$

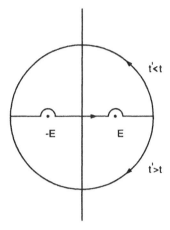

Figure 12.16: Integration contour for the retarded Green function.

$$\delta^4(x' - x) = \left(\frac{1}{2\pi}\right)^4 \int d^4p \, e^{-ip \cdot (x'-x)} , \qquad (12.123)$$

and (12.119) then implies

$$G(p) = \frac{1}{\not{p} - m} \qquad (12.124)$$

and therefore

$$G(x', x) = \int \frac{d^4p}{(2\pi)^4} \frac{e^{-ip \cdot (x'-x)}}{\not{p} - m} = \int \frac{d^4p}{(2\pi)^4} \frac{\not{p} + m}{p^2 - m^2} e^{-ip \cdot (x'-x)} , \quad (12.125)$$

where of course we must decide how to handle the singularity at $p^2 = m^2$.

How we choose to deal with the singularity amounts to a choice of boundary conditions. Suppose, for instance, that we want to have $G(x', x) = G_{\mathrm{ret}}(x'x)$, the *retarded* Green function satisfying $G_{\mathrm{ret}}(x', x) = 0$ for $t' < t$ (i.e., $x'_0 < x_0$). This might appear to be the natural choice, based on (12.120) and the analogous determination of the Green function in classical electromagnetic theory. To obtain $G_{\mathrm{ret}}(x', x)$ we choose the contour of integration in the complex p_0 plane such that there are no poles enclosed when $t' < t$ (Figure 12.16). The residue theorem then gives $G_{\mathrm{ret}}(x', x) = 0$ for $t' < t$.

Another choice of contour is shown in Figure 12.17. This choice is

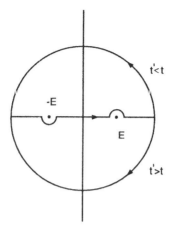

Figure 12.17: Integration contour for the Green function (Feynman propagator) $S_F(x', x)$.

equivalent to replacing the denominator in (12.125) by $p^2 - m^2 + i\epsilon$, $\epsilon \to 0^+$:

$$G(x', x) = \int \frac{d^4 p}{(2\pi)^4} \frac{\not{p} + m}{p^2 - m^2 + i\epsilon} e^{-ip\cdot(x'-x)} = S_F(x', x), \qquad (12.126)$$

where we recognize that this choice of contour gives a Green function identical to the Dirac (or Feynman) propagator defined by (10.198) or equivalently, in quantum field theory, (10.193).

Before addressing the question of which Green function to choose, let us pause to write $G_{\mathrm{ret}}(x', x)$ and $S_F(x', x)$ in more general forms. Using (10.195), (11.26), and (11.27), we can write $S_F(x', x)$ in terms of free-particle positive- and negative-energy wave functions $\phi_{p+}(\mathbf{x})$ and $\phi_{p-}(\mathbf{x})$, respectively:

$$
\begin{aligned}
iS_F(x', x) \ = \ & \theta(t' - t) \sum_p \phi_{p+}(\mathbf{x}')\overline{\phi}_{p+}(\mathbf{x})e^{-iE_p(t'-t)} \\
 & - \ \theta(t - t') \sum_p \phi_{p-}(\mathbf{x}')\overline{\phi}_{p-}(\mathbf{x})e^{iE_p(t'-t)} \ , \quad (12.127)
\end{aligned}
$$

where $E_p > 0$.[9] Consider now a positive-energy solution

$$\psi^{(+)}(x) = \sum_p c_p \phi_{p+}(\mathbf{x})e^{-iE_p t} \qquad (12.128)$$

---

[9] The summations are to be understood to include the spin components.

of the Dirac equation. We have

$$\int d^3 x\, i S_F(x', x) \gamma^0 \psi^{(+)}(x)$$

$$= \theta(t' - t) \sum_p \phi_{p+}(\mathbf{x}') e^{-i E_p (t' - t)} \sum_{p'} c_{p'} e^{-i E_{p'} t} \int d^3 x\, \phi_{p+}^*(\mathbf{x}) \phi_{p'+}(\mathbf{x})$$

$$- \theta(t - t') \sum_p \phi_{p-}(\mathbf{x}') e^{i E_p (t' - t)} \sum_{p'} c_{p'} e^{-i E_{p'} t} \int d^3 x\, \phi_{p-}^*(\mathbf{x}) \phi_{p'+}(\mathbf{x})$$

$$= \theta(t' - t) \sum_p \phi_{p+}(\mathbf{x}') e^{-i E_p t'} c_p\,, \tag{12.129}$$

or

$$\theta(t' - t) \psi^{(+)}(\mathbf{x}') = i \int d^3 x\, S_F(x', x) \gamma^0 \psi^{(+)}(x). \tag{12.130}$$

Similarly

$$\theta(t - t') \psi^{(-)}(x') = -i \int d^3 x\, S_F(x', x) \gamma^0 \psi^{(-)}(x). \tag{12.131}$$

According to the last two equations, $S_F(x', x)$ propagates positive-energy wave functions $\psi^{(+)}(x)$ forward $(t' > t)$ in time and negative-energy wave functions backward $(t > t')$ in time.

$G_{\text{ret}}(x', x)$, by contrast, is given by

$$\begin{aligned} i G_{\text{ret}}(x', x) &= \theta(t' - t) \sum_p \Big[\phi_{p+}(\mathbf{x}') \overline{\phi}_{p+}(\mathbf{x}) e^{-i E_p (t' - t)} \\ &\quad + \phi_{p-}(\mathbf{x}') \overline{\phi}_{p-}(\mathbf{x}) e^{i E_p (t' - t)} \Big], \end{aligned} \tag{12.132}$$

and propagates all wave functions forward in time.

Consider now the consequences of using $G_{\text{ret}}(x', x)$ or $S_F(x', x)$ in the first-order approximation to (12.120):

$$\psi(x') \cong \psi_0(x') + e \int d^4 x\, G(x', x) A(x) \psi_0(x). \tag{12.133}$$

Assume that $\psi_0(x')$ is a positive-energy plane-wave solution of the free-particle Dirac equation, and suppose we employ the retarded Green function $G_{\text{ret}}(x', x)$. Then

$$\psi(x') \cong \psi_0(x') - i \sum_p \phi_{p+}(\mathbf{x}') e^{-i E_p t'}$$

$$\times \left[e \int_{-\infty}^{t'} dt e^{iE_p t} \int d^3 x \overline{\phi}_{p+}(\mathbf{x}) A(x) \psi_0(x) \right]$$

$$- i \sum_{p} \phi_{p-}(\mathbf{x}') e^{iE_p t'} \left[e \int_{-\infty}^{t'} dt e^{-iE_p t} \int d^3 x \overline{\phi}_{p-}(\mathbf{x}) A(x) \psi_0(x) \right]$$

$$(12.134)$$

where we use the fact that $\int_{-\infty}^{\infty} dt \theta(t' - t)... = \int_{-\infty}^{t'} dt...$. This implies the amplitude

$$a_{p-}(t') \equiv -ie \int_{-\infty}^{t'} dt e^{-iE_p t} \int d^3 x \overline{\phi}_{p-}(\mathbf{x}) A(x) \psi_0(x) \qquad (12.135)$$

for the particle to be in a negative-energy state at time t', and the S-matrix element $a_{p-}(\infty)$. In other words, there is a nonvanishing probability for a transition from a positive-energy state to a negative-energy state, in contradiction to hole theory. For this reason the retarded Green function is physically unacceptable.

If instead we use the Feynman propagator $S_F(x', x)$ for $G(x', x)$ in (12.133), then (12.134) is replaced by

$$\psi(x') \cong \psi_0(x') - i \sum_{p} \phi_{p+}(\mathbf{x}') e^{-iE_p t'}$$

$$\times \left[e \int_{-\infty}^{t'} dt e^{iE_p t} \int d^3 x \overline{\phi}_{p+}(\mathbf{x}) A(x) \psi_0(x) \right]$$

$$+ i \sum_{p} \phi_{p-}(\mathbf{x}') e^{iE_p t'} \left[e \int_{t'}^{\infty} dt e^{-iE_p t} \int d^3 x \overline{\phi}_{p-}(\mathbf{x}) A(x) \psi_0(x) \right] .$$

$$(12.136)$$

As $t' \to \infty$ the second term vanishes, and so a positive-energy electron cannot make a transition to a negative-energy state. The use of the Feynman propagator therefore appears to be consistent with hole theory.

This satisfactory feature of the Feynman propagator is accompanied by something curious: $\psi(x') \neq \psi_0(x')$ as $t' \to -\infty$. According to (12.136), for $t' \to -\infty$,

$$\psi(x') \to \psi_0(x') + i \sum_{p} \phi_{p-}(\mathbf{x}') e^{iE_p t'} e \int_{-\infty}^{\infty} dt e^{-iE_p t} \int d^3 x \overline{\phi}_{p-}(\mathbf{x}) A(x) \psi_0(x).$$

$$(12.137)$$

To appreciate this result, suppose $\psi_0(x) = \sqrt{m/E_i V}\,u(\mathbf{p}_i, s_i)e^{-ip_i \cdot x}$, a plane wave for an electron of four-momentum p_i and spin s_i. Writing $\overline{\phi}_{p-}(\mathbf{x})e^{-iE_p t} = \sqrt{m/E_f V}\,\overline{v}(\mathbf{p}_f, s)e^{-ip_f \cdot x}$, we have

$$ie \int_{-\infty}^{\infty} dt\, e^{-iE_p t} \int d^3x\, \overline{\phi}_{p-}(\mathbf{x})\,\!\!\not{\!\!A}(x)\psi_0(x) =$$

$$ie\sqrt{\frac{m^2}{E_i E_f}}\frac{1}{V}\int d^4x\, \overline{v}(\mathbf{p}_f, s_f)\!\!\not{\!\!A}(x)u(\mathbf{p}_i, s_i)e^{-i(p_f + p_i)\cdot x}\,, \qquad (12.138)$$

which corresponds to the first-order S-matrix element for pair annihilation, Figure 12.8a, as the reader may easily verify using the quantum-field-theoretic formulation (see the following paragraph). In the present approach, where both the Dirac and Maxwell fields are *not* quantized, $v(\mathbf{p}_f, s_f)e^{ip_f \cdot x}$ is evidently interpretable as an *unoccupied* negative-energy state in hole theory, a state into which the electron can be scattered "backward in time" ($t' \rightarrow -\infty$) by the potential $A(x)$ (Figure 12.8a). The negative-energy electron scattered "backward in time," in other words, is the positron propagating forward in time in the present example of pair annihilation. This interpretation of positrons in the Feynman theory holds more generally: positive-energy electrons propagate forward in time, negative-energy electrons propagate backward in time, and the latter are equivalent to (positive-energy) positrons propagating forward in time. This is all a consequence of the choice $G(x', x) = S_F(x', x)$, a choice consistent with the Dirac hole theory.

In calculating the amplitude for the pair annihilation process of Figure 12.8a in quantum field theory, we specify the initial state $|i\rangle = |\mathbf{p}_i, s_i; \mathbf{p}_f, s_f; 0\rangle$ for an electron in state $(\mathbf{p}_i, s_i)$, a positron in state $(\mathbf{p}_f, s_f)$, and no photons, and the final state $|f\rangle = |0; 0; 1_{\mathbf{k}\lambda}\rangle$ of one photon in state $(\mathbf{k}, \lambda)$ and no electrons or positrons. Then, except for a physically insignificant minus sign,

$$\begin{aligned} S_{fi}^{(1)} &= i\int d^4x\, \langle f|h_I(x)|i\rangle \\ &= ie\int d^4x\, \langle f|\overline{\psi}(x)\!\!\not{\!\!A}(x)\psi(x)|i\rangle \\ &= ie\sqrt{\frac{m^2}{E_i E_f}}\frac{1}{V}\int d^4x\, \overline{v}(\mathbf{p}_f, s_f)\langle 1_{\mathbf{k}\lambda}|\!\!\not{\!\!A}(x)\psi(x)|0\rangle u(\mathbf{p}_i, s_i) \\ &\quad \times e^{-i(p_f + p_i)\cdot x}\,, \end{aligned} \qquad (12.139)$$

and obviously only the photon creation part of the operator $\!\!\not{\!\!A}(x)$ contributes:

$$\begin{aligned}
\langle 1_{\mathbf{k}\lambda}|A(x)|0\rangle &= \langle 1_{\mathbf{k}\lambda}|A^{(-)}(x)|0\rangle \\
&= \sqrt{\frac{1}{2\omega_k V}}e^{ik\cdot x}\epsilon(\mathbf{k},\lambda) \ .
\end{aligned} \qquad (12.140)$$

In order for (12.138) to reproduce the result (11.132) of quantum field theory, we must make the replacment

$$A(x) \rightarrow \sqrt{\frac{1}{2\omega_k V}}e^{ik\cdot x}\epsilon(\mathbf{k},\lambda) \qquad (12.141)$$

in (12.138) for pair annihilation. Similarly, for a process in which a photon is annihilated,

$$A(x) \rightarrow \sqrt{\frac{1}{2\omega_k V}}e^{-ik\cdot x}\epsilon(\mathbf{k},\lambda) \qquad (12.142)$$

in the Feynman approach. Doing this gives results in agreement with the full quantum field theory in which the Dirac and Maxwell fields are quantized. The appropriate replacement for processes in which the photon number does not change, such as those whose diagrams involve only internal photon lines, is discussed below.

Now in performing the integral over all $\mathbf{x}$ and t in equation (12.133) we get

$$\int d^4x\, e^{-i(p_f+p_i-k)\cdot x} = (2\pi)^4\delta(\omega_k - E_i - E_f)\delta^3(\mathbf{k}-\mathbf{p}_i-\mathbf{p}_f) \qquad (12.143)$$

for pair annihilation. The conditions $\omega_k = E_i + E_f$ and $\mathbf{k} = \mathbf{p}_i + \mathbf{p}_f$ for energy and momentum conservation cannot both be satisfied, as is obvious for the case $\mathbf{p}_i + \mathbf{p}_f = 0$: one-photon pair annihilation is forbidden. The rules (12.141) and (12.143) for photon creation and annihilation, however, are valid prescriptions for obtaining nonvanishing *matrix elements* in agreement with quantum field theory, even though the processes corresponding to these matrix elements might not be allowed by energy–momentum conservation.

Pair annihilation and creation processes are, of course, possible in an externally prescribed potential $A_{\text{ext}}^\mu(x)$ with time dependence such as to satisfy energy–momentum conservation. This was the case in Section 9.4, where we considered pair creation in a uniform electric field, corresponding to a vector potential $\mathbf{A}$ linearly proportional to time. For pair annihilation and creation processes in an external potential, the wavy (photon) lines in Figure 12.8 are deleted.

So the idea is to calculate S-matrix elements without direct recourse to quantum field theory. The fields $\psi(x)$ and $A(x)$ are not operators, but ordinary functions of $x = (\mathbf{x}, t)$, and the effect of the coupling of fermions to the electromagnetic field is studied by iteration of (12.113), using for the Green function the Feynman propagator $S_F(x', x)$. $S_F(x', x)$ is not a retarded Green function, so that there are contributions to $\psi(x')$ from $A(x)\psi(x)$ with $t < t'$ *and* $t > t'$. To calculate the amplitude for a given process, therefore, we must specify $\psi(x)$ in the remote *future* as well as the remote past. In particular, for positron scattering processes, $\psi(x')$ must have no positive-energy components in the remote past ($t' \rightarrow -\infty$), whereas for electron scattering processes $\psi(x')$ must have no negative-energy components in the remote future ($t' \rightarrow \infty$). This all comes about because the positive- and negative-energy electrons propagate forward and backward in time, respectively, and because of Feynman's interpretation of a positron as a negative-energy electron propagating backward in time.

We have one example thus far — pair annihilation in an external potential — of how this works. To calculate the pair annihilation amplitude, we require that $\psi_0(x')$ have no positive-energy components as $t' \rightarrow -\infty$, so that the amplitude is governed by the third term on the right side of (12.129). We can imagine the incoming electron to be scattered backward in time by the external potential, exactly as the diagram in Figure 12.8a suggests. We must therefore specify the final state at $t' \rightarrow -\infty$ as $\phi_{p_-}(x')e^{iE_p t'}$, i.e., corresponding to a *negative-energy* electron going *backward* in time.

Note that this approach allows us to describe pair annihilation as a single-particle process. We imagine the whole space-time history of the single particle (electron) to be laid out before us, the particle's history zigzagging backward and forward in time, as shown in Figure 12.8a. We are, in effect, following charge rather than individual particles themselves (Feynman, 1949a),

> ... considering [a] continuous world line as a whole rather than breaking it up into its pieces. It is as though a bombardier flying low over a road suddenly sees three roads and it is only when two of them come together and disappear again that he realizes that he has simply passed over a long switchback in a single road.
>
> This over-all space–time point of view leads to considerable simplification in many problems. One can take into account at the same time processes which ordinarily would have to be considered separately. For example, when considering the scattering of an electron by a potential one automatically takes into account the effects of virtual pair productions. The same equation, Dirac's, which describes the deflection of the world line of an electron in a field, can also describe the deflection (and in just as simple a manner) when it is

large enough to reverse the time-sense of the world line, and thereby to correspond to pair annihilation. Quantum mechanically, the direction of the world lines is replaced by the direction of propagation of waves.

This view is quite different from that of the Hamiltonian method which considers the future as developing continuously from out of the past. Here we imagine the entire space-time history laid out, and that we just become aware of increasing portions of it successively. In a scattering problem this over-all view of the complete scattering process is similar to the S matrix viewpoint of Heisenberg. The temporal order of events during the scattering, which is analyzed in such detail by the Hamiltonian differential equation, is irrelevant.

It is convenient to derive a general expression for the S-matrix element S_{fi}, basically following the example just given for the first-order approximation. Using the Feynman propagator $S_F(x', x)$ and the identity (12.127), we write (12.120) as

$$\begin{aligned}
\psi(x') &= \psi_0(x') + e \int d^4x \left[-i\theta(t' - t) \sum_p \psi_{p+}(x')\overline{\psi}_{p+}(x) \right. \\
&\quad \left. + i\theta(t - t') \sum_p \psi_{p-}(x')\overline{\psi}_{p-}(x) \right] \slashed{A}(x)\psi(x) \\
&= \psi_0(x') + \sum_p \psi_{p+}(x') \left[-ie \int d^4x\, \overline{\psi}_{p+}(x)\slashed{A}(x)\psi(x) \right] \\
&\qquad\qquad (\text{for } t' \to \infty) \\
&= \psi_0(x') + \sum_p \psi_{p-}(x') \left[ie \int d^4x\, \overline{\psi}_{p-}(x)\slashed{A}(x)\psi(x) \right] \\
&\qquad\qquad (\text{for } t' \to -\infty) ,
\end{aligned}$$
(12.144)

where $\psi_{p\pm}(x) = \phi_{p\pm}(\mathbf{x})e^{\mp iE_p t}$. Thus

$$S_{fi} = \delta_{fi} \pm ie \int d^4x\, \overline{\psi}_f(x)\slashed{A}(x)\psi_i(x)$$
(12.145)

is the S-matrix element for the transition $|i\rangle \to |f\rangle$. Here $\psi_f(x)$ is the final plane wave and the $-$ sign is used if it is a positive-energy solution propagating into the future, whereas a $+$ sign is used if $\psi_f(x)$ is a negative-energy solution propagating into the past. $\psi_i(x)$ is the initial wave, reducing as $t \to -\infty$ to an incident positive-energy solution or at $t \to \infty$ to an incident negative-energy solution. $\psi_i(x)$ for arbitrary t satisfies (12.120):

$$\psi_i(x) = \psi_0(x) + e \int d^4x'\, S_F(x, x')\slashed{A}(x')\psi_i(x').$$
(12.146)

Then

$$
\begin{aligned}
S_{fi} = \ & \delta_{fi} \pm ie \int d^4x \overline{\psi}_f(x) A(x) \psi_0(x) \\
& \pm ie^2 \int d^4x \int d^4x' \overline{\psi}_f(x) A(x) S_F(x,x') A(x') \psi_0(x') \\
& \pm \ \dots \ .
\end{aligned} \tag{12.147}
$$

In this example of pair annihilation the incident wave $\psi_0(x)$ corresponds to an electron propagating from $t = -\infty$ to the vertex of Figure 12.8a, and the final wave $\psi_f(x)$ corresponds to a positron propagating from $-\infty$ to the vertex, i.e., a negative-energy electron propagating backward in time to $t = -\infty$. We can think of the positive-energy electron as being scattered into a negative-energy state propagating backward in time.

In the case of pair creation (Figure 12.8b), the incident wave may be taken to correspond to a negative-energy electron propagating backward in time, i.e., a solution of the Dirac equation that reduces to a negative-energy plane-wave as $t \to \infty$. The final state $\psi_f(x)$, as indicated in the Feynman diagram in Figure 12.8b, is a positive-energy plane wave propagating forward in time.

Compton Scattering

To gain more familiarity with Feynman's perspective, let us consider again the example of Compton scattering, described in second-order perturbation theory by

$$
S_{fi}^{(2)} = \pm ie^2 \int d^4x \int d^4x' \overline{\psi}_f(x) A(x) S_F(x,x') A(x') \psi_0(x'). \tag{12.148}
$$

Using the initial and final states (12.64) and (12.65), we can consider the initial photon to be annihilated at x and the final photon to be created at x', corresponding to

$$
A(x) S_F(x,x') A(x') \to \sqrt{\frac{1}{4\omega_i \omega_f}} \frac{1}{V} \not{\epsilon}_i S_F(x,x') \not{\epsilon}_f e^{-ik_i \cdot x} e^{ik_f \cdot x'} , \tag{12.149}
$$

or the initial photon to be annihilated at x' and the final photon to be created at x, corresponding to

$$
A(x) S_F(x,x') A(x') \to \sqrt{\frac{1}{4\omega_i \omega_f}} \frac{1}{V} \not{\epsilon}_f S_F(x,x') \not{\epsilon}_i e^{-ik_i \cdot x'} e^{ik_f \cdot x} , \tag{12.150}
$$

when we use the rules (12.134) and (12.135). These two terms correspond to the diagrams $C2$ and $C1$, respectively, of Figure 12.2. For $\psi_o(x)$ we assume a plane wave for the initial electron, i.e., $\psi_0(x) = [(m/E_i)^{1/2}/V]u(\mathbf{p}_i, s_i) \times e^{-ik_i \cdot x}$. For $\overline{\psi}_f(x)$ we assume a plane wave $[(m/E_f)^{1/2}/V]\overline{u}(\mathbf{p}_f, s_f)e^{ik_f \cdot x}$. Then (12.141), with $A(x)S_F(x, x')A(x')$ equal to the sum of (12.142) and (12.143), gives exactly the S-matrix element $S_{fi}^{(2)}(C1) + S_{fi}^{(2)}(C2)$ derived from quantum field theory in Section 12.5.

Electron Self-Energy

The second-order amplitude (12.147),

$$S_{fi}^{(2)} = -ie^2 \int d^4x \int d^4x' \overline{\psi}_f(x)\gamma^\mu S_F(x, x')\gamma^\nu \psi_0(x')A_\mu(x)A_\nu(x'),$$
(12.151)

also applies for an electron *in the vacuum*. In this case, in which there are no external photon lines, $A_\mu(x)A_\nu(x')$ involves the emission and reabsorption of *virtual photons*, and so we replace $A_\mu(x)A_\nu(x')$ in the Feynman approach by the quantum field-theoretic expression

$$\langle 0|TA_\mu(x)A_\nu(x')|0\rangle = iD_{\mu\nu}(x, x'),$$
(12.152)

where $D_{\mu\nu}(x, x')$ is the photon propagator. Then

$$S_{fi}^{(2)} = e^2 \int d^4x \int d^4x' \overline{\psi}_f(x)\gamma^\mu S_F(x, x')\gamma^\nu \psi_0(x')D_{\mu\nu}(x, x'),$$
(12.153)

and when we take $\psi_0(x)$ and $\psi_f(x)$ to be wave functions for positive-energy electrons propagating forward in time, corresponding to the diagram in Figure 12.5, we recover exactly the results of Section 12.6 [compare (12.153) and (12.77)].

12.11   Discussion

When compared with the methods of "old-fashioned perturbation theory," Feynman's space–time approach is remarkable, to say the least, in its simplicity. We remarked earlier that the calculation of the electron self-energy in Section 12.6 is much simpler than the calculation of Section 11.4. In the Feynman approach the derivation of the basic self-energy S-matrix element [equation (12.151)] is simpler still. Note in particular that the use of the Feynman propagator automatically accounts for the effect of the negative-energy states, which are essential to the logarithmic rather than linear divergence of the electron self-energy (Section 11.4). With the "Feynman rules" for handling the vertices and internal and external lines of diagrams representing probability amplitudes, we can write expressions for the amplitudes without any operator algebra: neither the electron–positron field nor the electromagnetic field is quantized in Feynman's approach. Instead we regard the entire space–time history of a process to be laid out before us, and imagine that particles are scattered backward as well as forward in time. In the example of pair annihilation, we allow a positive-energy electron to be scattered into a "negative-energy" state propagating backward in time. The latter is equivalent to a (positive-energy) positron propagating forward in time. In this way we replace the two particles by a single particle that can zigzag backward or forward in time.

As is well-known, Stückelberg in the early 1940s proposed that a positron could be treated as a negative-energy electron propagating backward in time. Feynman, independently, went well beyond this in developing a general, relativistically covariant framework for all orders of perturbation theory. Unlike much of the previous work, furthermore, Feynman's did not involve a (noncovariant) splitting of the electromagnetic fields into transverse and longitudinal parts. This was an important part of the simplification he brought to the calculations.

As in the work of Tomonaga and Schwinger, Feynman's approach allowed for a systematic program of renormalization of mass and charge; the infinite values of the electromagnetic mass and charge were the source of all the nontrivial infinities of the theory. Feynman's approach made it easier to prove that renormalization of mass and charge could be carried out through all orders of the S-matrix expansion. This was done by Dyson (1949a), who used the counter term $\delta m \overline{\psi} \psi$ in the interaction Hamiltonian to renormalize the mass. Charge renormalization was handled by means of a counter term by Gupta (1951). We refer the reader to a paper by Matthews and Salam (1954) for a particularly clear discussion of the renormalizability of QED, and also to an excellent review by Gunn (1955).

Given its computational accessibility, it is no wonder that the covariant perturbation theory of Feynman – and the ubiquitous diagrams – provide a computational and conceptual basis for much of particle physics beyond QED, and have been usefully adopted in many-particle theory and other areas of theoretical physics.

A Feynman diagram really represents an amplitude that is a *sum over paths*, for the nth-order S-matrix element involves an n-fold integration over space and time. And of course the idea of a "negative-energy particle going backward in time" is only a mathematical artifice. One should not take the diagrams too literally.

In fact the physical interpretation suggested by the interaction picture can be different from that suggested by, say, the Heisenberg picture. In the interaction picture an electron in vacuum interacts with the *free* field, and its lowest-order self-energy arises from the free-field expectation values $(0|u_{\mathbf{k}\lambda}u^{\dagger}_{\mathbf{k}\lambda}|0) = 1$. Obviously the natural physical interpretation is that the electron self-energy is due to the vacuum fluctuations of the field, the fact that the averages of the *squares* of the electric and magnetic fields in the vacuum state are nonvanishing. These nonvanishing expectation values are associated with the photon propagator. But in the Heisenberg picture, as discussed in Chapter 11, this same self-energy can be attributed either to the vacuum field fluctuations or to the interaction of the electron with *itself* (radiation reaction), or some combination of the two, depending on the ordering of the field operators. The nonrelativistic theory of spontaneous emission and the Lamb shift in Chapter 4, similarly, can be worked out in the interaction picture, and the natural interpretation that then emerges is in terms of vacuum field fluctuations, as opposed to the possibility of associating these effects with radiation reaction in the Heisenberg picture.

Finally I think it is worth calling attention to the circumstances leading up to Feynman's space–time view of quantum electrodynamics. The basic ideas evolved from his efforts to solve the problem of infinite electromagnetic mass in *classical* electrodynamics, which he believed would provide a clue to the solution of the problem of infinities in QED (Feynman, 1966):

> I [gathered] from my readings ... that two things were the source of the difficulties with the quantum-electrodynamical theories. The first was an infinite energy of interaction of the electron with itself. And this difficulty existed even in the classical theory. The other difficulty came from some infinities which had to do with the infinite number of degrees of freedom in the field. As I understood it at the time (as nearly as I can remember) this was simply the difficulty that if you quantized the harmonic oscillators of the field (say in a box), each oscillator has a ground state energy of $\hbar\omega/2$, and there is an infinite number of modes in a box of ever increasing frequency ω, and therefore there is an infinite energy in the box. I now realize that that wasn't a completely correct statement of the central problem; it can be removed simply by changing the zero from which energy is measured. At any rate, I believed that the difficulty arose somehow from a combination of the electron acting on itself and the infinite number of degrees of freedom of the field.

The classical radiation reaction problem has never been "solved." It is now viewed by a majority of physicists, I suspect, as largely irrelevant, as discussed in Chapter 5. Regarding the infinities of QED, they have for the most part been "swept under the rug" by renormalization. Renormal izability is now thought to be a criterion that must be satisfied by any fundamental theory of physics. As discussed in Section 3.5, for instance, renormalization seems to be logically necessary regardless of whether masses and charges are finite or infinite. Whatever one thinks of the renormaliza tion program, its computational success has been little short of spectacular.

The two "sources of the difficulties" cited by Feynman — radiation re action and the infinite zero-point energy of the field — have figured promi nently in our physical interpretations of various QED effects in this book. The point has been made repeatedly that the radiation reaction and zero point fields are intimately related, that effects of the zero-point field are not eliminated by simply dropping the zero-point field energy from the Hamil tonian, and that very often one can interpret physical effects in terms of vacuum or source fields, this being largely a "matter of taste." These ideas are not entirely philosophical; they provide a useful intuitive framework even in such applied areas as the noise and coherence properties of lasers. It is my hope that this book may contribute to a better understanding of some of the simpler aspects of the quantum vacuum and different ways of thinking about them and also that it may provide the beginner with a gen tle introduction to the physics and the formalism of quantum optics and electrodynamics.

12.12   Bibliography

Bjorken, J. D. and S. D. Drell, *Relativistic Quantum Mechanics* (McGraw-Hill, New York, 1964).

Dyson, F. J., "The Radiation Theories of Tomonaga, Schwinger, and Feynman," *Phys. Rev.* **75**, 486 (1949a).

Dyson, F. J., "The S–Matrix in Quantum Electrodynamics," *Phys. Rev.* **75**, 1736 (1949b).

Feynman, R. P., "The Theory of Positrons," *Phys. Rev.* **76**, 749 (1949a).

Feynman, R. P., "Space-Time Approach to Quantum Electrodynamics," *Phys. Rev.* **76**, 769 (1949b).

Feynman, R. P., "Mathematical Formulation of the Quantum Theory of Electromagnetic Interaction," *Phys. Rev.* **80**, 440 (1950).

Feynman, R. P., "The Development of the Space-Time View of Quantum Electrodynamics," Nobel lecture reprinted in *Physics Today* (August 1966), 31.

Feynman, R. P., *QED: The Strange Theory of Light and Matter* (Princeton University Press, Princeton, New Jersey, 1985).

Gunn, J. C., "Theory of Radiation," *Rep. Prog. Phys.* **18**, 127 (1955).

Gupta, S., "On the Elimination of Divergencies from Quantum Electrodynamics," *Proc. Phys. Soc. Lond.* A64, 426 (1951).

Itzykson, C. and J.-B. Zuber, *Quantum Field Theory* (McGraw–Hill, New York, 1980).

Matthews, P. T. and A. Salam, "Renormalization," *Phys. Rev.* 94, 185 (1954).

Milonni, P. W., "Semiclassical and Quantum-Electrodynamical Approaches in Nonrelativistic Radiation Theory," *Phys. Rep.* 25, 1 (1976).

Milonni, P. W., "Wave-Particle Duality of Light: A Current Perspective," in *The Wave-Particle Dualism*, ed. S. Diner, D. Fargue, G. Lochak, and F. Selleri (Reidel, Dordrecht, 1984).

Pauli, W., "Relativistic Field Theories of Elementary Particles," *Rev. Mod. Phys.* 13, 203 (1941).

Sakurai, J. J., *Advanced Quantum Mechanics* (Addison–Wesley, Reading, Mass., 1976).

Scully, M. and M. Sargent, III, "The Concept of the Photon," *Physics Today* (March 1972), 38.

Schwinger, J. (editor), *Selected Papers on Quantum Electrodynamics* (Dover Books, New York, 1958).

Weinberg, S., "The Search for Unity: Notes for a History of Quantum Field Theory," *Daedalus* 106, 17 (1977).

Wick, G. C., "The Evaluation of the Collision Matrix," *Phys. Rev.* 80, 268 (1950).

Appendix A

Oscillator Equation and Absorption Rate

The Newton equation of motion for a particle of mass m and charge e, acted upon by an elastic restoring force $-m\omega_o^2 z$ and an external electric field $E_z(t)$, is

$$\ddot{z} + \omega_o^2 z = \frac{e}{m} E_z(t) + \frac{e}{m} E_{RR}(t). \tag{A.1}$$

For simplicity, and to follow Planck, Einstein, and Hopf, we assume the particle is constrained to one-dimensional motion.

The field $E_{RR}(t)$ in (A.1) is the field of radiation reaction, i.e., the electric field produced by the charged particle at the position of the particle. In other words, it is the electric field that the charge exerts on itself. For our purposes here a simplified derivation and expression for this field will suffice. A more detailed derivation is given in Appendix D.

We recall first the expression (1.8) for the rate at which an accelerating charge radiates electromagnetic energy. The energy radiated in the time interval from t_1 to t_2 is

$$W_{EM}(t_2, t_1) = \frac{2e^2}{3c^3} \int_{t_1}^{t_2} \ddot{z}(t)^2 dt = \frac{2e^2}{3c^3} [\ddot{z}(t)\dot{z}(t) \,|_{t_1}^{t_2} - \int_{t_1}^{t_2} \dddot{z}(t)\dot{z}(t)dt], \tag{A.2}$$

where the second equality follows from an integration by parts. We assume the motion of the charge is periodic and choose $t_2 - t_1$ to be an integral number of periods, in which case

$$W_{EM}(t_2, t_1) = -\frac{2e^2}{3c^3} \int_{t_1}^{t_2} \dddot{z}(t)\dot{z}(t)dt. \tag{A.3}$$

477

The change in energy of the charge, $-W_{DM}$, is attributed to the force $eE_{RR}(t)$ of radiation reaction:

$$- W_{EM}(t_2, t_1) = \frac{2e^2}{3c^3} \int_{t_1}^{t_2} \dddot{z}(t)\dot{z}(t)dt = \int_{t_1}^{t_2} eE_{RR}(t)\dot{z}(t)dt \qquad (A.4)$$

or

$$E_{RR}(t) = \frac{2e}{3c^3} \dddot{z}(t). \qquad (A.5)$$

Although this expression for the radiation reaction field was derived under the assumption of periodic motion, it actually holds more generally, as discussed in Chapter 5 and Appendix D. When it is used in (A.1), we obtain the equation (1.41) used by Planck, Einstein and Stern, and others.

For the case of a monochromatic applied field $E_z(t) = E_{z\omega} \cos(\omega t + \theta_\omega)$, equation (1.41) has the solution

$$z(t) = -\frac{e}{m} \text{Re} \left[\frac{E_{z\omega} e^{-i(\omega t + \theta_\omega)}}{\omega^2 - \omega_o^2 + i\gamma\omega^3} \right], \qquad (A.6)$$

so that the rate (force times velocity) at which the oscillator absorbs energy from the field is found after some simple algebra to be

$$\dot{W}_A = e\dot{z}(t)E_z(t) \rightarrow \frac{e^2}{2m} \frac{\gamma\omega^4 E_{z\omega}^2}{(\omega^2 - \omega_o^2)^2 + \gamma^2\omega^6}, \qquad (A.7)$$

where we have taken an average over the oscillations of the field, replacing $\cos^2(\omega t + \theta_\omega)$ by $1/2$ and $\sin(\omega t + \theta_\omega)\cos(\omega t + \theta_\omega)$ by 0.

Now suppose the applied field has a broad distribution of frequencies, with energy density in the interval $[\omega, \omega + d\omega]$ given by $\rho(\omega)d\omega = E_{z\omega}^2/8\pi$. In this case (A.7) is replaced by

$$\dot{W}_A = \frac{4\pi e^2}{m}\gamma \int_0^\infty \frac{\omega^4 \rho(\omega)d\omega}{(\omega^2 - \omega_o^2)^2 + \gamma^2\omega^6}. \qquad (A.8)$$

The time $\gamma = 2e^2/3mc^3 = 6.3 \times 10^{-24}$ sec is so short that, for natural oscillation frequencies ω_o of interest, $\gamma\omega_o \ll 1$. Furthermore $\rho(\omega)$ may be assumed to be flat compared with the sharply peaked function

$$\frac{\omega^4}{(\omega^2 - \omega_o^2)^2 + \gamma^2\omega^6} \cong \frac{\omega_o^4}{4\omega_o^2(\omega - \omega_o)^2 + \gamma^2\omega_o^6} \qquad (A.9)$$

in the integrand of (A.8), so that

$$\begin{aligned}
\dot{W}_A &\cong \frac{\pi e^2\gamma}{m}\omega_o^2\rho(\omega_o) \int_0^\infty \frac{d\omega}{(\omega - \omega_o)^2 + \gamma^2\omega_o^4/4} \cong \frac{\pi e^2\gamma}{m}\omega_o^2\rho(\omega_o)\left(\frac{2\pi}{\gamma\omega_o^2}\right) \\
&= \frac{2\pi^2 e^2}{m}\rho(\omega_o) = \frac{\pi e^2}{m}\rho(\nu_o) \rightarrow \frac{\pi e^2}{3m}\rho(\nu_0).
\end{aligned} \qquad (A.10)$$

In the last step we have replaced $\rho(\nu_o)$ by $\rho(\nu_o)/3$, where now the spectral energy density is defined by $\rho(\omega)d\omega = (E_{x\omega}^2 + E_{y\omega}^2 + E_{z\omega}^2)/8\pi = 3E_{z\omega}^2/8\pi$ for (isotropic and unpolarized) thermal radiation. We have thus arrived at equation (1.7) for the energy absorption rate.

By replacing e^2/m by $e^2 f/m$ in equation (1.7), where f is the oscillator strength of an atomic transition of frequency ω_o, we obtain the energy absorption rate given by quantum mechanics up to second order in perturbation theory.[1]

---

[1] See, for instance, M. Cray, M.-L. Shih, and P. W. Milonni, *Am. J. Phys.* **50**, 1016 (1982).

Appendix B

Force on an Atom in a Thermal Field

We shall follow Einstein's derivation of the force acting on an atom moving with velocity v in a thermal field. A classical derivation can be given along similar lines for the Einstein–Hopf force (1.42) acting on a classical dipole oscillator. Since the result differs from (1.92) only by simple multiplicative factors, we will not go through the classical derivation here.[1]

The field energy density in the frequency interval $[\omega, \omega + d\omega]$ and within the solid angle $d\Omega$ is $\rho(\omega)d\omega d\Omega/4\pi$, where $\rho(\omega)$ is independent of direction since thermal radiation is isotropic in the laboratory frame. Consider radiation propagating in a direction θ with respect to the axis defined by the atom's velocity. The frequency of radiation in the atom's frame is Doppler shifted to

$$\omega' \cong \omega(1 - \frac{v}{c}\cos\theta) \quad (v/c << 1). \tag{B.1}$$

The radiation appears to the atom to be directed at an angle θ' given by the aberration formula[2]

$$\cos\theta' \cong \cos\theta - \frac{v}{c}\sin^2\theta. \tag{B.2}$$

The field energy density $\rho'(\omega', \theta')d\omega'd\Omega'/4\pi$ in the frame of the moving atom can be obtained straightforwardly from the well-known transforma-

---

[1] See T. H. Boyer, *Phys. Rev.* **182**, 1374 (1969) for the classical derivation following Einstein and Hopf.

[2] See, for instance, A. P. French, *Special Relativity* (Nelson, Sunbury-on-Thames, Middlesex, 1979), p. 134.

tion properties of the electric and magnetic fields under Lorentz transfor mations of the coordinates. We simply write the result:

$$\rho'(\omega', \theta') d\omega' d\Omega' \cong (1 - \frac{2v}{c} \cos\theta) \rho(\omega) d\omega d\Omega, \qquad (B.3)$$

or

$$\rho'(\omega', \theta') \cong (1 - \frac{2v}{c} \cos\theta) \rho(\omega) \frac{d\omega}{d\omega'} \frac{d(\cos\theta)}{d(\cos\theta')} \cong (1 - \frac{3v}{c} \cos\theta') \rho(\omega), \quad (B.4)$$

where we have used (B.1) and (B.2) and continue to assume $v/c \ll 1$. From (B.1) it also follows that

$$\rho(\omega) \cong \rho(\omega' + \frac{v\omega'}{c} \cos\theta') \cong \rho(\omega') + \frac{d\rho(\omega')}{d\omega} \left(\frac{v}{c}\right) \omega' \cos\theta', \qquad (B.5)$$

so that (B.4) becomes

$$\rho'(\omega', \theta') \cong (1 - \frac{3v}{c} \cos\theta') \left[\rho(\omega') + \frac{d\rho(\omega')}{d\omega} \left(\frac{v}{c}\right) \omega' \cos\theta'\right]. \qquad (B.6)$$

Radiation in the solid angle $d\Omega'$ induces in the moving atom an average number

$$n_2 = B_{12} N_2 \rho'(\omega', \theta') d\Omega'/4\pi \qquad (B.7)$$

of stimulated emission transitions per unit time, and a number

$$n_1 = B_{12} N_1 \rho'(\omega', \theta') d\Omega'/4\pi \qquad (B.8)$$

of absorption transitions. The net momentum per unit time imparted to the atom due to stimulated emission and absorption of photons of momentum $\hbar\omega'/c$ is thus

$$F = \frac{dp}{dt} = (n_1 - n_2)\frac{\hbar\omega'}{c} \cos\theta', \qquad (B.9)$$

since absorption causes the atom to recoil in the same direction as the field propagation, whereas, from conservation of linear momentum, stimulated emission causes recoil in the opposite direction. From (B.7) – (B.9),

$$F = \frac{\hbar\omega'}{c} \frac{B_{12}}{4\pi} (N_1 - N_2) \rho'(\omega', \theta') \cos\theta' d\Omega'. \qquad (B.10)$$

Note that spontaneous emission adds no net momentum on average to the atom, since it is equally likely in all directions.

We now add up the forces associated with all directions of propagation of radiation of frequency ω :

$$
\begin{aligned}
F &= \frac{\hbar\omega'}{c}\frac{B_{12}}{4\pi}(N_1 - N_2) \int_0^{2\pi} d\phi' \int_0^{\pi} d\theta' \sin\theta' \rho'(\omega', \theta') \cos\theta' \\
&= \frac{\hbar\omega'}{2c} B_{12}(N_1 - N_2) \int_0^{\pi} d\theta' \sin\theta' \cos\theta' \rho'(\omega', \theta') \\
&\cong -(\frac{\hbar\omega}{c^2})(N_1 - N_2)B_{12} \left[\rho(\omega) - \frac{\omega}{3}\frac{d\rho(\omega)}{d\omega} \right] v, \qquad \text{(B.11)}
\end{aligned}
$$

where to lowest order in v/c we have dropped all primes. This is Einstein's equation (1.92) for the force on an atom moving in a thermal field.

Appendix C

Derivation of Equation (2.28)

The general identity $\nabla \cdot (\mathbf{F} \times \mathbf{G}) = \mathbf{G} \cdot \nabla \times \mathbf{F} - \mathbf{F} \cdot \nabla \times \mathbf{G}$, together with the Coulomb gauge condition $\nabla \cdot \mathbf{A}_\text{o} = 0$, implies

$$
\begin{aligned}
(\nabla \times \mathbf{A}_\text{o})^2 &= \nabla \cdot (\mathbf{A}_\text{o} \times \nabla \times \mathbf{A}_\text{o}) + \mathbf{A}_\text{o} \cdot \nabla \times (\nabla \times \mathbf{A}_\text{o}) \\
&= \nabla \cdot (\mathbf{A}_\text{o} \times \nabla \times \mathbf{A}_\text{o}) + \mathbf{A}_\text{o} \cdot [\nabla(\nabla \cdot \mathbf{A}_\text{o}) - \nabla^2 \mathbf{A}_\text{o}] \\
&= \nabla \cdot (\mathbf{A}_\text{o} \times \nabla \times \mathbf{A}_\text{o}) - \mathbf{A}_\text{o} \cdot \nabla^2 \mathbf{A}_\text{o} \\
&= \nabla \cdot (\mathbf{A}_\text{o} \times \nabla \times \mathbf{A}_\text{o}) + k^2 \mathbf{A}_\text{o}^2
\end{aligned}
\tag{C.1}
$$

when we use the fact that $\mathbf{A}_\text{o}$ satisfies the Helmholtz equation (2.24). Then the divergence theorem implies

$$
\begin{aligned}
\int d^3 r [\nabla \times \mathbf{A}_\text{o}(\mathbf{r})]^2 &= \oint dS \hat{\mathbf{n}} \cdot \mathbf{A}_\text{o} \times (\nabla \times \mathbf{A}_\text{o}) + k^2 \int d^3 r \mathbf{A}_\text{o}(\mathbf{r})^2 \\
&= k^2 \int d^3 r \mathbf{A}_\text{o}(\mathbf{r})^2 \ ,
\end{aligned}
\tag{C.2}
$$

since the surface integral vanishes as a consequence of the assumed periodic boundary condition on $\mathbf{A}_\text{o}(\mathbf{r})$.

Appendix D

Electric Field of Radiation Reaction

In the mode continuum limit (2.84) becomes

$$
\begin{aligned}
\mathbf{E}_{\mathrm{RR}}(t) &= -\frac{4\pi e}{V}\frac{V}{8\pi^3}\int d^3k \sum_\lambda \int_0^t dt'[\dot{\mathbf{x}}(t')\cdot\mathbf{e}_{\mathbf{k}\lambda}]\mathbf{e}_{\mathbf{k}\lambda}\cos\omega_k(t'-t) \\
&= -\frac{e}{2\pi^2}\int_0^t dt'\cos\omega_k(t'-t)\int_0^\infty dk\, k^2 \\
&\quad \times \int d\Omega_{\mathbf{k}}[\dot{\mathbf{x}}(t')-(\hat{\mathbf{k}}\cdot\dot{\mathbf{x}}(t'))\hat{\mathbf{k}}], \qquad (\mathrm{D}.1)
\end{aligned}
$$

where $\int d\Omega_{\mathbf{k}}$ denotes an integration over solid angles about $\mathbf{k}$ and we have used the identity $\dot{\mathbf{x}} = (\hat{\mathbf{k}}\cdot\dot{\mathbf{x}})\hat{\mathbf{k}} + \sum_\lambda(\dot{\mathbf{x}}\cdot\mathbf{e}_{\mathbf{k}\lambda})\mathbf{e}_{\mathbf{k}\lambda}$, where $\hat{\mathbf{k}}\equiv\mathbf{k}/k$. Now

$$
\int d\Omega_{\mathbf{k}}[\dot{\mathbf{x}}(t')-(\hat{\mathbf{k}}\cdot\dot{\mathbf{x}}(t'))\hat{\mathbf{k}}] = (4\pi - \frac{4\pi}{3})\dot{\mathbf{x}}(t'), \qquad (\mathrm{D}.2)
$$

and so

$$
\begin{aligned}
\mathbf{F}_{\mathrm{RR}}(t) &= -\frac{4e}{3\pi c^3}\int_0^t dt'\dot{\mathbf{x}}(t')\int_0^\infty d\omega\,\omega^2\cos\omega(t'-t) \\
&= \frac{4e}{3c^3}\int_0^t dt'\dot{\mathbf{x}}(t')\frac{\partial^2}{\partial t'^2}\delta(t'-t) \\
&= \frac{4e}{3c^3}[-\ddot{\mathbf{x}}(t)\delta(0) + \frac{1}{2}\dddot{\mathbf{x}}(t)], \quad t > 0, \qquad (\mathrm{D}.3)
\end{aligned}
$$

where the last line follows from two partial integrations. Here

$$\delta(0) \equiv \frac{1}{\pi} \int_0^\infty d\omega = \frac{1}{\pi\hbar} \int_0^\infty dE. \tag{D.4}$$

Thus

$$\mathbf{E}_{RR}(t) = \frac{2e}{3c^3} \overset{...}{\mathbf{x}}(t) - \frac{\delta m}{e} \ddot{\mathbf{x}}(t), \tag{D.5}$$

where

$$\delta m = \frac{4e^2}{3\pi c^3} \int_0^\infty d\omega = \frac{4e^2}{3\pi\hbar c^3} \int_0^\infty dE = \frac{4\alpha}{3\pi c^2} \int_0^\infty dE \tag{D.6}$$

is called the *electromagnetic mass*. We can then write (2.82) as

$$(m + \delta m)\ddot{\mathbf{x}} + \omega_o^2 \mathbf{x} = e\mathbf{E}_o + \frac{2e^2}{3c^3} \overset{...}{\mathbf{x}} . \tag{D.7}$$

According to this equation, δm is effectively a contribution to the mass and arises from the action on the dipole of its own field, i.e., from radiation reaction. (See also Appendix A for a simple derivation of the first term on the right side of (D.5).)

The calculation of the radiation reaction field (D.5) of a point dipole is valid both as a classical calculation and as a Heisenberg-picture, quantum-mechanical one.[1] Within the nonrelativistic approximation made in the calculation, the result for $\mathbf{E}_{RR}(t)$ applies also to a point charge. One way to see this in a crude sort of way is to consider the nonrelativistic Heisenberg equation of motion for an electron in a plane-wave field:

$$m\ddot{\mathbf{x}} = e\mathbf{E}_o e^{-i(\omega t - \mathbf{k}\cdot\mathbf{x})} , \tag{D.8}$$

where we drop terms of order $|\dot{\mathbf{x}}/c|$. Since $\mathbf{k} \cdot \mathbf{E}_o = 0, \mathbf{k} \cdot \dot{\mathbf{x}}$ is constant, the motion in the direction of field propagation is unaffected by the field in the nonrelativistic approximation. Thus we can effectively replace $e^{i\mathbf{k}\cdot\mathbf{x}}$ by 1 on the right side of (D.8). In other words, once the nonrelativistic approximation is made we are also making, in effect, the dipole approximation. Now if we extend this argument to all plane-wave modes of the field, we conclude that in the nonrelativistic approximation the expression (D.5) for $\mathbf{E}_{RR}(t)$ applies to a point charge as well as to a point dipole.

The nonrelativistic radiation reaction field for a rigid charge distribution is given, for instance, by Jackson.[2] We shall outline here a calculation using

---

[1] Recall the discussion in Section 4.6 about the formal correspondence between the classical and quantum-mechanical solutions of the Maxwell equations.

[2] J. D. Jackson, *Classical Electrodynamics*, 2nd ed. (Wiley, New York, 1975), Chapter 17.

the Coulomb gauge, which leads to some simplification. For a rigid charge distribution ρ the radiation reaction field is

$$\mathbf{E}_{RR}(t) = \frac{1}{e} \int d^3x \rho(\mathbf{x}) \mathbf{E}_s(\mathbf{x}, t), \tag{D.9}$$

where e is the total charge ($\int d^3x \rho(\mathbf{x}) = e$) and $\mathbf{E}_s(\mathbf{x}, t)$ is the self-field at $\mathbf{x}$. In the Coulomb gauge the transverse and longitudinal parts of $\mathbf{E}_s(\mathbf{x}, t)$ are, respectively,

$$\mathbf{E}_s^{\perp}(\mathbf{x}, t) = -\frac{1}{c}\frac{\partial}{\partial t}\mathbf{A}_s(\mathbf{x}, t), \tag{D.10}$$

$$\mathbf{E}_s^{\|}(\mathbf{x}, t) = -\nabla\phi_s(\mathbf{x}, t), \tag{D.11}$$

where

$$\mathbf{A}_s(\mathbf{x}, t) = \frac{1}{c}\int d^3x' \frac{\mathbf{J}^{\perp}(\mathbf{x}', t - |\mathbf{x} - \mathbf{x}'|/c)}{|\mathbf{x} - \mathbf{x}'|}, \tag{D.12}$$

$$\phi_s(\mathbf{x}, t) = \int d^3x' \frac{\rho(\mathbf{x}', t)}{|\mathbf{x} - \mathbf{x}'|}, \tag{D.13}$$

and $\mathbf{J}(\mathbf{x}, t)$ and $\rho(\mathbf{x}, t)$ are the current and charge densities, respectively.

It is clear that, for a spherically symmetric charge distribution, $\mathbf{E}_s^{\|}(\mathbf{x}, t)$ makes no contribution to $\mathbf{E}_{RR}(t)$. That is, the electrostatic radiation reaction field vanishes and

$$
\begin{aligned}
\mathbf{E}_{RR}(t) &= \frac{1}{e}\int d^3x \rho(\mathbf{x}, t)\mathbf{E}_s^{\perp}(\mathbf{x}, t) \\
&= -\frac{1}{ec^2}\int d^3x \rho(\mathbf{x}, t)\frac{\partial}{\partial t}\int d^3x' \frac{\mathbf{J}^{\perp}(\mathbf{x}', t - |\mathbf{x} - \mathbf{x}'|/c)}{|\mathbf{x} - \mathbf{x}'|} \\
&= -\frac{1}{ec^2}\sum_{n=0}^{\infty}\frac{(-1)^n}{c^n n!}\int d^3x \rho(\mathbf{x}, t)\int d^3x' |\mathbf{x} - \mathbf{x}'|^{n-1} \\
&\quad \times \frac{\partial^{n+1}}{\partial t^{n+1}}\mathbf{J}^{\perp}(\mathbf{x}', t),
\end{aligned}
\tag{D.14}
$$

where in the last step we have employed a Taylor expansion of $\mathbf{J}^{\perp}(\mathbf{x}', t - |\mathbf{x} - \mathbf{x}'|/c)$ about t. For a rigid charge distribution moving with velocity $\mathbf{v}(t) = \dot{\mathbf{r}}(t)$, $\rho(\mathbf{x}, t) = \rho(\mathbf{x} - \mathbf{r}(t))$.

According to (4.34) and (4.35),

$$
\begin{aligned}
J_i^{\perp}(\mathbf{x}, t) &= \int d^3x' \delta_{ij}^{\perp}(\mathbf{x} - \mathbf{x}') J_j(\mathbf{x}', t) \\
&= \frac{2}{3}J_i(\mathbf{x}, t) - \frac{1}{4\pi}\int d^3x' \frac{1}{R^3}\left(\delta_{ij} - \frac{3R_i R_j}{R^2}\right) J_j(\mathbf{x}', t),
\end{aligned}
\tag{D.15}
$$

where $\mathbf{R} = \mathbf{x} - \mathbf{x}'$. The current density $\mathbf{J}(\mathbf{x}, t)$ for a rigid charge distribution is

$$\mathbf{J}(\mathbf{x}, t) = \rho(\mathbf{x}, t)\mathbf{v}(t), \tag{D.16}$$

and for a spherically symmetric distribution the contribution to (D.14) from the second term on the righthand side of (D.15) vanishes. In this case we may effectively replace $\mathbf{J}^{\perp}(\mathbf{x}', t)$ by $(2/3)\mathbf{J}(\mathbf{x}', t)$ in (D.14):

$$\begin{aligned}
\mathbf{E}_{RR}(t) &= -\frac{2}{3ec^2} \sum_{n=0}^{\infty} \frac{(-1)^n}{c^n n!} \int d^3x \int d^3x' \rho(\mathbf{x}, t)|\mathbf{x} - \mathbf{x}'|^{n-1} \\
&\quad \times \frac{\partial^{n+1}}{\partial t^{n+1}}[\rho(\mathbf{x}, t)\mathbf{v}(t)].
\end{aligned} \tag{D.17}$$

If we ignore terms nonlinear in $\mathbf{v}(t)$, which is consistent with the nonrelativistic approximation, we may make the replacement

$$\frac{\partial^{n+1}}{\partial t^{n+1}}[\rho(\mathbf{x}, t)\mathbf{v}(t)] \rightarrow \rho(\mathbf{x}, t)\frac{d^{n+1}\mathbf{v}}{dt^{n+1}} \tag{D.18}$$

in (D.17) and write

$$\mathbf{E}_{RR}(t) = -\frac{2e}{3c^3} \sum_{n=0}^{\infty} \frac{A_n}{n!} \left(\frac{a}{c}\right)^{n-1} \frac{d^{n+1}\mathbf{v}}{dt^{n+1}}, \tag{D.19}$$

$$A_n \equiv \frac{(-1)^n}{e^2} \int d^3x \int d^3x' \left[\frac{|\mathbf{x} - \mathbf{x}'|}{a}\right]^{n-1} \rho(\mathbf{x})\rho(\mathbf{x}'), \tag{D.20}$$

and a is a length characterizing the extent of the (spherically symmetric) charge distribution $\rho(\mathbf{x})$.

The radiation reaction force $\mathbf{F}_{RR} = e\mathbf{E}_{RR}$. As discussed in Chapter 5, the electromagnetic mass for the extended charge distribution is

$$\begin{aligned}
\delta m &= \frac{2e^2}{3c^3}\left(\frac{a}{c}\right)^{-1} A_o = \frac{2}{3c^2} \int d^3x \int d^3x' \frac{\rho(\mathbf{x})\rho(\mathbf{x}')}{|\mathbf{x} - \mathbf{x}'|} \\
&= \frac{2}{3c^2} \int d^3x \rho(\mathbf{x})\phi(\mathbf{x}) = \frac{4}{3\pi c^2} \int_0^{\infty} dk \tilde{\rho}^2(k), \tag{D.21}
\end{aligned}$$

where $\nabla^2 \phi = -4\pi\rho$ and $\tilde{\rho}(k)$ is the Fourier transform of $\rho(\mathbf{x})$, a function only of k for the assumed case of a spherically symmetric charge distribution. In the point charge limit, $\rho(\mathbf{x}) = e\delta^3(\mathbf{x})$, $\tilde{\rho}(k) = e$, and

$$\delta m = \frac{4e^2}{3\pi c^2} \int_0^{\infty} dk = \frac{4e^2}{3\pi c^3} \int_0^{\infty} d\omega, \tag{D.22}$$

in agreement with (D.6).

In relativistic QED the linear divergence of δm is replaced by a logarithmic divergence (Chapters 11 and 12). A logarithmic divergence is also obtained nonrelativistically when retardation is included (Section 3.9). Historically, there were attempts to attribute all the mass of an electron to electromagnetic mass, but these attempts were not successful. Radiation reaction is reviewed briefly in Chapter 5, where we also mention theories constructed so that the electromagnetic mass, instead of diverging, is zero.

Appendix E

Photodetection and Normal Ordering

Devices used to measure light intensity nearly always do so by absorbing radiation and then converting the energy to another form. At the microscopic level the detection process involves, for instance, the promotion of bound electrons to continuum states. Thus a phototube operates on the basis of the photoelectric effect, involving a photoemissive surface and an anode that collects the photoelectrons to register a current proportional to the rate of absorption of photons.

Consider, to begin with, a highly idealized "detector atom" consisting of two states, $|a\rangle$ and $|g\rangle$, of energies E_a and E_g, $E_a > E_g$. We assume this atom interacts with radiation via the electric dipole interaction $-\mathbf{d} \cdot \mathbf{E}(\mathbf{r})$, where $\mathbf{d}$ is the electric dipole moment operator and $\mathbf{E}(\mathbf{r})$ is the electric field operator at the position of the (point) atom.[1] Suppose that at time $t = 0$ the atom is in the lower state $|g\rangle$ and the field is in the state $|I\rangle$. What is the probability amplitude that at time $t > 0$ the atom is in the upper state $|a\rangle$ and the field is in state $|F\rangle$? In perturbation theory this amplitude is

$$
\begin{aligned}
a_{fi}(t) &= -\frac{i}{\hbar} e^{-iE_F t/\hbar} \int_0^t dt_1 \langle f|h_I(t_1)|i\rangle \\
&= \frac{i}{\hbar} e^{-iE_f t/\hbar} \int_0^t dt_1 \langle f|\mathbf{d}(t_1) \cdot \mathbf{E}(\mathbf{r}, t_1)|i\rangle \\
&= \frac{i}{\hbar} e^{-iE_f t/\hbar} \int_0^t dt_1 \langle a|\mathbf{d}(t_1)|g\rangle \cdot \langle F|\mathbf{E}(\mathbf{r}, t_1)|I\rangle \quad \text{(E.1)}
\end{aligned}
$$

---

[1] See Section 4.4.

in the notation of Section 12.2. Here $d(t)$ and $\mathbf{E}(\mathbf{r}, t)$ are the dipole and electric field operators in the interaction picture.

The two-state model for an atom is discussed in Section 4.7. If we denote by $\mathbf{d}_{ag}$ the matrix element of $\mathbf{d}$ between the states $|a\rangle$ and $|g\rangle$, then the dipole operator

$$\mathbf{d}(t) = \mathbf{d}_{ag}\sigma^{\dagger}e^{i\omega_{ag}t} + \mathbf{d}_{ga}\sigma e^{-i\omega_{ag}t} \tag{E.2}$$

in the interaction picture, where $\hbar\omega_{ag} = E_a - E_g$ and σ and $\sigma^{\dagger}$ are the lowering and raising operators introduced in Section 4.7. Since $\langle a|\sigma|g\rangle = 0$ and $\langle a|\sigma^{\dagger}|g\rangle = \langle a|a\rangle = 1$, we have

$$
\begin{aligned}
a_{fi}(t) &= \frac{i}{\hbar}e^{-iE_f t/\hbar}\mathbf{d}_{ag} \cdot \int_0^t dt_1 \langle F|\mathbf{E}(\mathbf{r}, t_1)|I\rangle e^{i\omega_{ag}t_1} \\
&= \frac{i}{\hbar}e^{-iE_f t/\hbar}d_{ag,\mu} \int_0^t dt_1 \langle F|E_{\mu}(\mathbf{r}, t_1)|I\rangle e^{i\omega_{ag}t_1} ,
\end{aligned} \tag{E.3}
$$

where a sum over $\mu = 1, 2, 3$ is implicit.

The electric field operator in the interaction picture has the form

$$\mathbf{E}(\mathbf{r}, t) = -i\sum_{\alpha}(2\pi\hbar\omega_{\alpha})^{1/2}[a_{\alpha}e^{-i\omega_{\alpha}t}\mathbf{A}_{\alpha}(\mathbf{r}) - a_{\alpha}^{\dagger}e^{i\omega_{\alpha}t}\mathbf{A}_{\alpha}^{*}(\mathbf{r})], \tag{E.4}$$

where a_{α} and $a_{\alpha}^{\dagger}$ are (time-independent) photon annihilation and creation operators and the $\mathbf{A}_{\alpha}(\mathbf{r})$ are classically determined mode functions. We define the positive- and negative-frequency parts of the field by

$$\mathbf{E}^{(+)}(\mathbf{r}, t) = -i\sum_{\alpha}(2\pi\hbar\omega_{\alpha})^{1/2}a_{\alpha}e^{-i\omega_{\alpha}t}\mathbf{A}_{\alpha}(\mathbf{r}), \tag{E.5}$$

$$\mathbf{E}^{(-)}(\mathbf{r}, t) = i\sum_{\alpha}(2\pi\hbar\omega_{\alpha})^{1/2}a_{\alpha}^{\dagger}e^{i\omega_{\alpha}t}\mathbf{A}_{\alpha}^{*}(\mathbf{r}), \tag{E.6}$$

so that $\mathbf{E}^{(-)}(\mathbf{r}, t) = \mathbf{E}^{(+)}(\mathbf{r}, t)^{\dagger}$ and

$$\mathbf{E}(\mathbf{r}, t) = \mathbf{E}^{(+)}(\mathbf{r}, t) + \mathbf{E}^{(-)}(\mathbf{r}, t). \tag{E.7}$$

It is clear from (E.3) that only $\mathbf{E}^{(+)}(\mathbf{r}, t)$ will give rise to an energy-conserving transition amplitude:[2]

$$a_{fi}(t) = \frac{i}{\hbar}e^{-iE_f t/\hbar}d_{ag,\mu} \int_0^t dt_1 \langle F|E_{\mu}^{(+)}(\mathbf{r}, t_1)|I\rangle e^{i\omega_{ag}t_1} \tag{E.8}$$

---

[2] Recall the remarks following (12.18).

and

$$|a_{fi}(t)|^2 = \frac{1}{\hbar^2} d^*_{ag,\mu} d_{ag,\nu} \int_0^t dt_1 \int_0^t dt_2 \langle I|E^{(-)}_\mu(\mathbf{r}, t_1)|F\rangle$$
$$\times \langle F|E^{(+)}_\nu(\mathbf{r}, t_2)|I\rangle e^{i\omega_{ag}(t_2-t_1)} . \qquad \text{(E.9)}$$

We now sum over all possible final states $|F\rangle$ of the field, assuming that no observations are made to discriminate among possible final field states. Using the completeness relation $\sum_F |F\rangle\langle F| = 1$, we obtain

$$|a_{fi}(t)|^2 = \frac{1}{\hbar^2} d^*_{ag,\mu} d_{ag,\nu} \int_0^t dt_1 \int_0^t dt_2 \langle E^{(-)}_\mu(\mathbf{r}, t_1)E^{(+)}(\mathbf{r}, t_2)\rangle e^{i\omega_{ag}(t_2-t_1)}$$
$$\qquad \text{(E.10)}$$

where the expectation value refers to the initial field state $|I\rangle$.

What we have done thus far is nothing more than standard perturbation theory for the absorption of radiation by a two-state system. In the case of a practical photodetector there is effectively a *continuum* of final electron states $|a\rangle$. Not all the final states will have an equal probability of being counted by the device, and so we integrate (E.10) over final states $|a\rangle$, using some weighting function $P(a)$ characteristic of the device:

$$|a_{fi}(t)|^2 \rightarrow p^{(1)}(t) \equiv \frac{1}{\hbar^2} \int dE_a P(a) d^*_{ag,\mu} d_{ag,\nu} \int_0^t dt_1 \int_0^t dt_2 e^{i\omega_{ag}(t_2-t_1)}$$
$$\times \langle E^{(-)}_\mu(\mathbf{r}, t_1)E^{(+)}_\nu(\mathbf{r}, t_2)\rangle$$
$$= \int_0^t dt_1 \int_0^t dt_2 S_{\mu\nu}(t_2 - t_1)\langle E^{(-)}_\mu(\mathbf{r}, t_1)E^{(+)}_\nu(\mathbf{r}, t_2)\rangle,$$
$$\qquad \text{(E.11)}$$

where

$$S_{\mu\nu}(t) \equiv \frac{1}{\hbar^2} \int_0^\infty dE_a P(a) d^*_{ag,\mu} d_{ag,\nu} e^{i\omega_{ag}t} . \qquad \text{(E.12)}$$

The assumption of a continuum of final electron states means in effect that the possibility of an electron making a transition from a state $|a\rangle$ back to the bound state $|g\rangle$ is negligible as a practical matter. This justifies the use of (E.11) without accounting for higher order corrections in perturbation theory. In particular, the assumption of a continuum of final electron states allows us to ignore the possibility of temporally coherent atomic effects such as Rabi oscillations — we have gone beyond the simple two-state model we began with and constructed a more realistic model of photodetection.

The result (E.11) shows that the detector responds approximately to the integral over t_1 and t_2 of the normally ordered field correlation function $\langle E_\mu^{(-)}(\mathbf{r}, t_1) E_\nu^{(+)}(\mathbf{r}, t_2)\rangle$. The approximation lies in the step from (E.3) to (E.8), where we retain only the positive-frequency part of the field in order to obtain a nonvanishing, energy-conserving transition probability. We are assuming, in essence, that the observation time t, which might be controlled by a shutter shielding our detector, is large compared with times on the order of ω_{ag}^{-1}. For such time intervals the associated uncertainty in energy is $\Delta E \sim \hbar/t \ll \hbar\omega_{ag}$. Of course this is just the condition for the applicability of Fermi's golden rule. For shorter time intervals we cannot specify energy sufficiently precisely to impose the energy conservation condition that is part of the golden rule.[3]

We have not allowed for any complications associated with a real photodetector, but it is reasonable to suppose that (E.11) is accurate if we have an accurate expression for the response function $S_{\mu\nu}(t)$ of a real detector. All we have really assumed, after all, is that the detection of radiation is associated with an absorptive transition between a bound electron state and a continuum state.[4] It is useful to imagine an "ideal broadband detector" such that

$$S_{\mu\nu}(t) = s_{\mu\nu}\delta(t) \,, \qquad (\text{E}.13)$$

where $s_{\mu\nu}$ is a constant. Based on the simplified model leading to (E.12), we can see that such a detector is insensitive to the values of the frequencies ω_{ag} over a broad range. Actually, for a field of finite frequency bandwidth, this requires only that the detector response is insensitive to frequencies within the field bandwidth, no matter how it may vary with frequency outside this bandwidth.[5] Practical detectors can come very close to this ideal for optical frequencies. Assuming an ideal broadband detector, then,

$$p^{(1)}(t) = s_{\mu\nu} \int_0^t dt_1 \langle E_\mu^{(-)}(\mathbf{r}, t_1) E_\nu^{(+)}(\mathbf{r}, t_1)\rangle \,, \qquad (\text{E}.14)$$

---

[3] It is worth emphasizing that the appearance of the normally ordered field correlation function in (E.8) is an *approximation*, albeit an excellent one in practice. Thus, whereas the full electric field operator $\mathbf{E}(\mathbf{r}, t)$ is properly retarded, its separate positive- and negative-frequency parts are not. Provided we are not considering the detection of extremely short pulses of radiation, however, this is not of practical concern. "Energy-nonconserving" processes are known to be necessary for the formal demonstration of properly retarded interactions in related contexts, such as the resonant interaction of two atoms. See P. W. Milonni and P. L. Knight, *Phys. Rev.* A10, 1096 (1975); A11, 1090 (1975) for a discussion of the two-atom problem.

[4] The electric dipole approximation is not essential, and going beyond it does not change our results in any interesting ways.

[5] See the Les Houches lecture notes by R. J. Glauber in *Quantum Optics and Electronics*, ed. C. DeWitt, A. Blandin, and C. Cohen-Tannoudji (Gordon and Breach, New York, 1965), and R. J. Glauber, *Phys. Rev.* 130, 2529 (1963); 131, 2766 (1963).

and we can define the rate

$$R^{(1)}(t) = \frac{d}{dt}p^{(1)}(t) = s_{\mu\nu}\langle E_\mu^{(-)}(\mathbf{r}, t)E_\nu^{(+)}(\mathbf{r}, t)\rangle \ . \qquad \text{(E.15)}$$

The current registered by our ideal broadband detector should be proportional to this photon counting rate. For a single mode of the field, (E.15) is proportional to $\langle a^\dagger(t)a(t)\rangle$, the expectation value of the number of photons in the mode at time t.

A slightly different viewpoint can be taken to relate a field correlation function to a field spectrum. Suppose, for instance, that the field is statistically stationary in the sense that $\langle E_\mu^{(-)}(\mathbf{r}, t_1)E_\nu^{(+)}(\mathbf{r}, t_2)\rangle$ varies with t_1 and t_2 as some function of the difference $\tau = t_2 - t_1$:[6]

$$\begin{aligned}
\langle E_\mu^{(-)}(\mathbf{r}, t_1)E_\nu^{(+)}(\mathbf{r}, t_2)\rangle &= \langle E_\mu^{(-)}(\mathbf{r}, t_1)E_\nu^{(+)}(\mathbf{r}, t_1 + \tau)\rangle \\
&= \langle E_\mu^{(-)}(\mathbf{r}, t_2)E_\nu^{(+)}(\mathbf{r}, t_2 + \tau)\rangle \ . \quad \text{(E.16)}
\end{aligned}$$

Then, from (E.10),

$$\frac{d}{dt}p^{(1)}(t) = \frac{1}{\hbar^2}|d_{ag}|^2 \text{Re}\int_0^t d\tau\langle E^{(-)}(\mathbf{r}, 0)E^{(+)}(\mathbf{r}, \tau)\rangle e^{i\omega_{ag}\tau} \ , \qquad \text{(E.17)}$$

where for simplicity we restrict ourselves to a single polarization state of the field. Now if we imagine a detection process that responds to frequencies within an arbitrarily small range about ω_{ag}, and samples the field over times t much greater than the inverse of the field bandwidth, we can see that the number of photons counted at frequency ω_{ag} will be proportional to a long-time limit of (E.17). In other words, the measured spectrum of the field will be proportional to a Fourier transform of the correlation function $\langle E^{(-)}(\mathbf{r}, 0)E^{(+)}(\mathbf{r}, \tau)\rangle$. This is assumed in the calculation of the laser linewidth in Section 6.5.

It is useful to define the field correlation function

$$G_{\mu\nu}^{(1)}(\mathbf{r}_1, t_1; \mathbf{r}_2, t_2) = \langle E_\mu^{(-)}(\mathbf{r}_1, t_1)E_\nu^{(+)}(\mathbf{r}_2, t_2)\rangle \ , \qquad \text{(E.18)}$$

in terms of which various "first-order" field interference effects, including those measured in Michelson and Young interferometers, are described.[7]

---

[6] See, for instance, Glauber, ibid., or R. Loudon, *The Quantum Theory of Light*, 2nd ed. (Clarendon, Oxford, 1983).

[7] See Glauber, ibid., and Loudon, ibid., or L. Mandel and E. Wolf, *Rev. Mod. Phys.* **37**, 231 (1965). Many important papers on optical coherence are reprinted in *Selected Papers on Coherence and Fluctuations of Light*, Volumes 1 and 2, ed. L. Mandel and E. Wolf (Dover Books, New York, 1970). The theory of field coherence is discussed in an introductory fashion, mainly in classical terms, by P. W. Milonni and J. H. Eberly, *Lasers* (Wiley, New York, 1988).

In the theory of optical coherence it is in fact useful to define higher order correlation functions such as

$$G^{(2)}_{\mu\nu\sigma\delta}(\mathbf{r}_1, t_1; \mathbf{r}_2, t_2; \mathbf{r}_3, t_3; \mathbf{r}_4, t_4)$$
$$= \langle E^{(-)}_\mu(\mathbf{r}_1, t_1) E^{(-)}_\nu(\mathbf{r}_2, t_2) E^{(+)}_\sigma(\mathbf{r}_3, t_3) E^{(+)}_\delta(\mathbf{r}_4, t_4) \rangle. \qquad (\text{E.19})$$

In general the field will be described by a statistical mixture of states rather than a pure state, and the correlation functions are defined in terms of a density matrix ρ, e.g.,

$$G^{(1)}_{\mu\nu}(\mathbf{r}_1, t_1; \mathbf{r}_2, t_2) = \text{tr}[\rho E^{(-)}_\mu(\mathbf{r}_1, t_1) E^{(+)}_\nu(\mathbf{r}_2, t_2)]. \qquad (\text{E.20})$$

We were led to the first-order correlation function $G^{(1)}$ of the field by starting with a single detector atom at $\mathbf{r}$. Let us now imagine two detector atoms at $\mathbf{r}_1$ and $\mathbf{r}_2$, and consider the probability that each atom has absorbed a photon in the time interval from 0 to t. For this problem we calculate the *second*-order amplitude [see equation (12.16)]

$$a_{fi}(t) = (-i)^2 \int_0^t dt_1 \int_0^t dt_2 \langle f | h_I(t_1) h_I(t_2) | i \rangle , \qquad (\text{E.21})$$

with

$$h_I(t) = -\mathbf{d}_1(t) \cdot \mathbf{E}(\mathbf{r}_1, t) - \mathbf{d}_2(t) \cdot \mathbf{E}(\mathbf{r}_2, t). \qquad (\text{E.22})$$

The cross terms in $h_I(t_1) h_I(t_2)$ are responsible for the transition of interest. Following essentially the same procedure as that leading to (E.11), we obtain for a more realistic model of our two detectors the two-fold counting probability (see note 5)

$$\begin{aligned} p^{(2)}(t) &= \int_0^t dt_1' \int_0^t dt_2' \int_0^t dt_1'' \int_0^t dt_2'' S(t_1'' - t_1') S(t_2'' - t_2') \\ &\times G^{(2)}(\mathbf{r}_1, t_1'; \mathbf{r}_2, t_2'; \mathbf{r}_2, t_2''; \mathbf{r}_1, t_1'') \end{aligned} \qquad (\text{E.23})$$

when, for notational simplicity, we restrict ourselves to a single field polarization. Ideal broadband detectors used to measure the two–fold counting probability thus measure the time integral of the second-order correlation function:

$$p^{(2)}(t) = s^2 \int_0^t dt_1 \int_0^t dt_2 G^{(2)}(\mathbf{r}_1, t_1; \mathbf{r}_2, t_2; \mathbf{r}_2, t_2; \mathbf{r}_1, t_1). \qquad (\text{E.24})$$

The generalization to n-fold photon counting probabilities is straightforward (see note 5), but far less interesting as a practical matter.

For a slightly more general experiment in which a shutter before a detector at r_1 is open during a time interval $[0, T_1]$, and a shutter before a detector at r_2 is open during the interval $[0, T_2]$, one derives the joint counting probability (see note 5)

$$p^{(2)}(T_1, T_2) = s^2 \int_0^{T_1} dt_1 \int_0^{T_2} dt_2 G^{(2)}(\mathbf{r}_1, t_1; \mathbf{r}_2, t_2; \mathbf{r}_2, t_2; \mathbf{r}_1, t_1) \quad \text{(E.25)}$$

and the two-fold delayed coincidence rate

$$\begin{aligned} R^{(2)}(T_1, T_2) &= \frac{\partial^2}{\partial T_1 \partial T_2} p^{(1)}(T_1, T_2) \\ &= s^2 G^{(2)}(\mathbf{r}_1, T_1; \mathbf{r}_2, T_2; \mathbf{r}_2, T_2; \mathbf{r}_1, T_1). \quad \text{(E.26)} \end{aligned}$$

The second-order correlation function $G^{(2)}$ is measured in intensity correlation experiments of the type first performed by Brown and Twiss.[8] A simpler example occurs in two-photon absorption by a single atom, where the rate for transitions to the continuum is approximately proportional to $G^{(2)}(\mathbf{r}, t; \mathbf{r}, t; \mathbf{r}, t; \mathbf{r}, t)$. In the case of a single mode of the field this is proportional to $\langle a^\dagger(t) a^\dagger(t) a(t) a(t) \rangle$, as can be seen from (E.5), (E.6), (E.26), and the definition (E.19). In the special case of two-photon ionization by multimode laser radiation, which often approximates incoherent radiation in its temporal fluctuations, one calculates, under the assumption of Gaussian statistics, a two-photon ionization rate 2! times that calculated under the assumption of a perfectly coherent field. In the general case of n–photon ionization, the enhancement factor is $n!$. For instance, the 11–photon ionization rate of Xe in the (incoherent) multimode case has been observed experimentally to be about $10^7 \sim 11!$ times larger than in the single-mode case.[9]

Regarding multimode fields, note from (E.5) and (E.6) that

$$\begin{aligned} \langle \mathbf{E}^{(-)}(\mathbf{r}, t) \cdot \mathbf{E}^{(+)}(\mathbf{r}, t) \rangle &= \sum_\alpha \sum_\beta (2\pi\hbar)(\omega_\alpha \omega_\beta)^{1/2} \langle a_\alpha^\dagger a_\beta \rangle e^{i(\omega_\alpha - \omega_\beta)t} \\ &\quad \times \mathbf{A}_\alpha^*(\mathbf{r}) \cdot \mathbf{A}_\beta(\mathbf{r}), \quad \text{(E.27)} \end{aligned}$$

which is equal to the sum of the corresponding expectation values for the individual modes if $\langle a_\alpha^\dagger a_\beta \rangle = \langle a_\alpha^\dagger a_\alpha \rangle \delta_{\alpha\beta}$:

$$\begin{aligned} \langle \mathbf{E}^{(-)}(\mathbf{r}, t) \cdot \mathbf{E}^{(+)}(\mathbf{r}, t) \rangle &= \sum_\alpha (2\pi\hbar\omega_\alpha)\langle a_\alpha^\dagger a_\alpha \rangle |\mathbf{A}_\alpha(\mathbf{r})|^2 \\ &= \sum_\alpha (2\pi\hbar\omega_\alpha)\overline{n}_\alpha |\mathbf{A}_\alpha(\mathbf{r})|^2 , \quad \text{(E.28)} \end{aligned}$$

[8] See Section 2.11 and Glauber, ibid., Loudon, ibid., and Milonni and Eberly, ibid.
[9] C. Lecompte, G. Mainfray, C. Manus, and F. Sanchez, *Phys. Rev.* **A11**, 1009 (1975).

and similarly

$$
\begin{aligned}
\langle \mathbf{E}^{(+)}(\mathbf{r},t) \cdot \mathbf{E}^{(-)}(\mathbf{r},t) \rangle &= \sum_{\alpha} (2\pi\hbar\omega_\alpha)\langle a_\alpha a_\alpha^\dagger \rangle |\mathbf{A}_\alpha(\mathbf{r})|^2 \\
&= \sum_{\alpha} (2\pi\hbar\omega_\alpha)(\overline{n}_\alpha + 1)|\mathbf{A}_\alpha(\mathbf{r})|^2 \quad \text{(E.29)}
\end{aligned}
$$

when $\langle a_\alpha^\dagger a_\beta \rangle = \langle a_\alpha^\dagger a_\alpha \rangle \delta_{\alpha\beta}$. This absence of mode correlations is character-istic of thermal radiation, as assumed in Section 8.6, and of the vacuum state of the field. It is also approximately true of free-running multimode lasers, but not for mode-locked lasers.

Appendix F

Transverse and Longitudinal Delta Functions

Consider first the following vector field obtained from a vector field $\mathbf{F}(\mathbf{r})$:

$$\mathbf{A}(\mathbf{r}) \equiv \frac{1}{4\pi} \nabla \times \nabla \times \int d^3 r' \frac{\mathbf{F}(\mathbf{r}')}{|\mathbf{r} - \mathbf{r}'|} . \tag{F.1}$$

Using $\nabla \times \nabla \times \mathbf{C} = \nabla(\nabla \cdot \mathbf{C}) - \nabla^2 \mathbf{C}$ and $\nabla^2(1/|\mathbf{r} - \mathbf{r}'|) = -4\pi\delta^3(\mathbf{r} - \mathbf{r}')$, we have

$$
\begin{aligned}
4\pi\mathbf{A}(\mathbf{r}) &= \nabla \int d^3 r' \nabla \cdot \frac{\mathbf{F}(\mathbf{r}')}{|\mathbf{r} - \mathbf{r}'|} - \int d^3 r' \mathbf{F}(\mathbf{r}')\nabla^2 \frac{1}{|\mathbf{r} - \mathbf{r}'|} \\
&= \nabla \int d^3 r' \mathbf{F}(\mathbf{r}') \cdot \frac{1}{|\mathbf{r} - \mathbf{r}'|} + 4\pi\mathbf{F}(\mathbf{r}) \\
&= -\nabla \int d^3 r' \mathbf{F}(\mathbf{r}') \cdot \nabla' \frac{1}{|\mathbf{r} - \mathbf{r}'|} + 4\pi\mathbf{F}(\mathbf{r}) \\
&= \nabla \int d^3 r' \frac{\nabla' \cdot \mathbf{F}(\mathbf{r}')}{|\mathbf{r} - \mathbf{r}'|} + 4\pi\mathbf{F}(\mathbf{r}), \tag{F.2}
\end{aligned}
$$

where in the last line we have integrated by parts. Thus

$$
\begin{aligned}
\mathbf{F}(\mathbf{r}) &= \mathbf{A}(\mathbf{r}) - \frac{1}{4\pi}\nabla \int d^3 r' \frac{\nabla' \cdot \mathbf{F}(\mathbf{r}')}{|\mathbf{r} - \mathbf{r}'|} \\
&= \frac{1}{4\pi}\nabla \times \nabla \times \int d^3 r' \frac{\mathbf{F}(\mathbf{r}')}{|\mathbf{r} - \mathbf{r}'|} - \frac{1}{4\pi}\nabla \int d^3 r' \frac{\nabla' \cdot \mathbf{F}(\mathbf{r}')}{|\mathbf{r} - \mathbf{r}'|}
\end{aligned}
$$

$$= \mathbf{F}^{\perp}(\mathbf{r}) + \mathbf{F}^{\parallel}(\mathbf{r}) \tag{F.3}$$

This identity is called *Helmholtz's theorem*. We have defined

$$\mathbf{F}^{\perp}(\mathbf{r}) \equiv \frac{1}{4\pi} \nabla \times \nabla \times \int d^3r' \frac{\mathbf{F}(\mathbf{r}')}{|\mathbf{r} - \mathbf{r}'|}, \tag{F.4}$$

$$\mathbf{F}^{\parallel}(\mathbf{r}) \equiv -\frac{1}{4\pi} \nabla \int d^3r' \frac{\nabla' \cdot \mathbf{F}(\mathbf{r}')}{|\mathbf{r} - \mathbf{r}'|}. \tag{F.5}$$

Obviously $\nabla \cdot \mathbf{F}^{\perp}(\mathbf{r}) = 0$ and $\nabla \times \mathbf{F}^{\parallel}(\mathbf{r}) = 0$, and for this reason $\mathbf{F}^{\perp}(\mathbf{r})$ and $\mathbf{F}^{\parallel}(\mathbf{r})$ are called the *transverse* and *longitudinal* parts, respectively, of the vector field $\mathbf{F}(\mathbf{r})$. Thus Helmholtz's theorem says that any vector field can be decomposed uniquely into transverse and longitudinal parts.

Just as the delta function $\delta^3(\mathbf{r} - \mathbf{r}')$ has the property

$$\mathbf{F}(\mathbf{r}) = \int d^3r' \delta^3(\mathbf{r} - \mathbf{r}') \mathbf{F}(\mathbf{r}'), \tag{F.6}$$

the transverse and longitudinal delta function tensors, denoted $\delta_{ij}^{\perp}(\mathbf{r} - \mathbf{r}')$ and $\delta_{ij}^{\parallel}(\mathbf{r} - \mathbf{r}')$, have the properties

$$F_i^{\perp}(\mathbf{r}) = \int d^3r' \delta_{ij}^{\perp}(\mathbf{r} - \mathbf{r}') F_j(\mathbf{r}'), \tag{F.7}$$

$$F_i^{\parallel}(\mathbf{r}) = \int d^3r' \delta_{ij}^{\parallel}(\mathbf{r} - \mathbf{r}') F_j(\mathbf{r}'). \tag{F.8}$$

They are defined by

$$\delta_{ij}^{\perp}(\mathbf{r}) = \left(\frac{1}{2\pi}\right)^3 \int d^3k \left(\delta_{ij} - \frac{k_i k_j}{k^2}\right) e^{i\mathbf{k}\cdot\mathbf{r}}, \tag{F.9}$$

$$\delta_{ij}^{\parallel}(\mathbf{r}) \equiv \left(\frac{1}{2\pi}\right)^3 \int d^3k \frac{k_i k_j}{k^2} e^{i\mathbf{k}\cdot\mathbf{r}}, \tag{F.10}$$

and have the properties

$$\delta_{ij}^{\perp}(\mathbf{r}) + \delta_{ij}^{\parallel}(\mathbf{r}) = \delta_{ij}\delta^3(\mathbf{r}), \tag{F.11}$$

$$\delta_{ij}^{\perp}(\mathbf{r}) = \frac{2}{3}\delta_{ij}\delta^3(\mathbf{r}) - \frac{1}{4\pi r^3}\left(\delta_{ij} - \frac{3r_i r_j}{r^2}\right), \tag{F.12}$$

$$\delta_{ij}^{\parallel}(\mathbf{r}) = \frac{1}{3}\delta_{ij}\delta^3(\mathbf{r}) + \frac{1}{4\pi r^3}\left(\delta_{ij} - \frac{3r_i r_j}{r^2}\right), \tag{F.13}$$

all of which are easily derived. Note that $\delta_{ij}^{\perp}(\mathbf{r})$ and $\delta_{ij}^{\parallel}(\mathbf{r})$, unlike $\delta^3(\mathbf{r})$, do not vanish for $\mathbf{r} \neq 0$.

To prove (F.7), for instance, we need only show that $\nabla \cdot \mathbf{F}^{\perp}(\mathbf{r}) = 0$, i.e., $\partial F_i^{\perp}(\mathbf{r})/\partial x^i = 0$:

$$
\begin{aligned}
\frac{\partial F_i^{\perp}(\mathbf{r})}{\partial x^i} &= \frac{\partial}{\partial x^i} \int d^3 r' \delta_{ij}^{\perp}(\mathbf{r} - \mathbf{r}') F_j(\mathbf{r}') \\
&= \left(\frac{1}{2\pi}\right)^3 \frac{\partial}{\partial x^i} \int d^3 r' \int d^3 k \left(\delta_{ij} - \frac{k_i k_j}{k^2}\right) e^{i\mathbf{k}\cdot(\mathbf{r}-\mathbf{r}')} F_j(\mathbf{r}') \\
&= \left(\frac{1}{2\pi}\right)^3 \int d^3 r' \int d^3 k \left(\delta_{ij} - \frac{k_i k_j}{k^2}\right) i k_i e^{i\mathbf{k}\cdot(\mathbf{r}-\mathbf{r}')} F_j(\mathbf{r}') \\
&= i \left(\frac{1}{2\pi}\right)^3 \int d^3 r' \int d^3 k [\mathbf{k} \cdot \mathbf{F}(\mathbf{r}') - \mathbf{k} \cdot \mathbf{F}(\mathbf{r}')] e^{i\mathbf{k}\cdot(\mathbf{r}-\mathbf{r}')} \\
&= 0, \qquad\qquad\qquad\qquad\qquad\qquad\qquad\qquad\qquad\text{(F.14)}
\end{aligned}
$$

so that $\mathbf{F}^{\perp}(\mathbf{r})$ as defined by (F.7) is indeed the transverse part of $\mathbf{F}(\mathbf{r})$.

Appendix G

Lorentz-Invariant Measure

Equation (10.53) for the quantized Klein–Gordon field involves the phase-space measure $d^3k/[(2\pi)^3 2E_k]$. The factor $1/2E_k$ is introduced in order to have a Lorentz-invariant measure. The Lorentz invariance of d^3k/E_k may be checked directly, by making Lorentz transformations, or simply by noting the identity

$$\int \frac{d^3k}{(2\pi)^3} \frac{1}{2E_k} = \int \frac{d^4k}{(2\pi)^4} 2\pi\delta(k^2 - m^2)\theta(k^0), \qquad (G.1)$$

where $\theta(k^0)$ is the unit step function, equal to 1 for $k^0 > 0$ and 0 otherwise. The right side, involving $d^4k\delta^4(k^2 - m^2)$, is *manifestly* Lorentz invariant.

To establish (G.1) we recall that

$$\delta[y(x)] = \sum_i \frac{\delta(x - x_i)}{|y'(x_i)|} , \qquad (G.2)$$

where the x_i are the zeros of $y(x)$ and $y' = dy/dx$. Thus

$$\begin{aligned}
\delta(k^2 - m^2) &= \delta(k^{0\,2} - E_k^2) \\
&= \frac{1}{2E_k} \left[\delta(k^0 - E_k) + \delta(k^0 + E_k) \right]
\end{aligned} \qquad (G.3)$$

and

$$\delta(k^2 - m^2)\theta(k^0) = \frac{1}{2E_k}\delta(k^0 - E_k). \qquad (G.4)$$

Consequently

$$
\begin{aligned}
\int \frac{d^4k}{(2\pi)^4} 2\pi\delta(k^2 - m^2)\theta(k^0) &= \int \frac{d^3k}{(2\pi)^3}\frac{1}{2E_k}\int dk^0\delta(k^0 - E_k) \\
&= \int \frac{d^3k}{(2\pi)^3}\frac{1}{2E_k} \quad . \tag{G.5}
\end{aligned}
$$

Index

507

Printed and bound by CPI Group (UK) Ltd, Croydon, CR0 4YY

03/10/2024

01040425-0016